Paleokarst

N.P. James   P.W. Choquette

Editors

Paleokarst

With 277 Illustrations in 409 Parts

Springer-Verlag
New York Berlin Heidelberg
London Paris Tokyo

NOEL P. JAMES
Department of Geological Sciences, Queen's University,
Kingston, Ontario K7L 3N6, Canada

PHILIP W. CHOQUETTE
Department of Geological Sciences, University of Colorado,
Boulder, Colorado 80309-250, USA

Library of Congress Cataloging-in-Publication Data
Paleokarst.
    Based on a symposium convened at the 1985
Mid-year Meeting of the Society of Economic
Paleontologists and Mineralogists at Colorado
School of Mines.
    Bibliography: p.
    Includes index.
    1. Karst—Congresses.   I. James, Noel P.
II. Choquette, Philip W.   III. Society of Economic
Paleontologists and Mineralogists. Midyear Meeting.
(1985 : Colorado School of Mines)
GB599.2.P35   1987   551.4'47   87-13061

Typeset by David E. Seham Associates, Inc., Metuchen, New Jersey.
Printed and bound by Arcata Graphics/Halliday Lithograph, West Hanover, Massachusetts.
Printed in the United States of America.

9 8 7 6 5 4 3 2 1

ISBN 0-387-96563-7 Springer-Verlag New York Berlin Heidelberg
ISBN 3-540-96563-7 Springer-Verlag Berlin Heidelberg New York

Preface

Landscapes of the past have always held an inherent fascination for geologists because, like terrestrial sediments, they formed in *our* environment, not offshore on the sea floor and not deep in the subsurface. So, a walk across an ancient karst surface is truly a step back in time on a surface formed open to the air, long before humans populated the globe. Ancient karst, with its associated subterranean features, is also of great scientific interest because it not only records past exposure of parts of the earth's crust, but preserves information about ancient climate and the movement of waters in paleoaquifers. Because some paleokarst terranes are locally hosts for hydrocarbons and base metals in amounts large enough to be economic, buried and exhumed paleokarst is also of inordinate practical importance.

This volume had its origins in a symposium entitled "Paleokarst Systems and Unconformities—Characteristics and Significance," which was organized and convened by us at the 1985 midyear meeting of the Society of Economic Paleontologists and Mineralogists on the campus of the Colorado School of Mines in Golden, Colorado. The symposium had its roots in our studies over the last decade, both separately and jointly, of a number of major and minor unconformities and of the diverse, and often spectacular paleokarst features associated with these unconformities. The problems of correctly interpreting such paleokarst features were brought sharply into focus while we were preparing a detailed review paper on the alteration of limestones in the meteoric diagenetic realm for Geoscience Canada (James and Choquette, 1984). What struck us most forcefully then was that while the tempo of research on karst and karst-related diagenesis had increased dramatically over the last 20 years, much of the research was following parallel but separate pathways. Hydrologists, geomorphologists, geographers, and speleologists were documenting modern karst systems; petrologists and geochemists were unraveling the complex diagenetic textures, fabrics, and water chemistries of limestones in the meteoric realm; economic geologists were modeling processes associated with base metal deposits in subsurface paleokarst—but there seemed to be little interaction between the disciplines. We felt that the time was right to assemble specialists from these diverse fields to review the status of research, to look at paleokarst together, and to present papers outlining recent studies on a variety of paleokarst terranes.

This book represents the fruits of that effort. It brings together the ma-

jority of the symposium reports, along with several other manuscripts so-
licited or volunteered shortly after the meeting. In an introductory article
we have attempted to draw together the main threads of these contributions
in a brief synthesis of our understanding of the controls, processes, and
features of paleokarst. The main body of the book is devoted to presenting
well-documented examples which emphasize the sedimentological and
geochemical/geomorphological aspects of surface and subsurface paleokarst,
together with more general discussions of the processes, features, and geological signatures associated with karst systems.

We are indebted to all of our colleagues who contributed to the symposium and to this volume, for their cooperation, good humor, and fore-
bearance. To Judith V. James we extend special thanks for editing assistance,
text manipulation, and preparation of the subject index. We especially acknowledge Derek Ford and Dexter H. Craig for their counsel during the
preparation of the Introduction to this book. Richard H. De Voto and some
of his associates shared their knowledge of Mississippian paleokarst in central Colorado on a field trip with many of the symposium participants.
Finally, we are particularly grateful to the staff of Springer-Verlag New
York Inc. for the strong interest in this project that made possible the timely
publication of this book. Expenses incurred during the assembly, editing,
and processing of manuscripts were defrayed in part by the Natural Sciences
and Engineering Council of Canada (NPJ) and Marathon Oil Company
(PWC), to whom we express our gratitude.

Kingston, Ontario   NOEL P. JAMES
Littleton, Colorado   PHILIP W. CHOQUETTE
April 1987

Contents

Contributors

PHILIP W. CHOQUETTE
Department of Geological Sciences, University of Colorado, Boulder, CO
80309-250, USA

DEXTER H. CRAIG
Consulting Geologist, 6654 South Sycamore St., Littleton, CO 80120, USA

ANDRÉ DESROCHERS
Department of Geology, Faculty of Science, University of Ottawa, Ottawa,
Ontario KIN 6N5, Canada

RICHARD H. DE VOTO
Department of Geology, Colorado School of Mines, Golden, CO 80401,
USA

J. ALAN DONALDSON
Department of Geology, Carleton University, Ottawa, Ontario KIS 5B6,
Canada

DEREK FORD
Department of Geography, McMaster University, Hamilton, Ontario L8S
4K1, Canada

M. GARCIA-HERNANDEZ
Department of Stratigraphy and Paleontology, University of Granada,
18071 Granada, Spain

LUIS A. GONZALEZ
Department of Geological Sciences, University of Michigan, Ann Arbor,
MI 48109-1063, USA

NOEL P. JAMES
Department of Geological Sciences, Queen's University, Kingston, Ontario
K7L 3N6, Canada

CHARLES F. KAHLE
Department of Geology, Bowling Green State University, Bowling Green,
OH 43403-0218, USA

CHARLES KERANS
Bureau of Economic Geology, University of Texas at Austin, Box X, University Station, Austin, TX 78713-7508, USA

BARBARA H. LIDZ
U. S. Geological Survey, Fisher Island Station, Miami Beach, FL 33139, USA

KYGER C LOHMANN
Department of Geological Sciences, University of Michigan, Ann Arbor, MI 48109-1063, USA

WILLIAM J. MEYERS
Department of Earth and Space Sciences, State University of New York at Stony Brook, Stony Brook, NY 11794, USA

CHARLES J. MINERO
Pecten International Company, P.O. Box 205, Houston, TX 77001, USA

J. M. MOLINA
Department of Stratigraphy and Paleontology, University of Granada, 18071 Granada, Spain

ISABEL P. MONTANEZ
Department of Geological Sciences, Virginia Polytechnic Institute, Blacksburg, VA 24061, USA

WILLIAM J. MUSSMAN
Total Petroleum Inc., P.O. Box 500, Denver, CO 80201, USA

A. CONRAD NEUMANN
Curriculum in Marine Sciences, University of North Carolina, Chapel Hill, NC 27514, USA

R. J. PALMER
Department of Geography, University of Bristol, Bristol, BS8 1SS, England

KENNETH A. RASMUSSEN
Curriculum in Marine Sciences, University of North Carolina, Chapel Hill, NC 27514, USA

J. FRED READ
Department of Geological Sciences, Virginia Polytechnic Institute, Blacksburg, VA 24061, USA

P. A. RUIZ-ORTIZ
Department of Stratigraphy and Paleontology, University of Granada, 18071 Granada, Spain

WILLIAM J. SANDO
U. S. Geological Survey, U. S. National Museum of Natural History, Washington, DC 20560, USA

DONALD F. SANGSTER
Geological Survey of Canada, 601 Booth Street, Ottawa, Ontario K1A 0E8, Canada

EUGENE A. SHINN
U. S. Geological Survey, Fisher Island Station, Miami Beach, FL 33139, USA

PETER L. SMART
Department of Geography, University of Bristol, Bristol BS8 1SS, England

JUAN A. VERA
Department of Stratigraphy and Paleontology, University of Granada, 18071 Granada, Spain

F. WHITAKER
Department of Geography, University of Bristol, Bristol BS8 1SS, England

V. PAUL WRIGHT
Department of Geology, University of Bristol, Bristol BS8 1RJ, England

Introduction

PHILIP W. CHOQUETTE and NOEL P. JAMES

Karst is as dramatic a feature of the earth's surface as it is unique and complex. The manifold and convolute landforms, the complicated and delicately adorned caves, the bizarre drainage systems and collapse structures have few anologs in other kinds of terrains. But these features are only the most obvious in an array of surface and subsurface structures which range in size down to the submicroscopic and comprise systems that are only partly understood—systems that, quite uniquely, form almost entirely by dissolution.

Aims and Emphasis of this Volume

This book is devoted to the documentation and interpretation of karst in the geologic record. Understanding paleokarst is dependent, however, upon an appreciation of modern karst systems and how they are "fossilized." The volume is therefore in two sections. Part I consists of 7 papers dealing with aspects of the development, preservation, modification, and recognition of karst terranes. Part II consists of 11 papers documenting examples of paleokarst, Proterozoic to Cretaceous in age, from a wide variety of shelf and platform settings.

State of the Art

Karst terranes, with their cave systems and distinctive landforms, have for centuries intrigued students of earth processes. Most early studies were directed toward understanding the morphology, hydrology, and development of these features, which are well documented in numerous texts (e.g., Jennings 1971 and 1985, Sweeting 1973, Jakucs 1977, Bögli 1980, Trudgill 1985) that discuss theory as well as observations.

In the last 30 years, an awareness of ancient karst has developed among geologists whose primary interests lie in the stratigraphy and sedimentology of sedimentary carbonates (e.g., Roberts 1966, Roehl 1967, Bignot 1974, Quinlan 1972, Walkden 1974, Sando 1974, Meyers 1974 and 1978, Read and Grover 1977, Kobluk et al. 1977, Maslyn 1977, Wright 1982, Grover and Read 1983, Arrondeau et al. 1985). This awareness has come about because of a burgeoning of studies on meteoric-water diagenesis of carbonate sediments. Some threads of understanding between these two closely related fields have been drawn together in works by Bathurst (1971 and 1975) and more recently in major attempts at synthesis by Longman (1980), Esteban and Klappa (1983), and James and Choquette (1984). This growing appreciation of karst in the geologic record came about because of several factors: the detailed documentation of extant karst (e.g., Bögli 1980, Jennings 1985); refinement of our understanding of the hydrological and chemical processes that lead to chemical dissolution (e.g., Thrailkill 1968, Plummer et al. 1979, Hanshaw and Back 1980, Palmer 1984); substantial documentation of the diagenetic processes and features localized at the rock-soil-air interface (e.g., Esteban and Klappa 1983, and references therein); the growing recognition that paleokarst terranes, in addition to being destructive features, are also constructive sources of cations and carbonate ions for local speleothem deposition and for regional cementation (e.g., Meyers 1974 and 1978, Grover and Read 1983); and finally, a growing awareness that many puzzling fabrics and structures once thought to be meteoric in origin can be assigned to other diagenetic environments such as the sea floor and deep-burial realms (e.g., James and Choquette 1983 and 1984, Scholle and Halley 1985, Choquette and James 1987).

In spite of these encouraging trends, karst and paleokarst have received relatively little attention from carbonate sedimentologists and petrologists. Sedimentological studies have traditionally concentrated on documenting the makeup of carbonate deposits, deciphering their facies mosaics, and reconstructing their depositional and paleogeographic settings. Toward these ends, attention has focused on sedimentary structures, macroscopic features, and biotic constituents, and great effort has generally been made to record information from outcrops or drill cores. Studies of diagenesis, on the other hand, largely through thin-section petrography and geochemistry, have been directed toward unraveling the alteration history of carbonates and so have concentrated on microscopic cements and small-scale fabric-selective porosity.

The reason for this apparent neglect of larger-scale karst features may lie in the "negative," dissolutional nature of karst itself. Especially troublesome is intrastratal corrosion, which forms in the subsurface along lithologic boundaries and creates features that can be mistaken for surface karst. On a smaller scale, there is the problem of differentiating paleokarst surfaces from stylolites and other pressure-solution phenomena (Walkden 1974). It can also be quite difficult to separate local, more or less planar surfaces of a preserved paleokarst terrain from a paleokarst surface that was planed or corroded and bored during subsequent marine transgression. Finally, there is often uncertainty about when dissolution took place (Wright 1982)—did the observed features form soon after deposition of the host strata, are they the result of present-day processes, or were they fashioned at some intermediate time(s)? In short, the very *recognition* of paleokarst can be problematical.

Another reason may be that the natural laboratories of the carbonate petrologist, the mid-to-late Pleistocene carbonates of the tropics that have yielded most of our information on meteoric diagenesis, contain few accessible caves and little extant or former karst. This is because sealevel was much lower during most of the Pleistocene than it is today, so that even the most extensive Pleistocene karst systems are now drowned.

Lastly, the regional distribution and configuration of major karst unconformities have only recently become accessible with the development of high-resolution seismic profiling and reprocessing technology (recent summary and references in Fontaine et al. 1987). Until this development, attempts to reconstruct regional unconformities in any detail relied on information from large numbers of wells or outcrop sections. With the advent of sophisticated seismic technology has come the application of classical stratigraphic concepts and methods (e.g., Sloss 1963) to stratigraphic analysis using seismic-reflection profiles—the "new seismic stratigraphy" (e.g., Vail et al. 1977) in which regional and interregional unconformities are key elements (Schlee 1984).

Definitions

In this volume we use the term *karst* in the broad sense to include all of the diagenetic features—macroscopic and microscopic, surface and subterranean—that are produced during the chemical dissolution and associated modification of a carbonate sequence. By convention we also include the subsurface precipitates (speleothems) which may adorn dissolution voids, the collapse breccias and mechanically deposited "internal sediments" which may floor or fill the voids, as well as surface travertine. Evaporite karst is not considered in this volume.

Paleokarst is defined here (cf. Walkden 1974, Wright 1982) as ancient karst, which is commonly buried by younger sediments or sedimentary rocks and thus includes both *relict* paleokarst (present landscapes formed in the past) and *buried* paleokarst (karst landscapes buried by sediments) as defined by Jennings (1971) and Sweeting (1973). For many who use this book, the broad definition familiar to many

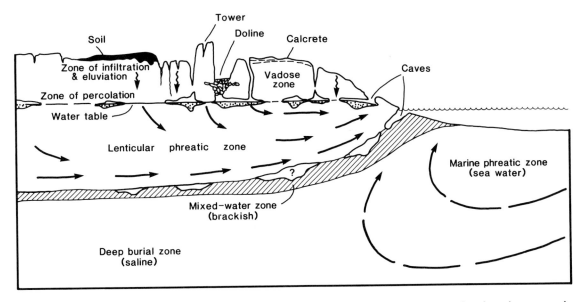

FIGURE 1. A diagram showing the general elements and hydrology of a karst terrane developed on recently deposited carbonates adjoining the sea.

carbonate petrologists and stratigraphers and expressed by Esteban and Klappa (1983, p. 11) may be helpful: "Karst is a *diagenetic facies* (our italics), an overprint in subaerially exposed carbonate bodies, produced and controlled by dissolution and migration of calcium carbonate in meteoric waters, occurring in a wide variety of climatic and tectonic settings, and generating a recognizable landscape." Karst and paleokarst are literally, in the words of Roehl (1967), *subaerial diagenetic terranes*, with an array of distinctive and generally interpretable features (Fig. 1).

For simplicity we also differentiate the types of calcite precipitated in a karst terrane into (1) *speleothems* for those precipitates deposited in spelean settings, or those cavities more than 50 cm in diameter (i.e., large enough to be explored), (2) *cements* for those precipitates which accumulate in smaller holes and are commonest in depositional, fabric-controlled (Choquette and Pray, 1970) pores, and (3) *surface travertines* for those carbonates precipitated from springs at the surface.

Controls of Karst Formation

The wide variety of karst features and the degree of karstification are the end results of in-

teracting processes governed by intrinisic and extrinsic factors (Table 1).

Intrinsic Factors

Most important among these are the *general lithology*, the "matrix" or *stratal permeability*, and the availability of *fractures* and other potential conduits for groundwater. It is well known that, all other rock properties being equal, limestones are several orders of magnitude more soluble

TABLE 1. Factors that influence the development of karst terranes.

Extrinsic	
Climate	Rainfall & evaporation
	Temperature
Base level	Elevation & relief
	Sealevel or local water bodies
Vegetation	
Time Duration	
Intrinsic	
Lithology	Mineralogy
	Bulk purity
	Fabric and texture
	Bedding thickness
	Stratal permeability
	Fractures
Structure &	Attitude of strata
stratigraphy	Confined or unconfined aquifers
	Structural conduits

than dolomites in meteoric water, and gypsum and anhydrite both are more soluble than either group of carbonates. In limestones the "maturity" or degree of stability as opposed to metastability of the $CaCO_3$ mineralogy is most important. Where carbonates with metastable forms of $CaCO_3$ are in contact with meteoric groundwaters, dissolution of aragonite will result in moldic and other forms of fabric-selective porosity, in addition to releasing Ca^{+2} and CO_3^{-2} which eventually precipitate as low-Mg calcite cement. Mineral-controlled alteration of this nature (James and Choquette 1984) may thus create new voids that form part of the karst diagenetic terrane, while concurrently occluding some original porosity (Harrison 1975).

Where poorly cemented carbonate sands or grainstones are subaerially exposed, high stratal permeability may cause groundwater flow to be diffuse, through the grain framework, bypassing or only partly using available fractures. Other processes, such as eluviation of sediment from the surface into the porous grainstone, and limited cave development above the water table, are also consequences of high permeability (Fig. 2). Low-permeability carbonates are likely to transmit groundwater chiefly by conduit flow through fractures and along bedding planes. Although most descriptions of extant karst emphasize the importance of large-scale voids and fracture-controlled dissolution, it is likely that small-scale dissolution and other alteration effects are widespread in the more permeable carbonates in karst terranes. Where dissolution cuts down deeply into the phreatically cemented "roots" of a karst terrane, stratal

LOW   Common fractures
          Mostly conduit flow
          Surface dissolution karst

HIGH   Rare fractures
           Mostly diffuse flow
           Subsoil rundkarren
           Much intergrain eluviation

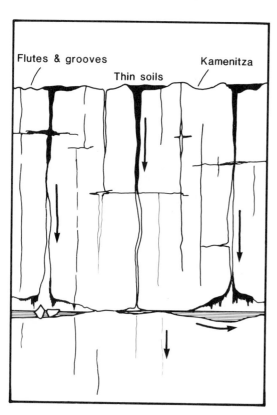
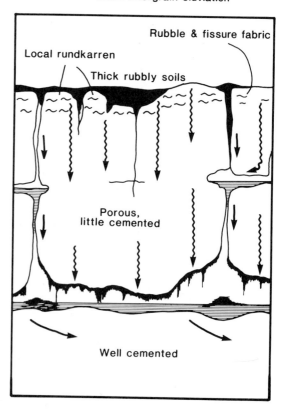

FIGURE 2. A sketch illustrating effects of contrasting stratal permeability on styles of surface and subsurface karst. Low-permeability carbonate might be a partly-lithified to well-lithified lime mudstone or tightly cemented grainstone. High-permeability limestone might be a little-cemented and/or leached, well-sorted lime sand or grainstone. A warm, temperate or humid climate is assumed.

permeability effects will be lessened and fractures will play a more important role in flow patterns.

In order to form intricate surface karst features such as grooves, flutes, and other karren, rainwater must run off rather than percolate into the rock. For this to happen low permeability is required, through either extensive cementation, high lime mud content, or a surface veneer of impermeable calcrete (Fig. 2).

Fractures are particularly important as water conduits in the development of caves. The role of fracturing is commonly "iterative," as small-fracture networks formed by dissolution-collapse become new conduits. Dissolution-enlarged fracture systems act as agents of mass transfer, transmitting soil and sediment downward from the surface and feeding vadose seepage waters that precipitate speleothems in caves.

Extrinsic Factors

Perhaps the most crucial extrinsic factor is *climate*, although *vegetation*, the relationship between initial subaerial relief and *diagenetic base level*, and, of course, the *time duration* of exposure are all important.

In areas of high rainfall and warm temperatures, alteration proceeds quickly, resulting in well-developed soil and terra rossa, abundant sinkholes (dolines), and subsurface dissolution-collapse breccias (Fig. 3). In some regions spectacular landforms evolve and may include pinnacles, jagged ridges, towers, and canyons with interspersed closed depressions that can be virtually impenetrable. In temperate or Mediterranean-type climates, karst and calcrete are common but their development is often seasonal or guided by longer-term cycles. On Caribbean islands built of Cenozoic calcarenite, it is common to find shallow sinkholes and other dissolution cavities veneered by calcrete (James 1972), or conversely, calcrete crusts that have been breached by dissolution. Deserts generally have little karst other than local surface karren, and in warm semiarid climates calcrete is common because of intense evaporation after occasional rains. In cold climates, karst is common even though reaction rates are slow; surface karst is well developed and subsurface karst forms at depth up to the continuous permafrost boundary. Calcrete is not present but calcite does precipitate onto clasts in the soil (D. Ford pers. comm. 1987).

The recognition that caliche and karst require somewhat different combinations of rainfall, evaporation, and to lesser extent temperature to form has led to the proposal that there are "karst facies" and "caliche facies" (Es-

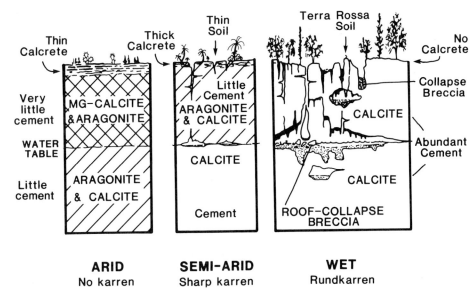

FIGURE 3. A diagram showing common karst features associated with different climatic conditions. Modified from James and Choquette (1984).

teban and Klappa 1983). At the same time, observations on relatively large islands at low latitudes, with marked orographic effects, have shown that both facies can develop not only in sequence, but synchronously, adjacent to one another, during the same season.

Climate also plays an important role in vadose and phreatic cementation beneath exposure surfaces. Recognition that cementation proceeds slowly in arid and semiarid regions such as the Persian Gulf, and that Jurassic carbonate reservoirs in the region commonly had very little cement, for example, led Illing et al. (1967) to suggest a climatic effect. It appears that different styles and abundances of meteoric cements are associated with paleokarst in arid as opposed to humid climates—and as such can be sensitive indicators of climate. Arid-climate diagenetic terranes tend to have little vadose cement other than needle-fiber crystals and sparse low-iron, blocky phreatic cement. Humid terranes, in contrast, tend to have extensive vadose and phreatic cements. Other factors, in particular, relief above water table and as a consequence hydraulic head, will come into play as well, since they too influence the throughput and vigor of groundwater.

The important concept of a base level for diagenesis in the meteoric zone, generally coincident with local drainages and/or the sea, appears in the landmark writings of Davis (1930) and Bretz (1942). Now that the effects of sea-level variations, the anatomy of broad, low-relief carbonate platforms standing only slightly above sealevel, and the contrasts between vadose and phreatic diagenetic alteration (e.g., Steinen and Matthews 1973, Longman 1980, James and Choquette 1984) are better understood, it is timely to reexamine this concept. The role of the water table per se in guiding the development of cave systems now seems strongly dependent on the nature of the conduit (pore) system offered by the host carbonates. In relatively "new" and/or little-cemented strata with high matrix or stratal porosity and permeability, the water table will be a general locus near which many caves first develop (Fig. 1). Cave systems also form in the vadose zone, where their prevailing elongation roughly normal to the water table betrays their origins, but these seem to make up only a small percentage

of known caves (Bretz 1942). In more "mature" and/or extensively cemented carbonates that have little or no stratal permeability, any conduits available will be fractures, faults, and bedding surfaces, and in these rocks the development of caves will determine where the water table will be, rather than vice versa (D. Ford pers. comm. 1987). Caves will thus develop to depths dictated by these larger conduits and topographic relief, and in the process will themselves create vadose and phreatic zones and intervening water tables. In general, whether stratal permeability is high or low, as cave systems develop groundwater will be diverted along them, bypassing deeper zones which then will become relatively stagnant. Karst landscapes erode down to a base level approximating the elevation of local water bodies, and the zone of maximum cave development encompasses the water table.

Cave systems also develop in sublenticular mixing zones along the coasts of some exposed carbonate platforms (e.g., Hanshaw and Back 1980, Back et al. 1984 and 1986), and may also form in inland brackish zones of extensive aquifers, such as the "boulder zone" in the Biscayne aquifer of south Florida (Vernon 1969). The importance of cave formation in the lower reaches of inland aquifers to depths of hundreds of meters below any extant water table has been demonstrated (e.g., Ford, this volume).

In general, the maximum relief possible on a karst landscape or paleolandscape will depend on initial elevation above local water bodies or the sea at the start of subaerial exposure (Fig. 4). This base level will control the depth of surface-karst erosion, but cave systems will generally develop to varying, often substantial depths (hundreds of meters) below it. On widespread carbonate platforms adjacent to the sea, relief will be limited by the elevation of the platform, which may be tens to hundreds of meters but is commonly a very few meters for very low-lying islands and coastal parts of such platforms. Subaerial exposure surfaces in the Pleistocene of Florida and the Bahamas (e.g., Perkins 1977, Beach 1982) now have relatively low relief (order of 10^{0}–10^{1} m) but probably represent drowned paleokarst terranes of higher relief. In strongly uplifted regions such

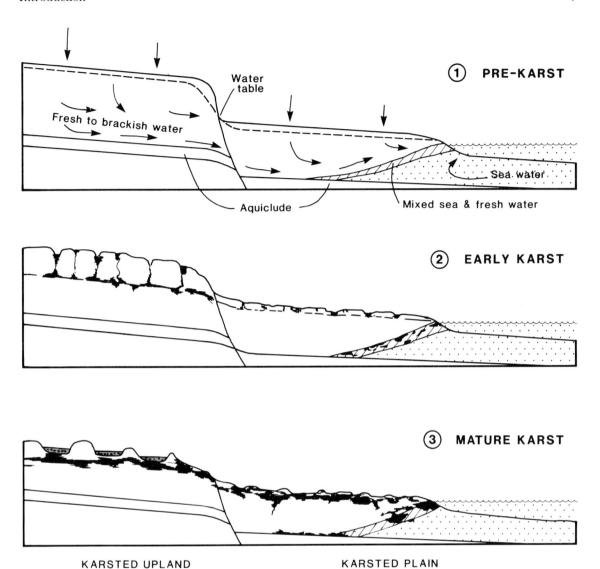

FIGURE 4. A sketch to suggest the kinds of contrasting karst landforms likely to develop because of differences in elevation above karst base level, in a slightly emergent carbonate shelf, now a broad coastal plain, adjoining an interior upland. Climate and other factors are assumed to be similar in both parts of the region. Caves are shown in black.

as the great karst mountains of eastern China, Mexico, and Belize, present topographic relief is much greater, in the hundreds of meters or more. The time durations recorded by such extant karst terranes probably range over two or three orders of magnitude—from 10^4 to 10^7 years or more, with their relief and elevation at any time a function of height above base level.

Biogenic CO_2 produced by decaying vegetation in soils and swamp deposits is of major importance in producing chemically aggressive meteoric groundwaters. The dissolution and precipitation of $CaCO_3$ are controlled in most natural situations by the flux of CO_2 in and out of water, respectively, dissolution increasing in proportion to the partial pressure of CO_2 (P_{co_2}) dissolved in water. It is well known that biogenic

CO_2 increases the P_{CO_2} of soil air, typically to about $10^{-2.0}$ atm., two orders of magnitude higher than atmospheric CO_2 (P_{CO_2} about $10^{-3.5}$ atm.), so that waters in equilibrium with soil air have a much greater capacity to dissolve calcium carbonate.

Synopsis of Part I: General Karst Features and Processes

Subsurface Karst

The one aspect of karst that is most commonly recognized in the geological record is extensive subsurface dissolution, in the form of caves and associated cavities. These features and the galleries connecting them show extraordinary complexity of form, size, disposition, and developmental sequence. It is here that almost all of the intrinsic and extrinsic factors operate in concert to create a myriad of different voids.

Early studies of these features emphasized the physical parameters which controlled their development. In recent years, as our knowledge of carbonate rock–water reactions has become more refined (Runnels 1969, Wigley and Plummer 1976m, Bögli 1980, Hanshaw and Back 1980, Back et al. 1984 and 1986) and exploration techniques have become more sophisticated, a clearer picture of cave formation has emerged.

The first paper in this volume, by D. Ford, entitled "Characteristics of Dissolutional Cave Systems in Carbonate Rocks," is a synthesis of subsurface dissolution. Most explorable cave systems have been formed, it appears, by meteoric waters circulating without geological confinement. Their geometry is not entirely predictable because their pattern of development is controlled by hydraulic gradients in penetrable fissures complicated by reorientation of gradients when initial conduits connect. Many caves are multiphase features with a sequence of levels reflecting vadose morphology or shallow, deep, or mixed phreatic morphology. Caves formed by CO_2-rich thermal waters display distributary dendritic or 2-D and 3-D maze forms while many honeycomb caves develop where fresh and salt waters mix in the coastal zone. Cave geometry, a critical element in determining the original genesis of a cave in

a paleokarst terrane, appears to be equivocal. Phreatic cross-sections tend to be elliptical but may be complicated by differing solubility or armoring of the floor; vadose cross-sections are canyon-like or trapezoidal; both may display dissolution scalloping or be modified or destroyed by breakdown.

Paleokarst is defined not only by caves and smaller dissolution features but also by the characteristic precipitates that adorn voids of all sizes. The complex sediment/rock–water reactions that characterize this system begin as soon as the materials come in contact with freshwater. If these are unaltered sediments composed of different carbonate minerals, some of which (aragonite and Mg-calcite) are unstable in freshwater, then processes are rapid and controlled by the relative solubilities of the minerals involved (*mineral-controlled alteration* of James and Choquette 1984). If the materials are all calcite or calcite limestone, then the saturation state of the circulating waters controls the style of dissolution and precipitation (*water-controlled alteration* of James and Choquette 1984). In his review paper entitled "Geochemical Patterns of Meteoric Diagenetic Systems and Their Application to Studies of Paleokarst," K.C. Lohmann outlines our current understanding of the chemistry of this system and focuses on trace element and isotopic signatures of the precipitated carbonates. Understanding the controls on incorporation of these elements into the calcite lattice in different parts of a meteoric terrane is fundamental if we are to make any sense of ancient karst-related precipitates. Lohmann reemphasizes the point that the $\delta^{18}O$ of meteoric groundwater is largely invariant in a given area during a particular interval of geologic time, while the $\delta^{13}C$ may vary greatly, dependent upon the amount of soil gas available to the system. He thus defines a *meteoric calcite line*, a pattern of essentially constant $\delta^{18}O$ and varying $\delta^{13}C$ which serves as a baseline against which chemical variations characteristic of different meteoric precipitates can be discriminated. The chemistry of the precipitates changes from those most in equilibrium with the waters to those more in equilibrium with the rock, both downdip along fluid flow lines and as the whole system matures and processes change from mineral-controlled to water-controlled. It seems that precipitates in the mixing zone will not display discrete hyperbolic mixing trends but rather offsets which parallel the meteoric calcite line. Spelean precipitates can be differentiated from typical vadose or phreatic cements on the basis of geochemistry but are disturbingly similar to some marine precipitates. As a final caution, Lohmann stresses the importance of examining all of the geochemical and fabric criteria before interpreting ancient precipitates.

Speleothems

Besides the familiar forms of flowstone and dripstone, cave popcorn, cave pearls, and the like (Moore and Sullivan 1978, James and Choquette 1984), smaller, less spectacular precipitates of aragonite and/or calcite are also common. These precipitates form on cave pool floors, on collapsed blocks, or on older speleothems in the pools as nodular branching masses, knobs, and nodules. They develop along pool surfaces as shelves, ledges, rims, and overhangs, or as floating crystal rafts that eventually settle to the floors. Speleothems in air-filled openings depend for their form on whether they precipitated from drip, spray, or capillary flow waters. Flowstone and some of the larger dripstone forms are recognized in paleokarst but the more delicate types are unlikely to be preserved unless they were quickly mantled and sealed by younger precipitates or sediment.

Utilizing geochemical and petrographic techniques, L.A. Gonzalez and K.C. Lohmann, in their paper "Controls on Mineralogy and Composition of Spelean Carbonates: Carlsbad Caverns, New Mexico," have analyzed precipitates from one of the world's largest and best-known caves. This report is a valuable integrated analysis of a cave system against which the geochemistry of paleokarst precipitates can be compared. Calcites in Carlsbad Caverns contain from 1.5 to 12.0 mole % $MgCO_3$, and the Mg^{2+} exhibits a nonlinear dependence on the fluid Mg/Ca ratio but a linear dependence on fluid $CO_3^=$ content. Calcite–aragonite polymorphism is controlled by elevated Mg/Ca ratios with the most enriched Mg-calcites coprecipitating with aragonite or from waters with higher $CO_3^=$ concentrations than those precipitating aragonite. Hydromagnesite and huntite precipitation are dependent upon extreme evaporation at elevated fluid Mg concentrations and Mg/Ca ratios, while primary dolomite precipitates at moderate Mg/Ca ratios, probably from fluids undersaturated with respect to both calcite and aragonite.

Cave Sediments

Studies in this volume illustrate the diversity of mechanically emplaced sediments found in ancient caves. The depth to which sediment may be eluviated into subsurface openings depends on a variety of factors, including: the vigor of the groundwater system, dependent in turn on the amount of recharge and the hydraulic head; the depth to "diagenetic base level"; the supply of sediment; and the locations relative to the zone of infiltration and zone of gravity percolation (Fig. 1). In the Permian paleokarst system described by Craig (this volume), caves in the upper 10 to 20 meters beneath a major unconformity are filled with greenish-gray, clayey carbonate mudstone, but below, dissolution cavities small enough to be sampled by cores are generally free of sediment. Although caves in extant karst terranes also show this pattern of near-surface filling, cavern systems in regions of high rainfall and high groundwater flux may be infiltrated to greater depths.

Karst Breccias

While some paleocaves, or parts of them, may be lined with precipitates and floored with sediments, many others are characterized by breccias, either alone or together with these features.

Karst breccias have such a wide variety of external forms and internal compositions as to defy more than general classification, but do show some system (Fig. 5). *Mantling breccias* associated with unconformities occur in irregular sheets or in patches filling topographic lows of virtually any scale and relief. Such breccias are usually mixtures of sharp-edged fragments of collapsed or residual, surface, or near-surface country rock, along with soil, eluviated soil, or sediment. Chert rubble is common. Similar breccias are likely to fill grikes and other openings at the surface and can have varied, irregular forms. *Breccia pipes and associated bodies* are more or less cylindrical to irregular masses that

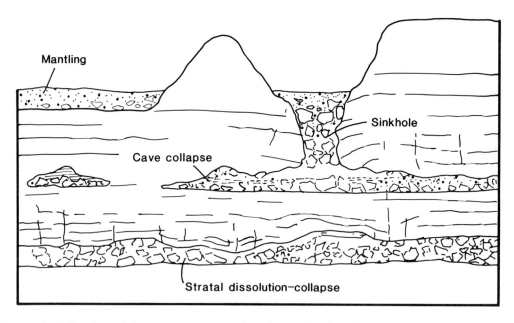

Figure 5. A sketch outlining common types of surface and subsurface breddia in karst terranes.

underlie sinkholes or dolines, generally contain higher proportions of displaced and jostled blocks, and are bordered by strata that are off-set as a result of dissolution-collapse. Such breccias may range in makeup from fitted "crackle" breccias to open-work breccias of ro-tated and displaced fragments, to soil-profile breccias with a matrix of soil or sediment. Fragments are usually sharply angular and show little sign of wear or displacement. Car-bonate fragments may all be the same lithology or may be mixed. Sediment or soil between fragments can be reddish or greenish in color.

Cave-roof collapse breccias have irregular roofs and regular floors with more or less tabular, conformable shapes in cross-sections, and gen-erally show extremely poor sorting, with large blocks many meters across as well as fine sed-iment carried downward from the surface. Breccia bodies that have complex, reticulate or dendroid forms in section or plan will be sus-pect cave-collapse features. Those which have no speleothems or sediments in a long- and well-exposed paleokarst breccia section are most probably intrastratal. *Evaporite-solution breccias* also show great variability of form and characteristics that depend in part on whether dissolution was vadose or phreatic. Phreatic

dissolution seems to have been responsible for the tabular "units" that are widespread near the top of the Madison Limestone (Mississippian) in Wyoming, described by Sando (this volume; see also Roberts 1966, McCaleb and Wayhan 1969, Sando 1974). Near surface vadose dis-solution in weathering profiles results in local-ized collapse features commonly found above base level in modern drainages through eva-poritic sequences.

Karst breccias are suggested as the hosts for several types of mineral deposits, particularly "Mississippi Valley-type" lead-zinc-sulfide de-posits (Ohle 1980, Anderson and Macqueen 1982, Kyle 1983, Kisvarsanyi et al. 1983, Rhodes et al. 1984). In his paper "Breccia Hosted Lead-Zinc Deposits in Carbonate Rocks," D.F. Sangster outlines the evidence for such assumptions and points out that current opinion favors one of two hypotheses: (1) cre-ation of shallow karst breccia and emplacement of ore into this breccia during burial, or (2) so-lution collapse induced by a slightly earlier phase of the ore solutions themselves (hy-drothermal karst) with no meteoric involve-ment. The geological evidence seems to suggest that the meteoric karst hypothesis is most widely applicable.

Surface Karst

A vast array of diagenetic features can be produced in karst terranes (see Roehl 1967, Esteban and Klappa 1983, James and Choquette 1984). At the upper end of the scale are the enormously varied landforms and the aberrant or nonexistent drainage systems. Large landforms can rarely be discriminated in the rock record, except where surface exposures are unusually good or in high-resolution seismic record sections where there are tens or hundreds of meters of paleotopographic relief, or in subsurface areas with exceptional well control. Large-scale landforms can be mapped with varying fidelity in well exposed terranes, using the familiar stratigraphic method of recording variations in the thickness of an interval immediately above or below the unconformity in question. The datum in either case should approximate a time surface and ideally should approximate a horizontal surface at deposition. This approach requires large numbers of control sections, whether mapping a paleokarst surface in broad outline across a region or in detail within a small area.

As outlined above, small-scale surface karst or solution sculpture is difficult to recognize in the rock record, because it is a dissolution phenomenon. In vertical sections it is manifest by irregular bedding contacts, so that only on well-exposed surfaces can the style and form of surface karst be determined with confidence.

The geomorphology of surface karst, with its myriad dissolution features, is another one of the keys to recognition of paleokarst. While these surfaces are readily apparent in modern karst terranes they may lose their distinctive attributes when buried (Wright 1982, Walkden 1974). Sediments overlying these surfaces, however, may signal the presence of an otherwise undetected paleokarst horizon. Among the deposits associated with peritidal and paleoexposure horizons, in rocks ranging in age from Ordovician to Holocene, are black pebbles (Strasser 1984). The origin of these clasts is a matter of some debate; although the blackening is generally ascribed to organic matter, the origin and incorporation of this material is controversial (Ward et al. 1970, Barthel 1974, Esteban and Klappa 1983). In their paper

"Blackened Limestone Pebbles: Fire at Subaerial Unconformities," E.A. Shinn and B.H. Lidz suggest an innovative possibility for the origin of these pebbles. From their studies of and experiments on Cenozoic and Holocene limestones of the Caribbean they conclude that selective blackening is caused by "instantaneous" heating in grass and forest fires. While proposing this as an alternative hypothesis they stress that blackening may also occur in subtidal sediments and that care should be taken when interpreting such clasts.

Drowned Karst

To be preserved, karst must be buried. Although initial deposits may be terrestrial, the terrane is commonly covered by newly deposited marine carbonate sediments. Since the terrestrial sediments are generally unlithified, marine transgression may extensively rework and remove them; the common result in the rock record is juxtaposed carbonates.

The nature of marine carbonates within drowned karst systems is poorly documented from modern seas. A step in the direction of documentation is presented by P.L. Smart, R.J. Palmer, F. Whitaker, and V.P. Wright in their paper "Neptunian Dikes and Fissure Fills: An Overview and Account of Some Modern Examples." They outline factors important in the initiation and development of and sedimentation in "neptunian" dikes and fissure fills. As a modern example they describe the results of preliminary studies of sediments in "blue holes" from the Bahamas.

From the same region K.A. Rasmussen and C.A. Neumann describe results of recent work on shallow buried karst in their paper "Holocene Overprints of Pleistocene Paleokarst: Bight of Abaco, Bahamas." Recent flooding of the dished bedrock surface of the Bight of Abaco, a shallow carbonate platform, has preserved three different, concentrically zoned Holocene/Pleistocene karst unconformities. The surface is best preserved in the deep center of the platform, where the doline karst is mantled by peats and paleosols. The slopes into the central depression are veneered with caliche and preserved because of the schizohaline conditions present during initial phases of sedimentation, resulting in a restricted bottom fauna. The karst surface of the wide upper margins, however, is extensively bioeroded, and much of the fine solution sculpture has been removed. This study dramatically demonstrates the subtle, yet important ways in which karst surfaces may be modified soon after formation and clearly indicates the need for more study of such surfaces.

Synopsis of Part II: Examples of Paleokarst Terranes

In this section are those papers which describe particular paleokarst terranes, often drawing heavily upon information from modern karst systems. We have arranged the papers in chronological order from Precambrian to Cretaceous.

Perhaps one of the most critical questions in any discussion of paleokarst is, how far back in geologic history can true karst be recognized? This question is clearly answered by C. Kerans and J.A. Donaldson in their paper "Proterozoic Paleokarst Profile, Dismal Lakes Group, N.W.T., Canada," in which they document a complete spectrum of Precambrian dissolution, precipitation, and depositional features. Particularly conspicuous are karst breccias with preserved roof pendants, cross-stratified chert gravel cave-floor sediments, grikes, fibrous flowstone, cave popcorn, and a complex overlying regolith. They interpret the karst profile to have developed in three stages, subaerial exposure and phreatic dissolution, vadose fill by clastic sediments and flowstone precipitation, and finally, cave collapse, all reflecting gradual lowering of the water table.

One of the most extensive interregional unconformities on the North American craton occurs near the base of the Middle Ordovician (Ross et al. 1982) where it separates the Sauk and Tippecanoe Sequences of Sloss (1963). The extent and timing of this karst unconformity has been discussed in detail by Mussman and Read (1986). This important paleokarst terrane is the subject of two papers.

A. Desrochers and N.P. James describe this and other paleokarst features in their paper "Early Paleozoic Surface and Subsurface Paleokarst: Middle Ordovician Carbonates: Mingan Islands, Quebec," near the northern margin of the Appalachaian Orogen. These carbonates are characterized by two major paleokarst unconformities and numerous local paleokarst surfaces. The lower paleokarst unconformity, which is correlative with the interregional post-Sauk break (Knox or Beekmantown unconformity), is an extensive karst plain developed on dolomite with remarkably well-preserved delicate solution features and small dolines. Associated subsurface karst is in the form of small caves and karst breccias. An upper paleokarst unconformity, within Middle Orodvician limestones, displays substantial karst relief which was modified by intertidal erosion preceding submergence. The surface is again sculpted into a variety of karren. Local paleokarst horizons are also present at the tops of meter-scale, shallowing-upward calcarenite cycles. These short-lived exposure events resulted in rapid lithification and the formation of surface karren. With time, however, the karren progressively widened and evolved into remarkably planar surfaces.

Building on their description of the karst unconformity at the top of the Knox–Beekmantown Group in the central Appalachians, W.J. Mussman, I.P. Montanez, and J.F. Read, in an integrated field, petrographic and geochemical study entitled "Ordovician Knox Unconformity, Appalachians," document a complete geologic history of this widespread terrane. The unconformity, locally representing a time gap of 10 m.y., has over 100 m of erosional relief in the southern part of the region, is most profound over syndepositional structures, and decreases to zero in depocenters. Paleokarst features include paleotopographic highs, sinkholes, and caves that extend more than 65 m below the surface, and intrastratal breccias down to 300 m. Sinkholes are filled with breccias and gravels in a fine detrital dolomite matrix. Bedding-concordant caves contain breccia, laminated fine dolomite sediment, and locally spar cement. Nonluminescent calcite cements fill particle molds and intergranular pores down to 200 m below the unconformity and are interpreted to have been precipitated from slowly moving oxidized meteoric waters undergoing diffuse flow in an unconfined aquifer. Bedding-concordant breccias are composed mainly of "fitted" dolomite clasts in a matrix of fine dolomite sediment and are thought to have been formed mainly as a result of dissolution by rapidly moving conduit flow cave waters, or by dissolution in the mixing zone. During burial, compaction further fractured the breccia beds, and in the Late Paleozoic warm, saline basinal brines caused further dissolution, formation of dolomitization haloes around breccias, and precipitation of saddle dolomite and associated sulphides within permeable horizons.

C.F. Kahle documents a wide variety of paleokarst features in the paper "Surface and Subsurface Paleokarst, Silurian Lockport and Peebles Dolomites, Western Ohio." This study, in which ancient karst features are matched with their modern counterparts, describes subsurface molds, vugs, in situ breccias with corroded clasts, collapse strata, solution-enlarged joints, internal sediment, boxwork, and caves within and surface paleosols and sinkholes at the top of the Peebles–Lockport. Maximum paleorelief may be as much as three meters but is generally less, and commonly the surface is a planar erosional paraconformity. A gley paleosol at the top of the Peebles–Lockport is remarkably similar to modern gley soils, which has implications for the formation of soils prior to the development of land plants. Syndepositional faulting during deposition of the Lockport is also interpreted to have resulted in local karst unconformities characterized by shallow subsurface vugs and caves and by scalloped and planar erosion surfaces mantled by terra rossa.

Mississippian time marks the appearance of land plants in abundance for the first time in the geologic record. This was accompanied by the development of true soils and thus the modification of freshwaters as they percolated through the biologically rich horizons. Coincidentally, rocks of this age exhibit numerous examples of paleokarst and caliche worldwide (Esteban and Klappa 1983, James and Choquette 1984). This volume contains several papers which concentrate on different aspects of Mississippian paleokarst.

Three of these papers focus on the interregional unconformity at the top of the Mississippian System in North America. In his paper

"Madison Limestone (Mississippian) Paleokarst: A Geological Synthesis," W.J. Sando integrates a vast amount of published and unpublished information to outline the development of this unconformity in Wyoming and adjacent states. Karst developed in the upper 120 m of the Madison during an estimated 34 m.y. spanning Late Mississippian to Early Pennsylvanian time. The paleokarst is differentiated from later and similar Cenozoic features by the age of contained late Paleozoic sediments. Madison paleokarst includes enlarged joints, sinkholes, caves, and two zones of karst breccia. Composite reconstruction of the karst topography reveals a mature karst landscape having a maximum relief of about 70 m and dominated by three major river systems and low rounded hills. An inferred paleoaquifer system illustrates bedding parallel flow radially away from an uplift in southeast Wyoming through conduits created by leaching of evaporites. The distribution of paleosolution features is interpreted to have been controlled by block faulting, which caused local structural relief.

The study entitled "Late Mississippian Paleokarst and Related Mineral Deposits, Leadville Formation, Central Colorado" by R.H. De Voto documents the same karst-related unconformity to the south in Colorado. The terrane is typified by caverns, sinkholes, solution-enlarged vertical joints, channels, and breccia-rubble soil zones. Subsurface channels are semiconcordant to bedding, and multiple cavern levels occur not only in the Mississippian, but also in underlying Devonian and Ordovician carbonates adjacent to 100-m to 200-m deep paleokarst valleys. Like karst on the Madison, the distribution of paleokarst features at the top of the Leadville is thought to have been controlled by Mississippian movement of local tectonic blocks.

Subsurface dissolution features are filled with breccia and carbonate sand or, if close to the unconformity, shale and clay of Pennsylvanian age. Hundreds of lead-zinc-silver-barite deposits occur within the karst solution features, and principal areas of mineralization, particularly in the Leadville and Aspen districts, occur immediately adjacent to deeply incised paleovalleys.

W.J. Meyers has concentrated on small-scale

features at the same unconformity further south in his paper "Paleokarstic Features in Mississippian Limestones, New Mexico." The effects of karst are most apparent in the upper few meters to few tens of meters of widespread crinoidal calcarenites, in contrast to the more common large-scale bedding and joint-controlled karst described in the previous two papers. This difference is interpreted to be due to the high original intergranular porosities and permeabilities of these sediments when karstified. Such karst, which develops on sediments that are only poorly lithified, commonly early in their history, has been called *syngenetic karst* by Jennings (1971). Meyers' careful documentation may be a good model for other such sequences.

A typical vertical profile, interpreted to have developed in the vadose zone and exhibiting progressively more dissolution and eluviated sediments upward, shows in ascending order: etching of freshwater phreatic syntaxial cements; micrite and microspar plugging of intergranular pores, filling in around crinoids, bryozoans, and earlier cements; clay and detrital quartz silt filling intergranular pores and surrounding etched crinoids and earlier cements; fragmented host limestone forming rubble and fissure fabrics surrounded by anastomosing veinlets and fissures filled with "weathering calcarenite," clay, and detrital quartz silt; and finally, at or near the surface, chert fragments in a matrix of clay, quartz silt, and sand-sized chert fragments.

V.P. Wright has chosen to concentrate on paleoclimate as a control in his paper "Paleokarsts and Paleosols as Indicators of Paleoclimate and Porosity Evolution: A Case Study from the Carboniferous of South Wales." Subaerial surfaces in these Mississippian limestones are of two types: (1) densely piped and rubbly solution horizons interpreted to have formed under humid karst conditions and in which the sediments are cemented by vadose, meniscus, and later phreatic/burial cements, and (2) poorly developed karst horizons with extensive calcretes containing rhizocretions and needle-fiber calcites indicative of a semi-arid climate and containing only minor early meteoric cements. This distribution is thought to be similar to Holocene profiles developed on Pleistocene calcarenites on the Yucatan Peninsula and re-

flects the modern distribution of karst and calcrete (Esteban and Klappa 1983). Extending the findings of this study, Wright reiterates Longman's (1980) observation that most major oolite petroleum reservoirs were cemented originally under arid climatic conditions, while comparable oolitic reservoirs cemented under humid conditions are virtually unknown because of extensive cementation.

Subsurface karst creates the extensive porosity necessary to pool large accumulations of hydrocarbons (McCaleb and Wayhan 1969, Fei Qi and Wang Xie Pei 1984, Loucks and Anderson 1985, Wilson 1985). Perhaps nowhere is this better demonstrated than in the Yates Field, as described by D.H. Craig in his paper "Caves and Other Features of Permian Karst in San Andres Dolomite, Yates Field Reservoir, West Texas." This 60-year-old field, located at the southern end of the Central Basin Platform, contained when discovered an estimated 4 billion barrels of oil in place and to date has produced just over 1 billion barrels. Because of the high well density it has been possible to reconstruct the paleokarst system on a very fine scale. The subsurface dissolution features are reflected by bit drops and sudden rushes of oil when drilling, extremely high flow rates recorded from early field wells (initially up to 205,000 barrels of oil per day), and fragments of dissolution rubble produced with the oil from some wells. The 285 unfilled caves discovered to date range up to 6.4 m in height and are most numerous in the eastern part of the field, where it is thought that permeable shelf-edge skeletal sands made the rock more susceptible to meteoric infiltration and dissolution. Cores contain speleothems, karst breccias, and several different kinds of internal sediment. The paleokarst system is interpreted to have developed beneath a cluster of low-relief limestone islands created by a eustatic or tectonically induced fall in sealevel during Late Permian time.

The development of paleokarst in an active tectonic region is described by J.A. Vera, M. Garcia-Hernandez, J.M. Molina, and P.A. Ruiz-Ortiz in their paper "Paleokarst and Related Pelagic Sediments in the Jurassic of the Subbetic (Southern Spain)." The exposure events, which occurred during collapse of the Liassic platform, were due to listric faulting and eustatic changes of sealevel. The karst features include caves, speleothems, collapse breccias with speleothems enveloping clasts, laminated internal terrestrial sediments, freshwater phreatic and vadose cements, and bauxites. The karst is buried by pelagic sediments.

The Cretaceous of Mexico is another region in which hydrocarbons are trapped in karst-enhanced reservoirs in the subsurface such as the Golden Lane trend, which has produced to date 1.6 billion barrels of oil. In his paper "Sedimentation and Diagenesis Along an Island-Sheltered Margin, El Abra Formation, Cretaceous of Mexico," C.J. Minero documents the multiphased karstification of an episodically exposed platform which is now largely exhumed and exposed again. The first phase of karstification took place as each meter-scale shallowing-upward sequence accreted to sea level and was exposed, resulting in dissolution beneath microkarst surfaces, calcrete formation, and mineral-controlled diagenesis. The second episode of karstification took place during late Cretaceous time and is interpreted to have resulted in minor surface dissolution but extensive precipitation of phreatic calcite cement. The final stage of karstification is occuring today as regional structure controls an evolving extensive subsurface and surface karst system.

Conclusion: General Themes and Directions

The examples of these paleokarst terranes, when considered together, illustrate several themes that could not have been predicted from the study of modern, extant karst alone. At the same time the results in these papers point up aspects of karst, paleokarst, and more generally meteoric-zone diagenesis in carbonates that remain poorly understood and deserve attention.

Diagnostic Criteria

It appears that some features of paleokarst—apart from the most dramatic landforms, caves, and cave deposits—are more common and diagnostic than others, as Esteban and Klappa

(1983) also showed. In Table 2 we list those criteria and highlight the attributes which recur most conspicuously in this collection of paleokarst studies. Meteoric diagenesis can be recognized on a microscopic scale by so many features that the presence of paleokarst may often be sensed or predicted where it cannot be directly confirmed. Thus, vadose processes can be inferred from the presence of needle-fiber cement, brownish or reddish micritized grains, gravity or meniscus cements, ^{13}C compositions of the bulk rock or cements that are exceptionally "light" (if the petrology precludes other origins besides surface biogenic CO_2), and so on. Phreatic cementation can commonly be inferred from the presence of iron-poor calcite spar, microspar, poikilotopic spar, cathodoluminescence that is "dark" with thin bright bands (see below), grain-rimming (isopachous) rhombohedral calcite cements, and the like.

TABLE 2. Features commonly associated with paleokarst.

Stratigraphic–geomorphic
*Karst landforms—towers, dolines, closed depressions, lack of fluviatile sediment
Unconformities—strata below are truncated, strata above onlap prominences
Shallowing-upward cycles—end abruptly at paleokarst surfaces

Macroscopic	
Surface karst	Subsurface karst
* Rundkarren	* Caves & smaller nonselective dissolution voids
* Other karren	
Kamenitzas & phytokarst	* In-place brecciated & fractured strata
Terra rossa & other soils	
* Caliche (calcrete)	* Collapse structures
Nonsedimentary channels	* Dissolution-enlarged fractures
Lichen structures	
Boxwork structure	* Rubble-and-fissure fabrics
Laminar brown or reddish fracture fillings	
	Sediment in nondepositional cavities
* Mantling nonsedimentary breccias	* Breccias in irregular bodies, conformable or not

Microscopic
* Eluviated soil in small pores
* Etched carbonate cements
Reddened & micritized grains
Meniscus, pendant, and needle-fiber vadose cements
* Extensive, dissolution, or enlarged, fabric-selective pores

* Features that seem especially diagnostic

What is missing in most studies of paleokarst and virtually all studies of extant karst, however, is attention to both the megascopic and microscopic products of karstification and their spatial and sequential relationships to one another. The studies in this volume by Meyers, Mussman et al., Wright, Craig, and Vera et al. illustrate the power of an integrated approach. As the microscale characteristics of paleokarst systems and their relationship to larger-scale features are more widely documented, it should be possible to construct more accurate models of the sequences of events, as well as the conditions, associated with the evolution of ancient karst terranes.

Stratigraphy of Paleokarst

Paleokarst develops at several different levels and in different major settings which deserve separate designation (Fig. 6). *Depositional paleokarst* is that karst which forms as a natural consequence of sediment accretion to sealevel and is to be expected within sediment packages that typify carbonate platforms. It is most commonly associated with meter-scale cycles, generally with topographic relief of centimeters to decimeters. Examples of depositional paleokarst are described in this volume by Desrochers and James, Wright, and Minero. The effects of exposure are usually restricted to surface solution sculpture, near-surface cementation, and minor subsurface dissolution because the processes act upon largely unlithified sediments with high porosity and permeability.

Local paleokarst forms when part of a carbonate shelf (or platform, ramp etc.) is exposed, generally because of tectonism, small drops in sealevel (e.g., exposure of the platform margin but not the platform interior), or synsedimentary block faulting. Depending on the length of time involved, the effects of exposure can vary from minor to extensive with the development of surface and subsurface karst. Such

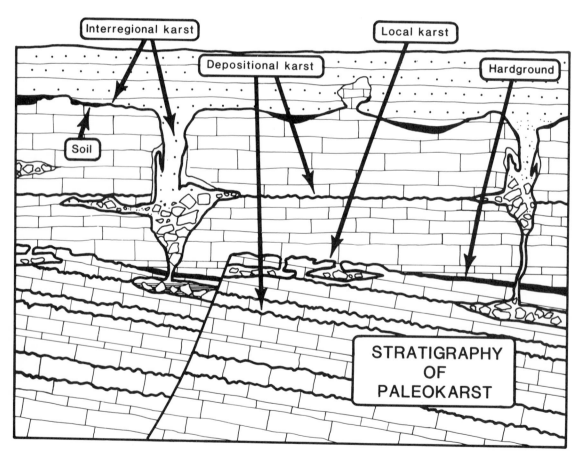

FIGURE 6. A diagram summarizing the different levels of karst to be expected within carbonate terranes.

localized paleokarst may be traced laterally into areas which show no exposure effects or record continuous deposition. Paleokarst terranes of this sort are documented by Kerans and Donaldson; Kahle; Craig; and Vera, Garcia-Hernandez, Molina, and Ruiz-Ortiz.

Interregional paleokarst is much more widespread, is related to major eustatic-tectonic events (e.g., Type 1 unconformities of Vail et al. 1984), and results in karst terranes that may illustrate profound erosion, a wide variety of karst features, and deep, pervasive dissolution. The extent and depth of karst development, including paleotopographic relief and cave systems, will usually be greater than in local paleokarst, although where local tectonics result in substantial uplift and long exposure, differentiation of the two scales of paleokarst may be difficult. This type of karst is illustrated in the papers by Desrochers and James; Mussman, Montanez, and Read; Sando; De Voto; and Meyers. Often the relief, stratigraphic truncation and onlap, and other large-scale features of the different paleokarst systems can be seen well in the amplitude patterns on seismic-reflection profiles (e.g., Vail, Mitchum, Todd and co-workers in Peyton 1977, Fontaine et al. 1987).

Paleokarst Profiles

The utility of describing paleokarst in terms of a profile, as suggested by Esteban and Klappa (1983), is well illustrated in these papers. This approach, used in the contributions by Kerans and Donaldson, Kahle, Meyers, and Wright, is useful in that it is stratigraphic in concept and thus allows different paleokarst sequences to be compared. Eventually, when a sufficient number of such paleokarst profiles have been documented it may be possible to erect a series of models that will allow the extraction of details such as paleoclimate, time of exposure, and the like from the rock record.

Cements and Cementation Associated with Paleokarst

Beginning with studies by Dunham (1969) and Steinen and Matthews (1973) and others on vadose diagenesis associated with various local

paleokarsts, and by Meyers (1974 and 1978) on vadose and phreatic cements associated with an interregional paleokarst, enough has been learned about the morphology, cathodoluminescence, and geochemistry of meteoric-zone cements to allow us to interpret with some confidence the general realms and waters in which these cements precipitate (references in James and Choquette 1984). In carbonate terranes beneath unconformities, and in most ancient carbonates in general, the great majority of cements seem to be of meteoric phreatic and deeper-burial origin (James and Choquette 1984, Scholle and Halley 1985, Choquette and James 1987). Shallower phreatic cements are typically clear, iron-poor, blocky or sparry calcite which is mostly nonluminescent with thin zones of bright luminescence. These are usually succeeded by a younger sequence, probably of deeper-burial origin, of iron-rich, commonly inclusion-rich, blocky calcite with dull, variably zoned cathodoluminescence.

Mussman, Montanez, and Read in this volume, expanding on earlier studies (Grover and Read 1983), observed that the dark nonluminescent cements with bright bands in Ordovician carbonates in Virginia fill dissolved-grain molds and interparticle pores down to 200 m below the post-Knox unconformity in Virginia. They interpret these cements to have formed from oxidizing meteoric groundwaters that moved slowly by diffuse flow through an unconfined aquifer (Fig. 7). The dully luminescent cements were precipitated at greater depths, further removed from recharge areas, out of more stagnant, reducing, subphreatic fluids. The depth to which nonluminescent cements were precipitated thus gives an indication of the hydrologic vigor of the system. It may be eventually possible to identify those cements precipitated in zones of mixed, brackish groundwater with more certainty, and to determine more specifically than is possible now the chemistry of subsurface fluids responsible for carbonate cements from the nature of fluid inclusions and the chemistry and cathodoluminescence of the cement. The approach of interpreting paleohydrological and paleochemical regimes from the cements in karst terranes, when done in the context of a carefully established stratigraphic and sedimentologic framework, has great power for unraveling the gen-

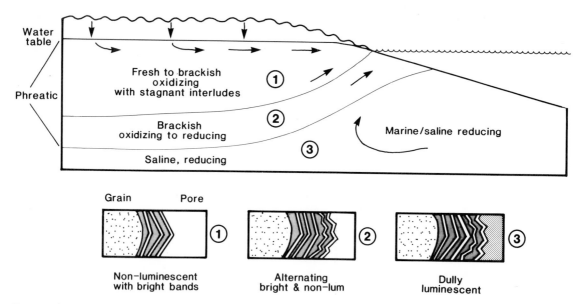

FIGURE 7. A diagram showing interpreted cathodoluminescent cement zones associated with a karst terrane being actively recharged. Groundwater in the phreatic should be dominantly oxidizing (nonluminescent cement), with interludes of stagnant/reducing conditions marked by bands of bright luminescence. Subsurface waters in the deeper burial realm should be more dominantly stagnant and reducing, with higher concentrations of Fe^{+2} and Mn^{+2} that cause luminescence to be generally dull and less sharply zoned. Modified from a diagram in this volume, by Mussman, Montanez, and Read.

eral diagenetic history of cement-bearing carbonates and the basins containing them.

Climate and Paleokarst

It is becoming more widely recognized that there is in some way a "semiarid climate" and a "humid climate" response of carbonate sequences to subaerial exposure in warm regions. This theme, discussed earlier as a general phenomenon by Esteban and Klappa (1983) and James and Choquette (1984), has been reaffirmed in the study by Wright and is implicit in many of the other studies reported here. It is likely to be more and more widely applied to the recognition of paleoclimates, to paleogeographic reconstructions, and to the prediction of cementation patterns and porosity in carbonates where the picture has not been complicated by subsequent deeper-burial diagenesis.

Relatively little is known about cementation and other microscale alteration of limestones and dolomites in karst terranes of cold climates. Modern marine carbonate sediments deposited outside the tropics seem to be entirely low-Mg calcite, so that upon exposure their alteration should be water-controlled (James and Choquette 1984) from the outset. This should slow the development of depositional karst, but its effects on local and regional or interregional, unconformity karst are more difficult to predict. In any event, there is a need for information about the profiles and the cementation and other microscale alteration that develop in modern karst in temperate climates.

Karst Through Geologic Time

These studies give strong indication that the larger-scale products of subaerial exposure have, on the whole, been similar since Middle Proterozoic time.

The presence of a true soil and the reservoir of CO_2 which it contains is critical for the development of surface and near-surface karst. True soils appear to have been an integral part of the terrestrial landscape since Carboniferous time, but not before. Early Paleozoic soils were probably more like the protosoils of today, dominated by algae and lichens. Thus, the amount of karst formation per unit time was probably less during the Cambro-Devonian.

The situation in the late Precambrian is more difficult to assess because it depends upon whether the $_pCO_2$ of the atmosphere was the same as at present or substantially greater (Holland 1984). If it was greater, e.g., 10^{-1} or more, then karst should be extensive. The role of soils in this case would be reversed, protecting the carbonates from acidic meteoric rainwaters (D. Ford pers. comm. 1987). Nevertheless, it is clear that dissolution, both fabric-selective and at larger scale, took place with enough intensity to produce recognizable dissolution karst in very old rocks, whether or not there were true soils to aid the process.

As other Proterozoic and early-middle Paleozoic karst terranes are studied it may be possible to decide whether caliche in such older terranes is rare, as suggested earlier (James and Choquette 1984) and by studies in this book, or instead has early analogs with different characteristics. Caliche is so climate-specific as to be expected only in certain paleoclimatic (and palegeographic) settings which may or may not be closely tied to worldwide climatic conditions. In any event, our ability to identify the imprints of climate in paleokarst from studies of extant karst will be increasingly important for the reconstruction of past climates and paleogeographic models.

References

Anderson, G.M. and Macqueen, R.W., 1982, Ore deposit models—6. Mississippi Valley type lead-zinc deposits: Geoscience Canada, v. 9, p. 108–117.

Arrondeau, J.-P., Bodeur, Y., Cussey, R., Fajerwerg, R., and Yapaudjian, L., 1985, Indices de pédogenèse et de karstification dans le Lias calcareo-dolomitique, Causse du Larzac (Languedoc, France): Bull. Centres Rech. Explor. Prod., Elf-Aquitaine, Pau, France, v. 9, p. 373–403.

Back, W., Hanshaw, B.B., Herman, J.S., and Van Driel, J.N., 1986, Differential dissolution of a Pleistocene reef in the groundwater mixing zone of coastal Yucatan, Mexico: Geology, v. 14, p. 137–140.

Back, W., Hanshaw, B.B., and Van Driel, J.N., 1984, Role of groundwater in shaping the eastern coastline of the Yucatan Peninsula, Mexico, in LaFleur, R.G., ed., Groundwater as a geomorphic agent (Binghamtom Smypos. in Geomorphology No. 13): Boston, Allen and Unwin, p. 281–294.

Barthel, K.W., 1974, Black pebbles, fossil and Recent, on and near coral islands: Second International Coral Reef Symposium Proceedings, v. 2, p. 395–399.

Bathurst, R.G.C., 1971, Carbonate sediments and their diagenesis: Amsterdam, Elsevier, 620 p.; also 1975, 2nd edition, 658 p.

Beach, D.K., 1982, Depositional and diagenetic history of Pliocene–Pleistocene carbonates of northwestern Great Bahama Bank, evolution of a carbonate platform: Ph.D. thesis, University of Miami, Coral Gables, FL, 447 p.

Bignot, G., 1974, Le paléokarst éocene d'Istrie (Italie et Yougoslavie) et son influence sur la sédimentation ancienne: Mémoires et Documents, nouvelle serie, v. 15, Phénomènes Karstiques, t. II, p. 177–185.

Bögli, J., 1980, Karst hydrology and physical speleology: Berlin/Heidelberg, Springer-Verlag, 285 p.

Bretz, J.H., 1942, Vadose and phreatic features of limestone caverns: Jour. Geology, v. 50, no. 6, pt. II, p. 675–811.

Choquette, P.W., and James, N.P., 1987, Diagenesis 12. Diagenesis in limestone—3. The deep burial diagenetic environment: Geoscience Canada, v. 14, no. 1, p. 3–35.

Choquette, P.W., and Pray, L.C., 1970, Geological nomenclature and classification of porosity in sedimentary carbonates: Amer. Assoc. Petroleum Geologists Bull., v. 54, p. 207–250.

Davis, W.M., 1930, Origin of limestone caverns: Geol. Soc. America Bull., v. 41, p. 475–628.

Dunham, R.J., 1969, Early vadose silt in Townsend mound (reef), New Mexico, in Friedman, G.M., ed., Depositional environments in carbonate rocks: Soc. Econ. Paleontologists & Mineralogists Spec. Publ. 14, p. 139–181.

Esteban, M., and Klappa, C.F., 1983, Subaerial exposure environments, in Scholle, P.A., Bebout, D.G., and Moore, C.H., eds., Carbonate depositional environments: Amer. Assoc. Petroleum Geologists Mem. 33, p. 1–54.

Fei, Qi, and Wang, Xie Pei, 1984, Significant role of fractures in Renquiu buried-hill oilfield in eastern China: Amer. Assoc. Petroleum Geologists Bull., v. 68, p. 983–993.

Fontaine, J.M., Cussey, R., Lacaze, J., Lanaud, R., and Yapaudjian, L., 1987, Seismic interpretation of carbonate depositional environments: Amer. Assoc. Petroleum Geologists Bull., v. 71, p. 281–297.

Grover, G.A., Jr., and Read, J.F., 1983, Paleoaquifer and deep-burial related cements defined by regional cathodoluminescence patterns, Middle Ordovician carbonates, Virginia: Amer. Assoc. Petroleum Geologists Bull., v. 67, p. 1275–1303.

Hanshaw, B.B., and Back, W., 1980, Chemical mass-wasting of the northern Yucatan Peninsula by groundwater dissolution: Geology, v. 8, p. 222–224.

Harrison, R.S., 1975, Porosity in Pleistocene grainstones from Barbados: some preliminary observations: Bull. Canadian Petroleum Geology, v. 23, p. 383–392.

Holland, H.D., 1984, The chemical evolution of the atmosphere and oceans: Princeton, NJ, Princeton University Press, 582 p.

Illing, L.V., Wood, G.V., and Fuller, J.G.C.M., 1967, Reservoir rocks and stratigraphic traps in non-reef carbonates, Seventh World Petroleum Congress Proceedings, Mexico City, 1967, v. 13, p. 487–499.

Jakucs, L., 1977, Morphogenetics of karst regions: New York, John Wiley & Sons, 284 p.

James, N.P., 1972, Holocene and Pleistocene calcareous crust (caliche) profiles: criteria for subaerial exposure: Jour. Sedimentary Petrology, v. 42, p. 817–836.

James, N.P., and Choquette, P.W., 1983, Diagenesis 6. Limestones—the sea floor diagenetic environment: Geoscience Canada, v. 10, p. 162–180.

James, N.P., and Choquette, P.W., 1984, Diagenesis 9. Limestones—the meteoric diagenetic environment: Geoscience Canada, v. 11, p. 161–194.

Jennings, J.N., 1971, Karst: Cambridge, MA, MIT Press, 252 p.

Jennings, J.N., 1985, Karst geomorphology: Oxford, Basil Blackwell, 293 p.

Kirsvarsanyi, G., Grant, S.K., Pratt, W.P., and Koenig, J.W., eds., 1983, International conference on Mississippi Valley-type lead-zinc deposits: Proceedings, University of Missouri–Rolla, 603 p.

Kobluk, D.R., Pemberton, S.G., Karolyi, M., and Risk, M.J., 1977, The Silurian–Devonian disconformity in southern Ontario: Bull. Canadian Petroleum Geology, v. 25, p. 1157–1186.

Kyle, J.R., 1983, Economic aspects of subaerial carbonates, in Scholle, P.A., Bebout, D.G., and Moore, C.H., eds., Carbonate depositional environments: Amer. Assoc. Petroleum Geologists Mem. 33, p. 73–92.

Longman, M.W., 1980, Carbonate diagenetic textures from near-surface diagenetic environments: Amer. Assoc. Petroleum Geologists Bull., v. 63, p. 461–487.

Loucks, R.G., and Anderson, J.H., 1985, Depositional facies, diagenetic terranes and porosity development in Lower Ordovician Ellenburger Dolomite, Puckett Field, west Texas, in Roehl, P.O., and Choquette, P.W., eds., Carbonate petroleum reservoirs: New York, Springer-Verlag, p. 19–39.

Maslyn, R.M., 1977, Recognition of fossil karst features in the ancient record: a discussion of several common karst forms: Rocky Mt. Assoc. Geologists Symposium, Denver, CO, 1977, p. 311–319.

Meyers, W.J., 1974, Carbonate cement stratigraphy of the Lake Valley Formation, (Mississippian), Sacramento Mts., New Mexico: Jour. Sedimentary Petrology, v. 44, p. 837–861.

Meyers, W.J., 1978, Carbonate cements: their regional distribution and interpretation in Mississippian limestones of New Mexico: Sedimentology, v. 25, p. 371–400.

McCaleb, J.A., and Wayhan, D.A., 1969, Geologic reservoir analysis, Mississippian Madison Formation, Elk Basin Field, Wyoming-Montana: Amer. Assoc. Petroleum Geologists Bull., v. 53, p. 2094–2113.

Moore, G.W., and Sullivan, G.M., 1978, Speleology: St. Louis, MO, Cave Books, 150 p.

Mussman, W.J., and Read, J.F., 1986, Sedimentology and development of a passive to convergent margin unconformity—Middle Ordovician Knox unconformity, Virginia Appalachians: Geol. Soc. America Bull., v. 97, p. 282–324.

Ohle, E.L., 1980, Some considerations in determining the origin of ores of the Mississippi Valley type, Part II: Econ. Geology, v. 75, p. 161–172.

Palmer, A.N., 1984, Geomorphic interpretation of karst features, in LaFleur, R.G., ed., Groundwater as a geomorphic agent: Winchester, MA, Allen and Unwin, p. 175–209.

Perkins, R.D., 1977, Depositional framework of Pleistocene rocks in south Florida: Geol. Soc. America Mem. 147, p. 131–198.

Peyton, C.E., ed., 1977, Seismic stratigraphy—applications to hydrocarbon exploration: Amer. Assoc. Petroleum Geologists Mem. 26, 516 p.

Plummer, L.N., Wigley, T.M.L., and Parkhurst, D.L., 1979, Critical review of the kinetics of calcite dissolution and precipitation, in Gould, R.F., ed., Chemical modelling in aqueous systems: American Chemical Society, Symposium Series No. 93, p. 537–577.

Quinlan, J.F., 1972, Karst-related mineral deposits and possible criteria for the recognition of paleokarst: a review of preservable characteristics of Holocene and older karst terranes: 24th Internat. Geol. Congr., Sect. 6, p. 156–168.

Read, J.F., and Grover, G.A., Jr., 1977, Scalloped and planar erosion surfaces, Middle Ordovician limestone, Virginia: analogues of Holocene exposed karst or tidal rock platforms: Jour. Sedimentary Petrology, v. 47, p. 956–972.

Rhodes, D.A., Lantos, E.A., Lantos, J.A., Webb, R.J., and Owens, D.C., 1984, Pine Point orebodies and their relationship to the stratigraphy, structure, dolomitization and karstification of the Middle Devonian barrier complex: Econ. Geology, v. 79, p. 991–1055.

Roberts, A.E., 1966, Stratigraphy of Madison Group near Livingston, Montana, and discussion of karst and solution-breccia features: U.S. Geol. Survey Prof. Paper 526-B, 23 p.

Roehl, P.O., 1967, Stony Mountain (Ordovician) and Interlake (Silurian) facies analogs of Recent low-energy marine and subaerial carbonates, Bahamas: Amer. Assoc. Petroleum Geologists Bull., v. 10, p. 1979–2032.

Ross, R.J. Jr., and others, 1982, The Ordovician System in the United States: International Union of Geological Sciences Publ. No. 13, 73 p.

Runnells, D.D., 1969, Diagenesis, chemical sediments, and the mixing of natural waters: Jour. Sedimentary Petrology, v. 39, p. 1188–1201.

Sando, W.J., 1974, Ancient solution phenomena in the Madison limestone (Mississippian) of north-central Wyoming: Jour. Research, U.S. Geol. Survey, v. 2, no. 2, p. 133–141.

Schlee, J.S., ed., 1984, Interregional unconformities and hydrocarbon accumulation: Amer. Assoc. Petroleum Geologists Mem. 36, 184 p.

Scholle, P.A., and Halley, R.B., 1985, Burial diagenesis: out of sight, out of mind, in Schneidermann, N., and Harris, P.M., eds., Carbonate cements: Soc. Econ. Paleontologists & Mineralogists Spec. Publ. 36, p. 309–334.

Sloss, L.L., 1963, Sequences in the cratonic interior of North America: Geol. Soc. America Bull., v. 74, p. 93–113.

Steinen, R.P., and Matthews, R.K., 1973, Phreatic vs. vadose diagenesis—stratigraphy and mineralogy of a cored borehole on Barbados, W.I.: Jour. Sedimentary Petrology, v. 43, p. 1012–1020.

Strasser, A., 1984, Black-pebble occurrence and genesis in Holocene carbonate sediments (Florida Keys, Bahamas and Tunisia): Jour. Sedimentary Petrology, v. 54, p. 1097–1109.

Sweeting, M.M., 1973, Karst landforms: London, Macmillan Publ. Co., 362 p.

Thrailkill, J., 1968, Chemical and hydrologic factors in the excavation of limestone caves: Geol. Soc. America Bull., v. 79, p. 19–46.

Trudgill, S., 1985, Limestone geomorphology, in Clayton, K.M., ed., Geomorphology texts, 8: New York, Longman, 196 p.

Vail, P.R., Mitchum, R.M. Jr., and Thompson, S. III, 1977, Seismic stratigraphy and global changes of sea level, part 4: Global cycles of relative changes of sea level, in Payton, C.E., ed., Seismic stratigraphy—applications to hydrocarbon exploration: Amer. Assoc. Petroleum Geologists Mem. 26, p. 83–98.

Vail, P.R., Hardenbol, R.G., and Todd, R.G., 1984, Jurassic unconformities, chronostratigraphy, and sea-level changes from seismic stratigraphy and biostratigraphy, in Schlee, J.S., ed., Interregional unconformities and hydrocarbon accumulations: Amer. Assoc. Petroleum Geologists Mem. 36, p. 129–144.

Vernon, P.D., 1969, The geology and hydrology associated with a zone of high permeability ("Boulder zone") in Florida: Soc. Mining Eng., Amer. Inst. Mining Engineers, Preprint 69-AG-12, Geochim. Cosmochim. Acta, v. 40, p. 989–995.

Walkden, G.M., 1974, Paleokarstic surfaces in Upper Visean (Carboniferous) limestones of the Derbyshire Block, England: Jour. Sedimentary Petrology, v. 44, p. 1232–1247.

Ward, W.C., Folk, R.L., and Wilson, J.L., 1970, Blackening of aeolianite and caliche adjacent to saline lakes, Isla Mujeres, Quintana Roo, Mexico: Jour. Sedimentary Petrology, v. 40, p. 548–555.

Wigley, T.M.L., and Plummer, L.N., 1976, Mixing of carbonate waters: Geochim. Cosmochim. Acta, v. 40, p. 989–995.

Wilson, J.L., 1985, Petroleum reservoirs and karst (abstr.): Soc. Econ. Paleontologists & Mineralogists Ann. Midyear Mtg., Aug. 11–14, 1985, Abstracts v. II, p. 97–98.

Wright, V.P., 1982, The recognition and interpretation of paleokarsts: two examples from the Lower Carboniferous of South Wales: Jour. Sedimentary Petrology, v. 52, p. 83–94.

PART I General Karst Features and Processes

1
Characteristics of Dissolutional Cave Systems in Carbonate Rocks

Derek Ford

Abstract

A dissolutional cave or cave system is defined as a solution conduit of 5 to 15 mm minimum diameter that extends continuously between groundwater input points and output points. Thousands that are of explorable dimensions are known; the greatest contain more than 100 km of accessible galleries or are more than 1000 m deep.

Approximately 80% of these caves were created by meteoric water circulating without unusual geologic confinement; these are common caves. Their plan pattern building is governed by hydraulic gradients in penetrable fissures, complicated by reorientation of gradients when initial conduits connect and by microfeatures of the fissures. Thus, patterns are not precisely predictable. On the long profile, caves may display drawdown or invasion, vadose morphology, and shallow, deep, or mixed phreatic morphology. Many caves are multi-phase features with sequences of levels.

Two-dimensional joint-guided mazes of passages develop as anomalous portions of common cave systems or as separate caves, due to artesian confinement or diffuse input or to rapid flooding. Caves formed by CO_2-rich thermal waters display distributary dendritic or 2-D or 3-D maze forms. Some large caverns may develop where H_2S-rich waters are oxidized. Many irregular honeycomb caves develop where salt and fresh water mix in the coastal zone.

Phreatic passage cross-sections tend to be elliptical but are complicated by differing solubility or armoring of the floor. Vadose cross-sections are canyon-like or trapezoid. Both may display solutional scalloping or a paleoflow indicator, or may be modified or destroyed by breakdown.

A variety of clastic deposits accumulate in cave interiors. Fluvial facies are dominant. More than 100 secondary minerals are precipitated in caves. Calcite is predominant and the most significant for paleoenvironmental reconstructions.

Introduction

The purpose of this review is to summarize the nature and supposed genesis of the kinds of natural dissolutional caverns in carbonate rocks that are explored by cave explorers and studied by speleologists, i.e., that are enterable instead of being filled or microfeatures inspected in outcrop, in a mine, or in a drill core. It is certain that these accessible caves do not encompass all types of solutional voids that are produced in carbonates. However, they do display a wide range of origins, morphology, and scale, as is shown in the discussion below. As consequences of submergence and/or burial (with or without infilling by detritus or precipitates), all the differing types of accessible caves may become deep paleokarst features.

Several tens of thousands of solutional caves have been mapped worldwide by cave explorers. The great majority are less than 1000 m in length or 100 m in depth, exploration and mapping being halted by some obstruction. However, the longest caves are multistory complexes of galleries aggregating more than 100 km each, and at least 24 cave systems in mountain regions have now been penetrated to depths greater than 1000 m. The largest numbers of caves are known in the most intensively explored regions, which are Europe (including the western USSR) and the Appalachian and midwestern United States. In recent years there have been major expeditionary finds in southeast Asia, New Guinea and neighboring islands, and in Mexico and the Caribbean region. The greatest potential for number and' variety of caves occurs in southern China, which contains the greatest modern karst terrains (see Zhang 1980, Waltham 1986).

Figure 1.1 places solutional cave genesis within the broader context of the karst system. The figure is drawn from the perspective of a karst geomorphologist. Thus, it emphasizes the process and landscape components and complexities occurring where carbonate solution prevails in the net erosion zone, in contrast to the net deposition zone that is the chief concern of carbonate sedimentologists or petrologists. Karst developed in the net erosion zone is principally a variant of the fluvial geomorphic system, a variant wherein part or all of the channels required to discharge runoff are located underground (i.e. are caves) because the rock is soluble and hydraulic gradients are sufficient. Most dissolution occurs at the input end of the system, creating surficial landforms at scales ranging from mm to km and an epikarst (or subcutaneous) zone of comparatively dense solutional fissuring that may extend downwards into the rock for a few meters. But caves are essential to remove the water from the surface or epikarst zone; their extension to spring points and their early expansion must precede surface karst development except at the smallest scales.

Dissolution by waters from juvenile, connate, or deep meteoric sources that are comparatively hot and thus ascend through the carbonate section is of much lesser quantitative significance in the karst net erosion zone in general, but may be locally or regionally significant. Mixing corrosion involving different proportions of meteoric and such exotic waters is also locally important in this zone.

Development of caves in the zones of modern (i.e., Holocene) carbonate deposition has been little studied by speleologists because of the grave problems of accessibility—most caves are too small to enter or are filled with water, or both.

Definition and Classification of Caves

Although we may only directly explore and study caves that are ~50 cm or more in diameter, much research suggests that once a solutional conduit has attained a minimum di-

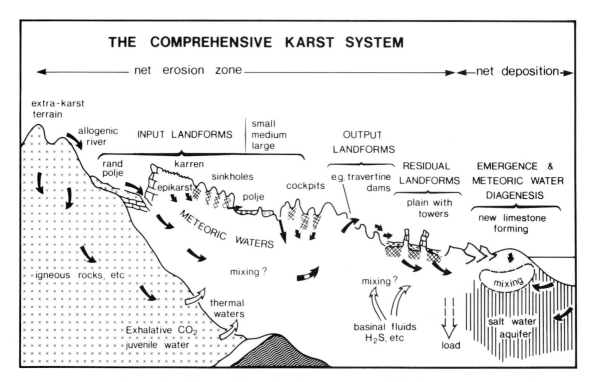

FIGURE 1.1. Diagram illustrating the components of the karst system.

ameter of between 5 and 15 mm, its form and hydraulic function need not change with further enlargement, even to the passage diameters of 30 m or more that are recorded in many caves (Ford and Ewers 1978, White 1984). This range of minimum diameters encompasses, for differing hydraulic gradients and channel roughnesses, the thresholds for turbulent flow, for optimum rates of dissolution, and for transport of clastic load. A *dissolutional cave system* is a conduit or connected sequence of dissolutional conduits of this minimum diameter or greater that extends continuously between input points (e.g., sinkholes) and output points (springs). Such systems are the subject of this review. They will be termed *caves* to avoid the repetition of a lengthy phrase. Dissolutional voids greater than 5 to 15 mm in diameter but not connected to water input or output points are types of isolated vugs. Dissolutional conduits less than ~5 mm in diameter but connected to an input or output point or both are *protocaves.*

When we apply these definitions we find that it is probably true that caves large enough to enter are in the minority. Most parts of most functioning cave systems have mean diameters less than ~50 cm but greater than 5 mm; they may be likened to First- and Second-Order

stream channels at the surface, which are generally difficult to study because they are too small to appear in maps, or in air photos if there is vegetation cover.

Table 1.1 presents a practical classification of the type of solution caves that are known. Its criteria are (1) type of solvent water, (2) hydrogeological setting, and (3) conduit patterns, applied in different combinations. The great majority of caves in the net erosion zone are created by meteoric waters circulating underground without any unusual geologic confinement, = common caves (Ford and Ewers 1978, Palmer 1984). They are the principal concern of this review. The other types are considered more briefly.

The Initial Conditions for Cave System Genesis

In most caves it is evident that the initial penetration by solvent waters occurred along fissures (bedding planes, joints, faults). Penetration via pores elsewhere in the rock is insignificant. In deep coal mines groundwater may "bleed" from fissures in limestone where the effective aperture is only ~0.5 μm, but much careful Hungarian work suggests that a minimum opening of about 10 μm is necessary for cave

TABLE 1.1. Generic classification of karst solution caves.

A.			
	NORMAL	Unconfined circulation in karst rocks	1. COMMON CAVES (~80% of known caves?)
	METEORIC WATERS	Confined circulation in karst rocks, or partial circulation in nonkarstic rocks	2. 2-D MAZE CAVES (artesian type) 3. BASAL INJECTION CAVES 4. COMBINATIONS OF TYPES 1 & 2
B.			
	DEEP	Enriched by exhalative CO_2 (normally, thermal waters)	5. HYDROTHERMAL CAVES (~10% of known caves?)
	ENRICHED WATERS	Enriched in H_2S, etc. (basin waters, connate waters)	6. CARLSBAD—TYPE CAVITIES
C.			
	MIXED WATERS	Chiefly marine & fresh water mixing	7. COASTAL MIXING ZONE CAVITIES
D.			
		Combinations of B or C with A, developing in sequence	8. HYBRID CAVES

propagation to occur in normal geologic circumstances and within reasonable timespans (Bocker 1969). White (1977) adopted a 25-μm minimum aperture for theoretical modeling. Probably most caves propagate through fissures where the connected apertures are between ~20 μm and 1.0 mm in width. The greater the initial aperture the faster is the propagation (extension) of the solutional protocave; it is the principal factor in determining the solvent penetration distances, discussed next.

Dissolution Processes in Meteoric Waters

In the pH range that is encountered in meteoric waters in karst terrains (broadly, pH between 5.5 and 9.5), the dissolution of calcite or dolomite occurs in two steps at the solid–liquid interface: dissociation of the cation and the anion from the solid, followed by the association of the latter with a proton in the liquid. For the simple system of calcite and pure water this is:

$$CaCO_3 + H_2O \rightleftharpoons Ca^{2+} + HCO_3^- + OH^- \quad (1)$$

This reaction equilibrates with 12 to 15 mg/l of $CaCO_3$ in the solution, which is a small amount. Thus, it has long been understood that most carbonate dissolution in normal meteoric waters is effected by the prior dissolution of CO_2 from an atmosphere; this creates carbonic acid which quickly releases protons:

$$CO_2 + H_2O \rightleftharpoons H_2CO_3^0 \quad (2)$$

$$CaCO_3 + H_2CO_3^0 \rightleftharpoons Ca^{2+} + 2HCO_3^- \quad (3)$$

The solubility of the gas is inversely proportional to temperature, approximately doubling between 20°C and 0°C.

Solute concentrations at equilibrium in this system are determined by two sets of conditions: (1) Whether the source atmosphere is the standard atmosphere or a CO_2-enriched enriched atmosphere in a soil. P_{CO_2} in the standard atmosphere is 0.033% at sealevel and diminishes a little with altitude. In soils P_{CO_2} values of 21% are conceivable (all O_2 is replaced) but in practice the upper limit appears to be ~7%. (2) Whether all three phases can react together: when rain falls onto bare rock or water seeps through a soil containing carbonate clasts and abundant air spaces, further

CO_2 is able to dissolve to replace that carbonic acid already complexed with the carbonate. This is the open (or coincident) system. In a closed (or sequential) system, the groundwater is first saturated with CO_2, then the liquid encounters carbonate rock and reacts with it in the absence of any further gas source. Effective carbonate solubility in an ideal closed system is only one-third to one-half that occurring in an ideal open system. The initiation of normal meteoric water caves and of thermal water caves takes place in ideal closed system conditions; later, the system may be opened (i.e., may become vadose).

Carbonate equilibria in meteoric waters are discussed comprehensively in Schoeller (1962), Garrels and Christ (1965), Langmuir (1971), Stumm and Morgan (1981), and Butler (1982). For discussion of the dissolution kinetics of calcite see Berner (1978), Plummer et al. (1978 and 1979), Bögli (1980), and Dreybrodt (1981); for dolomite, Busenberg and Plummer (1982) and White (1984). Drake (1980, 1983, 1984) has developed systematic and environmental models to explain the worldwide variations of solute concentrations that occur in bicarbonate karst waters sampled at equilibrium. Smith and Atkinson (1976) analysed a much larger data set where a majority of sampled waters were not at equilibrium.

It is found that the capacity of a meteoric water to dissolve calcite ranges from a minimum of 50 to 100 mg/L, where allogenic rivers flowing in the open air sink and the system closes at the limestone contact, up to a maximum of 350 to 400 mg/L, where waters pass through high P_{CO_2} soils containing abundant limestone fragments. Waters reported as containing more than 400 mg/L $CaCO_3$ will have been enriched by common ions such as Ca from sulfate rocks, or by foreign ions and mixing effects.

Net dissolution rates are determined primarily by the amount of runoff available. They are found to range from << 1 $m^3/km^2/yr$ in arid regions up to maxima of 100 to 150 $m^3/km^2/yr$ in some rain forests.

There have been many discussions of the length of time that is required for a given meteoric water to become saturated with respect to calcite (see, e.g., Bögli 1980). Under reasonable laboratory conditions this may be up to three weeks. Herman (1982, and see White 1984) was unable to saturate a solution with dolomite within a reasonable time despite the use of a rapid reaction experimental design.

More significant to cave genesis was Weyl's (1958) investigation of the effective penetration distance—how far a water will proceed through a fissure before it is effectively saturated and so incapable of enlarging its route further. Where a dissolutional cave system (as defined above) already exists, Weyl (1958) and White (1977) showed that the penetration distances to obtain 90% saturation are of the order of kilometers. The enlargement of existing caves is therefore readily understood.

The problem that remains is to understand the extension of protocaves from an input boundary through penetrable but very narrow fissures, as illustrated in Figure 1.2. Bögli (1964 and 1980) and Dreybrodt (1981) contend that simple penetration distances are infeasibly short, and they invoke mixing corrosion mechanisms to extend them. Ewers (1982) and the present author argue that such mixing is physically infeasible at the extending dissolution fronts unless there is high effective porosity

(i.e., the rock is diagenetically immature or fissure apertures are exceptionally large). Weyl (1958), White (1977 and 1984), and Palmer (1984) adopt solution kinetic models that will permit extension of protocaves without any mixing mechanisms, at rates between 10^3 and 10^5 yr/km. These receive support from field evidence (Ford 1980, Mylroie and Carew 1986) in the form of some postglacially initiated cave systems in glaciated areas.

Development of Common Caves

The patterns of conduits that are created where caves are formed by meteoric water circulating without any unusual geologic constraints in carbonate rocks may be very complex three-dimensional entities. Therefore it is convenient to divide description of the pattern building into three sections: (1) development of plan patterns in the first phase of cave genesis, (2) development of long sections (the dimensions of length and depth) in the first phase, and (3) pattern modifications introduced by later phases (multiphase development).

Development of Cave Plan Patterns

Case 1. The Single Input Cave in a Single Fissure. This is the simplest case of solution cave genesis. It is illustrated in Figure 1.2. The distance between input point and output boundary in the fissure may be no more than 1 m (the karren scale of development) or greater than 10 km. The pressure head on the groundwater ranges from the thickness of a single bed to hundreds of meters. Before any modification by dissolution has occurred the mode of flow is laminar, and anisotropic but homogeneous, as indicated by the flow envelope in Figure 1.2, i.e., it is Darcy-type flow.

From field observations (Ford 1968) and hardware solutional modeling (Ewers 1982), distributary patterns of protocave conduits are created at the solvent input and grow preferentially in the direction of the hydraulic gradient. Actual courses adopted within the fissure depend upon its geologic microfeatures and can be considered to be random and unpredictable. In electrical analog terms, all protocaves are connected in series with the solutionally unmodified fissure downstream of them. The lat-

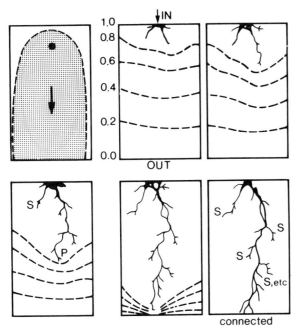

FIGURE 1.2. The propagation of a protocave pattern from a single input point towards an output boundary in a penetrable fissure. The flow envelope at the start of the experiment is shown top left; dashed lines = equipotentials; P = principal or victor tube; S = subsidiary tubes. From hardware experiments by Ewers 1982.

ter maintains very high resistance (the equipotential field in Figure 1.2 becomes deformed but maintains its initial value, or nearly so), so that the groundwater flow is slow and its discharge small.

One protocave (the principal tube) tends to grow ahead of others, deforming the equipotential field and so reducing the rate of solvent water supply at the solution fronts of its competitors (subsidiary tubes—Fig. 1.2). When the principal tube attains the output boundary the high resistance to flow is destroyed. Darcy flow conditions no longer apply because the connected tube is an inhomogeneity. The equipotential field in the remainder of the fissure is reoriented towards it.

The greater the initial aperture in a fissure, the simpler, straighter, and less branched is its pattern of protocaves. As principal tubes enlarge, much of the subsidiary branchwork is swallowed up; however, if a pressure head is maintained, surviving branchwork can continue to extend very slowly and may play an important role in pattern reorganization in later phases.

From experiments, principal tubes are known to have downstream diameters of about 1 mm when they first connect to the output boundary. Diameters rapidly enlarge thereafter because the flow rate is greatly increased upon connection. The longest time span in most speleogenesis is that required for principal tubes (the protocaves) to penetrate to outputs. As noted above, recent estimates suggest that most penetration times will fall between 10^3 and 10^5 yr/km. Penetration distances in a single fissure are sometimes greater than 10 km, but usually an output boundary (a spring point or a previously connected conduit) is encountered well within 1000 m.

Many thousands of such simple caves are known. The majority developed as groundwater cutoffs across the necks of incised river meanders, through the spurs at valley junctions, etc. (Jennings 1985) and so they are comparatively short. The plan pattern of their accessible passages is often complicated by the addition of a floodwater maze (see below) at the input end or by distributaries at the output end. This is the first type of cave to develop in many karst areas, and the most frequent type encountered in lowland glaciated regions (e.g.,

Quebec; Beaupre, 1975) because glacial disruptions have inhibited development of the more complex patterns.

Case 2. Multiple Inputs in a Rank. The basic situation is illustrated in Figure 1.3. In typical development principal tubes extend preferentially from one or more input points. When these connect to outputs, the reorientation of the equipotential fields towards them insures that protocaves extending from other, less successful input points will connect to them in sequence via the closest or most suitably oriented subsidiary tubes. Where fissure apertures are comparatively large (resistances are comparatively low) such lateral connections tend to be quite straight, i.e., close to the strike of the fissure. Where resistances are high the connections can be highly irregular.

The great example of these principles is Das Hölloch in Switzerland. It is the second longest cave system known in carbonate rocks (Bögli 1970 and 1980). Figure 1.4 shows the larger

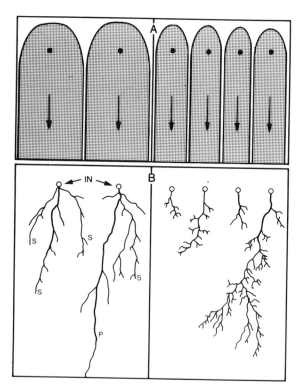

FIGURE 1.3. The competitive development of protocave patterns where there are multiple inputs in a rank, drawn for two and four inputs. Initial flow envelopes (A) indicate that input pressures are equal.

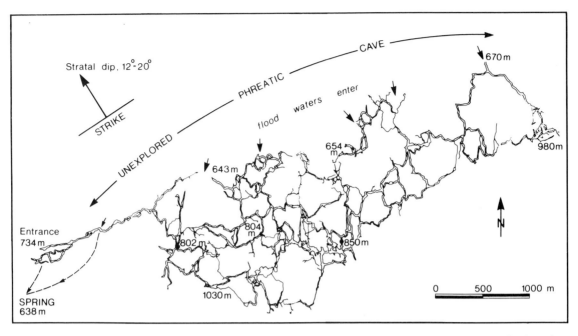

FIGURE 1.4. Simplified plan of the main system of galleries, Das Holloch, Muotatal, Switzerland. Elevations are quoted in meters above sea level. Adapted from Bögli 1970.

passages of its main section (about 80 km in aggregate mapped length), which is developed in a bedding plane that has also served as a thrust plane between two nappe sheets. The plane dips northwest at 12° to 25°. Inputs form a rank along its southern edge, and the output boundary is the western edge, i.e., to the strike. The cave is a multiphase sequence of trunk conduits that developed to drain the inputs westward at successively lower elevations. Their sinuous, highly irregular patterns occur because the trunk tubes connected in segments (west to east) that followed the closest protocave subsidiaries up, down, or aslant in the inclined, high resistance plane. Link lengths between adjacent subsidiaries are 50 to 250 m. Modern trunk conduits are typically phreatic in form and up to 8 m in diameter. Most of them are now relict. Modern waters flow westward below the water table via unexplored phreatic passages that are presumed to be of the same type. In the greatest recorded flood, water rose 180 m to overspill through the lower relict trunk routes.

Strike passages like these are common structural components of caves, though rarely are they as predominant as in Das Hölloch.

Case 3. Multiple Inputs in Multiple Ranks. The case sketched in Figure 1.5 is the one that most frequently occurs in karst. It is seen that where pressure heads are roughly equal, the rank of inputs that is closest to an output boundary will occupy most of the flowfield space approaching it. Principal and subsidiary tube competition proceeds in this rank as in Case 2, with propagation in further ranks being much slower. When the principal tubes of the nearest rank are connected, the hydraulic gradient to the second rank is greatly steepened so that its protocaves now propagate more rapidly, being directed toward connected tubes in the first rank. The process repeats itself in sequences of headward and lateral connections that are similar to the building of a Horton stream channel network. A priori, the particular patterns of connections that will occur cannot be predicted where there are many potential inputs in many ranks. If an input in a rank far from the output has a much greater pressure head than any others, it may connect before near ranks and thus distort the pattern, etc. Branchwork cave systems of this kind appear to predominate among the common caves. Because of geological and hydrological distortions of the flowfield

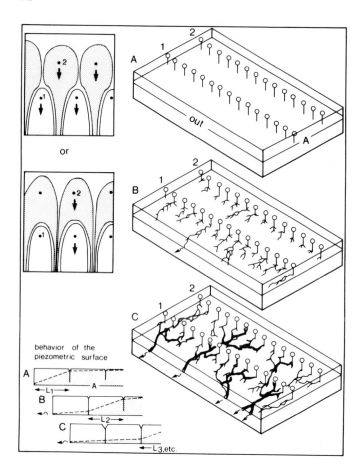

FIGURE 1.5. Competitive development of protocave patterns where there are multiple ranks of multiple inputs, drawn for the case of two ranks. Differing flow envelope configurations (equal pressures at all inputs) are shown. The multiple rank case is the most common found in karst and is normally the most complex in terms of interactions and resultant patterns.

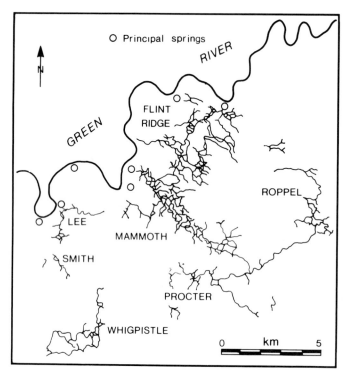

FIGURE 1.6. Illustrating the principal galleries of the Mammoth-Flint Ridge-Roppel-Procter cave system of Kentucky plus some nearby caves. All drain to springs along the Green River. All are multiphase, multilevel caves; the different levels are not discriminated here. The systems are excellent examples of multirank, multiple input cave genesis.

geometry during their propagation, very few examples conform well to the Horton laws or other empirical laws of channel geometry, and conduit position and scale is not predictable in detail. Nevertheless, the general principles of the pattern building are clear.

The Mammoth-Flint Ridge-Roppel-Procter system of interconnected conduits in Mississippian platform carbonates in Kentucky (Fig. 1.6) now exceeds 500 km in aggregate length and is the longest known system of caves of any kind. Its output points have shifted downwards and laterally many times in response to entrenchment of the Green River. As a consequence the conduit pattern is complex because much of it belongs to earlier phases. Earlier conduits are often truncated by surface erosion, while exploration and mapping remains incomplete in lower, younger levels. Despite these complications, it is possible to discern incomplete patterns of near and far ranks connecting to the springs. A recent lateral connection in far ranks (Procter and Roppel caves) beheads much drainage in Mammoth Cave. Detailed discussions are given in Palmer (1981) and Quinlan and et al. (1983).

Case 4. Inputs Restricted to a File. This case (Fig. 1.7) completes the range of significantly different karst input–output configurations that can occur. Inputs are restricted to a narrow zone where the carbonate rock is exposed, most often a valley floor (Fig. 1.7A). The spring point is at or close to one end of the zone. Flowfield geometry constrains the far inputs and so the cave is built in headward steps (Fig. 1.7B).

This type of cave is frequently encountered. The example given in Figure 1.7 is complicated because, once again, the network is multiphase and incompletely explored. Entrenchment in the Cumberland River (Fig. 1.7C) has exposed new, lower output points at C so that, in the latest phase, input water at point A has abandoned the subvalley course toward springs at point B, to develop a more direct route westward beneath the shale caprocks. This new route is not yet enterable in size.

Floodwater Mazes. In all four cases outlined above, the directions of protocave propagation and connection are dictated by hydraulic gradients and changes in their orientation. This implies that the initial skeletons of cave passages

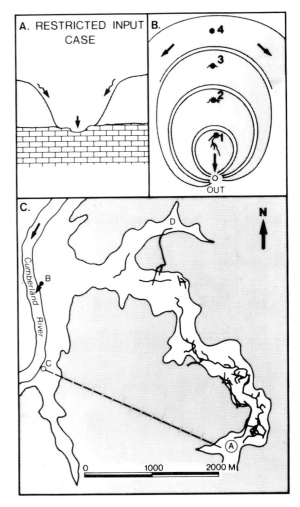

FIGURE 1.7. Competitive development of protocave patterns where input points are restricted to a file. This normally develops in a valley bottom, as shown. The example is of the Cave Creek caves in Kentucky (modified from Ewers 1982).

are developed under phreatic conditions, which is true in most instances (see below). With enlargement of their cross-sections, many caves later become partly or entirely vadose. If they are then subject to sudden flooding, local mazes of small passages may be created where trunk passages become obstructed, being generated by the very steep hydraulic gradients applied across rock around the obstruction during the flood (Palmer 1975).

Such floodwater mazes are a common superimposition onto the basic plan patterns of dissolution caves. They are most frequent in shallow caves such as those of the US Midwest, especially in the upstream parts (prone to damming by inwashed tree trunks) and where there is allogenic drainage to supply the sudden, large floods.

Development of Caves in Long Section (in Length and Depth)

For many years there was controversy concerning the question whether limestone caves should develop preferentially in the vadose zone (Martel 1921), or in the phreatic zone (Davis 1930, Bretz 1942), or in the water table–epiphreatic zone that separates them (Swinnerton 1932, Rhoades and Sinacori 1941); see Warwick (1953) for a comprehensive review.

The present author has shown that the question was irrelevant (Ford 1971, Ford and Ewers 1978). Common caves develop in all three zones, most of the larger cave systems being hybrids of two or more of the zonal types.

Types of Vadose Caves

Most limestones and dolomites exposed in the karst net erosion zone are diagenetically mature so that their effective porosity is generally confined to secondary fissures, as noted. The aggregate volume of such fissuring is normally very small when dissolution by circulating meteoric water begins, so that the water table is effectively at the surface of the rock. A vadose zone is then created by protocaves first connecting to springs (as described above), and then enlarging and progressively draining the rock that lies above the elevation of some minimal hydraulic gradient projected from those springs. This hydraulic gradient becomes the final, stable, and low-stage water table (Fig. 1.8A).

Two types of vadose caves occur. Drawdown caves are those developed as the vadose zone is created by dissolutional enlargement. Their courses follow the earlier phreatic skeleton of principal and connected subsidiary tubes and become entrenched below it as the genetic stream switches from a phreatic to a vadose state. Passages tend to be narrow river canyons, with rounded phreatic form preserved in their roofs.

Invasion caves occur where a new stream can

be delivered to a vadose zone that was created in an earlier phase. These may adapt some parts of the previous phreatic skeleton or create a rough and partial one of their own. In general they are steeper than drawdown vadose caves, exploiting any vertical fissures to form sequences of shafts linked by short canyons. Such caves may also develop where the effective porosity is primary (as in many young uplifted reef rocks) or where resistance in the fissure network is very low because it has been partly sprung open by rapid uplift in young mountain chains (Bögli 1980). As implied in Figure 1.8A invasion vadose caves are particularly common where glacial action disorders the allogenic drainage entering established karst terrains.

Differentiating Phreatic and Water Table Caves

The nature of the cave profile at or below the stabilized water table of a given phase is determined by the frequency and/or resistance (and to a lesser extent, array) of the fissures that are penetrable by groundwater (Ford and Ewers 1978). Simply stated, the principle is that the greater the frequency, the shorter and shallower will be the conduit connections of the type shown in Figures 1.3 and 1.5. Four distinct geometries are possible (Fig. 1.9A):

1) Where fissure frequency is very low the cave may constitute a single deep loop below the water table = *bathyphreatic cave*. Such caves are most difficult to explore, so that complete examples are few (Fig. 1.9B). Vaucluse Spring (France) and La Hoya de Zimapan (Mexico), are the outlet parts of an active and a fossil loop, respectively, where the vertical lift is greater than 300 m in solutional tubes of 20 to 30 m diameter. There are other great caves plus some of the world's largest karst springs in the Zimapan area (the Sierra de El Abra), and all are of this configuration because of low fissure frequency in a massive, vuggy reef (Fish 1977).

We do not know the maximum depth attained by bathyphreatic caves. Exploration drilling in many areas has tapped caves filled with young flowing water at depths down to −3000 m. Some may be bathyphreatic but it seems likely that many are shallower types of caves that have been lowered by tectonic activity.

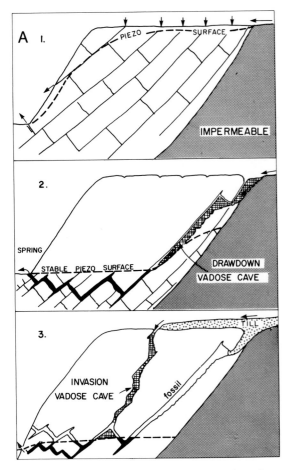

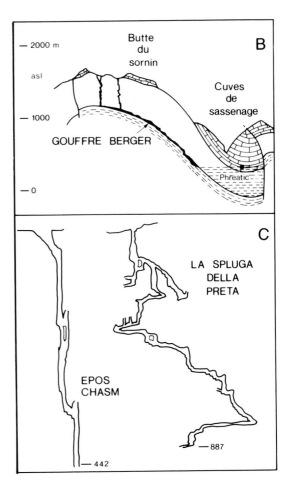

FIGURE 1.8A. The differentiation between drawdown and invasion types of vadose caves. *B.* Gouffre Berger, in the Forealps of France, has been explored to a depth of 1250 m below its higher entrance. It drains to the Cuves de Sassenage springs. Although it is a multiphase cave system with invasion vadose elements at its head, most of this great cave can be considered a drawdown vadose canyon at the contact between thick bedded, pure Urgonian limestones and underlying argillaceous limestones. *C.* Epos Chasm, Astraka Plateau, Greece, is an excellent example of a simple invasion vadose shaft. La Spluga della Preta, Veronese Alps, Italy, is a sequence of shafts and steep canyons that is more typical of invasion cave morphology.

2) Multiple-loop phreatic caves develop where fissure frequency is low. The tops of higher loops define the position of the stable water table, though this may later be lowered somewhat by entrenchment into them. Das Hölloch is a good example of this type of cave; the vertical amplitude of its phreatic loops generally ranges from 80 to 180 m.

3) Caves that are a mixture of shorter, shallower loops and quasi-horizontal canal (i.e., water table) passages represent a third category, of higher fissure frequency and diminishing resistance. The horizontal segments tend to exploit major joints or propagate to the strike in bedding planes.

4) Where fissure frequency is very high, low-gradient (most direct) routes are constructed via successive ranks of inputs behind the spring points. When sufficiently enlarged by dissolution, these may absorb all runoff; thus, the piezometric surface is lowered into them. They become *ideal water table caves*. Short examples that are wading or swimming canal passages with low roofs are very common, e.g., passing through residual limestone towers on alluvial plains in southern China, Vietnam, Malaysia,

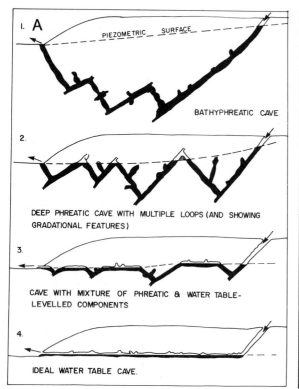

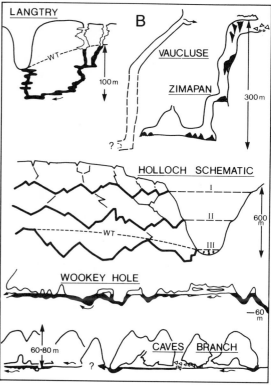

FIGURE 1.9A. The four-states model of phreatic and water table cave differentiation (from Ford and Ewers 1978). It is difficult to explore single, deep phreatic loops. B. Langtry Caves, Texas are shown in their middle phase of development with ~100 m of loop amplitude (Kastning 1983). Fontaine de Vaucluse, France, is the deepest known karst spring, −315 m. Lower parts were explored by a robot submersible. It is probably the ascending limb of a single loop in a syncline. La Hoya de Zimapan, Mexico, is a paleospring, an ascending limb in massive reef rocks that is partly infilled by stalagmite and choked with clay at −300 m (Fish 1978). Das Holloch, Switzerland is an outstanding example of multiple looping that is shown here in three successive phases (schematic figure adapted from Bögli 1980). Wookey Hole, England is a rapidly alternating sequence of loops and water table passages, in thick bedded and steeply dipping limestones. Chac Be Haabil System drains the Caves Branch polje, Belize. It is an ideal water table passage many kilometers in length that is interrupted by cockpits and breakdown (from the work of Miller 1982).

and Cuba. Large sectors of the lengthy Domica-Baradla system straddling the Czech-Hungarian border in highly fractured limestones (Jakucs 1977), are of this type, as are many allogenic river passages that traverse the dense sinkhole terrains (cockpit country) of Belize, Cuba, Jamaica, and Puerto Rico; see Miller (1981) for a detailed study of a Belize example. However, it is comparatively rare for water table passages to extend for more than ~1000 m without being interrupted by, at least, a shallow phreatic loop.

The division into four types (termed the *four states model* by Ford and Ewers 1978) is intended to recognize the four distinct geometries that might occur in phreatic–water table cave long sections. In practice, many caves display a mixture of states 2 and 3 or 3 and 4 in their composition. The longest system, Mammoth-Flint Ridge-Roppel-Procter, shows drawdown vadose plus later invasion vadose drains feeding into a mixture of states 3 and 4 passages. As a system ages, gradational processes tend to flatten its profile (Ford and Ewers 1978, p. 1791).

It follows that, with respect to the genesis of cave systems that are of explorable dimensions, there is a total of six states of significant fissure frequency and resistance. In state 0, frequency is so low or resistance so high that no cave can be generated in the available geomorphic timespans. This is true of many marble outcrops. States 1 to 4 are as defined. In state 5 fissure frequency is very high or a high primary porosity replaces it, so that a high density of tiny caves or protocaves is created instead of enterable caves. This is true of a majority of the chalks and much of the Jurassic oolitic limestone in Britain, for example, and for poorly lithified Holocene–Pleistocene limestones.

There can be no simple, single scale for the fissure frequencies required to generate the different geometries, because of differing resistances within fissures. Where frequency is low, a few interconnected, low-resistance fissures may permit state-3 caves to develop, or even state 4 if the route is short as in a subterranean meander cutoff. It should be noted that the frequency measured in natural outcrops or in quarries is generally a poor guide to what will be the effective frequency beneath the epikarst zone.

Where strata dip comparatively steeply (2° to 5° or more) and there are penetrable bedding planes, the planes tend to entrain groundwater to great depth because their gradients are steeper than most initial hydraulic gradients and there is a quasiartesian trapping effect. As a consequence, the deeper types of phreatic caves are particularly common. In contrast, where strata are nearly horizontal, conditions are most favorable for perching groundwater streams on aquitards such as some dolomite beds, thick shales, etc., so that shallow phreatic or water table caves are more common. In tightly folded rocks (i.e., where cave systems extend over several or many folds) effective fissure frequency is generally high, again favoring the shallower types of cave geometry.

To conclude this section, it is stressed that it is possible for a cave system, between the input points and the spring or springs, to consist entirely of one of the two vadose cave types or four phreatic–water table types. Many short caves are monotypic. However, most larger caves will display inlet passages of one or both vadose types draining into one or a mixture of the phreatic–water table types.

Multiphase Cave Systems

Cave systems can experience two different types of changes that will halt or radically alter their evolution. The first type occurs when the spring position is raised or lowered. The second occurs when, because of climatic change or some other cause, the cave is infilled by clastic or precipitate deposits. Raising the spring position inundates a cave and may also cause its aggradation by clastics. This has been the fate of the majority of caves reported in deep paleokarst. They are difficult or impossible to explore directly and so are little known to speleologists. Some principal features of clastic and precipitate deposits in caves are noted at the end of this review.

A majority of the longer limestone cave systems that are known display the effects of one or more lowerings of the spring position. The simplest effect is a vadose canyon entrenchment into the floor of the previous passage. This is common in short, simple caves such as cutoffs and in individual passages in larger systems. It implies that the spring point has merely descended, without there being any geographical (e.g., lateral) shift in its position.

Geographical shift of spring points occurs in most instances because the lowering will expose new fissures along which subsidiary protocaves are well extended. The consequence is that, in a series of steps proceeding headward through the system, a new cave is built by diversions from the old cave, which becomes partly or entirely abandoned by its formative streams and left as a relict upper story. Most speleologists term these upper passages *levels*, though it must be understood that (from Figs. 1.8 and 1.9) they rarely will be level in the strict sense.

The three-dimensional patterns of multiphase, multistory common cave systems are genetically perhaps the most complex forms that are preserved in all geomorphology. They certainly compose the most complex erosional networks. Passages may crisscross repeatedly on the plan without any direct past or present hydrological connection between them. In many systems it is evident that the effective fissure frequency increased in later phases, indicating that principal and subsidiary conduits continued to extend laterally and below the connected trunk conduits. The effect is that the later, lower stories have a greater proportion of level sections (water table cave) within them; these may originate at the bases of older loops, or pass beneath them from other inlets. Vadose shafts bisect older galleries of all types and plunge beneath them. Invasion vadose stream channels of recent phases may chance to intercept old phreatic passages, wander along them for a few hundred meters, and then leave again on a different bearing. These are all effects of downward plus lateral shifts of the springs. The Mammoth-Flint Ridge-Roppel-Procter cave system provides outstanding examples.

Development of Other Types of Caves

Meteoric water caves developed where there is confined circulation. The caves in this category are discussed below.

Two-Dimensional Maze Caves

These caves are quite frequently encountered, either as separate entities or as anomalous parts of common caves. They are reticulate mazes of small passages following a joint system or systems confined to one or a few adjoining beds. Passages have a rounded or tapered cross-section, and those that are parallel tend to be of similar dimensions. They are formed by slowly flowing phreatic waters, although late-stage transitions to floodwater maze form (sharper and rougher) are sometimes seen.

Such mazes perhaps occur most often where strata with a high frequency of penetrable fissures are confined in a state-1 or state-2 hydrodynamic environment by geological trapping effects. Local "sandwich" situations, where one or a few densely jointed beds are trapped between massive limestones, are the usual cause of maze sections within common caves; Lummelund Cave, Sweden, is an outstanding example (Engh 1980). Independent maze systems develop where the trapping is artesian, e.g., directly beneath a shale or other aquiclude or aquitard stratum. This is the explanation for the greatest such mazes, which occur in gypsum

A

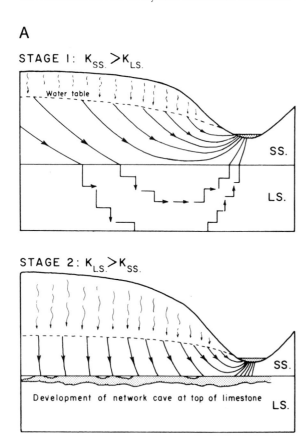

B

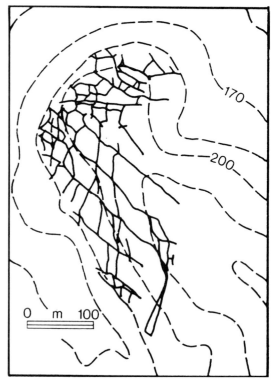

FIGURE 1.10A. A. N. Palmer's model (1975) for the development of 2-D rectilinear (joint) mazes in limestones directly beneath sandstone aquifers that function to diffuse the input waters; K = hydraulic conductivity. *B*. La Baume Cave, Missouri. (From A. N. Palmer, 1975.)

in Podolia, USSR (Jakucs 1977, Dublijansky 1979). Several examples there exceed 100 km in aggregate passage length.

Palmer (1975) made a careful investigation of maze caves in the United States and found that 86% were developed directly beneath permeable sandstones. He therefore proposed the alternative model given in Figure 1.10A. An equidimensional maze is created because water is introduced to the soluble rock by way of an insoluble homogeneous diffusing medium. No doubt each control applies in different cases.

Caves Created by Basal Injection of Normal Meteoric Water into Karst Rocks

Glennie (1954) considered this speleogenetic possibility and termed it *hypophreatic*. Only one example in carbonate rocks is known to the author. This is an excellent local study by Brod (1964) of small solutional pit and fissure caves scattered for 60 km along the denuded crest of the Rockwoods anticline in eastern Missouri. The caves descend steeply through 60 m of limestones and basal dolomites. There is local maze form at the bottom, where an underlying sandstone may be exposed, including sandstone blocks partly rounded in situ. The sandstone, St. Peter Formation, is an important regional aquifer 40 m in thickness. Brod proposes that it discharged upward through the carbonate aquitard where the latter is fractured and uncapped at the anticline.

Hydrothermal Caves Associated with CO_2-Enriched Waters

This is a frequent type of cave, accounting for roughly 10% of all that are known and explored. The principal studies have been made in eastern Europe and the Soviet Union, where recent reviews in English include Jakucs (1977), Muller and Sarvary (1977), and Dublijansky (1980).

The waters may be juvenile, connate, or meteoric in origin, or mixtures of these types. The common features are that they are significantly heated so that strong thermal convection may occur, and that they contain CO_2 gas from deep sources. As the waters cool during ascent this dissolves, thus permitting enhanced carbonate dissolution.

Hungarian specialists recognize three morphologically distinct types of hydrothermal caves. The first, which is certainly created by the hot waters alone, is represented by the remarkable Satorkopuszta Cave shown in Figure 1.11A. This comprises a basal chamber (the limit of exploration) that may be likened to a magma chamber, from which a branching pattern of rising passages has grown like the trunk and branches of a tree. Their form is modified by cupola-form ceiling dissolution pockets evidently created by convection currents in slowly flowing waters. Most branch passages terminate in the pockets; one gained the surface and discharged the hot water. Brod's Missouri fissure caves (noted above) are fundamentally similar but lack the branching and pocketing as well as any exotic thermal precipitates.

The second type of cave is a two-dimensional rectilinear maze. This may be created where water ascending fissures in carbonates is halted by an overlying aquiclude stratum and compelled to spread laterally or where the hot water gains the water table and spreads as a lens above meteoric water there, mixing with it to enhance the dissolutional capacity by mixing corrosion mechanisms.

The third type of cave is a three-dimensional rectilinear maze (i.e., a multistory cave) created within a single phase. This is the most common type of thermal cave. The hot springs and higher relict caves of Buda Hill, Budapest, are examples with a long history of use and study. The greatest examples are Jewel Cave (Fig. 1.11B)—118 km of galleries packed beneath an area of 2.7 km^2) and Wind Cave (70 km beneath 1.8 km) in the Black Hills of South Dakota. These are, respectively, the fourth and tenth longest caves that are known. They are developed in 90 to 140 m of well-bedded limestones and Ca-rich dolomites capped by sandstones and shales. There are modern hot springs nearby that discharge from the sandstones where the shale cap is breached by valleys.

Jewel Cave (Fig. 1.11B) is entirely relict. It has no relation to the modern surface topography and is entered where a gully chanced to intersect it. Its great features are big passages (up to 20 m in height) covered by 6 to 15 cm of euhedral calcite spar with a silica overgrowth (Deal 1968). Wind Cave is also a hydrologic rel-

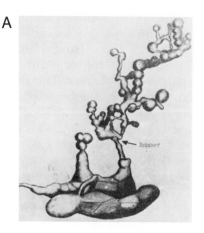

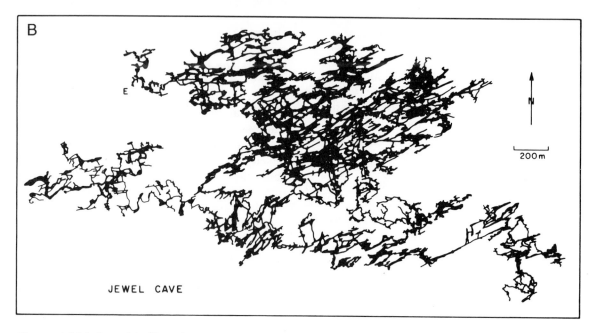

FIGURE 1.11*A*. A model of Satorkopuszta Cave, Hungary, an example of a dendritic thermal water cave with cupola-form (convectional) solution pocketing. (From Muller and Sarvary 1977.) *B*. Plan of Jewel Cave, South Dakota, a 3-D maze that is believed to be of hydrothermal origin. (From A. N. Palmer, 1975.)

ict displaying thermal water encrustations but its lowest accessible parts are filled by semistatic waters fed from below. Bakalowicz et al. (in press) use stable isotope ratios to show that the precipitates are of thermal origin and that wall rocks are altered toward the thermal ratios as well. They propose a model wherein regional heated groundwaters from the granitic core of the Black Hills converged and ascended through the cave zones en route to paleospring points in the overlying sandstones that are now largely stripped away. The morphological similarity of these caves to certain Mississippi Valley type (MVT) deposits such as Jefferson City Mine (Fulweiler and McDougal 1971) is evident.

Caves Formed by Waters Containing H_2S—Carlsbad Caverns and Other Caves of the Guadalupe Mountains, NM

The southern front of the Guadalupe Mountains is an erosional scarp in reef strata overlooking the Pecos evaporite basin (Fig. 1.12). The escarpment is ~50 km in length and con-

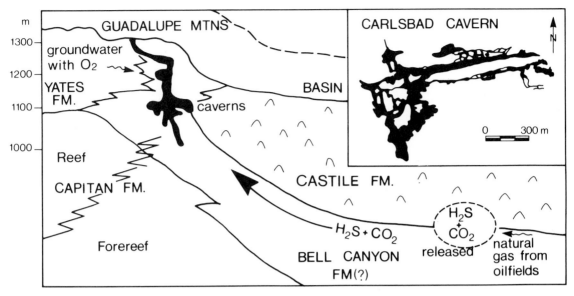

FIGURE 1.12. Geological section of the Guadalupe Mountains, New Mexico, at Carlsbad Caverns, with a plan of the cave in the inset. Drawn to illustrate genesis with CO_2 and H_2S migrating from oilfields. (Modified from C. A. Hill, 1987.)

tains more than 30 major caves, including the celebrated Carlsbad Caverns. These caves display a network form, with great rooms in the massive, vuggy reef rock being linked to higher passages or shafts in backreef strata and underlain by blind pits. Bretz (1949) proposed a complex meteoric water and backflooding model to explain Carlsbad Caverns, but subsequent explorations there and elsewhere have made it plain that the caves bear little or no relation to the modern topography or surficial hydrology and that they display only a slow-discharge phreatic morphology with the exception of a few late-stage and minor intrusions. Some big rooms and passages contain layered gypsum resting on thinner residual clays.

Queen et al. (1977) suggested a coastal zone mixing model for the genesis of the big rooms, in which gypsum replaced reef limestone in a reflux phase and was then preferentially dissolved in a later meteoric water phase. Davis (1980) and Hill (pers. comm., 1985) proposed instead that H_2S was produced by biogenic reduction in the adjoining gas fields (Figure 1.12) and migrated up-dip with other expelled basin fluids. Approaching the water table it became oxidized to form H_2SO_4, perhaps by some mixing with meteoric waters. The blind pits are the base of the oxidizing zone and the big rooms represent the areas of most effective oxidizing

and consequent dissolution; when local supersaturation occurred there gypsum was precipitated. Isotopic evidence shows that the sulphur has a gas field source rather than the marine source required in the model of Queen et al. (Hill, pers. comm., 1985).

Dublijansky (1980) has defined a chamber cave category of hydrothermal caves in the USSR, and the Guadalupe caves seem similar in many respects but larger. From work in progress, the present author accepts the ascending water proposals of Davis and Hill but suspects that gasfield CO_2 may play at least as important a role as does H_2S.

Caves in Limestone Formed by Gypsum Replacement and Solution

Egemeler (1981) describes a distinctive subtype of caves from the Big Horn basin of Wyoming. They are short, horizontal, vadose outlet passages in limestone that taper inward and terminate (after a few tens or hundreds of meters) in warm springs rising from narrow fissures. H_2S is liberated into the air from the water. It reacts to convert limestone to gypsum crusts which fall off due to their weakness and expansion forces, exposing fresh limestone to alteration. The fallen gypsum is removed in solution in the waters.

Seacoast Mixing Zone Caves

This final category of caves develops predominantly in young (late Tertiary and Quaternary) and diagenetically immature limestones within the karst net deposition zone. The mechanism is a salt–fresh water mixing corrosion effect discussed by many authors, e.g., Plummer (1975), Busenberg and Plummer (1982), and Back et al. (1984). Because there is high effective porosity via pores and vugs the pattern constructional role of bedding planes and joints (predominant in the other types of caves) is less significant. Spongework maze caves extending to strike along the mixing zone are anticipated. If large in size they will be partly collapsed.

Such caves do occur. Back et al. (1984) describe a "Swiss cheese zone" where there is modern mixing on the Yucatan coast. Dense honeycomb development within linear zones is described by Gregor (1981) in the British West Indies and by Ollier (1975) in the Trobriand Islands, Solomon Sea. Where very high cavernous porosity is created by this or other mechanisms the mixing zone migrates up close to sealevel so that the later cavities become sub horizontal.

Most explored caves in or close to modern mixing zones are more complex than this because they are polygenetic as a consequence of the rapid eustatic oscillations of sealevel that occurred during the Quaternary. Phreatic, then vadose, meteoric dissolution occurred when sealevels fell, together with much collapse of walls and roofs left unsupported by water. Vadose stalagmites have been recovered from −44 m in the Bahamas, and dated to ~140,000 yrs. B.P. by U series methods (Gascoyne et al. 1979). It is probably true to write that our best examples of mixing zone caves are to be found, not in the modern karsts at all, but in shelf-carbonate paleokarsts such as that of Pine Point, N.W.T., Canada (Rhodes et al. 1984) or West Texas (Craig, chap. 16, this volume).

Erosional Features within Cave Passages

When inside a cave we cannot see the planimetric and other gross structural characteristics discussed above, merely the local cross-sections and features that are indented into them. These are also the features most readily recognized in paleokarst caves exposed in outcrops and, with careful interpretation and some good fortune, in drill cores. This section of the review presents a brief summary of them. The very detailed discussion of phreatic and vadose features presented by Bretz in 1942 remains pertinent. Bögli (1980) and Jennings (1985) give modern reviews at greater length than is possible here.

There are four basic types of cave cross-section: (1) the phreatic type in which the solutional attack may be delivered uniformly around the perimeter or where the floor may be armored by clastic load; (2) the vadose type, where the erosional attack is confined to the bed and lower walls of a channel, or is straight down a vertical shaft; (3) the breakdown type, where rock fall has completely altered one of the previous types; (4) compound cross-sections, combinations of any two or all three of the previous types.

Phreatic Forms

Passage Cross-Sections. The ideal phreatic cross-section is circular (the minimum friction form) as in standard pipes. Caves of nearly perfectly circular cross-section do occur, and may be guided by horizontal or vertical fissures or the intercept of the two. "The Subway" in Castleguard Cave, Canada, is a celebrated example that maintains a diameter of 4 m and is quite straight for a distance of 500 m (Ford et al. 1983). It is guided by a bedding plane in dense, crystalline limestone and water flow was fast. However, it is more usual for the cross-section to be elongated along the host fissure to form an irregular ellipse (Fig. 1.13b, Fig. 1.14). Clastic armoring of the floors or their protection by chert bands, etc. causes the profile to extend preferentially upwards (Fig. 1.13c). Where flow is slow (as in most thermal water caves and other phreatic mazes) the dissolution process may pick out any minor variations in the solubility of successive beds or patches of rock, to create highly irregular profiles with projecting and recessive elements (Figs. 1.13d,e). At the extreme this produces honeycomb walls, as in "The Boneyard" of Carlsbad Caverns, which is in vuggy reefal strata.

FIGURE 1.13. *a–e* Schematic cross-sections for phreatic passages guided by a bedding plane fissure. *f* Cupola form solution pocket in the roof of Wind Cave, SD, believed to be a thermal convection dissolution feature. *g* Development of a standard solution pocket, triggered by mixing corrosion at a minor joint-trunk passage junction. *h* The enlargement of a paragenetic passage; under equilibrium conditions the open cross-sectional areas, X and X^1, remain equal.

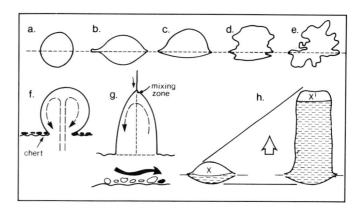

Solution Pockets. Solution pockets are among the most attractive features of phreatic caves and are those that most surprise geologists who are not cave specialists. They occur in walls or roofs and may extend upwards as much as 40 m. Many terminate in a tight joint that evidently guided their expansion. They may be single or multiple features (Figs. 1.13 and 1.15).

It is probable that most solution pockets are a "trigger effect" of mixing corrosion. Soil waters descending the guiding joint mix with allogenic trunk waters in the passage and create a small solutional depression at their juncture. This is then much enlarged by trunk waters at times of flood when they are flowing fast and are chemically aggressive; thus, the mixing zone depression recedes up the joint. As noted, very rounded or cupola-form pockets without guiding joints (Fig. 1.13f) are probably created by thermal convection.

Paragenesis, Ceiling Tubes, and Pendants. As defined by Renault (1968) and other French authors a *paragenetic* passage is any phreatic or water table passage where the dissolutional cross-section is partly attributable to accumulations of fluvial clastic detritus. Ford (1971) and other English-speaking authors have limited the term to passages that have an aggrading floor and a steadily rising solutional ceiling i.e., maintaining a roughly constant open cross-section (Fig. 1.13h). Pasini (1973) named this *erosione antigravitativa!*

Paragenetic passages originate in enlarged principal tubes. With enlargement, maximum groundwater velocities may be reduced, permitting the permanent deposition of part of any insoluble load. This armors the bed and lower walls; solution proceeds upwards on a thickening column of fill. The vertical amplitude of such paragenesis exceeds 50 m in some instances. Remarkably flat roofs that bevel dipping strata can be produced. The process terminates at the water table.

Ceiling half tubes are small, often meandering, channels that are indented into the roofs of some broader passages that have been completely filled with clastic debris, i.e., they develop to discharge the surviving flow through a conduit that has become choked. Although normally in the apex of a passage, horizontal examples are known in walls, while others climb from wall to apex.

Pendants, or *pendanting,* describes very complex anastomosing solutional channel patterns in walls or ceilings that most often develop where water is flowing slowly at the contact between rather impermeable (clay-rich) fill and the bedrock. It may be transitional to single half tubes.

Vadose Forms

Passage Cross-Sections. The form of vadose passages is that of net entrenchment, with or without channel widening. At its minimum development, the entrenchment is an underfit in the floor of a phreatic passage (Fig. 1.16a), suggesting that regional phreatic flow has been rerouted and the passage now conveys only local epikarst drainage. T-form or keyhole passages (Fig. 1.16b) suggest that the same stream has always occupied them, switching from phreatic to vadose conditions. Figure 1.16c is characteristic of drawdown vadose morphology; en-

FIGURE 1.14. The elliptical cross-section that typifies phreatic conduits formed in bedding planes. Clay banks masking the floor are from a later reflooding event.

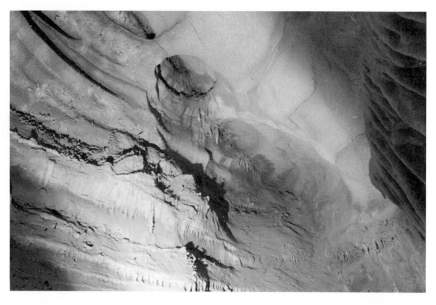

FIGURE 1.15. Oblique view of a line of small solution pockets developed along a joint in a relict phreatic cave. Since the trunk conduit waters were diverted to lower passages, subsoil waters saturated with re- spect to calcite have continued to descend the joint. They now precipitate stalactites where formerly they mixed with the trunk stream to dissolve the pockets.

FIGURE 1.16. *a–d* Standard types of vadose cross-sections. *e* McClung's Cave, W. Virginia. *f* Illustrates the mechanism of corrosion notching in a standing pool. *g* Corrosion beveling in Na Spicaku Cave, Czechoslovakia, where strata dip 70°.

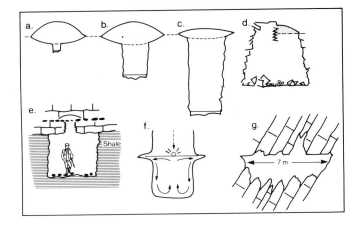

trenchment predominates but the phreatic part was first and fixes the plan position of the channel. Single trenches as deep as 100 m are known; they are "canyons with a roof on" (Fig. 1.17).

Many vadose streams establish rough equilibrium profiles and accumulate bedload that is moved only in floods. Central bedload deposits encourage solutional undercutting of the channel walls. There is collapse of large blocks from them into the channel, deflecting flow to an opposite wall where the process is repeated. By such undercutting and breakdown trapezoid cross-sections of stable width may be arrived at (Fig. 1.16d). These are common.

All these types of vadose channels can be created by dissolution acting alone but mechanical erosion (corrasion) is often important. This is attested by innumerable instances where the entrenchment is largely or entirely in insoluble rocks beneath the limestone where the initial solutional penetration was made (Fig. 1.16e).

Meandering Channels. Three types of meandering occur in vadose caves. The *entrenched* meandering canyon develops where a waterfall (a knick point) recedes along a meandering channel. The canyon has vertical walls and meanders only in plan view. This type is common where strata are flat lying and there is a deep vadose zone with many invasion streams, e.g., the Central Kentucky Karst.

Ingrowing meandering canyons are created by meandering entrenchment downwards, downstream and laterally, as in the famous "goosenecks" of the San Juan River canyon, i.e., there

FIGURE 1.17. A vadose canyon in Schoolhouse Cave, W. Virginia.

is no waterfall recession necessary. The form attains its extreme morphological development in well-bedded limestones in caves. Many examples are several tens of meters deep and some kilometers in length, but are too narrow for an explorer to pass through most parts; see Smart and Brown (1981).

Alluvial meanders develop where the channel is in sand, gravel, etc. on a broader passage floor. Deike and White (1969) showed that examples in Missouri caves behaved morphologically in the same manner as surface alluvial meanders. There is one major distinction, however; because the rock is soluble the alluvial channel may pass smoothly into the wall (becoming a bedrock channel) and back out again.

Vadose Shafts. Individual shafts created by falling water are known up to 400 m in depth. Their form ranges between two extremes: (1) the shaft created by a powerful waterfall, and (2) the *domepit*, created by a slow, steady film flow. In the first, the water mass itself will tend to create a circular or elliptical cross-section for its fall, but this is often modified. Breccia is swept out of any guiding fault to create parallel walls. A plunge pool undercuts the base, introducing irregular taper upwards, and spray at all levels may cause local block fall. Many shafts are highly irregular in form as a consequence. A domepit may form where local flow at the base of the epikarst or perched on a shale band, etc. is able to attack drained joints in the invasion vadose zone. In the ideal case the flow is always small enough to be retained against the vertical rock by surface tension. It disperses radially from an input point and carves a set of solutional flutings down the walls. The cross-section is circular. Domepits are best developed where strata are flat lying and the joints are few with high resistance, e.g., Mammoth Cave (Merrill 1960).

Many shafts display mixtures of the two forms, with waterfall features down the fall line and fluting of more distal parts of the perimeter.

Corrosion Notches, Bevels, and Facets. These features develop where water is nearly static, so that the slightest fluid density gradients can establish sharply delimited zones of accelerated erosion. In a standing pool in limestone, heavy solute ions and ion pairs sink, driving a cellular convection that carries fresh H^+ ion to the walls at the water surface (Fig. 1.16f). A sharp notch may develop there, extending as much as one

meter deep. It can signify a paleowater table level very precisely.

Notching becomes much larger in the foot caves of residual karst towers abutting alluvial floodplains (Fig. 1.1). Notching at the seasonal flood level and extending several meters into the rock is common, = *swamp notches*. Because they are more extensive and create a flat roof regardless of geologic structure, these features may be termed *corrosion bevels* (Ger. *laugdecke;* Kempe et al. 1975). The greatest notching and beveling occurs in the hyperacid conditions of some paragenetic massive sulfide emplacements, e.g., at Nanisivik Mine, Baffin Island, where the greatest notch is 400 m wide, 1 m deep, and filled with syngenetic layered pyrite.

If the water is *very* still, then in theory a linear taper should be created below the water line instead of a notch (Kempe et al. 1975). This is seen in many caves in the more soluble gypsum. It is most rare in limestone caves but Na Spicaku Cave, Czechoslovakia, is one spectacular exception (Skrivanek and Rubin 1978; and Fig. 1.16g).

Dissolution and Sublimation Scallop Patterns. Dissolution scallops are spoon-shaped scoops (Fig. 1.18). They occur in packed patterns so that individuals overlap and are incomplete. In homogeneous, medium- to fine-grained carbonates they may occupy channel floors and walls, and ceilings as well in phreatic caves. The steep face of the scoop points downstream i.e., they are paleoflow direction indicators. Most scallops in a given occurrence are of similar length. The length range is 0.5 to 20 cm, exceptionally up to 2 m. In the right conditions it is evident that patterns of scallops of a characteristic length extend to colonize all available surfaces, being the stable form on them. They develop (with greater lengths) where wind blows over firn fields or air flows through ice caves; here, erosion must be by sublimation.

Curl (1966) proposed an elegant theory of scallop formation in which there is detachment of the saturated fluid boundary layer in the subcritical turbulent flow regime. Detachment permits aggressive bulk fluid to dissolve or sublime the solid directly, rather than diffusing to it at the ionic scale. This theory has been exhaustively confirmed in laboratory experi-

FIGURE 1.18. Dissolutional scalloping in Lower Hughes' Cave, W. Virginia. This is a 2-D maze cave created by floodwaters beneath a calcareous shale aquitard (the bed comprising the roof). Scallops are well seen at midheight on the right-hand wall, which is in a thick micrite bed. They indicate that flow is toward the rear of the scene. They are absent at floor level, where abrasion processes destroy them. A less soluble calcareous siltstone (at the level of twigs jammed in the roof by floods) displays shallower, poor scallops. At the level of the person's head a shale band is protruding.

ments (Goodchild and Ford 1971, Allen 1972, Blumberg and Curl 1974) and in the field in caves (Gale 1984, see Lauritzen et al. 1986, who give the most useful current treatment). The length of scallops is inversely proportional to the velocity of flow. Lauritzen et al. (1986) show that the scallop-producing discharge in a Norwegian cave is the maximum 2 to 5% of discharge (i.e., annual peak floods). In any relict solution cave, therefore, scallops may be most useful paleodischarge indicators.

Features of Breakdown in Caves

Cave breakdown (collapse of bedrock walls and ceilings, in part or whole) may be due to a number of causes. The most important are (1) removal of the buoyant support of water when a phreatic cave is drained, (2) undermining of a wall or overwidening of a roof span by lateral vadose stream erosion, and (3) solutional attack upon fractures within the failing mass, reducing their bonding. Earthquake activity appears to be surprisingly unimportant, although most falls are clearly episodic rather than nearly continuous events. Preferred sites are at river passage junctions or where obstructions can accumulate in channels because these favor lateral undercutting, and where major vertical fractures intersect in a roof.

The extent of breakdown that is observed in caves ranges from zero in young or small caves or where rocks are exceptionally massive, to complete (all of the ceilings have fallen until temporarily stable tension dome boundaries are attained) in large, old caves in medium to thin bedded rocks. The largest unbroken solutional voids occur in massive reef rocks that lack any bedding planes. Carlsbad Big Room (Fig. 1.19) is the outstanding example, being generally 40 to 50 m wide and up to 80 m in height, yet with little breakdown. In the medium to thick bedded, flat-lying strata of the Mammoth Cave region of Kentucky, the oldest large passages (10 to 25 m wide) exhibit complete breakdown. Passages of intermediate age display "breakout domes," local tapering collapses in the tension dome, separated by tens to hundreds of meters of unbroken ceiling. Where strata dip more than a few degrees there is always preferential fall from the up-dip (hanging) wall. In addition to outright breakage and fall, elastic sagging of medium to thin, normally argillaceous beds is common. However, the *crackle breccia* type of partial displacement that is commonly reported in paleokarst outcrops, MVT deposits, etc. is not often seen in enterable caves, perhaps because the crackle is too small or too obviously dangerous to be penetrated!

The debris produced are (1) blocks, where the fragments consist of more than one bed remaining as a coherent unit, (2) slabs, where they are single, broken beds, and (3) chips, where beds are fragmented (White and White 1969).

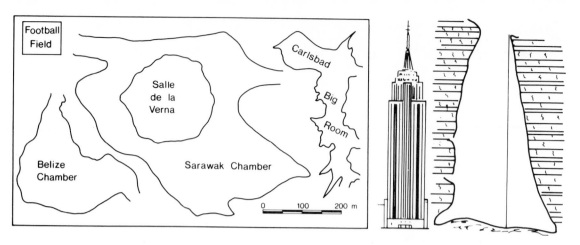

FIGURE 1.19. The largest rooms known in caves (at July 1986) and a vertical section through Sotano de los Golondrinas, Xilitla, Mexico. Salle de la Verna is in the Pierre St. Martin cave, which underlies the Franco-Spanish border in the central Pyrenees. The Sotano offers a free drop of 375 m on the short side (illustrated, right). Sarawak Chamber, Mulu National Park, Borneo, is approximately 20 million m^3 in volume.

Single blocks of greater than 25,000 m^3 are known, e.g., from a reef-backreef junction at Carlsbad Caverns. Slab breakdown is volumetrically predominant where strata are thin to thick bedded and flat lying, e.g., Mammoth Cave area. Chip breakdown is ubiquitous; its many forms depend on fabric and other properties of the rock.

The proportion of the fallen rock that is removed in solution from breakdown sites ranges from all to none. Complete removal may occur where there are large vadose streams. Part or all of the breakdown is conserved in relict passages. It may block them locally with "boulder chokes," which are very common. There is a gradation to the breccia pipe or geological organ type of feature.

The Largest Known Cave Voids. Large voids in caves are termed *rooms* by American speleologists and *chambers* by the British. The largest known examples (in 1986) are shown in Figure 1.19, together with one of the greatest open shafts. Sarawak Chamber, discovered in 1980 (Waltham and Brook 1980) has a volume of approximately 20 million m^3. Belize Chamber, Salle de la Verna, and Carlsbad Big Room are each in excess of one million m^3. Hundreds of rooms are known with volumes of 100,000 to 500,000 m^3.

A majority of these rooms are centered at vadose river passage junctions where under-cutting and solutional removal of debris produce repeated breakdown. Carlsbad Big Room and some large chambers in other hydrothermal caves probably owe most of their excavation to exotic mixing corrosion effects (as noted) and display relatively little breakdown morphology.

Principal Categories of Infilling in Dissolutional Caves

Classification of Cave Fills

Caves may function as giant sediment traps, accumulating samples of all clastic, chemical, and organic debris mobile in the natural environment during the life of the cave (Fig. 1.20).

A general classification of cave fills is given in Table 1.2. This is concerned with cave interior deposits only, and not the more complex facies of illuminated cave entrances that are so intensively explored by archaeologists, paleoanthropologists, etc. Entrance facies tend not to be preserved in buried paleokarst. The classification also ignores characteristic deposits where a major marine transgression terminates a karst erosional period and buries all karst features (surficial and underground) beneath later rocks, because the caves are not then explorable. The deposits of Table 1.2 can be considered merely transient within the karst net

FIGURE 1.20. This complex scene illustrates all the main categories of erosion and deposition seen in caves with the exception of breakdown. The original passage is a phreatic tube of circular cross-section, oriented down a 6° dip (toward rear of picture) in massive crystalline limestone. A later invasion vadose stream trench is seen in the floor (beyond the model). In several subsequent phases the passage was largely filled with layered silts and clays. 50% clastic filling is preserved in the rear of the picture. Fill in the foreground is being sapped into the invasion trench by renewed stream action there. Soda straws and quite massive carrot stalactites have grown since the last clastic filling event.

erosion zone. Nevertheless, their variety is considerable. Their extent ranges from none in many caves to complete infilling of large parts or the whole of others. Space permits only a very brief summary of some chief characteristics to close this review; see Bögli (1980) and Jennings (1985) for more detailed outlines.

Principal Clastic and Organic Facies

The principal clastic deposits are piles of breakdown (noted above), and fluvial debris laid down under open channel or pipeful flow conditions. Most kinds of depositional structures described from natural rivers, flumes, and pipe experiments can be seen in caves but broad floodplain sequences are absent.

Transport and deposition can be highly energetic. A steep alpine cave that is subject to daily meltwater floods is best likened to a flushing toilet. The extreme of transport is seen in such cases, where the maximum transported grain size may be determined by passage cross-sectional dimensions. Individual boulders firmly wedged between floor and roof are common. Vertical lifting of pebbles and cobbles in heterogeneous suspension up shafts as great

TABLE 1.2. Cave interior deposits.

A.

ALLOCHTHONOUS CLASTIC
1. FLUVIAL—many kinds—*dominant allochthone*
2. DEJECTA CONES, COLLUVIUM & MUDFLOWS—common
3. FILTRATES—cones of fines from seepage; minor in volume
4. LACUSTRINE—rare; clays, silts and fine sands
5. MARINE—rare; beach facies at entrances
6. EOLIAN—normally minor to negligible
7. GLACIAL & GLACI-FLUVIAL INJECTA—common in glaciated areas
8. SUBGLACIAL LACUSTRINE—often varve-like

ORGANIC
9. FLUVIAL OR EOLIAN TRANSPORTED—tree trunks to spores
10. CAVE-USING EXTERIOR FAUNA—bones, nests and middens, feces

B.

AUTOCHTHONOUS CLASTIC
11. BREAKDOWN
12. FLUVIAL—derived from breakdown
13. WEATHERING RINDS & EARTHS
14. EOLIAN—derivatives of all the above deposits

PRECIPITATES & EVAPORITES
More than 100 secondary minerals form in caves
PREDOMINANT
15. CALCITE—*most significant autochthone*
16. OTHER CARBONATES, HYDRATED CARBONATES
17. SULFATES, HYDRATED SULFATES, HALIDES
18. PHOSPHATES & NITRATES
19. SILICA & SILICATES
20. ORE-ASSOCIATED MINERALS
21. ICE—as glacial injecta, glaciers, water ice, frost

ORGANICS
22. TRACKS & REMAINS OF CAVE-ADAPTED (HYPOGEAN) & PHREATOPHYTE FAUNA

as 50 m in height is known (e.g., Schroeder and Ford 1983).

A facies that is frequently encountered is a poorly sorted to chaotic mixture of all sizes up to boulders, with patches that are grain-supported and others matrix-supported. The top is nearly flat, abruptly succeeded by a well-sorted sand or gravel. This appears to be a pipeful, sliding bed deposit (Acaroglu and Graf 1968). Well-sorted, imbricated shoals and bars develop downstream of constrictions in both pipeful and open channel caves; Gospodaric (1974) gives an excellent discussion.

Most types of sand forms (ripples, dunes, plain bed, etc.) are reported in caves. However, antidunes appear to be very rare indeed; evidently, the sliding bed substitutes for them in most instances.

Silts and clays are the most widespread deposits. They grade laterally from centers of passages and are finest grained in deep recesses. They may coat walls and ceilings as well as floors. They usually accrete in laminae that are parallel to the depositional surface. Each lamina is uniform in thickness and may display fining upwards (Bull 1977). It is deposited from a flood or other pulsed flow. Some well-sorted clays lack lamination and fining, suggesting decantation from a homogeneous suspension that was steadily renewed; see Bögli (1961).

The clay and silt fractions are normally composed of clay minerals and quartz fines. However, in areas that have been glaciated or exposed to glaciogenic loesses, carbonate grains typically compose 20% to 80% or more, being reworked "glacier flour" from the local rocks.

A wide variety of facies occur in phreatic caves but laminated silts and clays usually predominate. Paragenetic deposits may include sands and even pebbles, in fining upwards beds. Laminated fines in small amounts are the chief deposits in thermal water caves and artesian mazes.

The standard suite in gentle vadose, water table, and shallow phreatic channels is a series of cut-and-fill deposits with marked lateral fining. There may be a sliding bed deposit at the base of any given sequence. *Abandonment suites* are similar, but deposits of successive events become distinctly finer upwards as the passage is progressively abandoned by even large-magnitude, rare floods (Fig. 1.21).

Where a dam of breakdown is abruptly created the effects can be highly variable. Most often there is infilling with parallel beds or delta deposits upstream, fining upwards at a section. Downstream, there is reworking or even complete scouring of older deposits, plus a varying addition of finer grain sizes passed through the dam or winnowed from its detritus. If much of the dam is then removed, a *pond-and-sieve, winnow, cut-and-fill suite* is left. This is to emphasize that water-laid cave deposits can be very complicated and open to a variety of interpretations.

Caves everywhere display cones of dejecta where the particles have accumulated by a

FIGURE 1.21. An abandonment facies in a relict passage in a W. Virginia cave. Paleoflow was from left to right. Basal pebbles are succeeded by coarse sands and a prominent silt. Laminated clays displaying blocky fracture are at the top of the section. All of this material passed through an upstream screen of boulders (a boulder choke > 200 m in length) or was winnowed from it.

mixture of piecemeal falls and small slurries from steep openings to the surface. Tongues of colluvium extend from low gradient openings (e.g., Sorriaux 1982).

Glaciers bulldoze till or discharge meltwater and bedload directly into caves. Typical here is an *injection facies*; the passage entrance is blocked by wedged boulders, and a tail of gravels and sands extends inward, fining rapidly.

Waterborne vegetal material can be carried nearly everywhere in common caves. Where clastic detritus builds on or behind an organic barrier (such as a jam of tree trunks) that then decays and collapses, collapsed and jumbled fluvial sequences are created.

Animals, especially rodents, may nest in the deepest cave interiors. The nest and midden material is usually minor in volume and quite distinctive, but dung can accumulate to many meters and dry to an earth. Burrowing rodents can enormously disrupt clastic stratigraphy, intruding uppermost materials into the bottom layers. However, this is chiefly a problem in cave entrance zones. The deposits of the highly specialized, cave-adapted fauna (such as blind salamanders) that live all their lives in cave interiors are tiny.

Precipitates and Evaporites in Caves

More than 100 minerals are known to occur as secondary precipitates forming in caves. Some delicate polyhydrates can develop only in the ultrastable temperature and humidity regimes of deep cave interiors. Hill and Forti (1986) give a comprehensive review; see also Bull (1983) and White (1976).

Calcite

Cave calcite, also known as travertine, sinter, or speleothems, is the chief precipitate by volume and the most significant autochthone in terms of the paleo-environmental information that is to be gained from it. "Soft" calcite is evaporitic, composed of loose crystallite masses, earthy or pasty. It is associated with cave entrances, or seasonally or permanently arid caves. "Hard" calcite is crystalline. It is deposited as syntaxial overgrowths on previous crystals, with the c-axis usually oriented perpendicular to the growth surface (= *length-fast calcite*, cf. Folk and Assereto 1976; see also Broughton 1983). It is predominant in most cave interiors.

Hard calcite builds many different forms (Fig. 1.22). The chief ones are (1) dripstones, such as stalactites and stalagmites, the latter ranging up to 30 m in height and even more in girth; (2) flowstones or sheets that may extend for hundreds of meters and accumulate to 2 to 3 m in depth (both categories are vadose deposits); and (3) euhedral spar coatings on all passage surfaces and up to 1 m thick—these are pool wall or phreatic deposits, most often found in hydrothermal caves.

Hard calcite stalagmites and flowstones contain much paleo-environmental information (Fig. 1.23). They may be dated by ^{14}C, ESR (electron spin resonance decay), or U series methods, chiefly the last (Gascoyne and Schwarcz 1982). Their stable isotopes and fluid inclusions may yield paleotemperature or humidity data (Hendy and Wilson 1968, Schwarcz et al. 1976, Yonge 1982). Many carry detectable remanent magnetism (Latham et al. 1979). They trap and may beautifully preserve airborne pollen grains (Bastin 1979), and much of their color is attributable to varying concen-trations of humic and fulvic acids from soil sources.

Long-term rates of hard calcite growth are now being established by U series dating. Carrot stalactites probably grow up to 10 mm/yr, and soda straw stalactites several times faster. Stalagmites rarely extend more than 1 mm/yr, and mean rates as low as 0.001 mm/yr are established. Flowstone sheets probably thicken at a rate about one order of magnitude less than that of the local stalagmites.

The other precipitates are absent from many common caves and rarely amount to more than a few percent by volume of the calcite where they do occur. Aragonite is quite common as acicular clumps, or as massive layers within calcite. Hydromagnesite and other hydrated carbonates plus sulfates and hydrated sulfates such as thenardite or epsomite occur as thin pastes or crusts in evaporitic sequences. They are known from arctic to hot desert environments, so that they are poor environmental indicators. Gypsum, as flowers or thicker coatings, is common. It may be precipitated after bicarbonate

FIGURE 1.22. This small scene depicts the principal forms of calcite concretions in caves. Carrot and soda straw stalactites grow vertically downwards, with some "erratic" protuberances. A "drapery" forms where the ceiling is inclined. Flowstone sheets cover the floor, with stalagmites growing upwards from them where there are powerful drips overhead. White globular masses on the floor and wall to the right of the model are "moon milks"—pastes of calcite or aragonite crystallites with lesser hydromagnesite or sulfate minerals. (Photo by Dr. R. S. Harmon).

FIGURE 1.23. To illustrate the paleo-environmental evidence preserved in cave interior deposits. This cave was phreatic in origin. Laminated clays deposited as it was being abandoned are preserved at the junction of wall and roof (to left of the dark shadow); they are magnetically reversed ($> 700,000$ yr in age). Invading soil waters then deposited the massive "beehive" stalagmite in the relict passage. The chemical balance of these waters then reversed, so that they drilled a solutional domepit through the stalagmite and 3 m into the bedrock floor beneath it (beyond the model). New draperies and stalagmites are now growing in the domepit, filling it in again; some of these youngest deposits are older than 350,000 yr (from U series dating).

groundwaters are complexed with oxidized pyritiferous shale (e.g., Atkinson 1983) and thus is not necessarily indicative of the destruction of gypsum beds in a carbonate sequence. Salt (halite) stalactites and crusts are reported only from desert caves.

Cave phosphates and many nitrates are produced by urine (particularly bat urine) and other decaying animal products reacting with bedrocks or clastic sediments. Some nitrates are probably generated from soil-water solutions; see Hill (1981).

Aluminum and iron oxide coatings in caves are mostly carried in suspension from overlying bauxite or limonite deposits. Hematite is precipitated in thermal caves and some very dry caves. Black manganese coatings (chiefly birnessite) are common on gravels and channel walls, and indicate turbulent vadose conditions of deposition (Moore 1981).

A wide variety of seasonal and permanent ice deposits form in caves and glacier ice may be intruded. Ice rarely extends more than a few hundred meters from entrances, however. Some thick cave ice masses in Romania display growth rate curves that permit them to be correlated with the so-called Little Ice Age i.e., they began to build about 1700 A.D. and are now decaying (Racovitza 1972). Some very thick masses are most probably thousands of years in age (e.g., Marshall and Brown 1974) in the Rocky Mountains of Canada.

Conclusions

The great majority of explored limestone caves were created by dissolution in meteoric groundwaters circulating without unusual confinement. The principles governing their pattern building, in plan view and in depth, are now quite well understood. However, the particular pattern and, especially, details of location and passage dimensions within the host rock, can rarely be predicted because of the important role played by stochastic factors in the building. Multiphase effects greatly complicate the patterns of most larger caves. The longest known systems now exceed 100 km and the deepest are 1000 to 1500 m. One chamber as great as 20 million m^3 in volume is known.

Two-dimensional rectilinear mazes occur as anomalous parts of ordinary caves or as distinct caves. They are joint guided and develop due to local flooding, or to artesian or diffusion trapping. Caves formed by thermal, CO_2-rich waters make up approximately 10% of known caves and predominate in some regions. They display dendritic or 2- or 3-dimensional maze forms. Some very large caverns adjoining oil fields may be developed where H_2S-rich waters are oxidized. Many caves develop where salt and fresh waters mix in the coastal zone, but modern examples are complicated by effects of Quaternary sea level oscillations.

Cave passage cross-sections are those of phreatic or vadose action or of mechanical breakdown, or of combinations. Phreatic cross-sections are complicated by differing solubility, by armoring, pocketing, and paragenesis, vadose cross-sections by notching, beveling, and breakdown. Dissolution scallops develop where rocks are comparatively homogeneous and fine grained; they are important paleohydraulic indicators. A wide variety of clastic deposits accumulate in caves, fluvial deposits being dominant. The facies and hydraulic interpretations must be treated with care due to sliding bed, damming, and sieving processes that are much more significant than in surface channels. More than 100 secondary minerals are precipitated in caves. Calcite is the greatest of them by volume and is the most significant for paleoenvironmental reconstructions.

References

Acaroglu, E.R., and Graf, W.H., 1968, Sediment transport in conveyance systems, 2:the modes of sediment transport and their related bedforms in conveyance systems: Bull., Int. Assoc. Sci. Hydrol., v. 13, p. 123–135.

Allen, J.R.L., 1972, On the origin of cave flutes and scallops by the enlargement of inhomogeneities: Rass. Speleo. Ital., v. 24, p. 3–19.

Atkinson, T.C., 1983, Growth mechanisms of speleothems in Castleguard Cave, Columbia Icefields, Alberta, Canada: Arctic and Alpine Res., v. 15(4), p. 523–536.

Back, W., Hanshaw, B.B., and Van Driel, J.N., 1984, Role of groundwater in shaping the eastern coastline of the Yucatan Peninsula, Mexico, *in* LaFleur, R.G., ed., Groundwater as a geomorphic agent: London, Allen and Unwin, Inc., p. 280–293.

Bakalowicz, M., Ford, D.C., Miller, T.E., Palmer, A.N., and Palmer, M.V., Thermal genesis of solution caves in the Black Hills, South Dakota. Bull., Geol. Am. Soci, in press.

Bastin, B., 1979, L'analyse pollinique des stalagmites: une nouvelle possibilité d'approche des fluctuations climatiques du Quaternaire: Annales de la Société Géologique de Belgique, t. 101, p. 13–19.

Beaupré, M., 1975, Les regions karstiques du Quebec: Speleo-Quebec, v. 2(2), p. 22–51.

Berner, R.A., 1978, Rate control of mineral dissolution under earth surface conditions: Am. J. Sci., v. 278, p. 1235–1252.

Blumberg, P.N., and Curl, R.L., 1974, Experimental and theoretical studies of dissolution roughness: J. Fluid Mech., v. 65, p. 735–751.

Bocker, T., 1969, Karstic water research in Hungary: Internat. Assoc. Sci. Hydrol. Bull., v. 14, p. 4–12.

Bögli, A., 1961, Die Hohlenlehme: Mem. Rass. Spel. Ital., v. 5, p. 1–21.

Bögli, A., 1964, Mischungskorrosion, ein Beitrag zum Verkarstungs-problem: Erdkunde, v. 18(2), p. 83–92.

Bögli, A., 1970, Le Holloch et son karst: Neuchatel, Ed. La Baconniere.

Bögli, A., 1980, Karst hydrology and physical speleology: Berlin, Heidelberg, New York, Springer-Verlag, 284 p.

Bretz, J.H., 1942, Vadose and phreatic features of limestone caves: J. Geol., v. 50(6), p. 675–81.

Bretz, J.H., 1949, Carlsbad Caverns and other caves of the Guadalupe Block, New Mexico: J. Geol., v. 57, p. 447–463.

Brod, L.G., 1964, Artesian origin of fissure caves in Missouri: Bull., Nat. Speleo. Soc. Am., v. 26(3), p. 83–112.

Broughton, P.L., 1983, Environmental implications of competitive growth fabrics in stalactitic carbonate: Int. J. Speleo., v. 13(1–4), p. 31–42.

Bull, P.A., 1977, Laminations or varves? Processes of fine-grained sediment deposition in caves: Proc. of the 7th Internat. Speleo. Cong., p. 86–89.

Bull, P.A., 1983, Chemical sedimentation in caves, *in* Goudie, A.S., and Pye, K., eds., Chemical sediments and geomorphology: precipitates and residua in the near-surface environment: London, Academic Press, p. 301–320.

Busenberg, E., and Plummer, L.N., 1982, The kinetics of dissolution of dolomite in CO_2-H_2O systems at 1.5 to 65°C and 0 to 1 atm PCO_2: Am. J. Sci., v. 282, p. 45–78.

Butler, J.N., 1982, Carbon dioxide equilibria and their applications: Reading, MA, Addison-Wesley, 259 p.

Craig, D.H., this volume, Caves and other features of Permian karst in San Andres Dolomite, Yates Field reservoir, West Texas.

Curl, R.L., 1966, Scallops and flutes: Trans., Cave Res. Gp., G.B., v. 7(2), p. 121–160.

Davis, D.G., 1980, Cave development in the Guadalupe Mountains: a critical review of recent hypotheses: Bull., Nat. Speleo. Soc. Am., v. 42(3), p. 42–48.

Davis, W.M., 1930, Origin of limestone caves: Bull., Geol. Soc. Am., v. 41, p. 475–628.

Deal, D.E., 1968, Origin and secondary mineralization of caves in the Black Hills of South Dakota: Proc. of the 4th Internat. Speleo. Cong., Yugoslavia, v. 3, p. 67–70.

Deike, G.H., and White, W.B., 1969, Sinuosity in limestone solution conduits: Am. J. Sci., v. 267, p. 230–241.

Drake, J.J., 1980, The effect of soil activity on the chemistry of carbonate groundwaters: Water Resources Res., v. 16(2), p. 381–386.

Drake, J.J., 1983, The effects of geomorphology and seasonality on the chemistry of carbonate groundwater: J. Hydrol., v. 61, p. 223–236.

Drake, J.J., 1984, Theory and model for global carbonate solution by groundwater, in LaFleur, R.G., ed., Groundwater as a geomorphic agent: London, Allen and Unwin, p. 210–226.

Dreybrodt, W., 1981, Kinetics of the dissolution of calcite and its applications to karstification: Chem. Geol., v. 31, p. 245–269.

Dublijansky, V.N., 1979, The gypsum caves of the Ukraine: Cave Geol., v. 1(6), p. 163–183.

Dublijansky, V.N., 1980, Hydrothermal karst in the alpine folded belt of southern parts of the USSR: Kras i Speleologia, v. 3(12), p. 18–36.

Egemeier, S.J., 1981, Cavern development by thermal waters: Nat. Speleo. Soc. Bull., v. 43, p. 31–51.

Engh, L., 1980, Karstomradet vid Lummelunds bruk, Gotland: Lund Univ. Geog. Institute, 290 p.

Ewers, R.O., 1982, Cavern development in the dimensions of length and breadth: Ph.D. thesis, McMaster Univ., Hamilton, Ont., Canada, 398 p.

Fish, J.E., 1977, Karst hydrogeology and geomorphology of the Sierra de El Abra and the Valles–San Luis Potosi Region, Mexico: Ph.D. thesis, McMaster Univ., Hamilton, Ont., Canada, 469 p.

Folk, R.L., and Assereto, R., 1976, Comparative fabrics of length-slow and length-fast calcite and calcitized aragonite in a Holocene speleothem, Carlsbad Caverns, New Mexico: J. Sed. Pet., v. 46(3), p. 486–496.

Ford, D.C., 1968, Features of cavern development in central Mendip: Cave Res. Group, G.B., Trans., v. 10, p. 11–25.

Ford, D.C., 1971, Geologic structure and a new explanation of limestone cavern genesis: Cave Res. Group, G.B., Trans., v. 13(2), p. 81–94.

Ford, D.C., 1980, Threshold and limit effects in karst geomorphology, in Coates, D.L., and Vitek, J.D., eds., Thresholds in geomorphology: London, Allen and Unwin, p. 345–362.

Ford, D.C., and Ewers, R.O., 1978, The development of limestone cave systems in the dimensions of length and depth: Can. J. Earth Sci., v. 15, p. 1783–1798.

Ford, D.C., Smart, P.L., and Ewers, R.O., 1983, The physiography and speleogenesis of Castleguard Cave, Columbia Icefields, Alberta, Canada: Arctic and Alpine Res., v. 15(4), p. 437–450.

Fulweiler, R.E., and McDougal, S.E., 1971, Bedded-ore structures, Jefferson City Mine, Jefferson City, Tennessee: Econ. Geol., v. 66, p. 763–769.

Gale, S.J., 1984, The hydraulics of conduit flow in carbonate aquifers: J. Hydrol., v. 70, p. 309–324.

Garrels, R.M., and Christ, C.L., 1965, Solutions, minerals and equilibria: New York, Harper and Row, 450 p.

Gascoyne, M., Benjamin, G.J., Schwarcz, H.P., and Ford, D.C., 1979, Sea-level lowering during the Illinoisian glaciation: evidence from a Bahama blue hole, Science, v. 205, p. 806–808.

Gascoyne, M., and Schwarcz, H.P., 1982, Carbonate and sulphate precipitates, in Ivanovich, M., and Harmon, R.S., eds., Uranium series disequilibrium: applications to environmental problems: Oxford, Clarendon Press, p. 268–301.

Glennie, E.A., 1954, Artesian flow and cave development: Cave Res. Group, G.B., Trans., v. 3, p. 55–71.

Goodchild, M.F., and Ford, D.C., 1971, Analysis of scallop patterns by simulation under controlled conditions: J. Geol., v. 79, p. 52–62.

Gospodaric, R., 1974, Fluvial sediments in Krizna Jama: Acta Carsologica, v. 6, p. 327–363.

Gregor, V.A., 1981, Karst and caves in the Turks and Caicos Islands, B.W.I.: Proc. of the 8th Internat. Speleo. Cong., Lexington, KY, USA, p. 805–806.

Hendy, C.H., and Wilson, A.T., 1968, Paleoclimatic data from speleothems: Nature, v. 216, p. 48–51.

Herman, J.S., 1982, The dissolution kinetics of calcite, dolomite and dolomitic rocks in the CO_2-water system, Ph.D. thesis, Penn State Univ., College Park, Pennsylvania.

Hill, C.A., ed., 1981, Saltpeter: a symposium, Bull., Nat. Speleo. Soc., v. 43(4), p. 83–131.

Hill, C.A., 1987, Geology of Carlsbad Cavern and other caves in the Guadalupe Mountains, New Mexico and Texas: New Mexico Bureau of Mines and Mineral Resources, Bulletin 117, in press.

Hill, C.A., and Forti, P., 1986, Cave minerals of the world: Huntsville, AL, Nat. Speleo. Soc., 238 p.

Jakucs, L., 1977, Morphogenetics of karst regions: variants of karst evolution: Budapest, Akademiai Kiado, 284 p.

Jennings, J.N., 1985, Karst geomorphology: Oxford, Basil Blackwell, 293 p.

Kastning, E.H., 1983, Relict caves as evidence of landscape and aquifer evolution in a deeply dissected carbonate terrain: southwest Edwards Plateau, Texas, USA: J. Hydrol., v. 61, p. 89–112.

Kempe, S., Brandt, A., Seeger, M., and Vladi, F., 1975, "Facetten" and "Laugdecken," the typical morphological elements of caves developed in standing water: Ann. Speleol., v. 30(4), p. 705–708.

Langmuir, D., 1971, The geochemistry of some carbonate ground waters in central Pennsylvania: Geochim et Cosmochim Acta, v. 35, p. 1023–1045.

Latham, A.G., Schwarcz, H.P., Ford, D.C., and Pearce, G.W., 1979, Paleomagnetism of stalagmite deposits: Nature, v. 280, p. 383–385.

Lauritzen, S.E., Abbott, J., Arnesen, R., Crossley, G., Grepperud, D., Ive, A., and Johnson, S., 1986, Morphology and hydraulics of an active phreatic conduit: Cave Science, v. 12(3), p. 139–146.

Marshall, P., and Brown, M.C., 1974, Ice in Coulthard Cave, Alberta: Can. J. Earth Sci., v. 11(4), p. 510–518.

Martel, E.A., 1921, Nouveau traite des eaux souterraines: Paris, Editions Doin., 840 p.

Merrill, G.K., 1960, Additional notes on vertical shafts in limestone caves: Bull. Nat. Speleo. Soc., v. 22(2), p. 101–105.

Miller, T.E., 1981, Hydrochemistry, hydrology and morphology of the Caves Branch karst, Belize: Ph.D. thesis, McMaster Univ., Hamilton, Ont., Canada, 280 p.

Moore, G.W., 1981, Manganese speleothems, in Proc. of the 8th Internat. Speleo. Cong. p. 642–644.

Muller, P., and Sarvary, I., 1977, Some aspects of developments in Hungarian speleology theories during the last 10 years: Karsztes Barlang, Spec. Issue, p. 53–60.

Mylroie, J.E., and Carew, J.L., 1986, Minimum duration for speleogenesis: Communications, 9 Cong. Internat. Speleo., v. 1, p. 249–251.

Ollier, C.D., 1975, Coral Island geomorphology— the Trobriand Islands: Z. Geomorph., v. 19(2), p. 164–190.

Palmer, A.N., 1975, The origin of maze caves: Bull. Nat. Speleo. Soc., v. 37(3), p. 56–76.

Palmer, A.N., 1981, A geological guide to Mammoth Cave National Park: Teaneck, NJ, Zephyrus Press, 210 p.

Palmer, A.N., 1984, Geomorphic interpretation of karst features: in LaFleur, R.G., ed., Groundwater as a Geomorphic Agent: London, Allen and Unwin, p. 173–209.

Pasini, G., 1973, Sull'importanza speleogenetica dell' "erosione antigravitativa," Le Grotte d'Italia v. 4, p. 297–326.

Plummer, L.N., 1975, Mixing of seawater with calcium carbonate groundwater: quantitative studies in the geological sciences: Geol. Soc. Am. Mem. 142, p. 219–236.

Plummer, L.N., Parkhurst, D.L., and Wigley, T.M.L., 1979, Critical review of the kinetics of calcite dissolution and precipitation, in Jenne, E.A., ed., Chemical modeling in aqueous systems: Washington, DC, Am. Chem. Soc., p. 537–573.

Plummer, L.N., Wigley, T.M.L., and Parkhurst, D.L., 1978, The kinetics of calcite dissolution in CO_2-water systems at 5° to 60°C and 0.0 to 1.0 atm CO_2, Am. J. Sci., v. 278, p. 179–216.

Queen, J.M., Palmer, A.N., and Palmer, M.V., 1977, Speleogenesis in the Guadalupe Mountains, New Mexico: gypsum replacement of carbonate by brine mixing, in Proc. of the 7th Internat. Speleo. Cong., Sheffield, England, p. 333–336.

Quinlan, J.F., Ewers, R.O., Ray, J.A., Powell, R.L., and Krothe, N.C., 1983, Groundwater hydrology and geomorphology of the Mammoth Cave region, Kentucky, and the Mitchell Plain, Indiana, in Shaver, R.H., and Sunderman, J.A., eds., Field trips in midwestern geology: Bloomington, IN, Geological Society of America and Indiana Geological Survey, p. 1–85.

Racovitza, G.H., 1972, Sur la correlation entre l'évolution du climat et la dynamique des dépôts souterrains de glace de la grotte de Scarisoara: Trav. Inst. Speleo, "Emile Racovitza," v. 11, p. 373–392.

Renault, P., 1968, Contribution a l'étude des actions méchaniques et sédimentologiques dans la spéléogenèse: Ann. Speleol., v. 23, p. 529–596.

Rhoades, R., and Sinacori, N.M., 1941, Patterns of groundwater flow and solution: J. Geol., v. 49, p. 785–794.

Rhodes, D., Lantos, E.A., Lantos, J.A., Webb, R.J., and Owens, D.C. 1984, Pine Point ore bodies and their relationship to structure, dolomitization and karstification of the Middle Devonian barrier complex: Econ. Geol., v. 70, p. 991–1055.

Schoeller, H., 1962, Les eaux souterraines: Paris, Masson, 642 p.

Schroeder, J., and Ford, D.C., 1983, Clastic sediments in Castleguard Cave, Columbia Icefields, Alberta, Canada: Arctic and Alpine Res., v. 15(4), p. 451–461.

Schwarcz, H.P., Harmon, R.S., Thompson, P., and Ford, D.C., 1976, Stable isotope studies of fluid inclusions in speleothems and their paleoclimatic significance: Geochimica et Cosmochimica Acta, v. 40, p. 657–665.

Skrivanek, F., and Rubin, J., 1978, Caves in Czechoslovakia: Prague, Academica, 132 p.

Smart, C.C., and Brown, M.C., 1981, Some results and limitations in the application of hydraulic geometry to vadose stream passages, in Proc. of the 8th Internat. Speleo. Cong., Kentucky, USA, p. 724–725.

Smith, D.I., and Atkinson, T.C., 1976, Process, landforms and climate in limestone regions, in Derbyshire, E. ed., Geomorphology and climate: London, Wiley, p. 369–409.

Sorriaux, P., 1982, Contribution a l'étude de la sédimentation en milieu karstique: le système de

Niaux-Lombrives-Sabart, Pyrénées Ariegeoises: Thesis, 3rd cycle, Univ. Paul Sabatier, Toulouse, 255 p.

Stumm, W., and Morgan, J.J., 1980, Aquatic chemistry: an introduction emphasizing equilibria in natural waters, 2nd edition: New York, Wiley, 780 p.

Swinnerton, A.C., 1932, Origin of limestone caverns: Bull. Geol. Soc. Am., v. 43, p. 662–693.

Waltham, A.C., 1986, China caves '85: London, Royal Geographical Society, 60 p.

Waltham, A.C., and Brook, D.B., 1980, Geomorphological observations in the limestone caves of Gunung Mulu National Park, Sarawak: Trans., Brit. Cave Res. Assoc., v. 7(3), p. 123–140.

Warwick, G.T., 1953, The origin of limestone caves, in Cullingford, C.H.D. ed., British caving: London, Routledge and Kegan Paul, p. 41–61.

Weyl, P.K., 1958, Solution kinetics of calcite: J. Geol., v. 66, p. 163–176.

White, E.L. and White, W.B. 1969, Processes of cavern breakdown. Nat. Speleol. Soci. Am. Bulletin 31(4), p. 83–96.

White, W.B., 1976, Cave minerals and speleothems, Chap. 8 in Ford, T.D., and Cullingford, C.H.D., eds., The science of speleology: London, Academic Press, p. 267–327.

White, W.B., 1977, The role of solution kinetics in the development of karst aquifers, in J.S. Tolson, and Doyle, F.L., eds., Internat. Assoc. Hydrogeol., Mem. 12, p. 503–517.

White, W.B., 1984, Rate processes: chemical kinetics and karst landform development, in R.G. LaFleur, ed., Groundwater as a geomorphic agent: London, Allen and Unwin, p. 227–248.

Yonge, C.A., 1982, Stable isotope studies of water extracted from speleothems: Ph.D. thesis, McMaster Univ., Hamilton, Ont., Canada, 298 p.

Zhang, Z., 1980, Karst types in China: Geo. Journal, v. 4, p. 541–570.

2
Geochemical Patterns of Meteoric Diagenetic Systems and Their Application to Studies of Paleokarst

Kyger C Lohmann

Abstract

The isotopic and cation chemistry of meteoric waters changes in response to the effects of rock–water interaction, uptake of organically derived CO_2, and primary mineralogic differences among carbonate terranes. Moreover, variations in the dominance of these factors produce diverse chemical conditions within the meteoric systems which allow the subenvironments of vadose-phreatic, mixed-water, and spelean diagenesis to be distinguished. Therefore, geochemical patterns within the meteoric water system are examined to provide criteria for recognition of these subenvironments of meteoric diagenesis in ancient carbonate sequences.

The $\delta^{18}O$ composition of meteoric groundwater is largely constant at individual geographic sites. Variation in the amount of dissolved soil-gas CO_2 and in the extent of rock–water interaction produces a distinct geochemical trend for diagenetic alteration and precipitation products. This pattern of invariant $\delta^{18}O$ coupled with variable $\delta^{13}C$, termed here the *meteoric calcite line*, serves as the baseline relative to which chemical variations characteristic of vadose-phreatic, mixed-water, and spelean environments can be discriminated.

Interaction of meteoric water with country rock produces spatial and temporal changes in the chemistry of the water, and likewise in diagenetic products. Areas proximal to exposure surfaces exhibit low rock–water exchange, and diagenetic phases possess compositions in equilibrium with surface-derived water. In contrast, at stratigraphic positions distal from a recharge surface, diagenetic phases become progressively enriched in Mg^{2+} and/or Sr^{2+} and isotopic values converge toward country rock values due to increasing rock–water interaction along fluid flow lines. In response to the diminishing availability of dissolving metastable carbonate phases with time as the rock system matures, sequentially precipitated phases will record a history of decreasing rock–water exchange.

The $\delta^{18}O$ and $\delta^{13}C$ composition of water within the meteoric-marine mixing zone defines hyperbolic trends reflecting the relative proportions of intermixed marine and meteoric waters. Because calcite precipitation does not occur throughout the full range of mixing, and because the $\delta^{13}C$ value of meteoric waters vary through time, it is unlikely that discrete hyperbolic mixing trends can be deciphered in diagenetic products. Rather, the composition of replacive and precipitated phases formed in the mixed-water zone will define trends offset in $\delta^{18}O$ which parallel the meteoric calcite line.

Precipitation in spelean settings is induced by evaporation and/or CO_2 degassing, both of which can modify the isotopic and cation chemistry of vadose seepage waters. While the isotopic compositions of spelean carbonates overlap the range of coevally precipitated meteoric vadose and phreatic calcite values, combined effects of degassing and evaporation produce covariant trends which deviate from the meteoric calcite line. Such patterns of variation serve to distinguish carbonate precipitated in spelean settings from the more typical vadose-phreatic environments. However, similarities in mineralogy, fabric, and chemistry of spelean carbonates with primary cements precipitated in marine settings make it difficult to determine unambiguously a spelean origin for fibrous carbonates solely on the basis of geochemical or fabric critera.

Introduction

Peering into the darkness of the great caverns at Carlsbad, one is awed by the quantity of carbonate that has been removed by the chemical action of meteoric water. Simple calculations of the solubility of calcite emphasize the tremendous volume of water which must have flowed through these cavities. Clearly, karst forms in diagenetic environments in which the chemistry of water exerts a dominant role. Thus, the key for interpreting paleokarst systems lies in discerning and describing the temporal and spatial variations of present-day meteoric waters.

Karst refers to the physical structures formed through dissolution by meteoric waters. It is marked by macroscopic and microscopic features diagnostic of dissolution processes and meteoric water interactions with the host carbonate. Numerous criteria are utilized, in addition to obvious features of solution and collapse, to recognize this diagenetic environment. These include: gravity-controlled distribution of corrosional and accretionary structures, cavity formation by solution, and late-stage precipitation of speleothems. Mineralogically, karst-related mineralization is diverse, commonly comprised of aragonite, magnesian calcite, calcite, and dolomite (Moore 1961, Thrailkill 1971, Thrailkill 1968a, Gines et al. 1981). Troublesome, too, is the morphological similarity between fibrous and equant crystalline precipitates embodied by speleothems and the fibrous cements formed in cavernous porosity of marine origin (Folk and Asseretto 1976). At present, resolution of the karstic origin for cavernous porosity in ancient sequences requires recognition of numerous features associated with subaerial exposure such as calcrete horizons and the distribution of micro- and macroporosity relative to these surface disconformities or to paleowater tables (e.g., Wagner and Matthews 1982). These features are often absent from or difficult to recognize in ancient carbonate sequences. It is the purpose of this chapter, therefore, to examine the meteoric diagenetic system in which karst forms, not from a fabric, mineralogical, or geomorphic perspective, but as a rock–water geochemical system. The chemistry of the fluid and the complex interactions of the rock and water systems which lead to karst development are examined to provide criteria for recognizing paleokarst where other data are either lacking or ambiguous.

Environments of Karst Formation

Two conditions are necessary for the formation of karst: fluids must be undersaturated with respect to the country rock, and fluid flow must be available to transport products of dissolution away from the site of reaction. These conditions exist in the meteoric vadose and phreatic systems where undersaturated waters driven by gravity dissolve and transport vast quantities of carbonate.

Within the meteoric vadose and phreatic systems, several subenvironments can be identified where, because of differing physicochemical conditions, the style of karstification varies. For example, at subaerial surface exposures, rainwaters which are in equilibrium with atmospheric and soil-gas carbon dioxide are undersaturated with carbonate. Interaction of these with host rock results in extensive chemical erosion (Fig. 2.1). While precipitation may occur in response to local CO_2-degassing or evaporation, the dominant process in this setting is one of dissolution.

Intergranular flow within the vadose zone results in removal of calcite from grain tops and along other grain surfaces during episodes of meteoric water recharge. Yet, if concomitant precipitation is absent, dissolution may actually reduce porosity locally in this zone by processes of chemical compaction (Meyers and Hill 1983). Interestingly, cementation may strengthen the grain-pore framework, thus channeling fluid flow and leading to enlargement of the porosity network into open conduits for flow (Matthews 1974).

Development of fluid flow pathways which bypass the vadose zone (Thrailkill 1965 and 1968b, Matthews 1974, Bögli 1980) allows undersaturated surface waters, charged with CO_2, to directly infiltrate the meteoric phreatic zone, the fluid–rock environment defined by water-saturated pores. Here within the phreatic zone, the frequent recharge of meteoric waters undersaturated with respect to carbonate, the continuous subsurface flow of these fluids, and their long residence time compared to vadose waters, promote extensive dissolution (Matthews 1974, Bögli 1980, Jennings 1985). Near

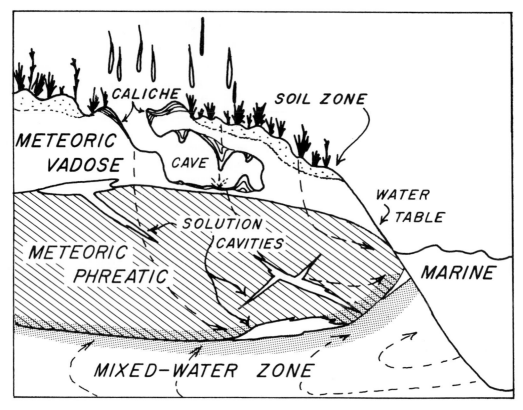

FIGURE 2.1. Environments of karstification within the meteoric diagenetic setting.

the water table, fractures and joints are enlarged and cavernous structures or other small openings develop. It is within this environment of active flow and repeated recharge by undersaturated water that most karstification develops (Fig. 2.1).

Along the basal margin of the meteoric phreatic system, dilute, calcite-saturated waters with elevated CO_2 partial pressures (P_{CO_2}) mix with low P_{CO_2} saline marine waters. Because the equilibrium P_{CO_2}'s and ionic strengths of these two waters differ, within a variable range the mixing waters can become undersaturated with respect to carbonate (Fig. 2.2) whereupon dissolution will occur (Plummer 1975, Back et al. 1979, Plummer and Back 1980). Such corrosion develops in areas spatially removed from the near-surface settings of vadose and proximal phreatic environments.

In general, since all karstification develops in the meteoric system, it should be possible to discriminate the vadose, phreatic, and mixed-water subenvironments in paleokarst, for each

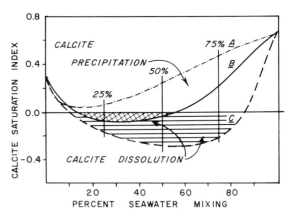

FIGURE 2.2. Variation in calcite saturation during mixing of dilute meteoric and saline marine waters. Mixtures of marine water with meteoric waters oversaturated with respect to calcite and in equilibrium with P_{CO_2} can result in undersaturation and calcite dissolution (ruled and crosshatched regions) (P_{CO_2} at: A = $10^{-2.0}$; B = $10^{-1.5}$; C = $10^{-1.0}$). Such mixed-water corrosion is envisioned as the principal mechanism for karstification in the distal portions of meteoric phreatic systems. Modified after Plummer (1975).

subenvironment is characterized by diverse yet distinct chemical conditions.

Chemical Considerations of the Fluid System

Meteoric water is born by evaporation as vapor from the ocean surface. During this stage, equilibration between the vapor and atmospheric CO_2 determines the initial chemical reactivity of condensing droplets of meteoric water. As droplets impinge on the terrestrial surface, interaction with organic and rock systems further modifies the chemical character of the water. Acidity is enhanced by addition of CO_2 from soil gases and is moderated by the dissolution of carbonate. Such water–rock–gas exchange profoundly alters concentrations of cations and abundances of isotopes in diagenetic waters and in the carbonate which precipitates from them. Therefore, if chemical variations of meteoric water are to be used to recognize subenvironments of the meteoric system, it is necessary to evaluate the nature and extent to which water–rock–gas interactions define the chemistry of diagenetic waters.

Fluid–Rock Saturation Relationships

Rainwaters are dilute solutions containing trace concentrations of dissolved solids, such as Mg^{2+}, Ca^{2+}, Na^+, Cl^-, and SO_4^{2-}, derived from airborne materials. More important, these waters are in equilibrium with atmospheric CO_2. While the dissolved phases are relatively insignificant in their effect on the reactivity of rainwater, dissolved CO_2 has a profound effect on the acidity of the water and thus its potential for carbonate dissolution. For example, under conditions of atmospheric equilibration the extent of CO_2 uptake by water is controlled by temperature and partial pressure of CO_2 (P_{CO_2}). This can be represented by two reactions and their equilibrium constants (values for 25°C):

$$CO_2(g) \leftrightharpoons CO_2(aq): K_1 = 10^{-6.4} \quad (1)$$

$$CO_2(aq) + H_2O \leftrightharpoons H_2CO_3: K_2 = 10^{-10.3} \quad (2)$$

Because the formation of H_2CO_3 is negligible ($K = 10^{-10.3}$) relative to the formation of $CO_2(aq)$ ($K = 10^{-6.4}$), the total carbon dioxide concentration in water is best represented by their sum, $H_2CO_3^*$ which is directly related to P_{CO_2} by K_H, Henry's constant (Lloyd and Heathcote 1985):

$$H_2CO_3^* \leftrightharpoons K_H \times P_{CO_2}: K_H = 10^{-1.5} \quad (3)$$

Dissociation of $H_2CO_3^*$ into HCO_3^- and H^+ results in a decrease in pH such that pure water, in equilibrium with atmospheric pressure ($P_{CO_2} = 10^{-3.5}$), will have a pH approximately equal to 5.7. Where organic-rich soil profiles are developed, production of CO_2 by oxidation of organic material and by biologic respiration locally increases the partial pressure of CO_2 in the soil zone (e.g., P_{CO_2}-soil gas $= 10^{-1.5}$) and underlying vadose and phreatic environments.

CO_2-saturated waters percolating into the carbonate sedimentary sequence react with $CaCO_3$, which acts as a base and increases the pH of the fluid:

$$H_2CO_3^* + CaCO_3 \leftrightharpoons Ca^{2+} + 2HCO_3^- \quad (4)$$

If the fluid is isolated from exchange with the gaseous soil-CO_2 reservoir (i.e., a *closed*-CO_2 system such as within the phreatic zone), calcite dissolution consumes $H_2CO_3^*$, and pH, HCO_3^- and CO_3^{2-} all increase until saturation with $CaCO_3$ is reached (K_s-calcite $= [Ca^2] \times [CO_3^{2-}] = 10^{-8.4}$). In a closed-system, further uptake of CO_2 is not possible and the concentration of $H_2CO_3^*$ decreases as pH is increased due to carbonate dissolution (Deines et al. 1974, Butler 1982, Lloyd and Heathcote 1985). The result is a decrease in the equilibrium P_{CO_2} of this fluid (Fig. 2.3).

In an *open*-CO_2 system (e.g., soil or vadose zone), CO_2 continuously reacts with water as calcite is dissolved, maintaining a constant concentration of $H_2CO_3^*$ in the fluid. Because CO_2 is added, the concentration of total dissolved carbon, C_T, increases. In contrast, C_T for a closed system is fixed, controlled by the P_{CO_2} prior to isolation of the fluid from gaseous exchange. As we will see in a following section, the elevated equilibrium P_{CO_2} of a water saturated with respect to calcite in an open-CO_2 system provides an important drive for precipitation of carbonate. When environments of lower P_{CO_2} are encountered, such as in ventilated caverns, degassing of CO_2 shifts the equilibrium to carbonate precipitation.

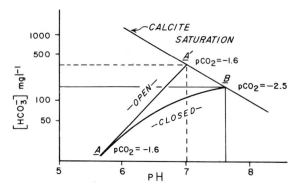

FIGURE 2.3. Contrast between *open-* and *closed*-CO_2 reaction pathways to calcite saturation. As calcite is dissolved, pH and HCO_3^- increase along different pathways depending upon the openness of the system to CO_2 gas. Both solutions begin in equilibrium with a $P_{CO_2} = 10^{-1.6}$ atm (A). Solution in a *closed*-system attains saturation with calcite at a lower $[HCO_3^-]$ and higher pH (B). *Open*-system reaction maintains equilibrium with CO_2 gas even at calcite saturation (A'). In contrast, consumption of $H_2CO_3^*$ under *closed* conditions significantly decreases the equilibrium P_{CO_2} of the solution.

For both the open- and closed-CO_2 reaction systems, in mature, monomineralic terranes where the sedimentary assemblage is comprised entirely of low magnesium calcite, calcite dissolution ceases when calcite saturation is reached at the ambient P_{CO_2}. Up to this point, rock–water reactions are controlled primarily by pH and concentrations of CO_3^{2-} and Ca^{2+}. It is within this regime in areas proximal to the fluid recharge surface that most dissolution in the phreatic zone of calcitic terranes occurs. Further dissolution (or precipitation) by the fluid will be negligible unless carbonate minerals of lower solubility (aragonite or magnesian calcite, as in the case of a polymineralic terrane) or contrasts in P_{CO_2} are encountered along the pathway of fluid flow.

An example of such a P_{CO_2} contrast is in the distal part of the meteoric system where marine and meteoric waters mix (Fig. 2.1). In the mixed-water zone, differences between the equilibrium P_{CO_2} and ionic strength of marine and meteoric water can result in undersaturation of the mixture of two saturated solutions (Fig. 2.2). The region of mixing where calcite dissolves depends on the initial saturation state of the meteoric water, and on differences in

the ionic strength and P_{CO_2} between the two waters and the proportion of meteoric and seawater in the mixture (Plummer 1975, Hanor 1978, Back et al. 1979, Plummer and Back 1980). Undersaturation of the mixed-water zone is responsible for macroporosity in the subsurface of the Florida Peninsula (Vernon 1969) and at sites along the east coast of the Yucatan (Back et al. 1979, Back et al. 1984). Moreover, chemical transformation of calcite to dolomite is favored within this zone of elevated Mg^{2+} availability and calcite undersaturation (Runnells 1969, Hanshaw et al. 1971, Badiozamani 1973, Ward and Halley 1985).

Cation Chemistry of Meteoric Waters

Implicit in the previous discussion has been the distinction between monomineralic carbonate assemblages and polymineralic assemblages in which metastable carbonate phases, aragonite or high magnesium calcite, coexist with stable, low-magnesium calcite. In monomineralic systems, changes in saturation are induced by *extrinsic* controls, such as temperature, P_{CO_2}, and evaporation; dissolution by meteoric waters proceeds until equilibrium is reached. While these factors also influence polymineralic systems, the effect of contrasting solubilities among coexisting phases superimposes an *intrinsic* drive for continued rock–water reaction (Matthews 1968, James and Choquette 1984); dissolution of more soluble phases oversaturates the solution relative to less soluble phases. Even though fluids are saturated with respect to calcite, rock–water reaction continues, progressively modifying the chemical composition of the fluid. It is within this context of exchange between water and rock systems that the major and minor cation chemistry of an evolving meteoric water can be envisioned. The extent of this rock–water reaction is ultimately controlled by initial differences in the mineralogy of the carbonate terrane.

As rainwater passes the threshold from the atmospheric into the lithospheric realm, the nature of its chemical interaction is dominated by rock–water exchange reactions. Trace, minor, and major cations in the carbonate are stoichiometrically transferred to fluid as carbonate rock is progressively dissolved. This increase in the concentration of cations within the fluid, in direct proportion to concentrations

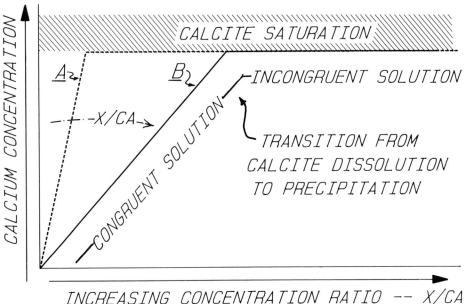

FIGURE 2.4. Patterns of increase in cation concentration during *congruent* and *incongruent* dissolution processes. Dissolution of a solid in the absence of precipitating phases results in a stoichiometric release of its constituents and a commensurate increase in fluid cation concentrations. In this example, calcite (A) and dolomite (B) dissolve congruently until a saturation threshold is reached. The slope of increase in Ca^{2+}/Mg^{2+} is directly proportional to the concentration ratio in parent phases. This relationship can be generalized for all elemental cations by relating their concentrations to that of calcium (X/Ca^{2+}). If saturation for an accompanying phase is exceeded, in this example calcite, continued dissolution of more soluble phases leads to calcite precipitation and an increase in the X/Ca^{2+} ratio of concentrations.

contained in the dissolving phases, is characteristic of a *congruent dissolution* process (Fig. 2.4). For example, dissolution of a calcite containing 15 mole percent $MgCO_3$ transfers 0.15 moles of Mg^{2+} to solution for every 0.85 moles of Ca^{2+}; thus the changes in Ca/Mg define a linear trend reaching a maximum at saturation. In a monomineralic system, this represents the climax of rock–water interaction, the point at which the chemical composition of the solution reaches steady state.

This pattern of chemical evolution contrasts with changes characteristic of polymineralic carbonate terranes. Where calcite coexists with aragonite and magnesian calcite, differing solubilities provide an intrinsic drive for continued rock reaction after saturation with respect to calcite is reached. Ca^{2+} is maintained at constant concentration by calcite precipitation. The onset of calcite precipitation (Fig. 2.4) reflects the change from *congruent* to *incongruent dissolution* (Wollast and Reinhart-Derie 1977). As

more carbonate is dissolved, minor elements which become enriched in the fluid are then partitioned between the fluid and the precipitating calcite.[1]

Unlike monomineralic systems, which rapidly reach a steady-state composition, incongruent dissolution continues in polymineralic systems until all soluble phases are exhausted, saturation of one mineral phase is exceeded relative to other phases, or the fluid exits the diagenetic system. Thus, the elemental composition of

---

[1]The ratio of molar concentrations of a minor element (mME: e.g., Mg, Mn, Fe, Sr, etc.) to the molar concentration of the major element mCa incorporated in a precipitating mineral can be related to the concentration ratio of the fluid by a distribution coefficient, D, such that: [mME/mCa]-mineral = [mME/mCa]-fluid × D. When D is greater than 1, the concentration ratio, mME/mCa, will be higher in the mineral relative to the fluid. When D is less than 1, the ME/mCa will be lower in the mineral than the fluid (Veizer 1983).

meteoric water in a polymineralic system is related to the composition and availability of soluble mineral phases, the duration of rock–water interaction, and the precipitation of secondary mineral phases (Pingitore 1978). This is effectively a measure of rates of reaction versus the rate of fluid flow.

Simplistically, the extent of rock–water interaction can be quantified as: the number of dissolution-precipitation reactions cycled through an individual volume of water. The water, acting largely as the medium for reaction, will change in chemistry as unstable phases are dissolved and secondary, diagenetic minerals are precipitated. Dissolution quantitatively adds the elemental components of dissolving mineral phases to the fluid via congruent dissolution (Fig. 2.4); subsequent partitioning of these elements, between the fluid and mineral phases during precipitation, further modifies water chemistry. For example, Mg^{2+} and Sr^{2+} are added to the fluid as magnesian calcite and aragonite dissolve. Because these elements have distribution coefficients less than unity, the Mg/Ca and Sr/Ca concentration ratios of diagenetic calcites will be less than their ratios in the fluid. Thus, via an incongruent dissolution process, the concentration of these elements will progressively increase in an individual volume of fluid as the extent of rock–water interaction increases.

Brand and Veizer (1980) refer to this feature of rock–water interaction in terms of the openness of the diagenetic system (Fig. 2.5). Alter-

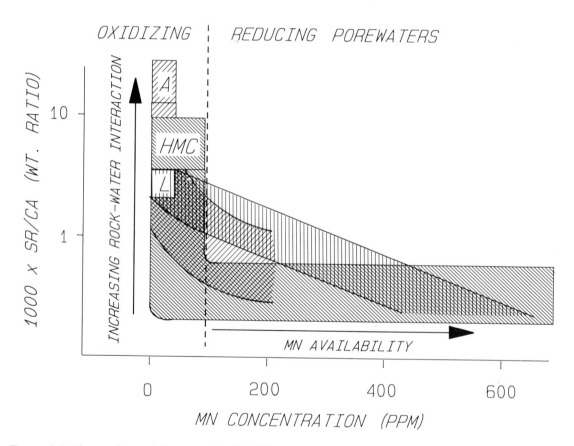

FIGURE 2.5. Diagenetic trends for aragonite (A), high-magnesium calcite (HMC), and low-magnesium calcite (L) which have been stabilized by meteoric water. In systems of low rock–water interaction, Sr/Ca ratio of diagenetic phases will decrease from primary compositions as Sr^{2+} is lost to the diagenetic water. In effectively closed systems, high rock–water interaction, Sr/Ca ratios of diagenetic phases approach values of primary components. Increases in Mn concentration of diagenetic calcite are possible only where the system is relatively open, porewaters are reducing, and a source of Mn is available in the sediment. Modified after Brand and Veizer (1980).

ation in an *open* or water-dominated system (low rock–water interaction) results in a decrease in cation concentrations in diagenetic carbonates for elements which have distribution coefficients less than unity. As the system becomes more *closed* (high rock–water interaction) these cations are retained in diagenetic phases because during incongruent dissolution their concentrations increase in the fluid. Such variations in cation abundance have been amply documented in Cenozoic carbonates presently undergoing diagenetic stabilization (Benson and Matthews 1971, Plummer et al. 1976, Back et al. 1979), in interstitial waters of pelagic carbonates stabilizing in marine burial phreatic environments (Matter and Perch-Nielson 1975, Sayles and Manheim 1975), and in ancient sequences of cratonic carbonates (Brand and Veizer 1980).

An additional class of cations, not solely controlled by rock–water reactions or by partitioning between the fluid and rock reservoirs, commonly exist in diagenetic carbonates. The solubilities in solution of cations such as Fe^{2+} and Mn^{2+} is predicated by redox equilibria which responds to the Eh and pH conditions of the solution. For example, if meteoric water entering a sedimentary sequence is an oxidizing solution, Fe^{2+} and Mn^{2+} released to solution from dissolving phases will be precipitated rapidly as insoluble oxides and hydroxides with cations in a higher valence state.

The oxidation–reduction (redox) system is controlled primarily by the decomposition (i.e., oxidation) of organic material. Dissolved oxygen in recharge waters oxidizes organic matter contained within the soil zone, a process which also elevates the P_{CO_2} of soil gas and increases the ambient P_{CO_2} of the underlying meteoric vadose and phreatic environments. In settings where the rate of organic decomposition surpasses the recharge of oxidized meteoric waters, dissolved oxygen can be rapidly depleted such that oxygen is then scavenged successively from oxyanions (NO_3^-, PO_4^{2-}) in the fluid and then from metal oxides (Mn_2O_4, Fe_2O_3, Fe_3O_4) contained within the sediment. This reduction sequence (cf. Froelich et al. 1979) can result in measurable concentrations of Fe^{2+} and Mn^{2+} in shallow diagenetic environments (Evamy 1969).

Diagenetic phases precipitating from water containing Mn^{2+} or Fe^{2+} will be enriched relative to the concentrations in the fluid because their distribution coefficient into carbonate is greater than one. On this basis, Brand and Veizer (1980) interpret that increased concentrations of these cations in diagenetic phases indicate increased openness of the rock–water system during alteration (Fig. 2.5). However, these cations are not derived from the dissolving carbonate phases, and thus cannot be considered like Mg^{2+} or Sr^{2+} to reflect the degree of rock–water interaction. Rather, their concentrations in the fluid require reducing conditions coupled with a local source for these elements within the sediment. From this, it follows that in a fully open system, alteration products may contain virtually no Mn^{2+} or Fe^{2+} if a source is not present or if the waters are oxidizing. In contrast, Mg^{2+} and Sr^{2+} are sourced directly from the mineral phases which are undergoing diagenesis; therefore, their bulk concentration, or gradients of concentration within individual crystal precipitates, provides a more reliable indication of the degree of rock–water interaction.

Isotopic Composition of Meteoric Waters

The isotopic composition of meteoric water reflects a combination of processes related to their generation by *evaporation* from the marine water reservoir to their modification by processes of vapor *condensation*, and to *rock–water interactions* within the sedimentary sequence (Seigenthaler 1977, Faure 1986). Isotopic components of particular interest in this examination are the stable isotopes of hydrogen (^{2}D-deuterium and ^{1}H), carbon (^{13}C and ^{12}C), and oxygen (^{18}O and ^{16}O).

Because of the small differences in natural abundance, measurement of isotopic composition is performed by determining the abundance ratio of the minor to the major isotope (i.e., $^{13}C/^{12}C$; $^{18}O/^{16}O$; and $^2D/^1H$). Isotopic enrichments are determined by comparison to standards and expressed in δ notation, in parts per thousand, ‰, or per mil.

$$\delta‰ = \frac{R\text{–(sample)} - R\text{–(standard)}}{R\text{–(standard)}} \times 10^3$$

where $R = {}^{18}O/{}^{16}O$, etc.

For waters, $\delta^{18}O$ and δD are compared to Standard Mean Ocean Water (SMOW); for carbonates, $\delta^{18}O$ and $\delta^{13}C$ are compared to the Pee Dee Belemnite (PDB) powdered carbonate standard.

Isotopes of Hydrogen and Oxygen. In evaluating the sources of waters, hydrogen and oxygen isotopes are of particular importance because they are the principal elemental constituents of water. During the processes of evaporation and condensation of meteoric water, isotopes of H and O are fractionated in response to mass differences between ^2D and ^1H and between ^{18}O and ^{16}O); the isotopes of lighter mass are preferentially removed in the vapor phase. The extent of this equilibrium fractionation is controlled primarily by temperature; however, at low relative humidities, kinetic fractionation diminishes the magnitude of the temperature-dependent, equilibrium fractionation. A direct relationship between mean annual air temperature and the annually averaged $\delta^{18}O$ of coastal rainwaters (Fig. 2.6) has been documented by Dansgaard (1964). This temperature effect, which is roughly correlated with latitude, can

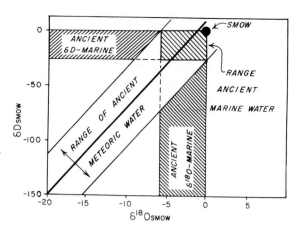

FIGURE 2.7. Isotopic variation of δD and $\delta^{18}O$ in meteoric waters. The isotopic compositions of hydrogen and oxygen in meteoric water vary due to fractionation during evaporation. Meteoric waters lie along the *meteoric water line* (heavy line) defined by the relationship $\delta D = 8\,\delta^{18}O + 10$, where the δD intercept is directly related to the isotopic composition of SMOW, present-day marine water (Craig 1961). With changes in the isotopic composition of marine water, *paleometeoric water lines* would form a range of parallel trends. The range of possible variation of ocean isotopic values can be estimated from observed variation of Phanerozoic marine carbonate (James and Choquette 1983; Lohmann 1983; Popp et al. 1986; Veizer et al. 1986) and proposed limits on variation of δD (Sheppard 1986).

be represented as a net fractionation between $\delta^{18}O$-*meteoric water* and $\delta^{18}O$-*marine water*. For example, at 15°C, Δ meteoric-marine = -4 per mil; if $\delta^{18}O$-marine = 0 per mil SMOW, then $\delta^{18}O$-meteoric = -4 per mil. The relationship between rainwater and groundwater compositions is relevant to this examination. Although seasonal variations of several per mil are observed at individual geographic sites, the composition of groundwater can be approximated by the annual average of rainwater compositions due to the time-averaging effect of mixing within the meteoric system.

Fractionation of the isotopes of hydrogen (D and H) accompanies the fractionation of oxygen isotopes during the genesis of meteoric water. The magnitude of fractionation is significantly greater for hydrogen because the relative mass contrast is larger (2/1 for D and H, relative to 18/16 for ^{18}O and ^{16}O). Craig (1961) determined that δD and $\delta^{18}O$ of meteoric waters (Fig. 2.7) derived by evaporation from marine water (δD and $\delta^{18}O$ = 0 per mil

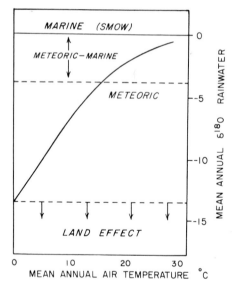

FIGURE 2.6. Relationship between $\delta^{18}O$ rainwater and mean annual air temperature provides a first approximation of the isotopic difference between meteoric and coeval marine waters (Δ meteoric–marine). Within a temperature range of 0° to 30°C, coastal water expected isotopic values exhibit a wide span $\delta^{18}O$ = 0 to –13‰. Additional fractionation due to "land effects" allows for increasingly more depleted compositions.

SMOW) covary. This covariance defines the *meteoric water line*. For present-day meteoric waters, where δD and δ¹⁸O equal 0 per mil SMOW, measurement of δD-water allows δ¹⁸O-water to be uniquely calculated (Fig. 2.7). This has been extensively applied in studies of paleoclimatology, where comparison of the δ¹⁸O-water (calculated from δD of fluid inclusions in speleothems) to δ¹⁸O-spelethem calcite is utilized to calculate temperatures of precipitation (Schwarcz et al. 1976). Application to studies of ancient meteoric systems, however, is hampered by secular variations in the isotopic composition of seawater (Fig. 2.8) (James and Choquette 1983, Lohmann 1983, Given and Lohmann 1986, Popp et al. 1986, Veizer et al. 1986), and thus by our uncertainty about the absolute values of paleometeoric waters.

Additional fractionation of meteoric water occurs as water vapor proceeds inland or to increasing altitudes. The condensation and removal of rainwater progressively depletes δ¹⁸O and δD of the remaining vapor (Faure 1986).

Through a Rayleigh fractionation process, this orographic or "land" effect can result in an additional depletion of 10 to 15 per mil in δ¹⁸O and 100 to 130 per mil in δD relative to coastal meteoric waters (Shepard et al. 1969, Seigenthaler 1977, Drever 1982).

When considering karst, it is possible to place some temperature limits on settings where karstification occurs. As a first approximation, active karstification is restricted to localities where mean annual air temperature lies between 0° and 30° C. In coastal settings, this temperature range will produce meteoric waters ranging in δ¹⁸O between 0 and −13 per mil relative to marine water. If inland localities are considered, the additional land effect significantly increases the range of possible meteoric water compositions. Clearly, the δ¹⁸O value of meteoric waters could range anywhere from 0 to less than −20 per mil relative to marine water; δD will vary even more, from +10 to −130 per mil. With such an extreme range of possible compositions, are predictions of paleometeoric

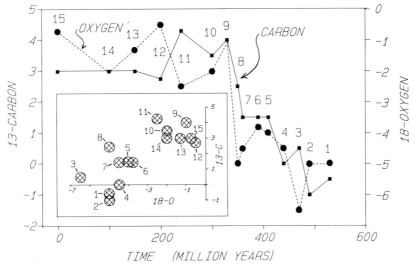

FIGURE 2.8. Secular variation of carbon and oxygen isotopic compositon of calcitic marine cements during the Phanerozoic. These data, derived from low latitude reefal sequences representing specific time intervals (enumerated), are estimates of primary marine *calcite* compositions after the effects of diagenesis have been accounted for and removed. The dramatic shift in both δ¹⁸O and δ¹³C, evident during the Devono-Carboniferous transition (7–9), should be matched by a comparable shift in composition of coeval meteoric waters. Comparison of the δ¹⁸O–δ¹³C crossplot, inset diagram, with Fig. 2.9 emphasizes the necessity of considering secular variation in marine water composition when estimating the composition of paleometeoric waters. Enumerated time intervals are: 15 = Holocene; 14 = Cretaceous (Aptian–Albian); 13 = Jurassic (Kimmeridgian); 12 = Triassic (Norian); 11 = Permian (Kazanian); 10 = Pennsylvanian (Moscovian); 9 = Mississippian; 8 = Late Devonian (Frasnian); 7 = Middle Devonian (Givetian); 6 = Late Silurian (Pridolian); 5 = Middle Silurian (Ludlovian); 4 = Late Ordovician (Ashgillian); 3 = Middle Ordovician (Llanvirnian); 2 = Early Ordovician (Tremadocian); and 1 = Early Cambrian (Data sources tabulated in Lohmann in press).

water compositions possible? If the paleolatitude, geographic setting and the isotopic composition of contemporaneous marine water were known, a much narrower range of possible values could be estimated.

Isotopes of Carbon. Carbonates precipitated in surface settings with open-system exchange with atmospheric CO_2 will have isotopic compositions which converge to a common value. This commonality reflects the equilibration of surface waters with carbon derived from the atmospheric CO_2 reservoir. Thus, coevally precipitated marine and meteoric carbonates may be indistinguishable on the basis of their carbon isotopic compositions. The isotopic composition of meteoric water, however, can be rapidly modified as the water percolates through a sediment soil zone containing CO_2 produced by the oxidation of organic matter (Salomons and Mook 1986). Organically derived CO_2 has an isotopic composition depleted in ^{13}C ($\delta^{13}C$ of -16 to -25 ‰), which is highly disparate from atmospheric carbon compositions. Equilibration with isotopically depleted CO_2 in the soil zone provides an isotopic signature for meteoric water which helps distinguish it from other diagenetic fluids.

The concept of congruent versus incongruent dissolution pathways is particularly applicable when evaluating isotopic variations of a fluid which occur in response to rock–water reactions. Such reactions result in isotopic exchange between the fluid and rock reservoirs. The magnitude of this effect is best represented by a general isotopic mass balance equation (Lawrence et al. 1975):

$$f_1[\delta_{c1}] + (1 - f_1)[\delta_{w1}]$$
$$= f_2[\delta_{c2}] + (1 - f_2)[\delta_{w2}] \quad (5)$$

where f_1 = mole fraction of element in dissolving mineral
f_2 = mole fraction of element in precipitating mineral
δ_{c1} and δ_{w1} = isotopic values of mineral and water prior to dissolution
δ_{c2} and δ_{w2} = isotopic value of precipitating mineral and water after precipitation

The simplest case is one where solution chemistry is modified solely by dissolution of carbonate, such as might occur in a monomineralic assemblage approaching carbonate saturation. Changes in fluid composition respond to the fluid/rock ratio of this reaction and the isotopic compositions of the fluid and the rock (Lohmann 1982, Given and Lohmann 1986, Martin et al. 1986). In the case of dissolution without precipitation, equation 5 simplifies to a stoichiometric addition of the dissolved mass to the fluid. For example, in equation 4, half the carbon is derived from dissolving calcite and half from $H_2CO_3^*$ of the fluid (Allan and Matthews 1977 and 1982). If the fluid begins with an isotopic composition $\delta^{13}C = -25$ ‰, addition of isotopically enriched carbon from $CaCO_3$ ($\delta^{13}C = +3$ ‰) modifies the fluid composition to an intermediate value ($\delta^{13}C = -11$‰). It is difficult to obtain significantly more depleted $\delta^{13}C$ values, except under extremely reducing conditions of methanogenesis. Because the equilibrium fractionation of calcite-CO_2 ($\Delta CaCO_3 - CO_2 = 10$‰) is quite large (Freidman and O'Neil 1977, Faure 1986), calcite precipitating in open-system equilibrium with soil-gas CO_2 is limited to about -15 per mil ($\delta^{13}C$-calcite = $\Delta CaCO_3 - CO_2 + \delta - CO_2 = +10 + (-25) = -15$‰).

The persistent dissolution and precipitation characteristic of polymineralic assemblages can result in a high degree of rock–water exchange.[2] Because of the small total concentrations of dissolved carbon in water ($C_T = 3$ to 8 millimoles/kilogram H_2O), the $\delta^{13}C$ rapidly converges to the value of the dissolving, metastable carbonate phases.

Combined Variation in Oxygen and Carbon Isotopes. When the simultaneous effects of rock–water exchange are considered for $\delta^{18}O$ and $\delta^{13}C$ using a mass balance approach (eq. 5), the relative rates at which carbon and oxygen in the fluid reach isotopic equilibrium with rock composition can be calculated. In contrast to $\delta^{13}C$, the $\delta^{18}O$-water will be modified only after extensive dissolution–precipitation has oc-

---

[2]The extent of rock–water interaction can be quantified as a rock to water volume ratio, V-rock/Y-water. This ratio represents the relative volume of rock cycled through a single volume of water in response to dissolution-precipitation reactions. Hence, a high rock–water ratio exemplifies a system in which fluid chemistry is dominated by the composition of dissolving rock phases.

curred (Fig. 2.9). When their combined variation is examined, a highly variable $\delta^{13}C$ coupled with invariant $\delta^{18}O$ is characteristic of environments where a source of dissimilar $\delta^{13}C$ is present (Lohmann 1982, Margaritz 1983, Meyers and Lohmann 1985). Although this setting is commonly observed in meteoric systems where organically derived, isotopically depleted CO_2 exists in the soil-zone, it is not exclusive to this environment. Any environment which contains sources for carbon with disparate $\delta^{13}C$ compositions can produce similar rock–water carbon and oxygen isotopic trends. For example, ^{13}C-depleted carbon derived from hydrocarbons under conditions of burial can produce similar isotopic trends (Hudson 1977, Budai et al. 1983).

Recognition of rock–water reaction trends can be extremely valuable in determining the $\delta^{18}O$ composition of a given diagenetic fluid. The highly variable carbon values coupled with invariant oxygen values serve to delineate a $\delta^{18}O$ of an individual diagenetic fluid. When considering the broad range of possible $\delta^{18}O$ values for meteoric waters (Fig. 2.6) large deviations in $\delta^{13}C$ provide for unique determination of the $\delta^{18}O$ for meteoric water at a specific site (Fig. 2.9). The usefulness of such trends in diagenetic studies stems from the preservation of these patterns of variation in precipitated phases.

The marine-meteoric, mixed-water zone of the phreatic system is the only environment in which $\delta^{18}O$ and $\delta^{13}C$ of waters covary. While $\delta^{13}C$ of meteoric water is characteristically variable, its $\delta^{18}O$ composition is largely invariant within a given hydrologic system. In comparison, both $\delta^{13}C$ and $\delta^{18}O$ of marine water are invariant. Mixing marine and meteoric waters at the distal margin of a phreatic system results in complex variations in water chemistry. In addition to changing saturation with respect to carbonate (Fig. 2.2), mixing alters both the cation and oxygen isotopic composition of the diagenetic water. Cation chemistry, such as $[Mg^{2+}]$ or $[Sr^{2+}]$, will exhibit a simple mixing relationship directly reflecting the proportion of intermixed marine water. In contrast, isotopic compositions define a range of covariant C-O trends which become hyperbolic (Fig. 2.10) as the difference in concentration of dissolved carbon increases between marine water ($C_T = 2.5$ mmoles/kg water) and meteoric water ($C_T = 2.5$ to >6.0 mmoles/kg H_2O). Because meteoric water typically contains higher concentrations of dissolved carbon (C_T), the $\delta^{13}C$ of meteoric water effectively dominates the isotopic composition of the mixture up to very high percentages of marine water Fig. 2.10A).

It is significant that a straight mixing line can develop only when C_T is equal for marine and meteoric waters, conditions that are not favorable for calcite dissolution. Therefore, carbon and oxygen isotopic values of karst-forming or dolomitizing mixed waters should parallel hyperbolic mixing curves (Fig. 2.10B). It is unlikely, however, that mixing will result in a single hyperbolic curve, because the $\delta^{13}C$ value of meteoric water should exhibit a range of compositions in response to varying degrees of rock–water exchange. In addition to the variation in $\delta^{13}C$ of meteoric water with time, local

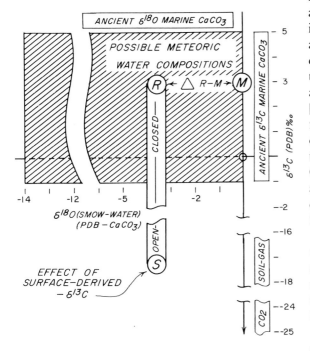

FIGURE 2.9. Range of possible $\delta^{18}O$ and $\delta^{13}C$ compositions of Phanerozoic meteoric waters. While the oxygen isotopic composition of meteoric waters can be variable due to the combined effects of equilibrium temperature fractionation and land effects, variation at one geographical site is quite limited. Identification of a characteristic $\delta^{18}O$ meteoric water is aided by the deviation of meteoric water $\delta^{13}C$ values from coeval marine values due to the interaction of highly depleted soil-gas CO_2. Such perturbations observed in $\delta^{13}C$ coupled with invariant $\delta^{18}O$ serve to mark the average composition of a meteoric water.

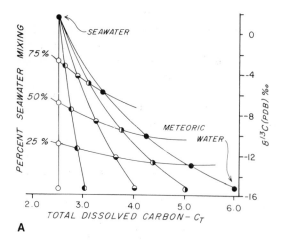

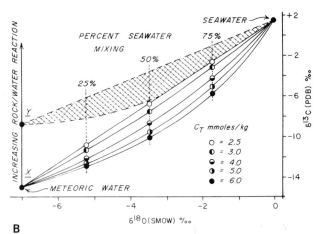

A

B

Figure 2.10A. Relationship of total dissolved carbon (C_T) and $\delta^{13}C$ during mixing of meteoric and marine waters. As C_T increases in meteoric water in response to equilibrium at higher P_{CO_2}, carbon isotopic composition becomes nonlinear. $\delta^{13}C$ composition of meteoric waters progressively dominate the mixed-water composition. B. Covariance of $\delta^{18}O$ and $\delta^{13}C$ during mixing of marine and meteoric waters. At equal total dissolved carbon (C_T = 2.5 mmoles/k H_2O) mixing follows a simple "straight line" mixing relationship between marine water and a meteoric water, (X) end member. As the contrast of C_T increases, deviation in $\delta^{13}C$ forms distinctive hyperbolic mixing lines. Local dissolution of carbonate in response to mixing (see Fig. 2.3) may modify the $\delta^{13}C$ end point composition of the meteoric water, (Y). Therefore, measured $\delta^{18}O$ and $\delta^{13}C$ values should define a field of wide variation converging at the marine water end-point.

dissolution in response to mixing further modifies its isotopic composition. The end-point value of meteoric water will define a range of $\delta^{13}C$ values with invariant $\delta^{18}O$. Thus, in response to rock–water reactions an array of mixing trends should be produced which diverge from the common and invariant marine water locus.

Applications to Studies of Paleokarst

Within the context of the variation observed in the chemistry of natural waters, it is possible to predict the chemical composition of alteration products and precipitates associated with meteoric systems. This, of course, requires that a mineral record of this diagenesis is preserved. Such mineral records include precipitated cements or replacive minerals resulting from fine-scaled processes of dissolution–precipitation. While dissolution of carbonate is dominant in karst-forming environments, mineral precipitation can occur locally as waters become oversaturated. Oversaturation may develop in

at least three ways (Salomons and Mook 1986): (1) transformation of metastable minerals such as aragonite or magnesian calcite to calcite, (2) degassing of CO_2 which modifies pH and CO_3^- or (3) evaporation which increases the total concentration of dissolved solids.

The initial step in evaluating the effects of meteoric alteration in ancient carbonate sequences is to determine that meteoric waters have actually been involved in the diagenetic history. Petrographic features have been proposed for identifying meteoric diagenesis of carbonate terranes (Halley and Harris 1979, Longman 1980, James and Choquette 1984). Perhaps most distinctive are alteration and cementation features associated with the vadose zone. Here, the presence of air- and water-filled pores results in meniscus or gravity-controlled distribution of cements. Macroscopic features at the surface of subaerial exposure may include soil-zone features of root casts, brecciation and corrosion, or carbonate accretion as caliche or calcrete (James and Choquette 1984). Although these diagenetic fabrics can be interpreted unambiguously, their preservation in ancient sequences is hampered by their volumetrically minor abundance, the thinness of the vadose zone, and the propensity for vadose sequences to be eroded during weathering.

In contrast, alteration and cementation are more extensive in the phreatic zone due to the ubiquitous water saturation of intergranular porosity and the comparatively long residence time of diagenetic waters in contact with reactive carbonate minerals. Despite the extent of alteration, petrographic fabrics characteristic of diagenesis in the meteoric phreatic zone are lacking. Textural criteria, such as equant calcite cements isopachously rimming voids or alternation of bright and dark cathodoluminescent zoning in early cements, have commonly been utilized to verify meteoric phreatic alteration (Halley and Harris 1979, Longman 1980, Grover and Read 1983). However, such fabrics can develop under a variety of diagenetic conditions during burial, with or without the catalyst of meteoric waters, and are thus ambiguous or inconclusive as criteria of meteoric diagenesis.

Most reliable are criteria which relate the timing of diagenetic alteration to the compactional history of a sequence (Meyers 1974). For example, dissolution or alteration of components with high reaction potential, such as aragonite or high magnesium calcite, prior to significant mechanical compaction is prime evidence for diagenesis in near-surface environments. An additional relationship which serves to identify meteoric diagenesis is the spatial distribution of fresh waters of a meteoric system, and thus the stratigraphic restriction of its diagenetic effects. An individual meteoric system has an upper boundary defined by the subaerial exposure surface at which meteoric waters are recharged. The thickness of a meteoric system is also finite, marked by the stratigraphic depth at which fresh waters buoyantly reside on denser marine or subsurface fluids. Thus, the spatial and stratigraphic restriction of diagenetic features and their relationship to known surfaces of exposure are paramount if a meteoric origin for diagenetic fluids is to be demonstrated (Given and Lohmann 1986).

Only after the spatial and temporal distribution of alteration and cementation have been resolved can geochemical tracers be integrated effectively to verify and subdivide the environments of meteoric diagenesis. It is important to note that the isotopic and cation chemistry of precipitating minerals will differ from the absolute composition of the fluid because of fluid–solid fractionation and partitioning effects. Despite this difference, variation present in the fluid will be mimicked by variation in diagenetic minerals. Therefore, spatial and temporal changes in chemistry of the meteoric water system will be recorded as trends in the chemical composition of diagenetic minerals.

The Meteoric Diagenetic Trend: Vadose and Phreatic Environments

As illustrated in Figures 2.7 and 2.9, the $\delta^{18}O$ value of meteoric water can lie within a broad range of possible values. This reflects a combination of numerous variables such as latitude, altitude, land effect, prevailing weather patterns, temperature, and the isotopic composition of coeval marine water. Although this range can be narrowed by considering these variables for ancient sequences, it is difficult to estimate accurately the isotopic composition of paleometeoric waters, and similarly to predict the composition of carbonate precipitated from them. For these reasons, recognition of meteoric diagenesis based solely on an absolute $\delta^{18}O$-calcite value is unjustified. Depleted oxygen isotopic compositions in carbonate can either result from an increase in temperature of precipitation (burial), or reflect isotopically depleted subsurface waters whose origin may be unrelated to meteoric processes. Thus, petrographic, stratigraphic and geochemical data must be combined to demonstrate that diagenetic alteration was indeed related to interaction with meteoric water.

It is first necessary to identify the $\delta^{18}O$ value characteristic of the diagenetic event. As discussed above, during rock–water interaction the $\delta^{18}O$ composition of the water remains largely unaffected while the $\delta^{13}C$ composition is highly variable, ranging between the value of dissolving host rock and the value of disparate carbon sources. When preserved in carbonate precipitates, this trend of variable $\delta^{13}C$ and invariant $\delta^{18}O$ identifies a characteristic $\delta^{18}O$-carbonate value for each diagenetic system (Allan and Matthews 1982, Lohmann 1982). In the case of meteoric diagenesis, this trend, here termed the *meteoric calcite line*, encompasses the majority of variations present in a single meteoric water system (Fig. 2.11A). Since the composition of meteoric water can vary geographically with individual systems possessing unique values, the $\delta^{18}O$ value of the meteoric calcite line must be determined individually for each sequence and locality studied.

Spatial changes in the isotopic composition of meteoric water, which occur in response to rock–water interaction, will be recorded in carbonate cements and alteration products. For example, meteoric waters at localities where an organic-rich soil is developed at the exposure surface will inherit a depleted carbon compo-sition from soil-gas CO_2. As these waters infil-trate the vadose zone and flow through the phreatic lens, rock–water interaction will pro-duce progressive shifts in the $\delta^{13}C$ composition related to the distance from this surface (Fig. 2.11A). Bulk compositions of alteration prod-ucts will be isotopically enriched and have more positive $\delta^{13}C$ with increasing stratigraphic depth relative to the exposure horizons (Allan and Matthews 1982, Given and Lohmann 1986).

Temporal changes in isotopic compositions occur as the carbonate system matures and the abundance of unstable mineralogies diminishes. Meteoric fluids will travel further along flow

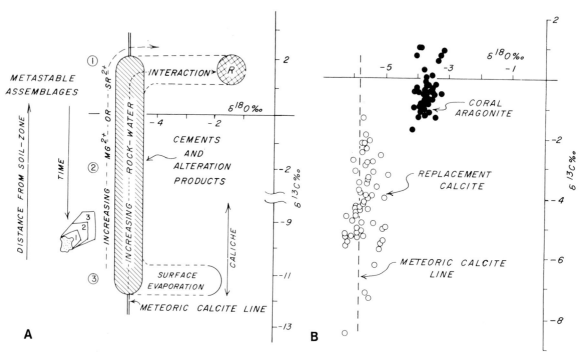

A

B

FIGURE 2.11A. Idealized plot of variation in $\delta^{18}O$ and $\delta^{13}C$ characteristic of meteoric vadose and phreatic carbonates. The isotopic composition of meteoric calcite cements and alteration products typically will define a trend of invariant oxygen coupled with highly variable carbon compositions. The constancy of $\delta^{18}O$, combined with this variable $\delta^{13}C$, defines a trend (termed the *meteoric calcite line*) which identifies the $\delta^{18}O$-calcite value characteristic of meteoric water alteration. Although this trend is typical of the dom-inant variation, deviations toward more enriched $\delta^{18}O$ are possible. In polymineralic assemblages at sites distal from the meteoric water recharge surface, increased rock–water interaction may lead to the precipitation of cements of equivalent isotopic com-position as dissolving mineral phases (R). Such dis-solution–precipitation reactions would be matched by increases in Mg^{2+} or Sr^{2+} in response to magne-sian calcite or aragonite stabilization. Similarly, at one site a progression toward more depleted composi-tions develops from temporally diminishing rock–water interaction as the carbonate terrane matures and the abundance of metastable components de-creases. At the recharge surface, where caliche is formed, intense evaporation associated with upward capillary flow produces enriched $\delta^{18}O$ coupled with depleted $\delta^{13}C$ derived from soil-gas CO_2. *B*. Meteoric alteration trend of the Key Largo Limestone, Florida. Open system alteration of aragonitic coralline skeletal carbonate by meteoric waters is recorded by the ver-tical limb of the characteristic meteoric calcite trend. After Martin et al. (1986).

pathways, and deeper in the phreatic lens, before rock–water reactions attenuate the depleted soil-gas carbon. Therefore, successively younger zones of cements, unlike bulk compositions, will have progressively more depleted $\delta^{13}C$ values (Fig. 2.11; Meyers and Lohmann 1985). To recognize such trends it is necessary to analyze microsamples which represent successive cement zones. While this can be difficult due to the thinness of cement zones, it is also possible to verify the meteoric calcite trend through probabilistic sampling of other components. Numerous microsized samples taken from micrites or altered skeletal grains generally exhibit a range in composition comparable to pure cement separates (Fig. 2.11B). This variation reflects differences in the timing of alteration, in response to differing diagenetic reactivity or intergranular porosities, during which time pore water chemistries are changing. Such an approach, however, cannot discern the temporal pattern of change which is most characteristic of the meteoric environment.

If paleoexposure surfaces can be found within a stratigraphic section, analysis of the chemistry of pervasively altered carbonate associated with these horizons provides a unique constraint for evaluating the spatial context of meteoric diagenesis. These surfaces, marked by caliches or extensive corrosional alteration, commonly define the open system end member of the meteoric calcite line (Fig. 2.11A). Comparison of the isotopic composition of cements and alteration products throughout the stratigraphic sequence allows meteoric vadose and phreatic zones to be uniquely correlated to specific subaerial exposure surfaces (Fernberg et al. 1986). Such correlations are particularly useful in sequences where multiple episodes of meteoric alteration are present and allow discrimination of the stratigraphic distribution and extent of alteration related to individual events of exposure.

The identification of a meteoric calcite line requires variation in the degree of rock–water interaction, either spatially or temporally. Therefore, in monomineralic calcite terranes, the isotopic values of meteoric calcites will not record a high degree of rock–water interaction because the intrinsic drive for continuing reaction and precipitation is absent. $\delta^{13}C$ values will typically cluster at the depleted end of the

meteoric trend. Variation observed in $\delta^{13}C$ is produced by the recycling of previously formed soil-zone carbonate (Salomons and Mook 1986) and by kinetic fractionation effects during the degassing of CO_2 (Hendy 1971).

The Mixed-Water Environment

Of all meteoric diagenetic environments, the fabric and chemistry of diagenetic components of the mixed-water zone are the most poorly documented and understood. This is a zone in which extensive corrosion can develop macroporosity; it is proposed as a site of replacive dolomitization of calcite. Documentation of such features, however, is sparse (Vernon 1969, Back et al. 1984, Ward and Halley 1985). In contrast, a great deal is known about the chemical variation present within the mixed water zone (see previous section). Therefore, it is possible to provide theoretical predictions of geochemical patterns which might be produced in a mixed, marine-meteoric environment. As discussed earlier, variable mixing of these waters produces, at any one point in time, a hyperbolic mixing curve reflecting differences in isotopic composition and total dissolved carbon, C_T, of the fluid end members. However, carbonate may be dissolved in response to mixing, or the $\delta^{13}C$ composition of the meteoric water may change temporally in response to evolving rock–water effects. Rather than a single mixing trend, an array of mixing curves (Fig. 2.12A) is more likely to be produced.

Although this pattern of chemical variation is distinctive, identification of mixed-water diagenesis in ancient sequences can be difficult because a carbonate record is generally not formed over the full range of mixing. If such mixed-water systems are to be identified in ancient sequences, a carbonate record must be precipitated. The drive for carbonate precipitation in the mixed-water zone is problematic; dissolution of carbonate seems more common than precipitation. Carbonate precipitation can occur either by CO_2-degassing (Hanor 1978) or by differential mineral solubility. When metastable phases are present calcite may precipitate, as aragonite and magnesian calcite dissolve (Matthews 1968 and 1974, Allan and Matthews 1982). However, where the contrast in P_{CO_2} is large, mixed waters will be undersaturated for

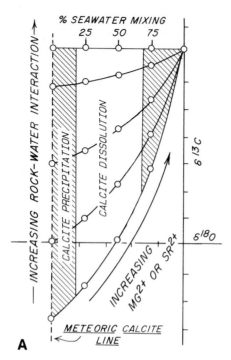

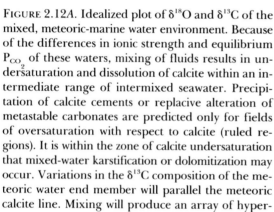

A

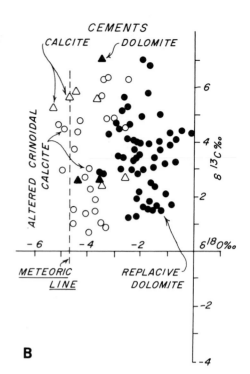

B

FIGURE 2.12A. Idealized plot of $\delta^{18}O$ and $\delta^{13}C$ of the mixed, meteoric-marine water environment. Because of the differences in ionic strength and equilibrium P_{CO_2} of these waters, mixing of fluids results in undersaturation and dissolution of calcite within an intermediate range of intermixed seawater. Precipitation of calcite cements or replacive alteration of metastable carbonates are predicted only for fields of oversaturation with respect to calcite (ruled regions). It is within the zone of calcite undersaturation that mixed-water karstification or dolomitization may occur. Variations in the $\delta^{13}C$ composition of the meteoric water end member will parallel the meteoric calcite line. Mixing will produce an array of hyper-

bolic mixing curves which diverge from the seawater end member composition. Thus, the carbonate record as cements or replacive phases should define vertical trends of variation parallel to the meteoric calcite line rather than a single hyperbolic mixed-water curve. B. Variation of $\delta^{18}O$ anc $\delta^{13}C$ in replacive calcite and calcite cement with cooccurring replacive dolomite and dolomite cement in the Madison Group, Wyoming. The covariation of these mineral phases, parallel to a meteoric calcite line, is compatible with alteration and dolomitization within a mixed, meteoric-marine water environment. After Budai et al. (in press).

both calcite and aragonite over most of the mixing range (Fig. 2.2). Diagenetic calcite, therefore, is likely to precipitate only during *partial mixing*—in regions dominated by either meteoric or marine waters. Because of these combined effects, the isotopic variation characteristic of mixed-water carbonates should parallel the meteoric calcite line and rarely mimic the hyperbolic compositional trends present in the water (Fig. 2.12A). Only under conditions where mixing does not induce calcite dissolution, for example where equilibrium P_{CO_2} values for marine and meteoric water are equal, will compositions of precipitated car-

bonate exhibit the full range of mixing present in the waters.

Another important feature of the mixed-water zone is the development of dolomite. Under conditions of calcite dissolution, replacive dolomite may form. As with calcite, dolomite values should reflect shifts in $\delta^{18}O$ in response to mixing, and the $\delta^{13}C$ values will vary in response to the rock–water effects of carbonate dissolution. Therefore, isotopic trends of replacive dolomite will exhibit dominant variation parallel to cooccurring mixed–water calcite; both will parallel the *meteoric calcite line* (Fig. 2.12B). Hyperbolic or linear trends will be produced only when the meteoric water composition is invariant and when diagenetic carbonate is directly precipitated rather than forming by replacement. This is a significant revision of the simplistic *straight-line* mixing relationship proposed in previous studies (Allan and Matthews 1982, Lohmann 1983, Buchbinder et al. 1984, Ward and Halley 1985).

Recognition of paleokarst associated with mixed–water environments cannot be accomplished without characterization of the overall meteoric vadose and phreatic system. Moreover, the spatial relationship between the surface of subaerial exposure and the stratigraphic distribution of meteoric alteration must be resolved if a mixed–water origin for corrosion or diagenetic alteration is to be confidently proposed. Patterns of chemical variation of mixed–water diagenesis can be interpreted only when the chemical patterns of diagenetic alteration in the meteoric system are well constrained, and when the composition of coeval marine waters is known.

The Spelean Environment

The dominant control on the variation of speleothem chemistry is the character of meteoric water entering caverns as vadose seepage along fractures and joints. Calcite precipitated from this water will define a meteoric calcite line equivalent to that of the vadose-phreatic trend (Fig. 2.11A and 2.13A). Speleothem chemistry can deviate significantly from the typical meteoric calcite trend, because the patterns of chemical variation can reflect the specific processes which drive speleothem precipitation (Hendy 1971). When CO_2-charged vadose waters contact a zone of lower P_{CO_2}, degassing and carbonate precipitation proceed until equilibrium is reached at the P_{CO_2} of cave air. Under conditions of rapid degassing, kinetic isotopic fractionation results in a marked enrichment of the $\delta^{13}C$ in the fluid as isotopically depleted CO_2 is released. Under conditions of gradual degassing the $\delta^{18}O$ of the fluid is not altered, and calcites precipitated in equilibrium at ambient cave temperatures will be indistinguishable from calcite typically forming in meteoric vadose and phreatic settings.

Where evaporation is significant, fluid compositions become more enriched in $\delta^{18}O$ and precipitating carbonates deviate significantly from the meteoric calcite line (Figs. 2.13A and 2.13B). CO_2-degassing can occur in the absence of evaporation, whereas evaporation almost always is accompanied or preceded by degassing. The coupled effect of evaporation and degassing produces distinctive trends with covariation of $\delta^{18}O$ and $\delta^{13}C$ (Fornaca-Rinaldi et al. 1968). Such variation has been documented along individual growth layers of speleothems, and the slopes of these trends reflect the relative dominance of evaporation versus degassing (Fantidis and Ehhalt 1970). In addition to dramatically increasing fluid saturation with respect to calcite, removal of Ca^{2+} from solution by calcite precipitation progressively enriches cation/Ca^{2+} ratios in the fluid. This combination of a high saturation state and a high Mg^{2+}/Ca^{2+} ratio results in aragonite and/or magnesian calcite precipitation (Holland et al. 1964); these define compositional end points of covariant trends (Figs. 2.13A and 2.13B) which are markedly enriched relative to typical meteoric values (Gonzalez and Lohmann, this volume).

Oxides are commonly associated with most speleothems. Iron and manganese, transported in meteoric waters, both as reduced Fe^{2+} and Mn^{2+} and as particulates, form distinctive color banding among growth bands. While dissolved iron and manganese are present in vadose seepage waters, little is incorporated into the calcite crystal lattice. Rather, upon contact with cave atmosphere, they are rapidly oxidized to form discrete concentrations which produce color banding along growth layers. The presence of such oxide banding may be a prime discriminator for spelean carbonates. Although some of the banding in speleothems is imparted by organic films (Schwarcz 1986), Mn- and Fe-oxides commonly form surface coatings on growing crystal faces. Thus, banding probably represents variation in climate, and likewise variation in the balance between organic production and fluid flow rates which control the oxidation state of meteoric groundwaters.

An important textural feature of speleothems is their coarse columnar and fibrous crystalline fabric (Kendall and Broughton 1978). This fabric traps abundant inclusions of the water which is precipitating these crystals. Sequestered meteoric fluids are the best record of the

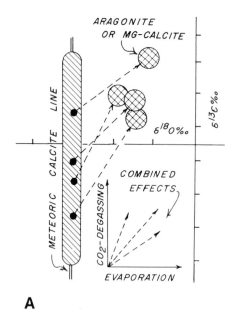

A

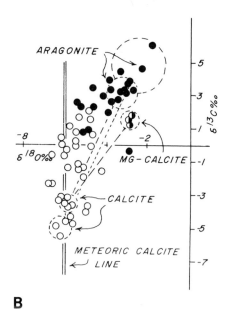

B

FIGURE 2.13A. Idealized plot of $\delta^{18}O$ and $\delta^{13}C$ variation associated with spelean environments. The isotopic composition of seepage waters defines the characteristic meteoric trend in response to variable degrees of rock–water reaction under differing degrees of open versus closed system equilibration with soil-gas CO_2. Precipitation of spelean carbonate is driven by a combination of CO_2-degassing and evaporation. In addition to modifying the isotopic composition of meteoric water at the site of deposition, increases in saturation state, alkalinity, and/or cation concentration may lead to the precipitation of metastable carbonate phases. *B*. Observed variation in mineralogy and isotopic composition of spelean carbonates from Carlsbad Caverns, New Mexico. Deviation from the meteoric calcite trend develops in response to the combined effects of evaporation and degassing, with aragonite or magnesian calcite precipitating at the enriched carbon–oxygen endpoint of covariant trends. After Gonzalez and Lohmann (this volume).

composition of paleometeoric waters (see review, Schwarcz 1986). Despite in situ exchange of oxygen and carbon isotopes between the inclusion waters and the carbonate host crystals, isotopic records of hydrogen remain unaffected unless inclusions are breached by later diagenesis. In speleothems of the recent past (at least through the Tertiary), comparison of the $\delta^{18}O$-calcite and δD-inclusion water values allows for unique verification of a meteoric water origin and estimation of paleotemperatures. At present, because we lack precise knowledge of paleo-ocean water compositions, *paleometeoric water lines* (Fig. 2.7) have not been determined. However, with future advances in this area, application of δD fluid inclusion approaches to paleokarst may be possible.

Because speleothems can differ considerably from the typical patterns of meteoric alteration and precipitation, in terms of both their chemistry and their occurrence as precipitational phenomena, they should be quite distinct and easily identified in the geologic record. An obvious complication, however, is the diversity of speleothem crystal fabric and mineralogy. Even though many speleothems are comprised of low-magnesium calcite, the development of fibrous and columnar crystal fabrics and the formation of metastable carbonate minerals complicate their recognition in ancient sequences. Speleothems, as primary precipitates, are quite similar texturally to the diagenetic products of calcitized marine cements.

In the case of marine cements, the columnar crystal fabric is a neomorphic product of a former metastable fibrous precursor. The increase in crystal size develops by dissolution of aragonite or magnesian calcite along intercrystalline boundaries; concomitant precipitation of low-magnesium calcite at sites of dissolution results in the coalescive growth of adjacent fibers (Given and Lohmann 1985). This replacement microfabric within columnar crystals, commonly visible as a mottled fabric under cathodoluminescence, is diagnostic of calcitized, fibrous precipitates initially comprised of metastable carbonate. In contrast, the majority of columnar crystals forming speleothems are directly precipitated as low-magnesium calcite. Thus, columnar crystal fabrics form as the speleothem is precipitated and do not possess replacive microfabrics.

Metastable fibrous carbonates precipitate in both marine and spelean environments and, regardless of their origin, these metastable carbonates will be calcitized during later diagenesis. It has been shown that isotopic analysis of calcitized marine cements can produce distinctive covariant $\delta^{13}C$ and $\delta^{18}O$ trends reflecting the replacive intermixture of carbonate recording diagenetic and primary compositions (Given and Lohmann 1985). However, a similar covariance of compositions is observed as a primary feature among cooccurring low-magnesium calcite and aragonite or magnesian calcite (Fig. 2.11B). While these metastable minerals will be transformed to low-magnesium calcite within the same meteoric environment which precipitated them and can inherit typical meteoric chemistries, it is possible that enriched isotopic compositions may be preserved in the diagenetic products. Thus, ancient speleothems can exhibit isotopic trends similar to calcitized marine cements. Because of the ambiguity of the chemical record, petrographic evidence that columnar calcites are primary and precipitated as low-magnesium calcite is needed to discriminate their spelean origin.

Conclusions

Meteoric water is the common denominator of all environments of karst formation such that overall trends in isotopic chemistry will focus about the meteoric calcite line and cation concentrations will reflect overall depletion from primary marine values. However, deviations from typical meteoric values occur as a natural part of the history of a given meteoric water. Variable degrees of rock–water interaction and the relative dominance of evaporation versus degassing which induce carbonate precipitation can modify the chemistry of meteoric waters. Thus, if diagenesis by meteoric waters is to be recognized in ancient carbonate sequences and paleokarst placed within this environmental context, stratigraphic, petrofabric, and geochemical data must be integrated to define the spatial and temporal evolution of the rock–water system. It is the combined observation of these features which provides for the unambiguous recognition of diagenesis by meteoric waters.

Acknowledgments. The author thanks E.R. Graber, L.A. Gonzalez, and B.H. Wilkinson for their stimulating discussions and assistance during all stages of the preparation of this chapter. Special thanks are due N.P. James and P.W. Choquette, for without their encouragement, persistence, and patience, this synthesis would not have been initiated. My appreciation continues for the many students and colleagues whose petrographic and geochemical data provide the foundation for my interpretive wanderings.

References

Allan, J.R., and Matthews, R.K., 1977, Carbon and oxygen isotopes as diagenetic and stratigraphic tools—data from the subsurface of Barbados, West Indies: Geology, v. 5, p. 16–20.

Allan, J.R., and Matthews, R.K., 1982, Isotope signatures associated with early meteoric diagenesis: Sedimentology, v. 29, p. 797–817.

Back, W., Hanshaw, B.B., Pyle, T.E., Plummer, L.N., and Weide, A.E., 1979, Geochemical significance of groundwater discharge and carbonate solution in the formation of Caleta Xel Ha, Quintana Roo, Mexico: Water Resources Res., v. 15, p. 1521–1535.

Back, W., Hanshaw, B.B., and Van Driel, J.N., 1984, Role of groundwater in shaping the eastern coastline of the Yucatan Peninsula, Mexico, *in* LaFleur, R.G., ed. Groundwater as a geomorphic agent: London, Allen and Unwin, p. 157–172.

Badiozamani, K., 1973, The *dorag* dolomitization model—application to the middle Ordovician of Wisconsin: Jour. Sed. Petrology, v. 43, p. 965–984.

Benson, L.V., and Matthews, R.K., 1971, Electron microprobe studies of magnesium distribution in carbonate cements and recrystallized skeletal grainstones from the Pleistocene of Barbados, West Indies: Jour. Sed. Petrology, v. 41, p. 1018–1025.

Bögli, A., 1980, Karst hydrology and physical speleology: Berlin, Springer-Verlag, 284 pp.

Brand, U., and Veizer, J., 1980, Chemical diagenesis of a multicomponent carbonate system—I: Trace elements: Jour. Sed. Petrology, v. 50, p. 1219–1236.

Buchbinder, L.B., Margaritz, M., and Goldberg, M., 1984, Stable isotope study of karstic-related dolomitization: Jurassic Rocks from the coastal plain, Israel: Jour. Sed. Petrology, v. 54, p. 236–256.

Budai, J.M., Lohmann, K.C, and Owen, R.M., 1983, Burial dedolomite in the Mississippian Madison limestone, Wyoming and Utah thrust belt: Jour. Sed. Petrology, v. 54, p. 276–288.

Budai, J.M., Lohmann, K.C, and Wilson, J.L., in press, Dolomitization of the Madison Group, Wyoming and Utah overthrust belt: Amer. Assoc. of Petroleum Geologists Bull.

Butler, J.N., 1982, Carbon dioxide equilibria and their applications: Reading, MA, Addison-Wesley, 259 pp.

Craig, H., 1961, Isotopic variations in meteoric waters: Science, 133, p. 1702–1703.

Dansgaard, W., 1964, Stable isotopes in precipitation: Tellus, v. 16, p. 436–468.

Deines, P., Langmuir, D., and Harmon, R.S., 1974, Stable carbon isotope ratios and the existence of a gas phase in the evolution of carbonate groundwaters: Geochim. Cosmochim. Acta, v. 38, p. 1147–1164.

Drever, J.I., 1982, The geochemistry of natural waters: Englewood Cliffs, NJ, Prentice-Hall, 388 p.

Evamy, B.D., 1969, The precipitational environment and correlation of some calcite cements deduced from artificial staining: Jour. Sedimentology, v. 39, p. 787–793.

Fantidis, J., and Ehhalt, D.H., 1970, Variations of the carbon and oxygen isotopic composition in stalagmites and stalactites: evidence of non-equilibrium isotopic fractionation: Earth Plan. Sci. Letters, v. 10, 136–144.

Faure, G., 1986, The principles of isotope geology: New York, John Wiley, 589 p.

Fernberg, R.S., Lohmann, K.C, and Wilson, J.L., 1986, Multiphase meteoric diagenesis of platform carbonates (abstr.): Ann. Meeting Geol. Soc. Amer. Abstr. with Programs, v. 18, no. 6, p. 600.

Folk, R.L., and Assereto, R., 1976, Comparative fabrics of length—slow calcite and calcitized aragonite in a Holocene speleothem, Carlsbad Caverns, New Mexico: Jour. Sed. Petrology, v. 46, p. 486–496.

Fornaca-Rinaldi, E., Panichi, C., and Tongiorgi, E., 1968, Some causes of variations of the isotopic composition of carbon and oxygen on cave concretions: Earth Plan. Sci. Letters, v. 4, 321–324.

Freidman, I., and O'Neil, J.R., 1977, Compilation of stable isotope fractionation factors of geochemical interest, in Fleischer, E.M., ed., Data of geochemistry: Washington, DC, US Government Printing Office.

Froelich, P.N., Klinkhammer, G.P., Bender, M.L., Luedtke, N.A., Heath, G.R., Cullen, D., Dauphin, P.L., Hammond, D., Hartman, B., and Maynard, V., 1979, Early oxidation of organic matter in pelagic sediments of the eastern equatorial Atlantic: suboxic diagenesis: Geochim. Cosmochim. Acta, v. 43, p. 1075–1090.

Gines, J., Gines, A., and Pomar, L., 1981, Morphological and mineralogical features of phreatic speleothems occurring in coastal caves of Majorca (Spain), in Beck, B.F., ed., Proc. 8th Internat. Speleological Congr., v. 2, p. 529–532.

Given, R.K., and Lohmann, K.C, 1985, Derivation of the original isotopic composition of Permian marine cements: Jour. Sed. Petrology, v. 55, p. 430–439.

Given, R.K., and Lohmann, K.C, 1986, Isotopic evidence for the early meteoric diagenesis of the reef facies, Permian Reef Complex of west Texas and New Mexico: Jour. Sed. Petrology, v. 56, p. 183–193.

Gonzalez, L.A., and Lohmann, K.C. (this volume), 1987, Controls on mineralogy and composition of spelean carbonates: Carlsbad Caverns, New Mexico, in Choquette, P.W., and James, N.P., eds., Paleokarst: New York, Springer-Verlag.

Grover, G.A., Jr., and Read, J.F., 1983, Regional cathodoluminescent patterns, middle Ordovician ramp, Virginia: Amer. Assoc. Petroleum Geologists Bull., v. 67, p. 1275–1303.

Halley, R.B., and Harris, P.M., 1979, Freshwater cementation of a 1,000-year-old oolite: Jour. Sed. Petrology: v. 49, p. 969–988.

Hanor, J.S., 1978, Precipitation of beachrock cements: Mixing of meteoric and marine waters vs. CO_2-degassing: Jour. Sed. Petrology, v. 48, p. 489–501.

Hanshaw, B.B., Back, W., and Deike, R.G., 1971, A geochemical hypothesis for dolomitization by groundwater: Econ. Geology, v. 66, p. 710–724.

Hendy, C.H., 1971, The isotopic geochemistry of speleothems—I. The calculation of the effects of different modes of formation on the isotopic composition of speleothems and their applicability as palaeoclimatic indicators: Geochim. Cosmochim. Acta, v. 35, p. 801–824.

Holland, H.D., Kirsipu, T.V., Huebner, J.S., and Oxburgh, U.M., 1964, On some aspects of the chemical evolution of cave waters: Jour. Geology, v. 72, p. 36–67.

Hudson, J.D., 1977, Stable isotopes and limestone lithification: Jour. Geol. Soc. London, v. 133, p. 637–660.

James, N.P., and Choquette, P.W., 1983, Diagenesis

6. Limestones—the seafloor diagenetic environment: Geoscience Canada, v. 10, p. 162–180.

James, N.P., and Choquette, P.W., 1984. Diagenesis 9. Limestones—the meteoric diagenetic environment: Geoscience Canada, v. 11, p. 161–194.

Jennings, J.N., 1985, Karst geomorphology: New York, Basil Blackwell Inc., 293 p.

Kendall, A.C., and Broughton, P.L., 1978, Origin of fabrics in speleothems composed of columnar calcite crystals: Jour. Sed. Petrology, v. 48, p. 519–538.

Lawrence, J.R., Gieskes, J., and Anderson, T.F., 1975, Oxygen isotope material balance calculations: Init. Reports Deep Sea Drilling Proj. Leg 35, p. 507–512.

Lloyd, J.W., and Heathcote, J.A., 1985, Natural inorganic hydrochemistry in relation to groundwater: Oxford, Clarendon Press, 294 p.

Lohmann, K.C, 1982, Inverted J carbon and oxygen isotopic trends—criteria for shallow meteoric phreatic diagenesis (abstr.): Ann. Meeting Geol. Soc. Amer. Abstr. with Program, p. 548.

Lohmann, K.C, 1983, Unraveling the diagenetic history of carbonate reservoirs, in Wilson, J.L., Wilkinson, B.J., Lohmann, K.C., and Hurley, N.F., New ideas and methods for exploration for carbonate reservoirs, short course, Dallas Geol. Soc.

Lohmann, K.C, in press, Carbon isotopic composition of Phanerozoic marine carbonates: implications for the global cycling of carbon and sulfur: Geology.

Longman, M.W., 1980, Carbonate diagenetic textures from near-surface diagenetic environments: Amer. Assoc. Petroleum Geologists Bull., v. 64, p. 461–487.

Margaritz, M., 1983, Carbon and oxygen isotopic composition of recent and ancient coated grains, in Peryt, T.M. ed., Coated grains: Heidelberg, Springer-Verlag, p. 27–37.

Martin, G.D., Wilkinson, B.H., and Lohmann, K.C., 1986, The role of skeletal porosity in aragonite neomorphism—*Strombus* and *Montastrea* from the Pleistocene Key Largo limestone, Florida: Jour. Sed. Petrology, v. 56, p. 194–203.

Matter, A., and Perch-Nielsen, K., 1975, Fossil preservation, geochemistry and diagenesis of pelagic carbonates from Shatsky Rise, northwest Pacific: Init. Repts Deep Sea Drilling Proj., v. 32, p. 891–919.

Matthews, R.K., 1968, Carbonate diagenesis: equilibration of sedimentary mineralogy to the subaerial environment coral cap of Barbados, West Indies: Jour. Sed. Petrology, v. 38, p. 1110–1119.

Matthews, R.K., 1974, A process approach to diagenesis of reefs and reef associated limestones, in Laporte, L.F., ed., Reefs in time and space: Soc.

Econ. Paleontologists Mineralogists Spec. Pub. 18, p. 234–256.

Meyers, W.J., 1974, Carbonate cement stratigraphy of the Lake Valley Formation (Mississippian), Sacramento Mountains, New Mexico: Jour. Sed. Petrology, v. 44, p. 837–861.

Meyers, W.J., and Hill, B., 1983, Quantitative studies of compaction in Mississippian skeletal limestones, New Mexico: Jour. Sed. Petrology, v. 53, p. 231–242.

Meyers, W.J., and Lohmann, K.C, 1985, Isotope geochemistry of regionally extensive calcite cement zones and marine components in Mississippian limestones, New Mexico, in Schneiderman, N., and Harris, P.M., eds., Carbonate cements: Soc. Econ. Paleontologists Mineralogists Spec. Pub. 36, p. 223–240.

Moore, G.W., 1961, Dolomite speleothems: Nat. Spel. Soc. News, v. 19, p. 82.

Pingitore, N.E., 1978, The behavior of Zn^{2+} and Mn^{2+} during carbonate diagenesis: theory and applications: Jour. Sed. Petrology, v. 48, p. 799–814.

Plummer, L.N., 1975, Mixing of seawater with calcium carbonate groundwater, in Whitten, E.H.T., ed., Quantitative studies in the geological sciences: Geol. Soc. Amer. Mem. 142, p. 219–236.

Plummer, L.N., and Back, W., 1980, The mass balance approach: application to interpreting the chemical evolution of hydrologic systems: Amer. Jour. Sci., v. 274, p. 61–83.

Plummer, L.N., Vacher, H.L., Mackenzie, F.T., Bricker, O.P., and Land, L.S., 1976, Hydrogeochemistry of Bermuda: a case history of groundwater diagenesis of biocalcarenites: Geol. Soc. Amer. Bull., v. 87, p. 1301–1316.

Popp, B., Anderson, T.F., and Sandberg, P.A., 1986, Brachiopods as indicators of original isotopic compositions in some Paleozoic limestones: Geol. Soc. Amer. Bull., v. 97, p. 1262–1269.

Runnells, D.D., 1969, Diagenesis, chemical sediments, and the mixing of natural waters: Jour. Sed. Petrology, v. 39, p. 1188–1201.

Salomons, W., and Mook, W.G., 1986, Isotope geochemistry of carbonates in the weathering zone, in Fritz, P., and Fontes, J.C., eds., Handbook of environmental isotope geochemistry, v. 2, Terrestrial environment: Amsterdam, Elsevier, p. 239–269.

Sayles, F.L., and Mannheim, F.T., 1975, Interstitial solutions and diagenesis in deeply buried marine sediments: results from the Deep Sea Drilling Project: Geochim. Cosmochim. Acta, v. 39, p. 103–127.

Schwarcz, H.P., 1986, Geochronology and isotopic geochemistry of speleothems, in Fritz, P., and Fontes, J. Ch., eds., Handbook of environmental isotope geochemistry, v. 2, Terrestrial environment B: Amsterdam, Elsevier, p. 239–269.

Schwarcz, H.P., Harmon, R.S., Thompson, P., and Ford, D.C., 1976, Stable isotope studies of fluid inclusions in speleothems and their paleoclimatic significance: Geochim. Cosmochim. Acta, v. 40, p. 657–665.

Seigenthaler, U., 1977, Stable hydrogen and oxygen isotopes in the water cycle, in Jaeger, E., and Hunziker, J.C., eds., Lectures in isotope geology: New York, Springer-Verlag, 329 p.

Sheppard, S.M.F., 1986, Characterization and isotopic variations in natural waters, in Valley, J.W., Taylor, H.P. Jr., and O'Neil, J.R., eds., Stable isotopes in high temperature geological processes: reviews in mineralogy; Mineral. Soc. Amer., v. 16, p. 165–183.

Thrailkill, J., 1965, Studies in the excavation of limestone caves and the deposition of speleothems: Part I. Chemical and hydrologic factors in the excavation of caves: Part II. Water chemistry and carbonate speleothem relationships in Carlsbad Caverns, New Mexico: Unpubl. Ph.D. dissertation, Princeton University, 193 p.

Thrailkill, J., 1968a, Dolomite cave deposits from Carlsbad Caverns: Jour. Sed. Petrology, v. 38, p. 141–145.

Thrailkill, J., 1968b, Chemical and hydrologic factors in the excavation of limestone caves: Geol. Soc. Amer. Bull., v. 79, p. 19–46.

Thrailkill, J., 1971, Carbonate deposition in Carlsbad Caverns: Jour. Geology, v. 79, p. 683–695.

Veizer, J., 1983, Chemical diagenesis of carbonates: Theory and application of trace element technique, in Stable isotopes in sedimentary geology: Soc. Econ. Paleontologists Mineralogists, Short course No. 10, Tulsa, Oklahoma.

Veizer, J. Fritz, P., and Jones, B., 1986, Geochemistry of brachiopods: oxygen and carbon isotopic records of Phanerozoic oceans: Geochim. Cosmochim. Acta, v. 50, p. 1679–1696.

Vernon, R.O., 1969, The geology and hydrology associated with a zone of high permeability (boulder zone) in Florida: Soc. Mining Engineers, 69-AG-12, 24 p.

Wagner, P.D., and Matthews, R.K., 1982, Porosity preservation in the upper Smackover (Jurassic) carbonate grainstone, Walker Creek Field, Arkansas: response of paleophreatic lenses to burial processes: Jour. Sedimentary Petrology, v. 52, p. 3–18.

Ward, W.C., and Halley, R.B., 1985, Dolomitization in a mixing zone of near-seawater composition, late Pleistocene, northeastern Yucatan Peninsula: Jour. Sedimentary Petrology, v. 55, p. 407–420.

Wollast, R., and Reinhart-Derie, D., 1977, Equilibrium and mechanism of dissolution of Mg-calcites, in Andersen, N.R., and Malahoff, A., eds., The fate of fossil fuel CO_2 in the ocean: New York, Plenum Press, p. 479–493.

3

Controls on Mineralogy and Composition of Spelean Carbonates: Carlsbad Caverns, New Mexico

Luis A. Gonzalez and Kyger C Lohmann

Abstract

Carbonate speleothems and precipitating fluids from Carlsbad Caverns, New Mexico, have been analyzed for their major and minor element and stable isotopic compositions in order to evaluate processes controlling the chemical evolution of cave water and factors determining the mineralogy and composition of cave carbonates. Chemistry and isotopic composition of fluids are determined by rates of CO_2 degassing, evaporation, and carbonate precipitation. Evaporation and calcium carbonate precipitation cause changes in the Mg/Ca ratio of fluids which, coupled with changes in $CO_3^=$ content, control the minor element chemistry and mineralogy of the precipitating phase.

A broad range of carbonate minerals precipitate from seepage cave fluids, with calcites containing 1.5 to 12.0 mole % $MgCO_3$; calcite Mg contents exhibit a nonlinear dependence on fluid Mg/Ca ratio, and a linear dependence on fluid $CO_3^=$ content, indicating dual control by both cation and anion concentrations. Calcite-aragonite polymorphism appears to be largely controlled by elevated fluid Mg/Ca ratios. Combined water and carbonate chemical and stable isotopic data suggest that the most Mg-enriched calcites (10 to 12 mole % $MgCO_3$) either coprecipitate with aragonite or precipitate from waters with higher $CO_3^=$ concentrations than those precipitating aragonite. The range of calcite compositions associated with aragonite suggest that it coprecipitates with Mg-depleted calcite at low fluid $CO_3^=$ concentrations, while Mg-enriched calcites form with aragonite at high $CO_3^=$ concentrations. Hydromagnesite and huntite precipitate under conditions of extreme evaporation at elevated fluid Mg concentrations and Mg/Ca ratios. Primary dolomite precipitates from waters of moderate Mg/Ca ratio, probably from fluids undersaturated with respect to calcite and aragonite.

Introduction

Factors which control calcium carbonate polymorphism and incorporation of magnesium and other minor elements have been an important focus in carbonate research in modern sedimentary environments. Most of this research, however, has been carried out in marine settings where carbonate precipitation occurs from water that is generally homogeneous with respect to major ions, but variable with respect to temperature and carbonate saturation state. While the morphology, crystallography, and chemistry of inorganic marine cements have been well documented, direct correlation of variations in these parameters to differences in water chemistry have been hindered by the difficulty of in situ sampling of ambient pore waters in which precipitation occurs.

In contrast, cave carbonate deposition occurs in a broad range of water chemistries, with variable ion concentrations, and at variable pCO_2. Moreover, within a specific cave system, temperature variation and biological interactions are virtually absent. Because of our ability to directly measure variability in these physiochemical settings by direct sampling of reactive fluids, spelean systems are ideal natural laboratories for evaluating controls on mineralogy and chemistry of carbonate phases. Despite this potential, comprehensive studies in spelean settings have been few and limited in scope.

The frequency with which both aragonite and calcite occur in the same caverns has made the study of the factors determining aragonite precipitation an important aspect of spelean research. Three factors have been suggested as leading to aragonite precipitation: (1) elevated temperature (Moore 1956, Siegel 1965); (2) elevated Mg/Ca ratios (Murray 1954, Thrailkill 1971); and (3) supersaturation from evaporation and/or degassing of CO_2 (Holland et al. 1964, Cabrol and Coudray 1982). Because aragonite precipitation has been documented in low-temperature caves (Rogers and Williams 1982, Lilburn Cave, California, 9° C; Harmon et al. 1983, Castleguard Cave, Canada, 2° to 4° C), and because temperature remains rather constant in cave areas removed from openings (with variation usually of less than 2° C), it is difficult to call on temperature as the primary control on carbonate mineralogy. Similarly, calcite and aragonite precipitate from waters of variable and overlapping Mg/Ca ratios and saturation states (Holland et al. 1964, Murray 1975). Thus, it is difficult to attribute carbonate polymorphism uniquely to either Mg/Ca ratio or saturation state. What is clear is that the relative importance of each of these factors is not well understood.

While several chemical controls have been implicated as influencing carbonate mineralogy, factors that may influence the incorporation of magnesium into calcite, such as fluid chemistry, Mg/Ca ratio, and saturation state, have been largely ignored. For example, the Mg content of calcite precipitated from fluids of moderate Mg/Ca ratio (2:5) in association with aragonite ranges from 2 mole % $MgCO_3$ (Siegel 1965, Murray 1975, Tietz 1981, Cabrol and Coudray 1982) to as high as 24 mole % $MgCO_3$ (from data in Zeller and Wray 1956). This broad range in calcite compositions occurring with aragonite argues for a coupled effect of fluid Mg/Ca ratio and saturation state as controlling both the mineralogy and composition of precipitating phases.

With this background, we have undertaken a study of cave carbonates in Carlsbad Caverns, New Mexico. Integration of isotopic, major, and trace elemental chemical data on carbonate minerals and co-occurring waters in areas of active speleothem deposition allows the relative effects of major cation chemistry and saturation state to be evaluated. Because of documented variability in water chemistry, speleothem composition, and mineralogy (Thrailkill 1965, 1968, 1971) Carlsbad Caverns is an ideal setting

in which to evaluate the relative effect of these variables over a narrow temperature range.

Carlsbad Caverns is located in the Guadalupe Mountains of southeastern New Mexico within the Permian Capitan Reef complex. This cave system consists of elongate passages and wide rooms trending NNW to ENE (Fig. 3.1) which extend to a depth of more than 300 m beneath the natural cave entrance (about 1300 m above MSL). The cave spans four Upper Permian formations, the Tansill, Yates, Seven Rivers, and Capitan Limestone, with passages and rooms showing no major relationship to bedding or formation contacts (Fig. 3.1).

Settings of Speleothem Formation

Sites of speleothem precipitation may be grouped into three general categories, each of which may relate to the composition of precipitating waters and to characteristic crystal morphologies. These are pools (Fig. 3.2 and 3.3), drips or drops (Figs. 3.4, 3.5, and 3.6), and sprays or capillary flow (Figs. 3.4, 3.5, and 3.6).

Pools

Growth of speleothems that are continuously bathed by water, in pools or by overflowing water, results in a variety of morphologies which relate to sites of deposition such as pool bottoms and pool edges (Fig. 3.2). The size of speleothem crystals from pools is typically large, with individual crystals easily distinguished in hand sample. Pool bottoms may be covered by more than one speleothem growth form, particularly when more than one mineralogy is present (Fig. 3.2). Speleothems along the bottom of a pool may take the form of nodules or knobs (Fig. 3.2A), which consist either of crystal arrays radiating from a single nucleation site, or of concentric layers of unoriented crystals (Fig. 3.3A). Calcite commonly grows as dendritic arrays of twinned scalenohedral crystals which resemble shrubs or as nodular branching masses (Fig. 3.2B), while aragonite may develop nodules consisting of crystal meshes (Fig. 3.3A) or radiating needle arrays (Fig. 3.3C).

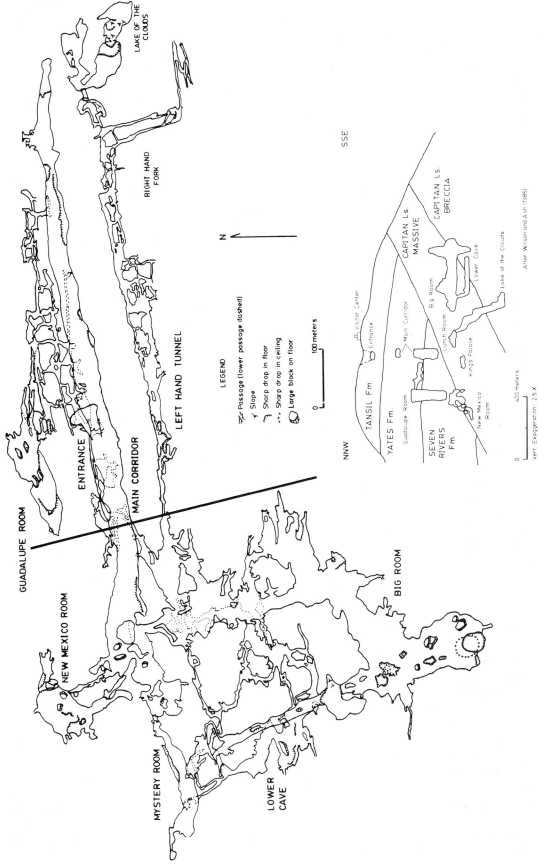

FIGURE 3.1. Floor map of Carlsbad Caverns, New Mexico. In general, rooms exhibit an elongation in ENE and NNW directions. The inset is a generalized cross-section of cavern geology. Cavern rooms are projected into the SSE traverse from the Guadalupe Room. (Map simplified from Cave Research Foundation 1979; cross-section simplified from Wilson and Ash 1985.)

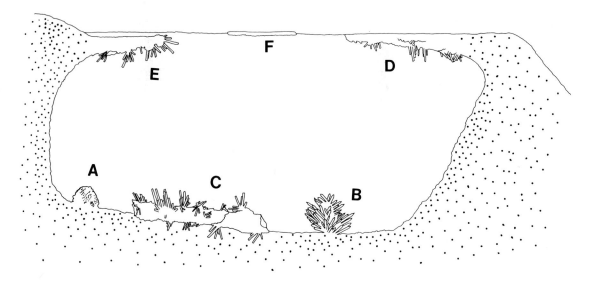

FIGURE 3.2. Schematic representation of sites of carbonate precipitation and resultant speleothem formations in pool setting. Carbonate precipitation occurs along pool floors as nodules or knobs (A), as nodular branching masses (B), or on older speleothem debris that has fallen into the pool (C), and along pool surfaces as rims and overhangs (D), as shelves or ledges (E), or as floating crystal rafts (F).

Borders of pools may be surrounded by either of two morphologically distinct growth formations, rim overhangs or shelf ledges (Fig. 3.2). Rim overhangs (Fig. 3.2D) develop where there is a continuous overflow of water. The surface over which water flows is usually smooth with no visible crystal terminations. Component crystals consist of fine coalesced calcite or (less frequently) aragonite, while the lower surface that faces into the pool usually develops larger crystals with well-defined terminations.

When water does not flow over the edges of pools, a ledge or shelf may be generated (Fig. 3.2E). Shelves whose upper sides are commonly bathed by overflow water but are periodically exposed to air may extend considerable distances into the pool. As with rim overhangs, the sizes of crystals on upper ledge surfaces are considerably smaller than on lower surfaces, but these may become progressively larger toward the pool center. Whereas lower surfaces commonly consist of a single mineralogy, ledge margins and undersides frequently consist of a mixed mineralogy (Fig. 3.3).

Throughout such pool settings, it is not uncommon to encounter more than one carbonate phase at any particular locality and, occasionally, even more than one calcite phase. In the few samples where two distinct calcite phases are present, one of these is commonly a needle calcite (Fig. 3.3F). In some pools it is also common to observe a thin film of minute calcite or aragonite crystal rafts covering the water surface (Fig. 3.2F). These nucleate at the surface of pools when the surface is not disturbed either by dripping or by water flow.

Drips

These speleothems originate when water flows along external or internal surfaces, forming droplets at speleothem tips which ultimately impact on growing speleothems on the cave floor (Fig. 3.4A–D). Common speleothem forms resulting from the downward flow of water are stalactites (Fig. 3.4A) and soda straws (Fig. 3.4B), although flow along cave walls can result in similar carbonate textures. The limiting factor for carbonate growth is the thickness of the water film bathing speleothem carbonate; precipitation in these settings produces crystals of such small size that morphologies are difficult if not impossible to discern even with the aid of a binocular microscope. Where aragonite precipitates from such water films, it consists of submicroscopic stubby crystals (Fig. 3.5A) with pseudohexagonal habits. Growth of calcite on soda straws and stalactites results in prismatic crystals oriented perpendic-

FIGURE 3.3. Textures of speleothems formed in association with cave pools. *A.* Thin section photomicrograph of a mixed mineralogy nodular speleothem from a pool floor (Fig. 3.2A). Clear rhombic crystals along the lower and right photo margin are calcite, whereas the meshlike arrays of needles toward the center and top of the photo are largely aragonite, but calcite rhombs may also occur nucleated on aragonite in some pool-bottom speleothems. *B.* Prismatic calcite crystals commonly found growing on shrublike speleothems at pool bottoms or on undersides of pool ledges (Fig. 3.2C and E). *C.* Radial aragonite arrays from a pool bottom branching nodular mass (Fig. 3.2B); length of individual needles is 1 to 2 mm. *D.* Large aragonite needles, up to 2mm in length, collected from a pool ledge (Fig. 3.2E). *E.* SEM image of a mixed mineralogy pool bottom speleothem (Fig. 3.2B) of aragonite needles overgrown by rhombic calcite crystals. Crystals are intimately mixed, suggesting either simultaneous growth of both phases or rapidly oscillating water chemistries. *F.* Closeup of a mixed mineralogy pool sample consisting of a robust needle of aragonite and thin needles of calcite.

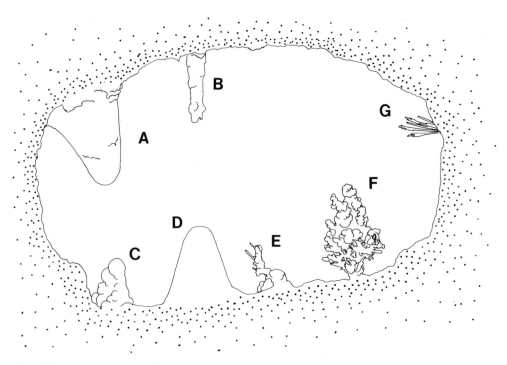

FIGURE 3.4. Schematic diagram of speleothems precipitated from drip, spray, and capillary flow fluids. These include forms related to drip waters as stalactites (A) and soda straws (B), to drip impact waters as nodular (C) and smooth (D) stalagmites, to capillary waters as spray trees (E) and bushes (F), and to impact spray waters as branching arrays (G).

ular to the speleothem surface, with terminations that do not extend above the water (Fig. 3.5B).

Speleothems that form in the impact zone of drip waters usually exhibit smooth surfaces. In these forms, calcite consists of a dense mass of submicroscopic rhombic crystals. However, aragonite may also occur as radiating fibers in well-defined bands, either as a relatively pure phase or mixed with small amounts of calcite. Mixed aragonite and calcite form a dense crystal mixture in which individual phases cannot be easily discerned (Fig. 3.5C).

Sprays and Capillary Flows

Speleothems which form from the spray of impacting drip waters, by capillary action (helictites or anthodites), and perhaps by water condensation (popcorn and frostwork) are the most interesting in terms of morphology because they exhibit a variety of shapes and forms in which individual crystals can be easily differentiated (Fig. 3.4E–G). The gross morphology of these speleothems is largely dictated by carbonate mineralogy and by the dominant water supply process.

Aragonite precipitation from capillary flow along a cave wall or stalactite results in elongated arrays of needles which may grow up to a few millimeters and branch from a nucleation single site (Figs. 3.4G, 3.6A). Calcite forming on the same walls, presumably by a similar process, consists of twisted or curved, elongated speleothem forms of considerably larger diameter (Fig. 3.6B), with individual crystals developing as either well-defined rhombs or blades. Speleothems formed in spray water from impacting drips usually show distinct orientations toward water sources (Fig. 3.4E). These may grow as branching arrays of microscopic to submicroscopic crystals and as small nodular masses (Fig. 3.4E and F). Crystal size in such speleothems varies considerably, and mixed mineralogy samples are frequently observed where aragonite grows as small needlelike crystals (Fig. 3.6C), while calcite occurs as minute rhombs (Fig. 3.6C and D).

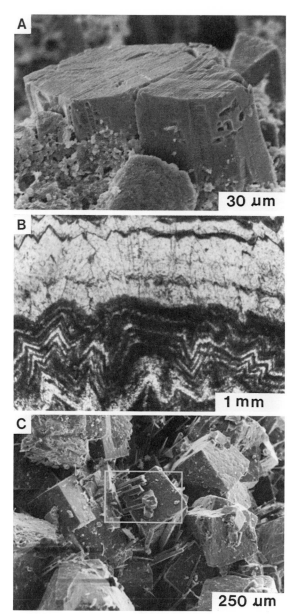

FIGURE 3.5. Spelean carbonates precipitated from drip waters. *A.* Stubby pseudohexagonal aragonite crystals which comprise the surface of some dripstones (Fig. 3.4A). *B.* Thin section photomicrograph of a soda straw (Fig. 3.4B) consisting of prismatic calcite crystals. In hand specimen, crystal terminations cannot be discerned. Internal dark bands consist of mixed mineralogy calcite–aragonite layers with approximately 10% aragonite. *C.* SEM image of such a mixed mineralogy speleothem that formed from drip impact waters (Fig. 3.4A). Note dominant rhombic calcite and minor elongate crystals of aragonite in the central box.

Cave Fluids

Small specimens of carbonate and associated waters were sampled during August and September of 1984 and May of 1985 from speleothems undergoing active carbonate precipitation. Two water samples per site, each less than 30 ml, were collected: water temperature and pH were measured in situ prior to and immediately following collection. One sample was immediately treated with HgCl and sealed for later isotopic analyses. A 10-cc aliquot was extracted from the second water sample and titration alkalinity determined within 12 hours of collection; the remainder was filtered using a 0.45 μm millipore filter and treated with 5 cc of 10% HNO_3 for trace element analysis. All in situ field and laboratory pH measurements agreed within 0.1 pH unit. Air samples were also collected from each cavern room to evaluate isotopic composition of cave air CO_2.

Water oxygen isotopic compositions were determined by CO_2 equilibration (Epstein and Mayeda 1953). Cave air CO_2 was stripped by flowing samples through a vacuum line with three sequential liquid nitrogen cooled gas traps. All stable isotope analyses were carried out with a VG 602E Micromass ratio mass spectrometer and all were corrected for contribution of ^{17}O (Craig 1957). Carbon values are reported relative to the PDB carbonate standard while oxygen values of waters are reported relative to SMOW. Precision of water analyses was monitored by daily analysis of two water standards (V-SMOW and SLAP) for oxygen, with a measured precision better than 0.2 ‰. Precision for cave air CO_2 was monitored by daily analysis of a laboratory dry CO_2 standard; measured precision was better than 0.05 ‰. Trace element compositions of water samples were determined by atomic absorption spectrophotometry using a Perkin-Elmer 2380 Spectrophotometer.

Water and air temperature in the cave ranges from 13.5° to 15.5° C throughout most of the cave, with the exception of the Left Hand Tunnel area (Fig. 3.1), where temperature increases progressively from the Right Hand Fork to a maximum of 20° C at Lake of the Clouds, the deepest section of the cave. Relative humidity ranges from 86% in open and dry rooms to 100% in closed rooms and constricted pas-

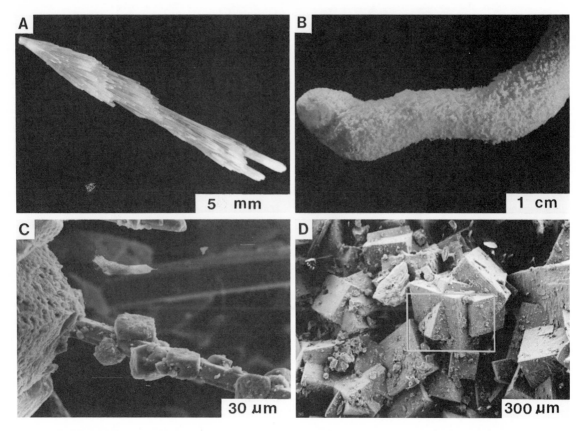

FIGURE 3.6. Speleothems from spray and capillary waters. *A.* Aragonite anthodite consisting of clusters of aragonite needles that grew from water supplied by capillary flow. *B.* Calcite anthodite from the same wall as that in A. *C.* Intimately associated elongate aragonite and rhombic crystals of calcite from a speleothem that precipitated from spray waters. *D.* Well-defined submicroscopic calcite rhombs in dripstones and speleothems bathed by spray waters. Note twinned crystals in the central box.

sages where condensation of water is observed on cave walls.

Cave waters exhibit a broad spectrum of compositions, with Ca and Mg concentrations displaying the greatest variability. The pH of water samples ranges from 7.5 to 8.6. Magnesium concentrations range from 0.5 to 5.4 mmoles, while Ca concentrations range from 0.3 to 3.3 mmoles (Fig. 3.7). Potassium concentrations range from 0.012 to 0.35 mmoles and Na concentrations from 0.067 to 1.09 mmoles (Fig. 3.8). Oxygen isotopic compositions of water range from -8.2 ‰ to -4.9 ‰ $\longrightarrow$

FIGURE 3.7. Ca and Mg concentrations in sampled cave waters. Three analyses of water in contact with hydromagnesite–huntite have been omitted as they have Mg concentrations up to five times higher than other cave waters shown here.

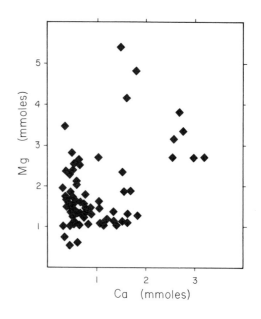

FIGURE 3.8. Na and K concentrations in sampled cave waters. As in Figure 3.7, analyses of waters in contact with hydromagnesite–huntite have been omitted as the concentration of both Na and K cations in these waters is up to five times higher than other water samples.

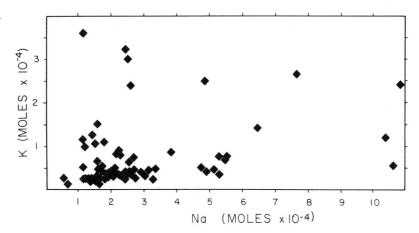

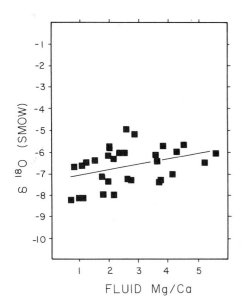

FIGURE 3.9. Water Mg/Ca ratios and $\delta^{18}O$ values of water samples. Note a progressive ^{18}O enrichment with higher Mg/Ca ratio; the regression line has an $r^2 = 0.6$.

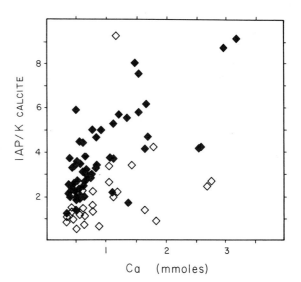

FIGURE 3.10. Ca concentrations and calcite saturation of cave waters collected during 1984 (open diamonds) and 1985 (closed diamonds).

and covary with Mg/Ca ratio (Fig. 3.9). The observed variability in water chemistry bears no systematic relationship to source such as pool water versus drip water.

Carbonate saturation state of waters varied between 1984 and 1985 (Fig. 3.10), a variation that reflects an increase in water seepage during 1984 due to increased rainfall during months preceding sampling. All 1985 water samples were saturated with respect to aragonite, calcite, and dolomite, whereas approximately 15% of the 1984 samples were undersaturated with respect to both calcite and aragonite. In the most

extreme case, 1985 samples were as much as five times more saturated than 1984 samples.

Cave Carbonates

Small carbonate specimens associated with cave waters were also collected during 1984 and 1985. In the laboratory, the outermost 500 μm of external surfaces was sampled using a microscope-mounted dental drill with 500- to 1000-μm bits to produce homogenized powders of approximately 20 mg. Similarly, internal surfaces of hollow soda straws, the undersides of pool ledges, and internal layers of contrasting crystal morphology or mineralogy were also sampled. In such mixed mineralogy samples, individual crystals of calcite and aragonite were separated using surgical forceps. Powdered samples were analyzed by X-ray diffraction to determine mineralogy, and a 0.2- to 0.5-mg portion was analyzed for carbon and oxygen isotopic composition. Carbonate samples for stable isotopic analysis were roasted under vacuum for one hour at 200° C to eliminate volatile components. Following roasting, samples were reacted in anhydrous phosphoric acid at 55° C in an on-line extraction system connected to the inlet of a VG 602E Micromass ratio mass spectrometer. All isotopic analyses were corrected for contribution of ^{17}O (Craig 1957); all carbonate isotopic values are reported relative to the PDB carbonate standard. Precision for carbonates was monitored by daily analysis of a powdered calcite standard (NBS-20) and was better than 0.10 ‰ for both oxygen and carbon. Minor and trace element compositions of carbonate samples were determined by atomic absorption spectrophotometry using a Perkin-Elmer 2380 Spectrophotometer.

Mineralogically, Carlsbad Cavern samples consist of calcite, aragonite, hydromagnesite, huntite, dolomite, gypsum, and quartz, with calcite (ranging from 0 to 12 mole % $MgCO_3$) and aragonite making up the bulk of any particular speleothem (Table 3.1). In samples from many sites, particularly pools, more than one mineralogy is present, either as separate speleothem forms or mixed within a single speleothem. In many instances, it is common to encounter calcites of differing $MgCO_3$ content, particularly between forms occurring in the bottom of a pool versus those occurring as ledges or rims.

Cave carbonates exhibit a large variation in both $\delta^{13}C$ and $\delta^{18}O$ composition (Fig. 3.11). Carbon values range from -6.1 to $+6.0$ ‰, while the oxygen values range from -7.0 to -1.4 ‰. In addition, carbon and oxygen isotopic values show an overall positive covariance with an apparent separation by mineralogy. However, the slopes of covariant trends are not always the same and are characteristic of individual cavern rooms (Fig. 3.11). Depleted endpoints for covariant trends correspond to

TABLE 3.1. Mineralogy of all speleothem samples grouped by sites of precipitation.

Mineralogy	Pools	Drips	Spray	Other	Total
Calcite	33	22	6	1	62
Aragonite	12	4	5	3	24
Calcite and aragonite	22	10	7	1	40
Hydromagnesite	1	2	0	5	8
Hydromagnesite and huntite	1	1	0	0	2
Hydromagnesite and aragonite	0	0	0	2	2
Dolomite and aragonite	1	0	0	0	1
Dolomite and calcite	3	0	0	1	4
Gypsum and calcite	1	0	0	0	1
Calcite, dolomite and quartz	1	0	0	0	1
Quartz and aragonite	2	0	0	0	2
Total	77	39	18	13	147

samples with low $MgCO_3$ contents, while aragonite and high-magnesium calcites (about 8 mole % $MgCO_3$) exhibit enriched compositions. In speleothems of mixed mineralogy, calcites with lower Mg contents are isotopically more depleted than cooccurring aragonite and calcites with a higher mole % $MgCO_3$ (Fig. 3.12).

Chemical Evolution of Meteoric Vadose and Cave Seepage Water

In order to evaluate those factors controlling the composition of spelean carbonates, it is first necessary to understand the role of those processes that control cave water chemistry. Changes in water composition associated with the precipitation of carbonate minerals should be reflected in the isotopic composition of those carbonate phases. If, for example, carbonate mineralogy or composition reflects water degassing and/or evaporation, then carbonate compositions should be noticeably different for each set of conditions because water carbon and oxygen isotopic composition is highly sensitive

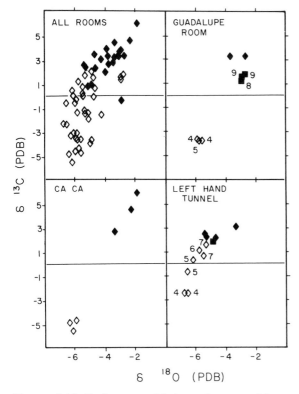

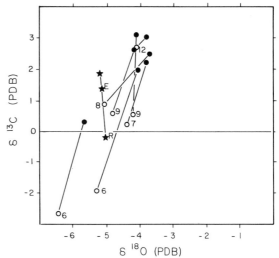

FIGURE 3.12. Stable isotopic compositions of associated aragonite and calcite crystals within individual samples. In all cases, the heaviest carbon isotopic composition corresponds to aragonite. Relative to aragonite, those calcites which exhibit the largest carbon isotopic difference are those with 6 to 9 mole % $MgCO_3$, while those with the smallest difference have compositions of 10 to 12 mole % $MgCO_3$. Points as solid stars are from one sample with co-existing rhombic calcite (R), elongated bladed calcite (E), and aragonite; due to the fine nature of this mixture it is extremely difficult to analyze the individual calcite phases; the average composition for these two calcites is 10 mole % $MgCO_3$.

FIGURE 3.11. Carbonate stable isotopic compositions of water-bathed monomineralic speleothem surface crystals; calcites are open diamonds and solid squares; aragonites are solid diamonds. All Rooms: Note a strong nonlinear covariance of isotopic composition for all samples; calcites exhibit the most depleted compositions relative to aragonites. Ca Ca: Stable isotopic composition of samples from along the Lower Main Corridor and the Big Room (Fig. 3.1). Note the marked difference in isotopic composition of calcites which contain 2 to 4 mole % $MgCO_3$ and aragonites. Guadalupe Room and Left Hand Tunnel: Calcite Mg contents shown as numbers beside open diamonds. Note that calcites show marked variance in isotopic composition that correlates with mole % $MgCO_3$; those with depleted isotopic compositions have compositions from 4 to 5 mole % $MgCO_3$, whole those with the enriched compositions contain 8 to 9 mole % $MgCO_3$ (solid squares).

to these two processes. On the other hand, cave water chemistry could possibly reflect different amounts of rock–water interaction as seepage fluids move through overlying rocks. In this case, speleothem isotopic compositions should be different than those reflecting changes in water composition in the cavern itself.

Rock–Water Interactions

As rainwater seeps into the soil, it is charged with CO_2 from the soil zone, increasing its capacity to dissolve carbonate. Since soil CO_2 has $\delta^{13}C$ compositions ranging from -20 to -25 ‰ (Deines et al. 1974), dissolved CO_2 becomes extremely negative. In a pure carbonate terrain, the Ca and Mg concentration of water will be a function solely of the amount of carbonate rock dissolved, which is determined by the amount of CO_2 dissolved into the water while in transit through the soil zone.

Water Mg/Ca ratio will reflect the proportion of dolomite to calcite and Mg content of the limestone. In the most extreme case, country rock will be all dolomite, and a maximum Mg/Ca ratio of about 1.0 will be attained. Conversely, dissolution in a pure limestone terrain would produce a Mg/Ca ratio approaching zero. If the seepage water equilibrates under closed system conditions, and there are no external controls on the dissolution process, the carbon isotopic composition of the water will be determined by both the carbon composition of the country rock and that of the soil CO_2. Assuming relatively heavy values for these carbon reservoirs of $+4$ ‰ for limestone and -20 ‰ for dissolved CO_2, the carbon isotopic composition of dissolved CO_2 will be approximately -10 ‰, and the carbon isotopic composition of any precipitated spelean carbonate will be approximately -9.0 ‰ at $15°$ C.

Barring the presence of metastable carbonates in the country rock (a reasonable assumption for these Permian carbonates), only three scenarios can be envisioned that would cause deviation of present-day seepage water compositions from those of the closed-system setting. The first is the dissolution of host rock in the presence of a gas phase (Deines et al. 1974). In this case, dissolution is enhanced and the amount of carbonate rock dissolved is increased. Such processes would increase the total concentration of Ca and Mg (as well as other cations) in the fluid, yet are incapable of producing a Mg/Ca ratio greater than unity. Since the solution remains in contact with a gas phase, there is continuous exchange of CO_2 between fluid and host rock, and the $\delta^{13}C$ of dissolved CO_2 cannot attain values as heavy as those under closed system dissolution.

The other two scenarios involve pH buffering, result in different cation chemistries, and do not involve external sources of CO_2. These are pyrite oxidation and cation exchange with clays (Sears 1976). Both mechanisms enhance the dissolution of country rock and, since they do not provide additional $CO_3^=$ to the solution, can result in a greater shift of water carbon isotopic composition toward that of the country rock than simple dissolution in a closed system. Pyrite oxidation will result in an increase in cation concentration, but in proportion to those present in the country rock; as in previous scenarios, this mechanism is incapable of producing Mg/Ca ratios greater than one. Cation exchange processes, on the other hand, may result in a decrease in the concentration of major cations. If removal of Ca and Mg is proportional to their concentration in the fluid, then we would expect the Mg/Ca ratio to re-

main constant at the value determined by the host rock dolomite/calcite ratio. If there is selective removal of one of the cations, on the other hand, then a decrease of both cations would yield a linear relationship whose slope would be dependent on the relative amounts of each cation removed. In most cases (see Sears 1976), preferential removal of Mg would be the case and, hence, exchange processes would be unable to produce Mg/Ca ratios greater than unity.

In most caves, the oxygen isotopic composition of seepage water reflects the mean annual composition of the precipitation in the area (Harmon 1979, Yonge et al. 1985). Due to the short residence time of water in the vadose zone, significant shifts in the oxygen composition of water seeping into caves cannot be induced by any of the above processes. The apparent rapid response of Carlsbad Cavern waters to increased precipitation during the summer of 1984 gives an indication of the short residence time of vadose zone waters.

CO_2 Degassing and Evaporation

Upon entrance into caverns, the two most important processes which could affect the chemical composition of cave water are degassing and evaporation. Each of these processes will change water cation chemistry and isotopic composition in a characteristic and unique way which, in turn, will be reflected in the cation chemistry and isotopic composition of precipitated carbonate phases.

If water has attained saturation with respect to calcium carbonate prior to its entrance to the cave, then degassing will cause a marked increase in $CO_3^=$ concentration, the precipitation of a carbonate mineral, and a marked decrease in the Ca concentration of the fluid. Otherwise, degassing will bring the water closer to saturation, and carbonate precipitation will commence only if saturation is attained before the water comes to equilibrium with cave air pCO_2. During rapid degassing, kinetic disequilibrium occurs, and the isotopic composition of dissolved CO_2 will not be in equilibrium with that of the atmospheric pCO_2. The escaping gas will have a negative $\delta^{13}C$ while the remaining dissolved CO_2 will be enriched in ^{13}C (Fantidis and Ehhalt 1970, Fornaca-Rinaldi et al. 1968, Inoue and Sugimura 1985). With progressive degassing, dissolved CO_2 becomes progressively heavier, and, thus, any carbonate precipitated in response to degassing should be enriched in $\delta^{13}C$. The major effect of evaporation will be to concentrate all ions while maintaining initial cation ratios (Thrailkill 1965), and to enrich water ^{18}O values by preferential removal of light oxygen (Craig et al. 1963, Lloyd 1966).

With respect to these processes, plots of important fluid components aid in conceptualizing paths followed by a water during its evolution (Figs. 3.13–15). Evaporation without carbonate precipitation will produce a linear increase in both Ca and Mg while maintaining original fluid Mg/Ca ratio, while precipitation of pure calcium or magnesium mineral phases will produce linear decreases in either of these cations. When both cations are removed simultaneously, a linear trend with a negative slope (dictated by the amount of either ion removed) will result (Fig. 3.13).

If, on the other hand, one ion is preferentially removed before the other (such as calcium removal by carbonate precipitation during degassing followed by evaporative concentration),

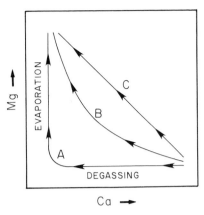

FIGURE 3.14. Possible changes in Ca and Mg concentration during degassing, evaporation, and carbonate precipitation. Precipitation induced by fluid degassing results in decreasing Ca concentrations; evaporation after degassing then gives rise to an increase in Mg concentrations (line A). If degassing and evaporation occur simultaneously and at a constant rate during carbonate precipitation, trend slope is determined by relative amounts of evaporation and degassing (e.g., line C). If degassing rate decreases at a constant rate of evaporation, all trends will lie under line C (e.g., line B). The unlikely alternative, that evaporation rate decreases at a constant rate of degassing, will produce trends similar to those displayed by A and B but symmetrical above line C.

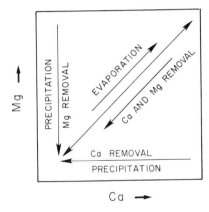

FIGURE 3.13. Possible changes in Ca and Mg concentration during degassing and evaporation of a cave water. Degassing and $CaCO_3$ precipitation will result in a decrease in Ca concentration at constant Mg while $MgCO_3$ precipitation will give rise to a decrease in Mg concentration at constant Ca. Simultaneous removal of Ca and Mg results in a linear decrease in both cations with the slope of the trend depending on rates of removal. Simple evaporation results in an increase in the concentration of both ions, with trend slope determined by initial Mg/Ca ratio.

then the resultant trend will be a linear Ca decrease at constant Mg followed by an increase in Mg at constant Ca (Fig. 3.14). In the context of caves, perhaps the most likely scenario is initial rapid degassing with concomitant evaporation, with evaporation becoming the dominant process as water approaches equilibrium with cave pCO_2. Such processes will result in a curvilinear (hyperbolic) trend whose general shape is dictated by (relatively constant) evaporation rate and the rate at which degassing decreases. It will lie between a trend generated by simultaneous degassing and evaporation and a trend generated by degassing followed by evaporation. It will exhibit an initially rapid decrease in Ca and slight increase in Mg, followed by a rapid increase in Mg and a slight decrease in Ca (Fig. 3.14).

Similarly, varying rates of degassing and evaporation should also give rise to distinctive isotopic trends. If both processes take place simultaneously, carbon and oxygen isotopic

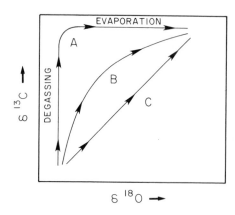

FIGURE 3.15. Changes in stable isotopic compositions due to degassing, evaporation, and carbonate precipitation. Fluid degassing results only in enrichment in ^{13}C; if degassing is followed by evaporation, oxygen isotopic compositions are then enriched at constant carbon values (line A). At the opposite extreme, simultaneous (but constant) degassing and evaporation gives rise to a trend whose slope is determined by the relative importance of these two processes (line C). Trends generated by simultaneous degassing and evaporation, with the former decreasing with time, all lie above line C (e.g., line B). Arrows along the X and Y axis indicate relative enrichment.

Mineralogy and Carbonate Isotopic Composition

In addition to the above considerations, speleothem isotopic composition may also reflect the mineralogy of precipitated phases. When calcite and aragonite precipitate in equilibrium from the same water, experimental data (Rubinson and Clayton 1969, Tarutani et al. 1969) and empirical observations (Gonzalez and Lohmann 1985) suggest they should exhibit different isotopic compositions. Relative to aragonite, calcite $\delta^{13}C$ values should be 1.0 to 1.6 ‰ more depleted. Low-Mg calcite $\delta^{18}O$ values also should be lighter by about 0.6 ‰, with differences decreasing by 0.06 to 0.08 ‰ per calcite mole % $MgCO_3$.

Carlsbad Cavern Waters

With this background in mind, it is now possible to discuss the composition of cave waters and speleothem carbonates. Carlsbad Caverns water compositions exhibit a large variability which cannot be explained by the simple closed system reaction of waters with overlying rock. Nor can the observed variability be accounted for by any of the external mechanisms noted above which might control dissolution process, since changes in Ca and Mg are not linear as would be expected if they merely reflected rock–water interactions. The Mg/Ca ratio of most water samples is in excess of unity. The only attribute of the cave waters which is dictated by interaction with overlying rock is initial (as it enters the cave) ion concentration. Since most changes in Ca concentration seem to correlate inversely with changes in Mg concentration (Fig. 3.16), it is likely that variation in water chemistry is due solely to degassing, evaporation, and carbonate precipitation.

Relations between fluid Mg/Ca ratios and fluid Mg and Ca concentrations allow for the extrapolation to initial water compositions by projecting trends to the lowest Mg and highest Ca concentration (Fig. 3.12). These define two populations, one with an initial composition of 1.9 mmoles Ca and 1.1 mmoles Mg, and one with an initial composition of 3.1 mmoles Ca and 2.8 mmoles Mg. The former comprise a relatively large number of samples, whereas the latter is a smaller group of samples from a constricted section in the Left Hand Tunnel (Fig. 3.1). The apparent difference in their compositions probably reflects differences in amounts of rock–water interaction and evaporation rate.

Comparison of Ca and Mg concentrations in sample waters with those expected from degassing, evaporation, and carbonate precipitation (e.g., Fig. 3.14) aids in understanding the

compositions of waters and precipitating carbonates should exhibit a linear trend (Fig. 3.15). If degassing is followed by evaporation, an abrupt enrichment in ^{13}C with no change in $\delta^{18}O$ will be followed by an increase in $\delta^{18}O$ at constant $\delta^{13}C$ (Fig. 3.15). If, as is likely in caves, rapid degassing and evaporation is followed by reduced degassing as waters reach equilibrium with the cave pCO_2, the resultant trend will be a hyperbolic curve similar to that generated by Ca and Mg; its shape will reflect the invariant evaporation rate and the rate at which degassing decreases (Fig. 3.15).

dominant processes controlling water chemistry. The majority of waters (those whose initial Ca concentration is about 1.9 mmoles) have evolved primarily in response to CO_2 degassing because Ca concentrations decrease with a small increase in Mg. The smaller group of waters (initial Ca concentration about 3.1 mmoles), however, lie along a linear path where Ca decreases and Mg increases at comparable rates (slopes; Fig. 3.16). This suggests that both degassing and evaporation processes operated simultaneously and with invariant intensity during the evolution of this water. These samples follow a path where evaporation exerts a stronger influence than degassing (Fig. 3.16). These trends are consistent with those observed by Thrailkill (1965 and 1971). The Na and K data are consistent with evaporative concentra-

TABLE 3.2. Carbon isotopic composition of ambient cave air CO_2 by rooms or cave areas, and that of air at the ground surface over the cave. All air samples were collected in May 1985.

Room or area	$\delta^{13}C$ (PDB)
Lower Cave	−15.52
Guadalupe Room	−15.47
Ca Ca	−14.94
Mystery Room	−14.89
New Mexico Room	−14.85
Left Hand Tunnel	−14.12
Cave average	−14.97
Carlsbad air	−10.74

tion of all samples, as most values lie along linear trends of constant Na/K (Fig. 3.8).

The stable isotopic composition of waters (Fig. 3.9) and, in turn, that of the carbonates (Fig. 3.11), also records the effects of degassing and evaporation. Enriched ^{18}O fluid compositions exhibit a significant correlation, with elevated fluid Mg/Ca ratio (Fig. 3.9) indicating that the main process controlling $\delta^{18}O$ is evaporation. With enrichment in $\delta^{18}O$ values, carbonate $\delta^{13}C$ compositions also exhibit a marked enrichment which can only be attributed to rapid water degassing under disequilibrium conditions (Fig. 3.11). Isotopic equilibrium with cave pCO_2 is achieved by few if any waters, since the mean $\delta^{13}C$ composition of cave atmospheric CO_2 is about −15 ‰ (Table 3.2). Carbonate precipitated from water at equilibrium should not exhibit values heavier than −5 ‰, and should exhibit lighter values if precipitation is mediated by kinetic effects.

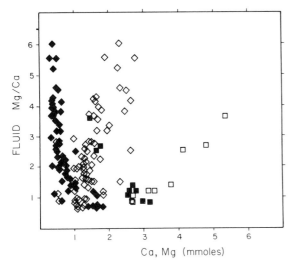

FIGURE 3.16. Cave water Ca and Mg concentrations plotted against water Mg/Ca ratio; open symbols are Mg; solid symbols are Ca. Trends displayed by the data define two evolution paths for waters from which an estimate of initial cation concentration can be made. Data for the dominant group (open and solid diamonds) appears to have had a starting compositon of about 1.8 mmoles Ca and 1.1 mmoles Mg. The smaller group (open and solid squares) had an initial Ca concentration of 3.1 mmoles and an initial Mg concentration of about 2.8 mmoles. This latter group is composed entirely of Left Hand Tunnel samples (Fig. 3.1); higher cation concentration reflects greater rock–water interaction while the slope of Mg concentration is due to a different evaporation rate.

Controls on Carbonate Mineralogy and Chemistry

Given the range in water chemistries and carbonate mineralogies in Carlsbad Caverns, what are the factors determining the mineralogy of the precipitating carbonate (particularly the dominant minerals, aragonite and calcite), and what is controlling the Mg content of calcites? From the composition of cave fluids, it is apparent that the two major controls on the chemical evolution of waters are degassing and evaporation. Variables which respond strongly to these two processes are Ca and Mg concentrations (thus controlling Mg/Ca ratio) and the amount of dissolved CO_2 (and its dissociation products as HCO_3^- and $CO_3^=$. Mineralogy and minor element incorporation must be dictated by the changes in these parameters.

Magnesium Incorporation in Calcite

Which of these variables is controlling the amount of Mg incorporated in calcite? Mole % $MgCO_3$ exhibits a nonlinear dependence on Mg/Ca ratio of the fluid (Fig. 3.17) but there is a considerable scatter in these data. However, oxygen isotopic composition of carbonate exhibits a strong relationship to carbonate Mg content (Fig. 3.18). The variation in calcite isotopic composition is greater than that which could be attributed to $MgCO_3$ content and, indeed, correlates well with water isotopic composition. Since the $\delta^{18}O$ water is dependent only on evaporation, which, in turn, controls the Mg/Ca ratio, then correlation between calcite mole % $MgCO_3$ and $\delta^{18}O$ composition demonstrates that the relationship between carbonate Mg content and water Mg/Ca ratio is not accidental. But then why the scatter of data in Figure 3.17? Is the poor correlation between carbonate Mg content and fluid Mg/Ca ratio reflecting comparison of water samples that were not responsible for the precipitation of the analyzed carbonate, or is there yet another variable influencing the Mg content of calcites?

It has been suggested that calcite Mg content is dependent to some extent on rates of precipitation as determined by $CO_3^=$ concentration in ambient fluids (Berner 1975, Given and Wilkinson 1985), even though some experimental work (e.g., Mucci and Morse 1983) seems to suggest no such dependence. In Carlsbad Caverns, the Mg content of calcites exhibits a linear relationship with the $CO_3^=$ concentration of associated fluids (Fig. 3.19). Furthermore, these data define two linear trends with similar slopes. Although not as clearly separated, these two groups are also distinct populations in Mg/Ca ratio–mole % $MgCO_3$ space (Fig. 3.17). Difference between the two may reflect the fact that one group comprises disequilibrium fluid and/or carbonate samples where water chemistry does not directly correspond to carbonate composition, or may reflect the effects of some unmeasured water parameter such as the presence of sulfate

or phosphate ions. The presence of calcite phases of differing Mg content in many samples suggests that speleothem formation may occur from oscillating water chemistries which are responding to periodic variation in the vadose water system.

The most probable origin for these two groups of samples, which exhibit similar but displaced trends, is the effect of seasonal changes in water chemistry and the fact that one of the collected carbonate phases is not undergoing precipitation from the sampled water. Since at least half of these samples also belong to that group of waters which had higher initial Mg and Ca concentrations (Fig. 3.16), it is possible that absolute Mg concentrations may also influence Mg incorporation. This would be expected if distribution coefficients Mg in calcite where dependent on fluid Mg concentration as well as on water Mg/Ca ratio. Even with scatter in these data, it is evident that there is a significant dependence of calcite Mg content on the $CO_3^=$ concentration of the precipitating fluid. This dependence is also supported by correlation of carbonate ^{13}C compositions and Mg contents (Fig. 3.20). Magnesium incorporation in calcite from Carlsbad Caverns is not a simple function of fluid Mg/Ca ratios or $CO_3^=$ concentrations. Both parameters exert a strong influence on calcite compositions, with seasonal variation in both leading to the wide range of compositions observed within particular areas and even specific sites.

Controls on Calcite–Aragonite Polymorphism

Data on carbonate oxygen and carbon isotopic composition indicate that degassing is the dominant control on the precipitation of isotopically depleted calcite, whereas evaporation becomes the dominant process during the precipitation of isotopically enriched aragonite. It is clear that the two factors controlling carbonate mineralogy are fluid $CO_3^=$ concentration and Mg/Ca ratio. However, it is not clear which of these two factors exerts the greatest influence on mineralogy. At Mg/Ca ratios of about 1, aragonite does not precipitate regardless of $CO_3^=$ concentration (Fig. 3.21), even though all waters (1985 samples) saturated with respect

FIGURE 3.17. Mg contents of cave calcites and Mg/Ca ratios of associated
fluids. Note that, despite *some* scatter, these data define two general
trends (as solid and open squares)
of increasing MgCO₃ with increasing Mg/Ca ratio ($\tau^2 = 0.6$).

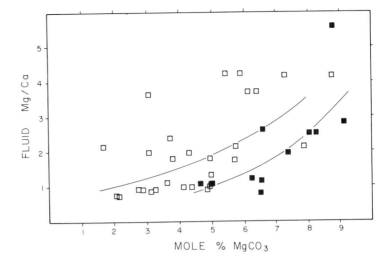

FIGURE 3.18. Mg contents and $\delta^{18}O$ values of
cave calcites. Note the good correlation (r^2
≥ 0.80) between these two parameters.
Carbonate $\delta^{18}O$ values and those of associated waters are also strongly correlated
($r^2 \geq 0.7$).

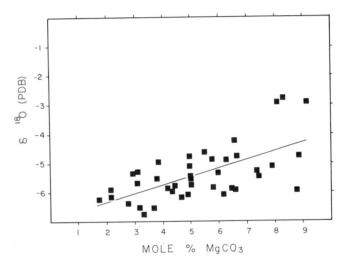

FIGURE 3.19. Mg content of cave calcites and
$CO_3^=$ contents of associated fluids. As in
Figure 3.17, these data define two distinct
populations (as solid and open squares),
both of which display the same linear dependence of calcite Mg content and $CO_3^=$
concentration. The population with generally lower $Co_3^=$ concentrations is composed mostly of samples containing two
calcite phases ($r^2 \geq 0.6$, for the upper data
set; $r^2 \geq 0.4$ for the lower data set).

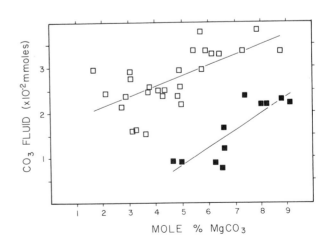

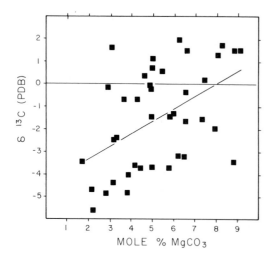

FIGURE 3.20. Mg contents and δ^{13}C compositions of cave calcites ($r^2 \geq 0.60$). Despite the scatter in the data there is a significant correlation between Mg and δ^{13}C ($r^2 \geq 0.6$).

to calcite are also saturated with respect to aragonite. Aragonite precipitation commences only after fluid Mg/Ca ratios exceed values of about 1.5.

However, elevated Mg/Ca ratio does not preclude calcite precipitation, and at high $CO_3^=$ concentrations, calcite is precipitated in the presence of moderate Mg/Ca ratios. When Mg/Ca ratios exceed 2.5 and at higher $CO_3^=$ concentrations, aragonite becomes the dominant phase (Fig. 3.21). If, as suggested by Given and Wilkinson (1985), the precipitation of aragonite was simply initiated by a $CO_3^=$ controlled calcite–aragonite threshold, the occurrence of calcite at higher $CO_3^=$ centrations should not be observed at any fluid Mg/Ca.

The most Mg-enriched calcites occur in monomineralic samples (7 to 9 mole % MgCO₃), and in mixed mineralogy samples (up to 12 mole % MgCO₃), from the Left Hand Tunnel, the New Mexico Room, and the Guadalupe Room. The oxygen isotopic composition of these calcites overlaps those of cooccurring aragonite, and carbon isotopic compositions differ by 0.2 to 1.5 ‰, with calcites being lighter. Since these differences are of the magnitude expected for equilibrium calcite–aragonite precipitation, aragonite and high-Mg calcite are probably precipitating from a common fluid in these areas. If so, then coprecipitation of aragonite and calcite, rather than the

precipitation of either one polymorph or the other, may be a common process in these (and possibly other) settings. In addition, some samples (particularly those with the most Mg-enriched calcites) yield smaller δ^{13}C calcite–aragonite differences than would be expected for equilibrium precipitation from a common fluid. This suggests that high-Mg calcite precipitation may occur even when $CO_3^=$ concentrations are extremely high.

The role of Mg in enhancing the precipitation of aragonite is not through direct interaction with aragonite crystals (aragonite precipitation rate is not enhanced), but through its inhibition of calcite nucleation and growth (Bischoff 1968, Berner 1975). As Mg concentration or Mg/Ca ratio increases, calcite growth rate decreases and aragonite becomes the dominant phase. Plummer and Mackenzie (1974) suggest that the solubility of magnesian calcites decreases with increasing Mg content. If solubility constants derived from such experimental dissolution data are valid expressions for magnesian calcite equilibrium reactions and stoichiometric saturation (as defined

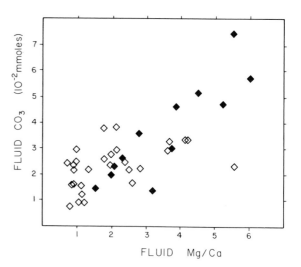

FIGURE 3.21. Fluid Mg/Ca ratios and $Co_3^=$ concentrations for waters associated with monomineralic cave samples. Aragonites are solid diamonds; calcites are open diamonds. Aragonite precipitation seems to occur only at Mg/Ca ratios higher than 1.5, indicating the importance of elevated Mg concentration or Mg/Ca ratio for aragonite precipitation to occur. Initial Mg/Ca values for Carlsbad Caverns water are 0.9 to 1.0; hence the lack of data at lower ratios.

by Thortstenson and Plummer 1977), they indicate that (at 25° C) calcite with 7 to 8 mole % $MgCO_3$ has a stability equivalent to that of aragonite (Walter 1985). This implies that precipitation of more Mg-enriched calcites requires elevated $CO_3^=$ concentrations.

In the case of Carlsbad Caverns, as waters degass, $CO_3^=$ concentrations increase at the same time that Ca removal begins to increase Mg/Ca ratios. As waters approach equilibrium with ambient pCO_2, degassing rates decrease and evaporation becomes the dominant process. In response to evaporation and carbonate precipitation, Mg/Ca ratios increase while $CO_3^=$ concentrations change less rapidly than in waters of lower Mg/Ca. At these elevated Mg/Ca ratios, $CO_3^=$ concentration is generally not sufficiently high to produce calcite with over 9 mole % $MgCO_3$. The presence of aragonite coexisting with two calcite phases of differing Mg contents and morphologies, and the fact that differences in calcite–aragonite isotopic compositions are smaller than expected, suggests that precipitation progresses from low-Mg calcite to aragonite to high-Mg calcite.

Precipitation of Hydrated Carbonates

Hydromagnesite and huntite in pools precipitate from waters with significantly elevated Mg concentrations and Mg/Ca ratios. Hydromagnesite also occurs as crusts on some dripstones (soda straws and popcorn) coexisting with aragonite. The precipitation of hydrated Mg and Ca–Mg carbonates results from evaporative concentration of waters and the development of extreme fluid Mg/Ca ratios and elevated pH (Thrailkill 1965 and 1971). The presence of hydromagnesite and aragonite on the same speleothems is further indication of the important role of high fluid Mg/Ca ratios as controls on the mineralogy of carbonate phases.

Dolomite Precipitation

Unlike occurrences of dolomite documented by Thrailkill (1968), dolomite in Carlsbad Cavern samples occurs on speleothem surfaces and is not an apparent alteration product of hydromagnesite, huntite, or aragonite. The dolomite is also not detrital, as the range of dolomite Mg contents is 42 to 46 mole % $MgCO_3$, while that from the country rock ranges from 46 to 50 mole % $MgCO_3$ and shows excellent ordering reflections. Moreover, in two sites dolomite is found on the undersides of pool shelves where detrital dolomite could not accumulate. In addition to speleothem surfaces, dolomite has also been detected in internal layers of pool ledge carbonate in association with another phase that is either a high-Mg calcite or a dolomite phase with approximately 38 mole % $MgCO_3$. From the present data it is difficult to determine the conditions under which dolomite has formed in these settings. High variability in cave water chemistry suggests that it may form under extreme conditions. Because samples containing dolomite contain no associated hydrated magnesium carbonates, dolomite precipitation is probably not a product of extreme evaporation. An alternative explanation is that dolomite may form when pool waters attain equilibrium with ambient pCO_2 and become undersaturated with respect to calcite and aragonite. The occurrence of dolomite and documentation of associated water compositions in these settings deserve careful study, as such information may prove to be a valuable key to understanding dolomite formation under other sedimentary conditions.

Conclusions

The chemical composition of water in Carlsbad Caverns is controlled by the cave-related processes of CO_2 degassing and evaporation. Progressive degassing increases the $CO_3^=$ concentration of fluids, results in carbonate precipitation and removal of Ca ions, and, thus, an increase in the Mg/Ca ratio. As these changes occur, precipitated calcites contain higher amounts of Mg as dictated by the Mg/Ca ratio and $CO_3^=$ content of the fluid. Where $CO_3^=$ concentrations are not sufficiently high, Mg inhibition of calcite growth is not overcome, and either aragonite or a Mg-deficient calcite precipitates. As degassing decreases in importance and as evaporative concentration of Mg increases the fluid Mg/Ca ratio, aragonite becomes the dominant phase because $CO_3^=$ concentrations are not sufficient to precipitate high-Mg calcite. Because experimental and empirical data indicate that aragonite precipitation is favored at higher temperatures (>40° C), the lack of cave aragonite precipitation from low Mg/Ca solutions at elevated $CO_3^=$ may be a function of the relatively low temperatures of these systems. It is clear that extensive documentation in natural systems, particularly in spelean settings, is needed to further assess relationships between carbonate precipitation and ambient water chemistry.

Acknowledgments. Funds for this study were provided by grants to the senior author by the University of Michigan Scott Turner Fund, by a University of Michigan Rackham Dissertation Grant, and by the North Central Section of the Geological Society of America. Scanning electron photomicrographs were taken with a Hitachi SEM obtained with NSF grant BSR-83-14092. All samples were collected with the permission and under supervision of the National Park Service. Sampling at Carlsbad Caverns was guided by Park Naturalist Ronald C. Kerbo, to whom we are grateful. Field assistance was provided by R. Kevin Given and William G. Zempolich. Scott J. Carpenter, Elizabeth A. Finkel, and Bradley N. Opdyke assisted in various aspects of the preparation of this manuscript. David Dettman provided invaluable technical assistance in the stable isotope laboratory. Final preparation of this manuscript was made possible by the extraordinary efforts of Bruce H. Wilkinson, to whom we are extremely grateful. The senior author wishes to express his extreme gratitude to Sahudi A. Gonzalez for her assistance with illustrations and her infinite patience.

References

Berner, R.A., 1975, The role of magnesium in the crystal growth of calcite and aragonite from sea water: Geochim. Cosmochim. Acta, v. 39, p. 489–504.

Bischoff, J.L., 1968, Kinetics of calcite nucleation: magnesium ion inhibition and ionic strength catalysis: Jour. Geophys. Res., v. 73, p. 3315–3322.

Cabrol, P., and Coudray, J., 1982, Climatic fluctuations influence the genesis and diagenesis of carbonate speleothems in southwestern France: Nat. Spel. Soc. Bull., v. 44, p. 112–117.

Cave Research Foundation, 1979, Map of Carlsbad Caverns National Park: Cave Research Foundation, Washington, DC, 1 p.

Craig, H., 1957, Isotopic standards for carbon and oxygen and correction factors for mass-spectrometric analysis of carbon dioxide: Geochim. Cosmochim. Acta, v. 12, p. 133–149.

Craig, H., Gordon, L.I., and Horibe, Y., 1963, Isotopic exchange effects in the evaporation of water: Jour. Geophys. Res., v. 68, p. 5079–5087.

Deines, P., Langmuir, D., and Harmon, R.S., 1974, Stable carbon isotope ratios and the existence of a gas phase in the evolution of carbonate ground waters: Geochim. Cosmochim. Acta, v. 38, p. 1147–1164.

Epstein, S., and Mayeda, T., 1953, Variation of O^{18} content of waters from natural sources: Geochim. Cosmochim. Acta, v. 4, p. 213–224.

Fantidis, J., and Ehhalt, D.H., 1970, Variations of the carbon and oxygen isotopic composition in stalagmites and stalactites: evidence of nonequilibrium isotopic fractionation: Earth Planet. Sci. Lett., v. 10, p. 136–144.

Fornaca-Rinaldi, G., Panichi, C., and Tongiori, E., 1968, Some causes of the variation of the isotopic composition of carbon and oxygen in cave concretions: Earth Planet. Sci. Lett., v. 4, p. 321–324.

Given, R.K., and Wilkinson, B.H., 1985, Kinetic control of morphology, composition, and mineralogy of abiotic sedimentary carbonates: Jour. Sed. Petrol., v. 55, p. 109–119.

Gonzalez, L.A., and Lohmann, K.C, 1985, Carbon and oxygen isotopic composition of Holocene reefal carbonates: Geology, v. 13, p. 811–814.

Harmon, R.S., 1979, An isotopic study of groundwater seepage in the central Kentucky karst: Water Resources Res., v. 15, p. 476–480.

Harmon, R.S., Atkinson, T.C., and Atkinson, J.L., 1983, The mineralogy of Castleguard Cave, Columbia Icefields, Alberta, Canada: Arctic Alpine Res., v. 15, p. 503–516.

Holland, H.D., Kirsipu, T.V., Huebner, J.S., and Oxburgh, U.M., 1964, On some aspects of the chemical evolution of cave waters: Jour. Geol., v. 72, p. 36–67.

Inoue, H., and Sugimura, Y., 1985, Carbon isotopic fractionation during the CO_2 exchange process between air and seawater under equilibrium and kinetic conditions: Geochim. Cosmochim. Acta, v. 49, p. 2453–2460.

Lloyd, R.M., 1966, Oxygen isotope enrichment of seawater by evaporation: Geochim. Cosmochim. Acta, v. 30, p. 801–814.

Moore, G.W., 1956, Aragonite speleothems as indicators of paleotemperature: Amer. Jour. Sci., v. 254, p. 746–753.

Mucci, A., and Morse, J.W., 1983, The incorporation of Mg++ and Sr++ into calcite overgrowths: influences of growth rate and solution composition: Geochim. Cosmochim. Acta, v. 47, p. 217–233.

Murray, J.W., 1954, The deposition of calcite and aragonite in caves: Jour. Geol., v. 62, p. 481–492.

Murray, J.W., 1975, Additional data on the mineralogy of the New River Cave: Nat. Spel. Soc. Bull., v. 37, p. 79–82.

Plummer, L.N., and Mackenzie, F.T., 1974, Predicting mineral solubility from rate data: application to the dissolution of magnesian calcites: Amer. Jour. Sci., v. 274, p. 61–83.

Rogers, B.W., and Williams, K.M., 1982, Mineralogy of Lilburn Cave, Kings Canyon National Park, California: Nat. Spel. Soc. Bull., v. 44, p. 23–31.

Rubinson, M., and Clayton, R.N., 1969, Carbon-13 fractionation between aragonite and calcite: Geochim. Cosmochim. Acta: v. 33, p. 997–1002.

Sears, S.O., 1976, Inorganic and isotopic geochemistry of the unsaturated zone in a carbonate terrain: Ph.D. thesis, Pennsylvania State University, 236 p.

Siegel, F.R., 1965, Aspects of calcium carbonate deposition in Great Onyx Cave, Kentucky: Sedimentology, v. 4, p. 285–299.

Tarutani, T., Clayton, R.N., and Mayeda, T., 1969, The effect of polymorphism and magnesium substitution on oxygen isotope fractionation between calcium carbonate and water: Geochim. Cosmochim. Acta, v. 33, p. 987–996.

Thorstenson, D.C., and Plummer, N.L., 1977, Equilibrium criteria for two-component solids reacting with fixed composition in an aqueous phase-example: the magnesian calcites: Amer. Jour. Sci., v. 277, p. 1203–1223.

Thrailkill, J., 1965, Studies in the excavation of limestone caves and the deposition of speleothems: Part I. Chemical and hydrologic factors in the excavation of limestone caves. Part II. Water chemistry and carbonate speleothem relationships in Carlsbad Caverns, New Mexico: Ph.D. thesis, Princeton University, 193 p.

Thrailkill, J., 1968, Dolomite cave deposits from Carlsbad Caverns: Jour. Sed. Pet., v. 38, p. 141–145.

Thrailkill, J., 1971, Carbonate deposition in Carlsbad Caverns: Jour. Geol., v. 79, p. 683–695.

Tietz, G.F., 1981, Hollow calcite crystals on surfaces of small pools in the Liethühle/Sauerland, West Germany, in Beck, B.F., ed., Proceedings of the Eighth International Congress of Speleology: v. 1, p. 362–363.

Walter, L.M., 1985, Relative reactivity of skeletal carbonates during dissolution: implications for diagenesis, in Schneidermann, N., and Harris, P.M., eds., Carbonate cements: Soc. Econ. Paleontologists and Mineralogists, Spec. Publ. 36, p. 3–16.

Wilson, W.L., and Ash, D.W., 1985, Stratigraphy of the New Mexico and Guadalupe Rooms in Carlsbad Caverns, New Mexico, in Lindsley, K.B., ed., Annual report: Cave Res. Found. 1984, p. 13–15.

Yonge, C.J., Ford, D.C., Gray, J., and Schwarcz, H.P., 1985, Stable isotope studies of cave seepage water: Chem. Geol., v. 58, p. 97–105.

Zeller, E.J., and Wray, J.L., 1956, Factors influencing precipitation of calcium carbonate: Amer. Assoc. Petroleum Geologists Bull., v. 40, p. 140–152.

4
Breccia-Hosted Lead–Zinc Deposits in Carbonate Rocks

Donald F. Sangster

Abstract

Although several deposit types have either been shown or suggested to be associated with paleokarst features in carbonate rocks, lead–zinc sulfide deposits of the "Mississippi Valley-type" (MVT) have received the most research and exploration.

MVT deposits are not simply sulfide-filled open caves; rather, they are emplaced in carbonate breccias and consist of sulfides and carbonates, as either matrix or cement, filling spaces between angular carbonate fragments. Evidence of replacement is characteristically minimal, except in high-grade portions of deposits.

Two main hypotheses have been proposed to account for the origin of the breccias. One school advocates solution collapse brought about by the incursion of meteoric water into the carbonate strata (i.e., meteoric karst) during a period of emergence. Emplacement of ore into the breccias is envisaged as a separate and later event. The other school proposes solution collapse induced by a slightly earlier phase of the ore solutions themselves (hydrothermal karst) with no meteoric involvement.

Circumstantial evidence, such as the nearly universal presence of an overlying unconformity, the morphological similarity of the ore-bearing breccia zones to known paleokarst solution collapse breccias, the presence of solution-thinned strata below the breccia zones, and the relative paucity of wallrock alteration, has led to a majority opinion favoring the meteoric karst hypothesis.

Introduction

Those lead–zinc deposits which are hosted by carbonate rocks and contained in structures widely accepted as paleokarst phenomena are commonly referred to as "Mississippi Valley-type" (MVT) deposits (Ohle 1959 and 1980). They are so named because they were originally recognized as a distinct deposit-type in the Mississippi Valley region of the United States almost a hundred years ago. Since then, several symposia have summarized our knowledge of these deposits, and the reader is referred to these for details (Bastin 1939, Brown 1967, Kisvarsanyi et al. 1983).

Displayed in Figure 4.1 are the MVT districts of Upper Mississippi Valley, Southeast Missouri, and Tri-State districts of the USA, now regarded as the classical type of localities. The Central and East Tennessee districts, as well as Pine Point in Canada, are also recognized as significant districts on a world scale. Outside North America, only two major MVT districts are known, these being in the Silesian district of Poland and a narrow belt in the southern Alpine region of Italy, Austria, and Yugoslavia (Fig. 4.2).

MVT deposits tend to be relatively low grade, averaging perhaps 4% to 8% combined lead–zinc, although several examples of higher-grade ore bodies, or portions of ore bodies, are known. Mineralogy is typically simple, consisting predominantly of sphalerite, galena, pyrite, marcasite, and dolomite. Silver content of the ore is negligible and the metal is not normally recovered. Lead–zinc ratios of MVT deposits range from Pb-dominant to Zn-dominant with a majority of deposits displaying $Zn/(Zn + Pb)$ ratios around 0.7. Minerals are usually coarse-grained.

From studies of fluid inclusions in the ore minerals, depositional temperatures of MVT

FIGURE 4.1. Locations of major MVT districts in North America. District name abbreviations as follows: E.Tenn. = East Tennessee; C.Tenn. = Central Tennessee; SE Mo. = Southeast Missouri; UMV = Upper Mississippi Valley.

deposits were between 100° and 150° C; salinities of the same inclusions are high, ranging from 15 to over 23 wt% NaCl equivalent. Sulfur isotopes in ore minerals suggest derivation of sulfur by reduction, either bacterial or nonbacterial, of seawater sulfate (Anderson and Macqueen 1982).

These features, and many other lines of evidence, have led investigators to propose a genetic model involving transport of metals by compactional expulsion of metal-bearing connate brines toward basinal margins (Jackson and Beales 1967). Precipitation of sulfides is envisaged to have occurred by reaction between these brines and hydrogen sulfide-bearing gas trapped in paleokarst breccias developed beneath unconformities at the basin margins (Anderson 1983, Cathles and Smith 1983). Other authors have proposed that both reduced sulfur and metals were carried to the depositional site in the same fluid; in this case, precipitation was probably effected by cooling (e.g., Sverjensky 1984).

Deposits are discordant but stratabound; that is to say, in any one district, one or two for-

FIGURE 4.2. Locations of major MVT districts in Europe. E. Alpine = Eastern Alpine; U. Silesia = Upper Silesia.

mations are the preferred hosts. Of the diagnostic features of MVT deposits, perhaps the two most relevant to this volume are (1) the open-space filling of carbonate breccias as the main mineralizing process, and (2) the nearly universal presence of an unconformity above the ore district.

Ore-Hosting Breccias

Abundant evidence of sulfide minerals filling or partly filling open spaces is perhaps the single most characteristic feature of MVT deposits. In some high-grade zones, massive replacement of carbonate host seems an inevitable conclusion, but for the deposit-type as a whole, open-space filling is the diagnostic texture. In some instances, these open spaces are primary, such as in biomicrites, where examples of brachiopod molds, partially or completely filled with sulfide minerals, are not uncommon (Akande and Zentilli 1984). Similarly, primary sedimentary porosity seems to have been the main ore control in the Old Lead Belt of the Southeast Missouri district (Snyder and Gerdemann 1968).

These examples notwithstanding, however, secondary porosity, occurring as voids between angular carbonate fragments, is by far the most widespread type of open space observable in MVT deposits. No examples are known of world-class MVT deposits occurring as simple open-cave fill; all appear to occupy zones of carbonate breccias. A proposal by Olson (1977 and 1984) that the Nanisivik deposit, hosted by Proterozoic dolomites in northern Baffin Island, is an open-cave fill has been challenged by Ford (1982, p. 49): "95% of the volume of the Main Ore cavity was created by ore fluids dissolving dolomite as sulphide was emplaced." If Ford's thesis is correct, then Nanisivik is not a MVT deposit at all but a carbonate replacement body.

A candid discussion of breccias in MVT deposits has recently been presented by Ohle (1985), in which he reviews the evidence for a solution collapse origin for the breccias. Features supporting this concept are:

1. Removal of host rock carbonate as evidenced by thinning of the favored ore horizons near the ore bodies.

2. The observation that, within the breccia zones, identifiable carbonate fragments have dropped tens of meters from their original stratigraphic position.
3. The morphological similarity between MVT ore deposits and patterns of modern cave systems.

What has not been mentioned specifically by Ohle, but could be considered parallel evidence, is the nearly universal presence of an erosional hiatus immediately above virtually all MVT deposits. In some districts a well-developed karst morphology, with tens of meters of relief, has developed on the unconformity (see discussion in Sangster 1983, p. 10). The presence of an unconformity overlying MVT districts is important to the paleokarst concept because subaerial exposure is necessary to recharge the subsurface aquifer with meteoric waters to effect the solution thinning of soluble units. An important exception is the southeast Missouri district where no regional unconformity has been recognized. The absence of an unconformity in this major MVT district may call into question the meteoric origin of collapse breccias as a universal genetic concept for all MVT districts.

Internal Features

Ore-bearing breccias vary considerably, from one MVT district to the other, in the nature and abundance of their internal features. There are, however, three which are common to many districts; these are crackle breccia, rock-matrix breccia, and ore-matrix breccia (Figs. 4.3 to 4.5).

Crackle Breccia. Occurring most commonly in the uppermost part of the ore zones, crackle breccia consists of highly fractured host dolostone, generally fine-grained. The fragments are only slightly separated, along bedding planes and other fractures, without rotation or collapse of fragments (Figs. 4.3 and 4.4).

Rock-Matrix Breccia. This breccia is composed of matrix-supported dolostone fragments, ranging from blocks to silt-sized, sealed within a clastic matrix of even finer-grained carbonate fragments (Fig. 4.6). The large fragments are typically rotated, and in cases of distinctive

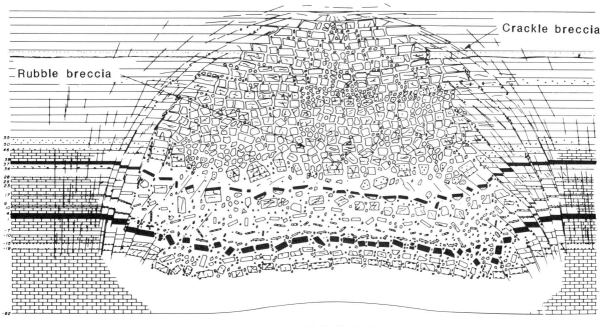

FIGURE 4.3. Diagrammatic representation of a collapsed, high-dome, ore-bearing breccia body in East Tennessee. Unpatterned horizontal ruling represents dolostone. Note: (1) the positions of crackle breccia and rubble (ore-matrix) breccia; (2) the structural arch configuration; and (3) the outward-dipping bounding fractures (from Ohle 1985, reproduced from Economic Geology with permission).

stratigraphy can be demonstrated to have dropped relative to their original stratigraphic position.

Ore-Matrix Breccia. This variety of breccia is characterized by dolostone fragments cemented by white or pink crystalline sparry dolomite with or without sulfides (Fig. 4.7). As in the rock-matrix breccia, fragments are typically rotated and have dropped from their original stratigraphic position (Figs. 4.3–4.5). Ore-matrix breccia differs from rock-matrix breccia in that, in the latter, the matrix is a clastic sediment, whereas in the former it is a chemical precipitate. Consequently, although it should more properly be referred to as a cement, the term *ore-matrix* is well engrained in the scientific literature of MVT deposits (e.g., Hoagland et al 1965, Hill 1969). A very common feature of ore-matrix breccias is the virtual absence of fine-grained material, either siliceous or carbonate, in the interstices between the fragments (Fig. 4.8).

Trash Zones. During dissolution of the host carbonate, much insoluble material accumulated at the bottom of the breccia bodies (Fig. 4.9). These accumulations, which can range up to a few meters in thickness, are referred to as *trash zones* (Hill et al. 1971). They consist of poorly sorted grains of a variety of minerals, depending on the host-rock lithology, and black, carbonaceous material. Most common detrital minerals are dolomite, clay, chert, quartz, and the ore sulfides.

Internal Sediment. In several deposits, space between breccia fragments is partly or entirely filled with well stratified material consisting mainly of dolomite but commonly containing a large component of detrital sand-size sulfide grains and angular sulfide fragments; the sulfide is most commonly sphalerite. The sand is deposited in alternating light and dark layers a few mm thick; graded bedding is not uncommon. Significantly, laminations in these sands are parallel to the dip and strike of the country

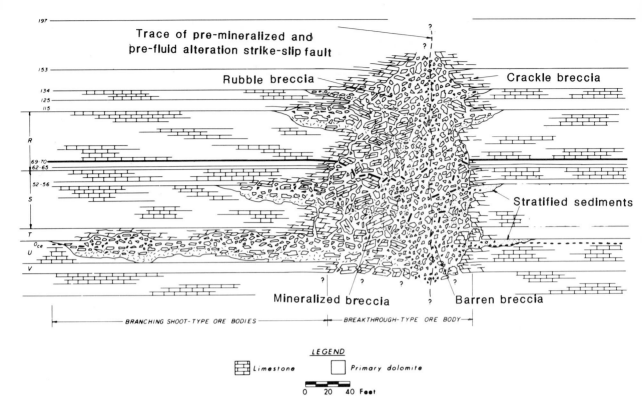

FIGURE 4.4. Schematic representation of a columnar, ore-bearing, breccia body in east Tennessee. Note: (1) that not all breccia is mineralized, (2) the presence of conformable breccia zones occurring as offshoots of, but connected to, the discordant breccia body; (3) the presence of internal, stratified sediments; (4) a narrow alteration zone of dolomite (with minor silicification) surrounding both the discordant and concordant breccias; and (5) all fragments within the breccia are dolomite even though significant amounts of limestone occur in the host rocks (from Ohle 1985, reproduced from Economic Geology with permission).

rock. Internal sands in ore-hosting breccia zones have been described in the East Tennessee (Figs. 4.4 and 4.10, Kendall 1960), Pine Point (Figs. 4.5 and 4.11, Rhodes et al. 1984), and Upper Silesia (Sass-Gustkiewicz et al. 1982, Bogacz et al. 1973, Sass-Gustkiewicz 1975) MVT districts. In the last two districts, the internal sediments are found most extensively on or near the floor of the breccia bodies.

Snow-on-Roof. First described by Oder and Hook (1950), this, in the author's opinion, is the most diagnostic texture in MVT deposits because it has not been reported in any other type of mineral deposit. The texture is most commonly displayed by sphalerite occurring as a layer of coarse crystals preferentially coating the *tops* of the dolomite fragments in ore-matrix breccias (Figs. 4.12 and 4.13). Undersides of the same fragments are coated either not at all or to a much lesser degree. A variety of snow-on-roof texture is sphalerite lining the *bottoms* of cavities (Fig. 4.14). In either case, the feature can be regarded as a geopetal indicator because it can be used to indicate "right side up."

Morphological Shape

As with internal features, MVT ore-bearing breccias display a variety and abundance of forms both within and between deposits. Those which serve as the main control to mineralization in some of the major districts shown in Figure 4.1 are briefly described below.

Probably the most spectacular form of ore-hosting breccias are discordant, stratabound domes or columns. The dome-shaped bodies (Fig. 4.3), with their characteristic bounding shears, have been compared to "pressure arches" which have reached a state of equilibrium (McCormick et al. 1971). Breccia bodies of this type have been described in the East Tennessee (McCormick et al. 1971), southeast Missouri (Mouat and Clendenin 1977, Rogers and Davis 1977, Sweeney et al. 1977), and Upper Silesia (Sass-Gustkiewicz et al. 1982) MVT districts.

The columnar bodies, referred to as *break-through* (McCormick et al. 1969) or *prismatic* (Rhodes et al. 1984) type (Fig. 4.15), can affect up to 100 m of section. Compared with the domelike bodies, the columnar breccias display vertical walls which impart a much more cylindrical form to the bodies rather than the beehive shape of the domelike bodies (Figs. 4.4 and 4.5).

In plan view, the discordant bodies produce a sinuous, linear pattern which can be up to 8 km long in individual mines (e.g., the Buick mine, Viburnum trend, southeast Missouri; Rogers and Davis 1977) or as much as 50 km

for the entire Viburnum trend. The Tri-State District "runs" can be straight, up to 30 m high, 150 m wide, and several kilometers long; they can also occur in clusters of arcuate trends up to 500 m across (Brockie et al. 1968). In other districts such as the Jefferson City mine, East Tennessee district, the discordant bodies occur as an interconnected, reticulate, boxwork pattern dominated by northwest-trending linear breccia zones up to several hundred meters in length (Crawford et al. 1969, Fulweiler and McDougal 1971). A similar reticulate pattern is present in the Elmwood deposit, central Tennessee (Gaylord and Briskey 1983). Rectilinear patterns such as these have been compared to similar patterns in present-day solution caves in limestone (Harris 1971, Ohle 1985). At Pine Point, in contrast, the prismatic breccias are typically elliptical in plan with the longer dimension averaging 120 to 180 m.

In some instances, the discordant type breccia

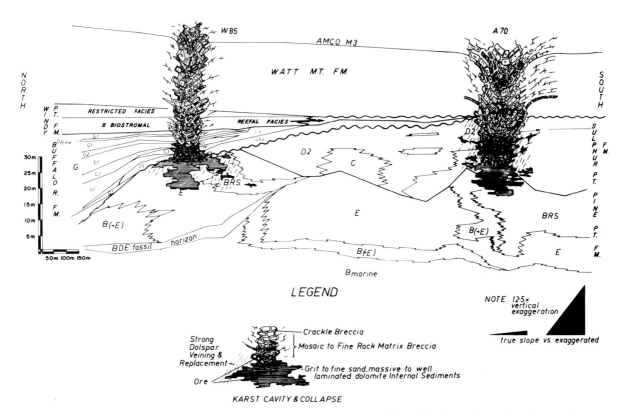

FIGURE 4.5. Prismatic ore-bearing breccias, Pine Point. Note: (1) the vertical exaggeration; (2) that the breccias pierce the unconformity at the base of the Watt Mountain Formation; (3) the position of internal sediments at the base of the breccia bodies; (4) position of crackle breccias and rock-matrix breccias; and (5) the halo of dolomite veining and replacement (alteration?) (from Rhodes et al. 1984, reproduced from Economic Geology with permission).

4.6

4.7

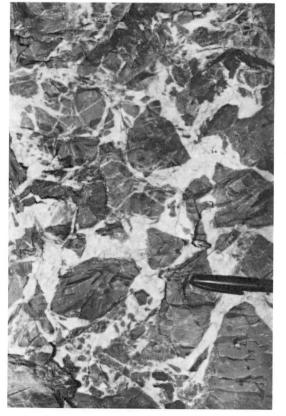

4.8

FIGURE 4.6. Rock-matrix breccia, Elmwood mine, Central Tennessee district. Note the relative absence of sparry carbonate cement (contrast with Figs. 4.7 and 4.8).

FIGURE 4.7. Ore-matrix breccia, Jefferson City mine, East Tennessee district. Photograph of stope back; size of large dark fragment approximately 25 cm.

FIGURE 4.8. Ore-matrix breccia, Jefferson City mine, East Tennessee district. White material is sparry dolomite cement. Note: (1) the absence of fine-grained fragments in the matrix: and (2) several of the larger fragments appear to "float" in the dolomite cement without touching other fragments. Length of pen visible is 6 cm.

FIGURE 4.9. Floor of breccia body, Gordonsville mine, Central Tennessee district. Note: (1) the breccia is of the rock-matrix type; (2) the rotated nature of the fragments; (3) the presence of a thin, clay-rich layer of insoluble residue ("trash zone") between the breccia and the floor (dark discordant layer in photo); and (4) truncation of the host rock bedding beneath the trash zone layer. Marker is 16 cm.

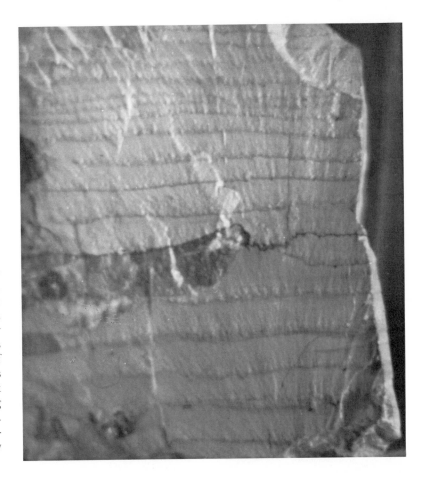

FIGURE 4.10. Internal sediment, Jefferson City mine, East Tennessee district. Note: (1) the presence of breccia fragments in lower left corner of specimen; (2) darker layers consist almost entirely of sphalerite grains, light layers are dolomite grains; (3) not visible is the graded bedding in the sphalerite-rich layers. Specimen is oriented right-way-up and is approximately 15 cm high.

FIGURE 4.11. Internal sediment in K-57 prismatic orebody, Pine Point district. Note: (1) cross-bedded nature of the layers; (2) at base of infill zone, layering "sags" downward, presumably between adjacent breccia fragments, only one of which is visible immediately to the right of the sag; and (3) sediment consists of detrital sand-size dolomite grains. Knife is 8 cm.

FIGURE 4.12. Snow-on-roof texture, K Zone, Newfoundland Zinc mine, Newfoundland. Sphalerite is the grey material within the white sparry dolomite cement and on top of the dark fragment under the pen. Note the sphalerite is absent underneath the same fragment but is present again on top of the next fragment, lower right corner. Pen, left corner, is 6 cm.

FIGURE 4.13. Snow-on-roof texture, K Zone, Newfoundland Zinc mine, Newfoundland. Sphalerite is the coarse-grained material lying on top of the three large fragments on either side of the knife and the large, lenticular fragment in the center left of the photo. Note the absence of sphalerite on the undersides of the same fragments. Knife is 8 cm.

bodies are linked by relatively concordant breccia zones (Fig. 4.4) referred to as *bedded* (which alludes to their conformability rather than the presence of layering) (Crawford et al. 1969), *manto* (McCormick et al. 1971), or *tabular* (Rhodes et al. 1984) ore bodies. In east Tennessee, these concordant breccia bodies are more common than the breakthrough type and are developed over a considerable lateral extent, forming large, tabular deposits (Crawford et al. 1969). At Pine Point the tabular breccias also link the prismatic breccias but do so in northeast-trending linears, 2 to 3 km wide, extending for more than 50 km along strike

(Rhodes et al. 1984). Tabular breccia "is the most common, widespread and typical karstic phenomenon at Pine Point" (Rhodes et al. 1984 p. 1014).

In both concordant and discordant types, the breccia bodies are completely filled with breccia material, including matrix, cement, and scattered open cavities between fragments. Breccia extends right to the walls and top of the body (Figs. 4.3 and 4.4). No examples are known where the lower portion of an ore zone consists of the usual sulfide-filled breccia and the upper portion is a large, formerly open cave filled with sulfides only.

FIGURE 4.14. Coarse-grained sphalerite (dark) lying on bottom of cavity; remainder of void filled with carbonate. Elmwood mine, Central Tennessee district. Marker is 16 cm.

FIGURE 4.15. Portion of 0–28 prismatic orebody, Pine Point district. Note: (1) beds on left of photo bend downward as they approach the breccia body outlined by short dashes in the photo; (2) disappearance of bedding within the breccia; and (3) host rock bedding passes over the breccia body with only slight evidence of sag.

Solution Thinning

Thinning, or complete destruction, of carbonate units beneath the ore-hosting breccias has been documented in several MVT districts, most notably in east Tennessee. Evidence for this dissolution lies, for example, in the presence of extensive "trash zones," such as in the Flat Gap mine (Hill 1969), or the "shalification" of carbonates in the Upper Mississippi Valley district (Heyl et al. 1959), both of which represent the insoluble residue following carbonate dissolution. McCormick et al. (1971) report volume reduction of as much as 80% in limestones beneath the breccias. Solution thinning has also been reported in several deposits in the Viburnum trend, Southeast Missouri district (Mouat and Clendenin 1977, Sweeney et al. 1977).

Solution thinning of selected stratigraphic units is generally acknowledged to be the process which ultimately leads to collapse of overlying strata. Development of the ore-hosting breccias is envisaged as a more or less continuous process which stopes *upward* from the initial aquifer until an equilibrium "pressure arch" configuration is attained, the breccia reaches a lithologic unit of sufficient strength that it does not collapse, or the hole is completely filled with breccia and presumably thereby becomes self-supporting. In some instances, the upward-stoping breccia zones have actually pierced the unconformity thought to

be responsible for ingress of the corrosive meteoric solutions and the breccias contain fragments of lithologies *above* the unconformity. Such features are known, for example, in the Central Tennessee (Gaylord and Briskey 1983) and Pine Point (Fig. 4.5) districts. They could be taken as evidence of continued subterranean dissolution by meteoric waters long after submergence of the erosional surface and/or that meteoric recharge was not involved in the breccia-forming process at all (see discussion below).

Alteration

Extensive haloes of host rock visibly altered by the ore solutions are notably absent in MVT deposits. Silicification, for example, is virtually absent from a majority of districts; a major exception is the Tri-State district, which contains widespread silicification ("jasperoid") of carbonate adjacent to the ore zones (Brockie et al. 1968). In other districts which contain only minor amounts of secondary silicification, the hosting stratigraphic units contain primary silica as chert nodules, concretions, etc., and it appears that the relatively minor ore-related silicification may be the result of nothing more than local mobilization of silica already present in the vicinity.

Dolomitization is a common phenomenon in MVT districts, but in most cases it is not obvious whether dolomitization was a direct result of the ore-forming process or was produced by regional-scale diagenetic processes unrelated to ore. The widespread development of Presqu'ile dolomite at Pine Point is an example of the latter (Rhodes et al. 1984). In some instances, however, local dolomitization appears to be directly related to the brecciation process if not the ore-forming process itself. In East Tennessee, for example, almost all fragments within the breccia zone consist of dolostone in spite of the fact that the collapse zone penetrates both dolostone and limestone strata (Figs. 4.3 and 4.4). A narrow zone of secondary dolomite also surrounds the breccia body (Fig. 4.4).

Discussion

The ore-bearing carbonate breccias have generally been attributed to a carbonate dissolution origin on the basis of four main lines of evidence:

1. All known deposits, except in southeast Missouri, occur below an unconformity or disconformity.
2. The spatial distribution of breccia bodies forms an interconnected reticulate to rectilinear coalescing network closely resembling modern dissolution patterns in limestone terrain.
3. Solution-thinned carbonate units occur below the ore-bearing discordant carbonate breccias.
4. Individual fragments in the breccias can be shown to have dropped as much as tens of meters relative to their proper stratigraphic position.

Solution-induced collapse as the main process responsible for the ore-hosting breccias in MVT deposits was first proposed by Ulrich (1931) and has received considerable support since then (Ohle 1985). Two main hypotheses have been proposed to account for the dissolution. Both recognize the presence of collapse breccia but differ in the process by which the breccia is formed.

One school advocates solution collapse brought about by the incursion of meteoric water into the carbonate platform during a period of emergence indicated by the nearly ubiquitous unconformities above the ore deposits. Paleokarst formed in this manner is referred to herein as *meteoric karst*. Emplacement of ore into the karst breccias is envisaged as a separate and later event. Proponents of the meteoric school of breccia formation point to the nearly universal presence of an overlying unconformity in MVT districts the world over. The implication is that the association is too marked to be attributed to coincidence. A meteoric origin for at least the preore breccias in the East Tennessee district was the major conclusion of an interdisciplinary symposium in 1969 (reported in Economic Geology, v. 66, no. 5, 1977) at which, among other topics, the Middle Ordovician paleohydrodynamics were examined and comparisons drawn with a modern carbonate aquifer in Florida (LeGrand and Stringfield 1971, Harris 1971). Morphological comparisons have also been made between MVT breccia bodies and modern cave systems (Harris 1971, Ohle 1985). Such comparisons are useful but it must be kept in mind that the features being compared are, in the one instance, breccias with relatively minor open space and, in the other, open caves with only minor breccia.

The other school, advocated chiefly for the Polish deposits, proposes that "the emplacement of ores and the formation of underground karst were parts of the same formative processes and were essentially simultaneous" (Sass-Gustkiewicz et al. 1982, p. 392). A meteoric component is not perceived to be involved at any stage and, consequently, breccia formed in this manner is considered to have formed by "hydrothermal karst" processes (Dzulynski 1976, Bogacz et al. 1970). An explicit documentation of hydrothermal karst is that of Sass-Gustkiewicz et al. (1982) for the Upper Silesia district of Poland. These authors recognized five stages of brecciation, each of which is characterized by its own specific assemblage of ore minerals; they also noted the absence of any indication of oxidizing aqueous solutions between the stages of sulfide mineralization. From such evidence, it was concluded that "the breccia formation and the emplacement of sulfide ores were parts of the same formative process rather than a repeated coincidental superposition of two different and genetically unrelated events" (Sass-Gustkiewicz et al. 1982,

p. 404). Further evidence precluding a meteoric aqueous phase in the development of MVT paleokarst breccias is the virtual total absence of carbonate speleothems such as stalactites, stalagmites, cave pearls, dripstone, etc. in ore breccias. The present author has never seen any such features whatsoever in MVT districts and Sass-Gustkiewicz et al. (1982) report them as "rare" in Silesian deposits. Similarly, the scarcity of red muds (terra rosa) in MVT breccias suggest little or no connection with surface oxidation (Ohle 1985). Although wallrock alteration is minimal in most MVT districts, Ohle (1985) notes the almost complete absence of limestone blocks in the breccias, even in deposits such as the Jefferson City mine, where the stratigraphic section pierced by the breccias contains limestone beds. This is regarded as "further indication of the intimate relationship of alteration, solution, brecciation, and ore deposition" (Ohle 1985, p. 1743); that is to say, all features present in the deposit were the result of one (ore-forming) process. The absence of "fines" in the ore-matrix breccias has been attributed to the preferential dissolution of the finer-grained material (Sass-Gustkiewicz et al. 1982). This proposal, however, seems at variance with the typical angular nature of the breccia fragments. If corrosive solutions attacked the fines, should not the sharp edges of the larger fragments exhibit some evidence of rounding or partial replacement?

In conclusion, it has been demonstrated that by far the most typical host for MVT deposits is secondary breccia zones and *not* open cave systems. With respect to the origin of the ore-hosting breccias, judging by current literature, proponents of the meteoric origin for MVT paleokarst breccias appear to be in the majority. Examination of the evidence, however, reveals that support for this hypothesis is largely circumstantial and little or no direct evidence of a preore meteoric component, such as oxygen isotopic composition of carbonates or oxidized red earth in ore breccias, has yet been recognized. Similarly, the complex alternation between brecciation and mineralization documented so admirably for the Polish Upper Silesian district has not yet been recognized as a dominant feature in other MVT districts, leaving open the possibility that this process may not be generally applicable in the development of MVT ore-hosting breccias. The truth, indeed, may lie somewhere between these two extremes with both processes contributing, in differing proportions, to most or even all MVT districts. The presence of these two hypotheses, however, can be used to advantage in directing future research into both MVT deposits and modern karst systems.

Acknowledgments. The author takes this opportunity to thank the organizers of the 1985 SEPM Symposium on Paleokarst in Golden, Colorado, for extending an invitation to participate in the program. The symposium and the associated Leadville Formation field trip led by Dick DeVoto provided a welcome and suitable venue to discuss MVT paleokarst features with hydrogeologists and speleologists. In particular, the author is grateful to Derek Ford, Jim Quinlan, and John Thrailkill for sharing their extensive experience in modern karst systems.

A very great debt is also acknowledged to the staff of the many North American and European mining and exploration companies who, over two decades, have shared their knowledge of MVT deposits with the author.

Several perceptive comments and suggestions by the author's colleagues, Hugh Dunsmore and Murray Duke, significantly improved an earlier version of this manuscript. Similarly, constructive suggestions by E.L. Ohle and the editors of this volume further refined the text and figures. To all these people, the author extends his appreciation for their time and effort.

References

Akande, S.O., and Zentilli, M., 1984, Geologic, fluid inclusion, and stable isotope studies of the Gays River lead–zinc deposit, Nova Scotia, Canada: Economic Geology, v. 79, p. 1187–1211.

Anderson, G.M., 1983, Some geochemical aspects of sulfide precipitation in carbonate rocks; *in* Kisvarsanyi, G., Grant, S.K., Pratt, W.P., and Koenig, J.W., eds., Proceedings of the International Conference on Mississippi Valley-Type Lead–Zinc Deposits: University of Missouri–Rolla, Missouri, p. 61–76.

Anderson, G.M., and Macqueen, R.W., 1982, Ore deposit models—6. Mississippi Valley-type lead–zinc deposits: Geoscience Canada, v. 9, p. 108–117.

Bastin, E.S., ed., 1939, Contributions to a knowledge of the lead and zinc deposits of the Mississippi Valley region: Geological Society of America, Special Paper 24, 156 p.

Bogacz, K., Dzulynski, S., and Haranczyk, C., 1970, Ore-filled hydrothermal karst features in the Triassic rocks of the Cracow-Silesian region: Acta Geologica Polonica, v. 20, p. 247–265.

Bogacz, K., Dzulynski, S., and Haranczyk, C., 1973, Caves filled with clastic dolomite and galena mineralization in disaggregated dolomites: Annales de la Société Géologique de Pologne, v. 43, p. 59–76.

Brockie, D.C., Hare, Jr., E.H., and Dingess, P.R., 1968, The geology and ore deposits of the tri-state district of Missouri, Kansas, and Oklahoma, in Ridge, J.D., ed., Ore deposits of the United States, v. 1: New York, The American Institute of Mining, Metallurgical, and Petroleum Engineers, Inc., p. 400–430.

Brown, J.S., ed., 1967, Genesis of stratiform lead–zinc–barite–fluorite deposits (Mississippi Valley-type deposits): Economic Geology, Monograph 3, 443 p.

Cathles, L.M., and Smith, A.T., 1983, Thermal constraints on the formation of Mississippi Valley-type lead–zinc deposits and their implications for episodic basin dewatering and deposit genesis: Economic Geology, v. 78, p. 983–002.

Crawford, J., Fulweiler, R.E., and Miller, H.W., 1969, Mine geology of the New Jersey Zinc Company's Jefferson City mine, in Papers on the stratigraphy and mine geology of the Kingsport and Mascot Formations (Lower Ordovician) of east Tennessee, State of Tennessee, Department of Conservation, Division of Geology, Report of Investigations No. 23, p. 64–75.

Dzulynski, S., 1976, Hydrothermal karst and Zn–Pb sulfide ores: Annales de la Société Géologique de Pologne, v. 46, p. 217–230.

Ford, D.C., 1982, Karstic features of the zinc–lead Main Ore deposit at Nanisivik, Baffin Island (abstr.): Geological Association of Canada–Mineralogical Association of Canada, Annual Meeting, Program with Abstracts, v. 7, p. 49.

Fulweiler, R.E., and McDougal, S.E., 1971, Bedded-ore structures, Jefferson City mine, Jefferson City, Tennessee: Economic Geology, v. 66, p. 763–769.

Gaylord, W.B., and Briskey, J.A., 1983, Geology of the Elmwood and Gordonsville mines, central Tennessee zinc district: Tennessee zinc deposit field trip guide book; Virginia Polytechnic Institute, Blacksburg, VA, Guide book 9.

Harris, L.D., 1971, A Lower Paleozoic paleoaquifer—the Kingsport Formation and Mascot dolomite of Tennessee and southwest Virginia: Economic Geology, v. 66, p. 735–743.

Heyl, Jr., A.V., Agnew, A.F., Lyons, E.J., and Behre, Jr., C.H., 1959, The geology of the Upper Mississippi Valley zinc–lead district: US Geological Survey, Professional Paper 309, 310 p.

Hill, W.T., 1969, Mine geology of the New Jersey Zinc Company's Flat Gap mine at Treadway in the Copper Ridge district, in Papers on the stratigraphy and mine geology of the Kingsport and Mascot Formations (Lower Ordovician) of east Tennessee; State of Tennessee, Department of Conservation, Division of Geology, Report of Investigations No. 23, p. 76–90.

Hill, W.T., Morris, R.G., and Hagegeorge, C.G., 1971, Ore controls and related sedimentary features at the Flat Gap mine, Treadway, Tennessee: Economic Geology, v. 66, p. 748–756.

Hoagland, A.D., Hill, W.T., and Fulweiler, R.E., 1965, Genesis of the Ordovician zinc deposits in east Tennessee: Economic Geology, v. 60, p. 693–714.

Jackson, S.A., and Beales, F.W., 1967, An aspect of sedimentary basin evolution: the concentration of Mississippi Valley-type ores during the late stages of diagenesis: Bulletin of Canadian Petroleum Geology, v. 15, p. 393–433.

Kendall, D.L., 1960, Ore deposits and sedimentary features, Jefferson City mine, Tennessee: Economic Geology, v. 55, p. 985–1003.

Kisvarsanyi, G., Grant, S.K., Pratt, W.P., and Koenig, J.W., eds., 1983, Proceedings of the International Conference on Mississippi Valley-Type Lead–Zinc Deposits, University of Missouri–Rolla, Missouri, 603 p.

LeGrand, H.E., and Stringfield, V.T., 1971, Tertiary limestone aquifer system in the southeastern states: Economic Geology, v. 66, p. 701–709.

McCormick, J.E., Evans, L.L., Palmer, R.A., Rasnick, F.D., Quarles, K.G., Mellon, W.V., and Riner, B.G., 1969, Mine geology of the American Zinc Company's Young mine, in Papers on the stratigraphy and mine geology of the Kingsport and Mascot Formations (Lower Ordovician) of east Tennessee; Nashville, State of Tennessee, Department of Conservation, Division of Geology, Report of Investigations No. 23, p. 45–52.

McCormick, J.E., Evans, L.L., Palmer, R.A., and Rasnick, F.D., 1971, Environment of the zinc deposits of the Mascot–Jefferson City district, Tennessee: Economic Geology, v. 66, p. 757–762.

Mouat, M.M., and Clendenin, C.W., 1977, Geology of the Ozark Lead Company mine, Viburnum Trend, southeast Missouri: Economic Geology, v. 72, p. 398–407.

Oder, C.R.L., and Hook, J.W., 1950, Zinc deposits in the southeastern states, in Snyder, F.G., ed., Symposium on mineral resources of the southeastern United States: Knoxville, Tennessee, University of Tennessee Press, p. 72–87.

Ohle, E.L., 1959, Some considerations in determining the origin of ores of the Mississippi Valley-type: Economic Geology, v. 54, p. 769–789.

Ohle, E.L., 1980, Some considerations in determining the origin of ores of the Mississippi Valley-type, Part II: Economic Geology, v. 75, p. 161–172.

Ohle, E.L., 1985, Breccias in Mississippi Valley-type deposits: Economic Geology, v. 80, p. 1736–1752.

Olson, R.A., 1977, Geology and genesis of zinc–lead deposits within a late Proterozoic dolomite, northern Baffin Island, N.W.T.: Unpub. Ph.D. thesis, University of British Columbia, Vancouver, 371 p.

Olson, R.A., 1984, Genesis of paleokarst and stratabound zinc–lead sulfide deposits in a Proterozoic dolostone, northern Baffin Island, Canada: Economic Geology, v. 79, p. 1056–1103.

Rhodes, D.A., Lantos, E.A., Lantos, J.A., Webb, R.J., and Owens, D.C., 1984, Pine Point orebodies and their relationship to the stratigraphy, structure, dolomitization, and karstification of the Middle Devonian barrier complex: Economic Geology, v. 79, p. 991–1055.

Rogers, R.K., and Davis, J.H., 1977, Geology of the Buick mine, Viburnum Trend, Southeast Missouri: Economic Geology, v. 72, p. 372–380.

Sangster, D.F., 1983, Mississippi Valley-type deposits: a geological mélange, in Kisvarsanyi, G., Grant, S.K., Pratt, W.P., and Koenig, J.W., eds., Proceedings of the International Conference on Mississippi Valley-Type Lead–Zinc Deposits: University of Missouri–Rolla, Missouri, p. 7–19.

Sass-Gustkiewicz, M., 1975, Stratified sulfide ores in karst cavities of the Olkusz mine (Cracow-Silesian region, Poland): Annales de la Société Géologique de Pologne, v. 45, p. 63–68.

Sass-Gustkiewicz, M., Dzulynski, S., and Ridge, J.D., 1982, The emplacement of zinc–lead sulfide ores in the Upper Silesian district—a contribution to the understanding of Mississippi Valley-type deposits: Economic Geology, v. 77, p. 392–412.

Snyder, F.G., and Gerdemann, P.E., 1968, Geology of the southeast Missouri lead district, in Ridge, J.D., ed., Ore deposits of the United States: New York, The American Institute of Mining, Metallurgical, and Petroleum Engineers, Inc., p. 327–358.

Sverjensky, D.A., 1984, Oil field brines as ore-forming solutions: Economic Geology, v. 79, p. 3–37.

Sweeney, P.H., Harrison, E.D., and Bradley, M., 1977, Geology of the Magmont mine, Viburnum trend, southeast Missouri: Economic Geology, v. 72, p. 65–371.

Ulrich, E.O., 1931, Origin and stratigraphic horizon of the zinc ore in the Mascot district of east Tennessee (abstr.): Washington Academy of Science Journal, v. 21, p. 31–32.

5
Blackened Limestone Pebbles: Fire at Subaerial Unconformities

EUGENE A. SHINN and BARBARA H. LIDZ

Abstract

Irregularly shaped blackened limestone pebbles mixed with similar but unblackened material at unconformities have long presented a mystery to geologists examining Tertiary and Holocene limestones in the Caribbean. How can irregularly shaped limestone pebbles showing no signs of lateral transport be mixed together, especially when a distant or underlying source is invariably absent?

The blackened pebbles generally are composed of soilstone crust, lightly lithified grainstone, or multicomponent limestones, and may occur at subaerial unconformities in marine or subaerial (eolian) limestones. The most common examples occur as multicolored breccias in karst potholes, which are abundant throughout the Caribbean.

We propose that selective blackening is caused by "instantaneous" forest fire heating. Simple experiments showed that thorough blackening can occur in one-half hour at temperatures between 400° and 500° C. Heating experiments showed that only those limestones that are commonly black in nature blackened when heated to temperatures similar to those of forest fires. Blackening, restricted mainly to individual sand-size skeletal grains, also occurs under subtidal conditions, and the product of this process should not be confused with fire-blackened limestone pebbles. Correct identification can be useful for distinguishing between submarine diastems and unconformities and subaerial unconformities in ancient limestones.

Introduction

Dark grey to black, irregularly shaped limestone pebbles of millimeter to centimeter size occur at subaerial disconformities and unconformities in limestones of Tertiary and Holocene age throughout the Caribbean. The origin of these pebbles has long been a puzzle to local geologists. In south Florida, many black pebbles are fragments of soilstone crusts often called *calcrete*,[1] but they also include blackened coral, lightly lithified grainstones, or multicomponent limestones. Most blackened pebbles occur in a matrix of similar but unblackened irregularly shaped pebbles. These multicolored breccias are generally concentrated in karst potholes. The intimate mixture of blackened and unblackened irregularly shaped pebbles is puzzling, because an underlying or adjacent grey or black limestone source is always absent, and irregular shapes argue against extensive lateral transport. Although it has been argued that the black pebbles are erosional remnants of a missing overlying bed, we believe they are produced essentially in situ by forest fire-induced heating. Such an interpretation may have useful implications for limestone containing black pebbles in ancient limestone, especially in the Mesozoic, where they are extremely common yet often go undescribed. We point out, however, that there is also a well-documented subtidal blackening process which affects molluscs and produces the familiar "salt-and-pepper" sands phenomenon. Subtidal blackened fossils and sand-size car-

---

[1]The term *soilstone crust* is preferred because of its formation beneath a peaty soil. Petrographically, such crusts are similar to calcrete or caliche, which is normally associated with arid climates and lack of soil (see Multer and Hoffmeister 1968, and Robbin and Stipp 1979).

bonate grains often accumulate at disconformities and unconformities (either submarine
or subaerial unconformities subsequently
drowned by rising sealevel) and can easily be
confused with those of subaerial fire origin.
Correct distinction may therefore provide useful clues for identification of subaerially exposed unconformities and disconformities in
the geologic record.

In this chapter, we will describe some observations and simple experiments relating to the
origin of black pebbles. Later, we will discuss
the origin of subtidal blackened grains and fossils, show how they are often mixed with subaerially formed black pebbles, and discuss the
dangers of misidentifying their origin. Furthermore, we note that geochemical analysis is
usually equivocal for distinguishing between
subtidal and subaerially produced black pebbles, grains, and fossils.

Field Observations and Some Experimental Results

Black pebbles can be found along subsurface
Pleistocene unconformities in south Florida
(Perkins 1977) and the Bahamas (Beach and
Ginsburg 1980). They also occur on the Holocene subaerial unconformity[2] presently
forming on the surface of Pleistocene limestones throughout south Florida and much of
the Caribbean. These unconformities, easily
mistaken for diastems, invariably are marked
by the presence of soilstone crusts or calcretes
(Kornicker 1958, Multer and Hoffmeister 1968,
Barthel 1974, Perkins 1977, Robbin and Stipp
1979, Beach and Ginsburg 1980, Pierson 1982,
Strasser 1984, and Williams 1985).

The most common black pebbles consist of
angular fragments of soilstone crusts (Figs.
5.1A, B, C). They often occur cemented within
unblackened in situ soilstone crusts, as shown
in Figure 5.1B, or just above such crusts (Fig.
5.1A). The more striking examples occur in so-

---

[2]We recognize that this unconformity is part Pleistocene and part Holocene. Subaerial erosion occurred during the last Pleistocene sealevel drop;
however, the distinguishing features—soilstone crust
and blackened pebbles—developed mainly during
the past 5,000 to 10,000 years (Robbin and Stipp
1979).

lution holes where they are imbedded in a cemented multicolored breccia composed of unblackened soilstone pebbles, both blackened
and unblackened limestone fragments, corals,
and mollusc shells (Figs. 5.2A, B). The vertical
walls of these solution pits are often lined with
laminated reddish-brown soilstone crust 1 to 3
cm thick, which is usually continuous with adjacent horizontal crusts. Most black pebbles occur in association with thin soilstone crusts often
sandwiched in marine limestone sequences.
They also occur in young eolian carbonates 10
to 20 m above sealevel. Figure 5.3A shows black
pebbles of cemented pelletal and ooid grainstone in a soilstone-lined solution pit approximately 20 m above sealevel. Figure 5.3B and
C are photomicrographs showing the contact
between the black pebbles and unblackened
matrix in which they are imbedded. Brown in
situ soilstone crusts sometimes contain discontinuous blackened laminae 1 to 2 m thick (Fig.
5.4). Significance of these black laminae and
their relationship with black pebbles will be
discussed later.

Although there are also rounded and bored
black pebbles that have been reworked in a
marine environment, the main thrust of this
chapter is to explain how irregularly shaped
fractured pebbles, which clearly have experienced little transport, can become intermixed
with unblackened counterparts.

Chance observations around outdoor fire-
places at campgrounds throughout the Florida
Keys and the Bahamas suggested the fire hypothesis. Native limestones used for campfire
containment are generally blackened and show
a distinct color gradient and selectivity. They
are darkest toward the fire and grade over a
distance of several centimeters to their normal
color. Soilstone crust fragments show the most
darkening, and some limestones resist blackening.

To test whether fires could explain the mul-
titude of shades and colors in nature, we first
performed a simple experiment. Various pebbles of Pleistocene limestone, modern corals,
shells, and soilstone crusts were placed in an
open oil drum and covered with a 30-cm-thick
layer of seaweed which was set afire. When the
fire cooled a few hours later, we examined the
pebbles and found that, as in multicolored
breccia-fills, only certain pebbles blackened.

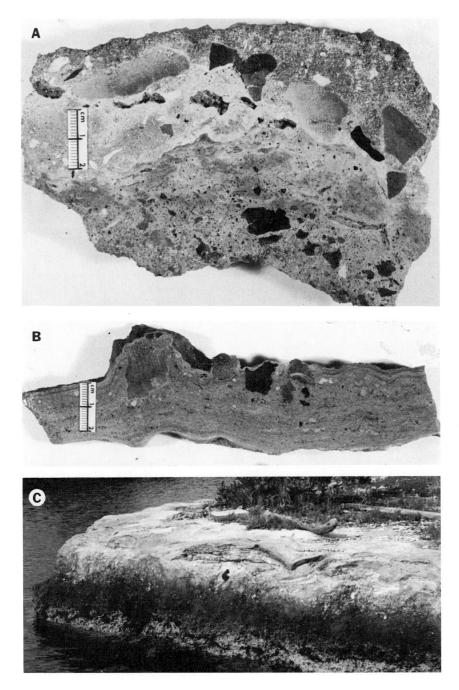

FIGURE 5.1A. Poorly cemented soilstone crust with blackened lithoclasts from Ramrod Key, Florida. Note shading from dark to light in lithoclast above scale. Matrix is angular, light tan to rusty brown in color. 5.1B. Angular blackened lithoclasts in brown laminated soilstone crust from Ramrod Key, Florida.

5.1C. Laminated soilstone crust exposed at edge of artificial canal on Ramrod Key where overlying soil has been stripped away. Thickness of crust at center of photo is about 8 cm. Top of outcrop to water level is about 45 cm.

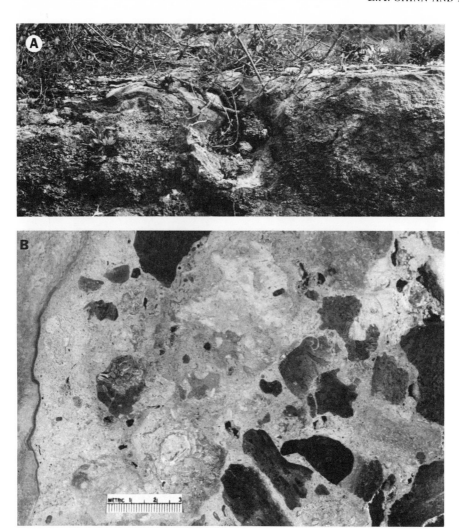

Figure 5.2A. A soilstone crust-lined solution pit exposed alongside artificial canal on Ramrod Key, Florida. Note soil and poorly cemented breccia in solution pit. Depth of solution pit is approximately 40 cm. 5.2B. Well-cemented multicolored breccia in solution pit from top of Q3 Unit of Perkins (1977) approximately 20 ft (6 m) below sealevel (from quarry tailings on Big Pine Key, Florida). Note soilstone crust lining at far left of specimen. Blackened fragments include soilstone crust and fossiliferous Key Largo Limestone.

Soilstone crust blackened most thoroughly, followed by coral, shells, and those poorly cemented grainstones which still contain considerable aragonite. Well-cemented Pleistocene limestones with essentially no aragonite did not blacken. To test the idea further, a series of tests was run in a jeweler's kiln. The results are summarized in Table 5.1, and selected examples of experimentally blackened material are shown in Figures 5.5 and 5.6. Thin sections of experimentally blackened soilstone crust appear identical to natural blackened soilstone crust and pebbles (Fig. 5.7). It is impossible to distinguish any difference between normal reddish-brown and blackened soilstone crusts in black-and-white photographs. Laminae of darker (reddish-brown) cryptocrystalline material blacken more than the light brown cryptocrystalline matrix.

The data in Table 5.1 show that blackening

begins first in modern ooids heated 2 h between 232° and 288° C. Coral, both modern and Pleistocene, began blackening after 3 h of exposure at 343° C. After 4 h at 399° C, all specimens began blackening but curiously not the soilstone crusts. After 6 h at 510° C, samples attained maximum blackening. Temperatures above 510° C caused lightening of the outer surface, and after 8 h at 566° C, all samples had calcined. Visual aspects of the blackening process are shown in Figures 5.5A, B, and C. In the oil-drum and experiments subsequent to those listed in Table 5.1, we found that maximum darkening could be obtained in just one-half hour. Although temperature was not measured in the oil-drum experiment, additional experiments in the kiln showed that maximum darkening could be obtained at around 400° C in just one-half hour. Results of these later experiments are shown in Figures 5.5A, B, and C.

Naturally blackened and unblackened Pleistocene limestone and experimentally blackened Holocene crust were subjected to LECO and Rock-EVAL pyrolysis, as were subtidal salt-and-pepper sands from Florida Bay near Crane Key (1.5 m of water) and from the base of an 8-m-thick Holocene sediment section (30 cm below sealevel) (Table 5.2).

Another experiment considered significant to our hypothesis was performed on a cemented breccia sample. A slab of the same breccia shown in Figure 5.2B was photographed (Fig. 5.6A), wrapped in aluminum foil (to retard oxidation), and heated in an oven for one-half hour at approximately 400° C. The experiment caused blackening of some previously unblackened breccia pebbles and the in situ soilstone crust which lined the breccia-filled pit (Fig. 5.6B).

Discussion

The material which causes blackening is difficult to discern in thin section. Figure 5.7 is a thin section of artificially blackened soilstone crust. It is identical in appearance to its unblackened counterpart except that what were darker brown areas have turned dark grey to black. Finely disseminated organic matter which causes the brown color in soilstone crust (Multer and Hoffmeister 1968) apparently chars to dark grey or black. Thin sections of naturally blackened soilstone crust and pebbles look identical to the artificially blackened samples. Even in thin sections of heat-blackened coral, discrete black particles were impossible to discern. Scanning electron microscopy might reveal individual charred particles; however, its use was beyond the scope of this study.

Figures 5.3B and C are photomicrographs of two different black pebbles from the breccia shown in Figure 5.3A. Note in Figure 5.3B that blackened ooid grains surrounded by clear calcite cement give the pebble its overall black color. The darkened pebble in Figure 5.3C, on the other hand, is a fragment of soilstone crust. We think it significant that the black pebbles in the breccia shown in Figures 5.3A, B, and C occur in late Pleistocene eolian deposits situated approximately 20 m above sealevel. These 20- to 30-m-thick cross-bedded eolian dunes were deposited shortly before or during the last Pleistocene glacial lowering of sealevel, and there is no evidence of a higher sealevel stand in this area since sealevel rose again during the Holocene. The age and position of these black pebbles therefore are compatible with our forest fire hypothesis, and blackening mechanisms involving subtidal or hypersaline conditions can be ruled out.

The data from LECO and Rock-EVAL analyses of experimentally blackened material are interesting but equivocal (Table 5.2). The control sample contained 0.53 wt % organic carbon by LECO and 0.89 wt % organic carbon by Rock-EVAL. The heat-blackened sample, however, contained slightly less, 0.46 wt % and 0.19 wt %, respectively. Surprisingly, the naturally blackened Pleistocene pebbles from the breccia shown in Figure 5.2B contained more carbon than its unblackened counterpart, although the values are less than for both the control and experimentally blackened Holocene crusts. It should be pointed out, however, that these Pleistocene pebbles are not of soilstone crust but instead are coralline grainstone, so a direct comparison cannot be made.

Surprisingly, the data for unblackened subtidal skeletal grains collected in Florida Bay show significantly more total organic carbon than the blackened grains analyzed from the same sediment sample. The Marquesas Keys samples, which were older (buried beneath 8 m of lime mud and silt) and at a depth of 30 m below sealevel, showed more carbon in the black sample than in their unblackened counterpart. We do not know how to explain these few data adequately but do not find it surprising that heated crust samples apparently contain less carbon than unheated samples. Carbon was probably lost by vaporization during heating.

We do not know why darkening was obtained so quickly at only 400° C in the later experiments, when the previous work listed in Table 5.1 indicated that 510° C for 6 h was required to obtain maximum blackening. There are at least two possibilities: (1) All samples listed in Table 5.1 were small sawed samples approximately 1 cc in size. The later experiments (see Figs. 5.5A, B, C) utilized larger hand specimens. (2) The thermometer and heat control devices, which were later changed, were defective. At best, temperature control was no better than

5% and likely as low as 10%. Regardless of the precision, it seems likely that temperature is more critical than time.

Although all our experiments and observations support the fire hypothesis, we nevertheless feel that it would be fruitful to determine more precisely the temperature–time relationships. To obtain the data, more expensive and sophisticated temperature control, not now available to us, will be required. In addition, more geochemical analyses should be aimed at identifying the blackening agent. If there is a unique chemical signature, then it might lead to development of a simple geochemical test that could distinguish subaerial from submarine blackened pebbles and grains.

All our observations, simple experiments, and carbon analyses support the fire hypothesis. We thus find it interesting that Barthel (1977), who also noted anthropogenic blackening of limestone in the Florida Keys, thought fire-blackening a hindrance to discovering the true cause of blackening. Certainly there has been much burning associated with land-de-

Figure 5.3A. Blackened lithoclasts in karst solution pit breccia in cross-bedded eolian grainstone approximately 20 m above sealevel. Eolian host rock is late Pleistocene and has never been submerged. Note fractures in black pebble, indicating fracturing took place after emplacement in breccia. Length of key 6 cm. 5.3B. Thin section of contact between oolitic black pebble and reddish-brown eolian grain

stone to left. Blackening of the oolitic grainstone pebble is due to blackening of individual ooid grains. Cement is clear blocky calcite spar of vadose origin. 5.3C. Another pebble from the same exposure showing contact between blackened soilstone crust pebble at right and reddish-brown carbonate grainstone (highly altered by soil processes) to left. Scale bar 400 μm.

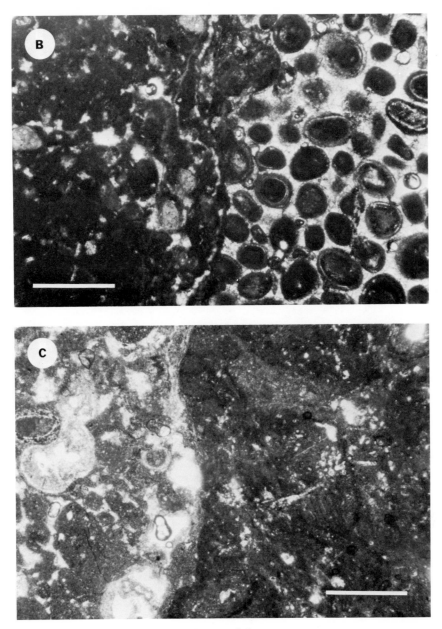

FIGURE 5.3

velopment projects. One of us (EAS) observed many forest fires in the Florida Keys during the late 1940s and early 1950s, when that was a common land-clearing technique. There is evidence, however, of natural forest fires. Figure 5.8 is a photomicrograph of a charred twig or wood fragment imbedded within soilstone crusts from Ramrod Key, Florida. The C^{14} work of Robbin and Stipp (1977) showed that these crusts range in age from about 5000 y to the present. Inclusion of charcoal throughout the crusts therefore indicates that natural fires were occurring in the area a few thousand years ago, before influx of modern man. We are concerned, however, that we often see charred wood fragments in unblackened laminated soilstone crust. If our hypothesis is correct, then this material is probably the result of minor grass fires which did not generate sufficient heat to blacken thoroughly the crust in which they became incorporated. Some crusts in this area, however, do contain blackened laminae,

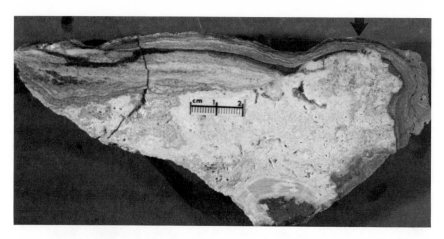

FIGURE 5.4. Detail of soilstone crust from quarry on Long Key (Florida Keys) showing normal brown laminations between two blackened laminae. Top black layer (arrow) is discontinuous. Basal black layer grades to normal brown color at right where layer enters depression in underlying rock. Unblackened portion of this layer was protected from fire in depression.

TABLE 5.1. Time, temperature heating experiments in jeweler's kiln

Time (h)	Temperature[b] (°C)	Materials tested[a]						
		Holocene			–	Pleistocene		
		Coral	Crust	Ooids		Coral	Crust	Oolite
1	177	NC	NC	NC		NC	NC	NC
2	288	NC	NC	◐		NC	NC	NC
3	343	◐	NC	◐		◐	NC	NC
4	399	◐	NC	◐		◐	◐	◐
5	454	◐	◐	◐		◐	◐	◐
6	510	●	●	●		●	●	●
7	566	◑	◑	◑		◑	◑	◑
8	621	○	○	○		○	○	○

NC = No change
◐ = Beginning to darken
● = Maximum darkening
◑ = Beginning to lighten
○ = Almost all white

[a]New samples used for each timed test (total 48 samples).
[b]Oven temperature stabilized before each timed test.
Note: Ooids were the first to begin darkening (after 2 h at 288° C) and Holocene soilstone crust, which is the most common black pebble in nature, was the last to begin darkening (5 h at 454° C). All samples attained maximum blackening after 6 h at 510° C. It should be noted that a jeweler's kiln as used here has imprecise temperature control, and accuracy is probably between 5% and 10%.

FIGURE 5.5A. Modern ooids heated for one-half hour in kiln. (a) Control, not heated. (b) Heated to 176° C. (c) heated to 400° C. (d) heated to 625° C. Note in (d) that outer surface of ooids is calcined (white color) but inner parts of individual ooids are dark grey to black. 5.5B. Fragment of well-cemented coralline limestone (a) and *Porites* coral (b) heated together at approximately 400° C for one-half hour. Note only the coral blackened. 5.5C. Soilstone crusts: (a) heated to approximately 400° C for one-half hour. (b) Unheated.

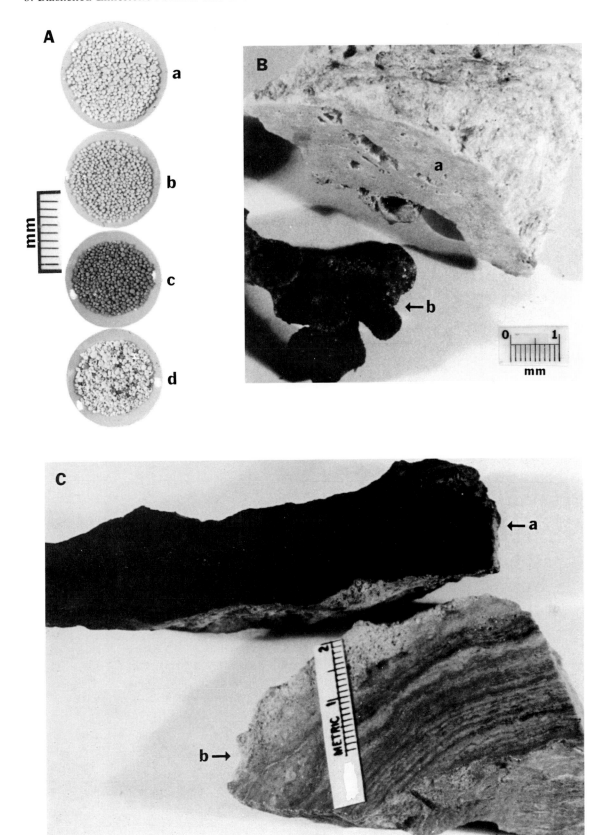

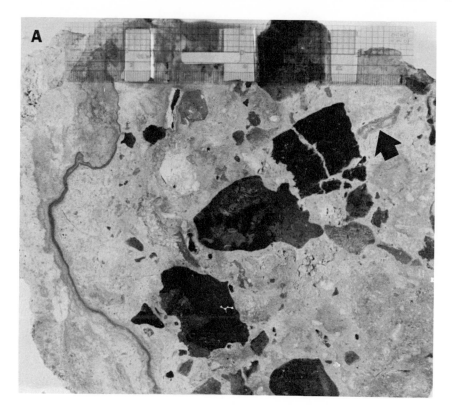

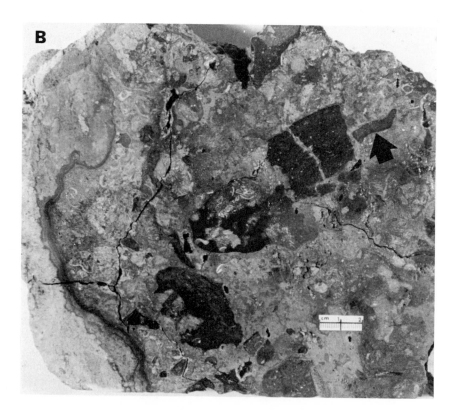

which we think were caused by fire before deposition of subsequent unblackened layers (Fig. 5.4).

We consider the experiment shown in Figures 5.6A and B to be particularly significant. The experiment shows that even though breccia components experienced the same temperature conditions, only some of the previously unblackened pebbles became blackened. Soilstone crust lining the breccia accumulation also turned black. The fractured black pebble in Figure 5.6A is consistent with the in situ forest fireheating hypothesis. The pebble was clearly fractured in place, possibly during the same heating event that blackened it. It is probable, however, that it was blackened before emplacement and was subsequently fractured by pedogenic processes. If this was not the case, the pebble next to it which blackened experimentally would have already been blackened by the heat that colored the fractured pebble. Although probably transported after blackening, its angularity argues against significant transport.

Figure 5.1A shows two pebbles which display a pronounced color gradient. The one at the upper left indicates a heat source from below, whereas the one to the right of center indicates a source from above. We believe the gradient was caused by heat, more or less in situ, but that the upper left pebble has been rotated by pedogenic processes, possibly roots, or burrowers, or some other process of which we are unaware.

Ward et al. (1970) described blackened soilstone crust pebbles mixed with unblackened crust pebbles in and around a hypersaline pond on Isla Mujeres off Mexico, and concluded that the color was caused by organic matter, probably algae, which had been darkened by reducing hypersaline waters. Analyses showed that the blackened fragments contained no more iron, manganese, or sulfur than nonblackened fragments (Ward et al. 1970). Analyses by LECO carbon analyzer showed a slightly higher (2000 to 3000 ppm) concentration of organic carbon (versus 1000 to 2000 ppm) in unblackened soilstone crust pebbles. Ward et al. (1970) also concluded that there is no evidence of crusts being blackened today but that blackening occurred during the early Holocene, and the crusts were subsequently fractured by desiccation to form angular pebbles. We suggest that the blackening was caused by forest fires and that the recent relative rise in sealevel created the ponds which inundated both blackened and unblackened crusts and pebbles with hypersaline waters. Our analyses in Table 5.2 would suggest little relationship between total organic carbon and color. The total organic carbon data of Ward et al. (1970) for blackened and unblackened crusts overlap and are not considered sufficient evidence that blackening was caused by reducing hypersaline waters. Although reducing waters can cause blackening, our simple experiments and occurrences of black pebbles in young eolian deposits demonstrate that hypersaline water is not a necessary condition.

Although we may be the first to propose an "instantaneous" fire blackening origin, we are not the first to implicate fire as a cause of limestone blackening. Strasser (1984) noted the possible role of fire in Florida and the Bahamas but concluded that "the blackening substances are transported by percolating waters" (p. 1103). He concluded from his analyses that traces of organic matter in the 0.1 to 0.5% range caused blackening. Although Strasser also listed blackening by anoxic bottom water, he concluded that in ancient limestones blackened pebbles indicate subaerial exposure and the former presence of islands and coastlines. We would like to stress that indeed there are grains

◁

FIGURE 5.6A. Well-cemented multicolored breccia from solution pit in Pleistocene Key Largo Formation in Florida Keys. Note blackened angular lithoclasts. Also note in situ fracturing of black pebble. Laminated soilstone crust to left was part of solution pit lining. Soilstone crust is rusty brown in color. 5.6B. Same rock as in A after heating for one-half hour at 400° C. Note darkening of fragment shown by arrow. Soilstone lining at left has turned dark grey. (Note that in black-and-white photographs, rusty brown appears the same as dark grey, thus color change from A to B cannot be fully appreciated.) Notice that some darkened lithoclasts, such as one in center of specimen, became lighter upon heating.

FIGURE 5.7. Thin section of experimentally blackened soilstone crust (one-half hour at 400° C). Crude laminations of originally reddish-brown cryptocrystalline calcite (color imparted mainly by organic matter) turned black, while less darkly stained portions between dark laminae only turned grey. In hand spec- imen, as shown in Figure 5.5C, the rock looks com- pletely black. A thin section photo of unheated crust (control) is not shown because in black-and-white photography no differences could be seen (plain light). Scale bar 400 μm.

TABLE 5.2. Organic carbon content and pyrolysis assay of related pairs of light- and dark-colored carbonate crusts, lithoclasts, and salt-and-pepper carbonate sands

Sample description/ alteration state	LECO combustion organic carbon wt %	Rock-EVAL TOC estimate wt %	S_1 mg/g	S_2 mg/g	S_3 mg/g	Tmax °C
Subaerial						
Soilstone crust						
Not heated	0.53	0.89	0.07	2.00	6.42	435
Heated	0.46	0.19	0.01	0.07	1.29	474
Breccia pebbles						
Brown	0.04	0.0	0.0	0.01	0.73	—
Black	0.15	0.14	0.0	0.0	0.27	—
Subtidal						
Crane Key salt-and- pepper sands						
Not dark	n.a.	1.24	0.45	4.15	3.62	432
Dark	n.a.	0.41	0.04	0.71	1.17	426
Marquesas salt-and- pepper sands						
Not dark	n.a.	0.13	0.04	0.11	1.75	400
Dark	n.a.	0.24	0.02	0.06	1.39	415

n.a. = not analyzed
— = insufficient S_2 yield for Tmax determination

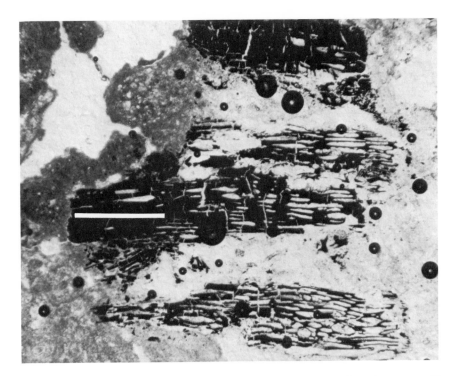

FIGURE 5.8. Thin section of unblackened soilstone crust from Ramrod Key, Florida showing charred twigs (charcoal). Charcoal scattered throughout 5000-year-old crust indicates natural forest fires occurred be-fore the invasion of modern man. Circular objects are bubbles created during thin section preparation. Scale bar = 400 μm.

blackened under subtidal conditions, and they should not be confused with subaerially blackened pebbles.

Submarine Blackening and Accumulation of Salt-and-Pepper Sands

Intertidal and subtidal blackening of fossils and other carbonate materials has been reported from numerous areas of Holocene deposition (for example, Van Straaten 1954, Ginsburg 1957, Houbolt 1957, Bush 1958, Sugden 1966, Macintyre 1967, Maiklem 1967, Shinn et al. 1969, Pilkey et al. 1969, Shinn 1973, Wagner and van der Togt 1973). On the basis of equivocal data, these investigators have cited both organic matter and/or iron or manganese sulfide as the blackening agent. Pilkey et al. (1969) proved the necessity of reducing conditions in an experiment in which modern and fossil shells from the beach were buried in a North Carolina salt marsh. In just three weeks, most modern shells turned black. Fossil shells did not blacken. Shells placed in distilled water through which H_2S was bubbled also turned black. Color was thought to be caused by reduced organic matter.

Subtidally blackened grains and fossils often occur together as a kind of salt-and-pepper accumulation that may or may not include cross-bedding. Salt-and-pepper sands commonly accumulate just above a contact, whether it be a diastem produced by submarine hardground formation or a drowned subaerial unconformity. Laterally migrating tidal channels erode blackened grains from their outer banks and redeposit them with unblackened counterparts, usually as a lag deposit above an underlying unconformity. Some channels that do not reach underlying unconformities produce their own diastems as they migrate.

Migrating mudbanks also cause mixing of blackened and unblackened materials. Fischer (1961) and Pilkey et al. (1969) showed how blackened grains concentrate as a result of landward migration of barrier islands along the eastern United States.

The blackened grains in Table 5.2 (Florida Bay near Crane Key and the Marquesas core samples) are considered to be of subtidal origin. Recognizing that blackening can occur both above and below the water, it seems important, then, that the two different kinds be distinguishable. Correct identification of subaerial fire-blackened pebbles on an unconformity could be useful for recognizing subaerial unconformities. Recognition of marine black grains may accurately distinguish submarine diastems. Difficulties are likely to arise, however, when a subaerial unconformity is inundated by a marine transgression, thus mixing two kinds of blackened material together. The following criteria are considered useful for separating the two.

Subtidal Grains

1. Most grains are rounded and predominantly in the sand-size range.
2. Grains are often whole and skeletal, such as foraminifera, or they may be broken fossil fragments, such as molluscan pieces, all imbedded in a light-colored matrix.
3. Blackened grains occur with unblackened counterparts, as do subaerial pebbles, but they are in the sand-size range and produce a mixture resembling salt and pepper.
4. They frequently occur above an unconformity or diastem. Subaerial black pebbles also occur above unconformities but not diastems.
5. Accumulations are graded (fining upward) and the ratio of black to white grains decreases upward.
6. Black grains are current-sorted and sometimes cross-laminated.

Subaerial Fire-Blackened Pebbles

1. Black material is usually in the pebble-size range.
2. Blackened lithoclasts are generally angular and some show evidence of having fractured in situ.
3. Black pebbles often occur in multicolored breccias with no particular grading.
4. Black pebbles may occur at the top of pedogenic sequences, especially soilstone crusts.

Crusts may contain patchily distributed, millimeter-thick black layers.
5. Some individual blackened pebbles may show gradation of blackening from white through grey to black.

Many of these criteria overlap; therefore, the entire sedimentary sequence should be considered when making an environmental interpretation. Since present chemical analyses are of little use in correctly identifying the environment or even the cause of blackening, it is proposed that more detailed study is needed. A simple chemical technique, once developed, could be invaluable for distinguishing subaerial and subtidal environments.

Conclusions

Simple experiments and observations have shown that pebbles of limestone (soilstone crust, coral, poorly cemented grainstone, and molluscs) can blacken almost instantaneously during natural forest fires. Temperatures between 400° and 500° C readily cause blackening and can account for the bulk of angular black pebbles observed in Pleistocene and Holocene limestones throughout the Caribbean.

The fact that fire-blackened pebbles occur in young rocks where their processes of formation can be observed suggests that they should be equally abundant in ancient rocks. Once observed and analyzed, the presence of ancient black pebbles could provide valuable clues to the environment of deposition and the nature of underlying unconformities.

Acknowledgments. The authors thank J. Harold Hudson, Jack L. Kindinger, and Robert B. Halley for aid in collecting samples and for in-field discussion of their significance. Randolph P. Steinen, Nealy Bostick, and Judith Parrish kindly reviewed the manuscript and offered comments and suggestions which greatly improved it. Appreciation is extended to Daniel M. Robbin for conducting the majority of the heating experiments, Renato Diaz for making thin sections and printing photographs, George E. Claypool and Ted Daws for LECO carbon analysis, and Charles W. Holmes for constructive discussions. Finally, the authors are indebted to André Strasser for having stimulated our interest in the subject.

References

Barthel, K.W., 1974, Black pebbles, fossil and Recent, on and near coral islands: Proceedings, Second International Coral Reef Symposium, v. 2, Great Barrier Reef Committee, Brisbane, p. 395–399.

Beach, D.K., and Ginsburg, R.N., 1980, Facies succession of Pliocene–Pleistocene carbonates, northwestern Great Bahama Bank: American Association of Petroleum Geologists Bulletin, v. 64, no. 10, p. 1634–1642.

Bush, J., 1958, The foraminifera and sediments of Biscayne Bay, Florida, and their geology: Unpubl. Ph.D. dissertation, University of Wisconsin, Madison, WI, 127 p.

Fischer, A.G., 1961, Stratigraphic record of transgressing seas in light of sedimentation on Atlantic Coast of New Jersey: American Association of Petroleum Geologists Bulletin, v. 45, p. 1713–1721.

Ginsburg, R.N., 1957, Early diagenesis and lithification of shallow-water carbonate sediments in southern Florida, in LeBlanc, R.J., and Breeding, J.G., eds., Regional aspects of carbonate deposition: Society of Economic Paleontologists and Mineralogists Special Publication 5, 178 p.

Houbolt, J.J.H.C., 1957, Surface sediments of the Persian Gulf near the Qatar Peninsula: Unpubl. Ph.D. dissertation, University of Utrecht, the Hague, Mouton and Company, 113 p.

Kornicker, L.S., 1958, Bahamian limestone crusts: Transactions, Gulf Coast Association of Geological Societies, v. 8, p. 167–170.

Macintyre, I.G., 1967, Recent sediments off the west coast of Barbados, W.I.: Unpubl. Ph.D. dissertation, McGill University, Montreal, Canada, 169 p.

Maiklem, W.R., 1967, Black and brown speckled foraminiferal sand from the southern part of the Great Barrier Reef: Journal of Sedimentary Petrology, v. 37, no. 4, p. 1023–1030.

Multer, H.G., and Hoffmeister, J.E., 1968, Subaerial laminated crusts of the Florida Keys: Geological Society of America Bulletin, v. 79, p. 183–192.

Perkins, R.D., 1977, Depositional framework of Pleistocene rocks in south Florida, in Enos, Paul, and Perkins, R.D., eds., Geological Society of America Memoir 147, p. 131–198.

Pierson, B.J., 1982, Cyclic sedimentation, limestone diagenesis and dolomitization in Upper Cenozoic carbonates of the southeastern Bahamas: Unpubl. Ph.D. dissertation, University of Miami, Coral Gables, FL, 343 p.

Pilkey, O.H., Blackwelder, B.W., Doyle, L.J., Estes, E., and Terlecky, M.P., 1969, Aspects of carbonate sedimentation on the Atlantic Continental Shelf off the southern United States: Journal of Sedimentary Petrology, v. 39, no. 2, p. 744–768.

Robbin, D.M., and Stipp, J.J., 1979, Depositional rate of laminated soilstone crusts, Florida Keys: Journal of Sedimentary Petrology, v. 49, no. 1, p. 0175–0180.

Shinn, E.A., 1973, Carbonate coastal accretion in an area of longshore transport, NE Qatar, Persian Gulf, in Purser, B.H., ed., The Persian Gulf: New York, Springer-Verlag, p. 179–191.

Shinn, E.A., Lloyd, R.M., and Ginsburg, R.N., 1969, Anatomy of a modern carbonate tidal flat, Andros Island, Bahamas: Journal of Sedimentary Petrology, v. 39, no. 3, p. 1202–1228.

Strasser, A., 1984, Black-pebble occurrence and genesis in Holocene carbonate sediments (Florida Keys, Bahamas and Tunisia): Journal of Sedimentary Petrology, v. 54, no. 4, p. 1097–1109.

Sugden, W., 1966, Pyrite staining of pelletly debris in carbonate sediments from the Middle East and elsewhere: Geology Magazine, v. 103, p. 250–256.

Van Straaten, L.M.J.U., 1954, Composition and structure of Recent marine sediments in the Netherlands: Leidse Geol. Mededel., Leiden, deel 19, p. 1–110.

Wagner, C.W., and Van der Togt, C., 1973, Holocene sediment types and their distribution in the southern Persian Gulf, in Purser, B.H., ed., The Persian Gulf: New York, Springer-Verlag, p. 123–155.

Ward, W.C., Folk, R.L., and Wilson, J.L., 1970, Blackening of eolianite and caliche adjacent to saline lakes, Isla Mujeres, Quintana Roo, Mexico: Journal of Sedimentary Petrology, v. 40, no. 2, p. 548–555.

Williams, S.C., 1985, Stratigraphy, facies evolution, and diagenesis of Late Cenozoic limestones and dolomites, Little Bahama Bank, Bahamas: Unpubl. Ph.D. dissertation, University of Miami, Coral Gables, FL, 216 p.

6
Holocene Overprints of Pleistocene Paleokarst: Bight of Abaco, Bahamas

Kenneth A. Rasmussen and A. Conrad Neumann

Abstract

Unconformable facies relationships at the Holocene/Pleistocene interface in the Bight of Abaco lagoon, northern Bahamas, have been investigated by high-resolution seismic profiling, rock coring, and excavation of buried bedrock surfaces. Lateral variation in the expression of a single lagoonal unconformity can be explained in terms of different genetic and preservational processes affecting the Pleistocene subaerial exposure surface during Holocene sealevel rise.

The dished bedrock surface beneath the Bight of Abaco has been divided geomorphically into a central depression, a transitional slope, and a marginal terrace. Upon Holocene sealevel rise, irregular karstic central depressions (dolines) are first wet by freshwater, preserving them beneath peats or paleosols. Along the transitional bedrock upslope, pedogenic caliche crusts are buried and preserved intact during schizohaline (mixed salinity) conditions. Continued sealevel rise ultimately results in the stable marine environment present within the lagoon today. These conditions initiate marine deposition and bioerosion at the widening lagoon margins, both of which drastically change the burial environment and preservation of the subaerial exposure surface there. A seismically observed break in bedrock slope characterizes the marginal terrace. Bedrock excavation suggests that it is a result of the initiation of coastal marine bioerosion and consequent truncation of the original subaerial exposure surface. Holocene flooding of a silled interior bedrock depression has created three distinct preservational environments, resulting in the burial of three demonstrably different, concentrically zoned Holocene/Pleistocene unconformities at the contact.

Differences in the final expression of paleokarstic unconformities (i.e., karstic, caliche, and bioeroded) are the combined product of the exposure environments which produced them and the burial environments which altered or preserved them. Temporal and spatial differences in the genetic and taphonomic environments of paleokarstic unconformities must surely complicate attempts at stratigraphic correlation, especially when glimpsed in intermittent outcrop. An actualistic model is proposed for central to marginal changes in substrate preservation beneath "dished" carbonate lagoons as a function of sealevel change. Predictable changes in the preservation of ancient subaerial unconformities can yield useful information on antecedent platform topography, sealevel history, basin geometry, and the paleoceanography of the evolving lagoon.

Introduction

A considerable body of petrologic literature has emerged on the paleoenvironmental, diagenetic, and stratal significance of karst and caliche development—both in Quaternary and more ancient limestone terranes (Multer and Hoffmeister 1968, James 1972, Read and Grover 1977, Harrison and Steinen 1978, Klappa 1979, Sweeting 1979, Isphording 1976, Coniglio and Harrison 1983, Esteban and Klappa 1983, Semeniuk and Searle 1985, James and Choquette 1984 and this volume, and others). Subaerial exposure surfaces identified by karst/caliche features are particularly common to the Quaternary of Florida and the Bahamas, serving as useful stratigraphic horizons (Perkins 1977, Beach and Ginsburg 1980, Wilber 1981, Mylroie 1982, Pierson 1983, Garrett and Gould 1984). Paleokarstic horizons figure prominently in the record of ancient carbonate lagoons development as well, as exemplified by studies of the Alpine and Italian Triassic "löfer cycles" (Fischer 1964, Carannante 1971, Catalano et al. 1974a and 1974b). These stacked lagoon sequences are bounded by paleokarstic unconformities which serve as identifiable cycle boundaries. The stratigraphic utility of ancient subaerial unconformities, as well as their genetic relationship to relative sealevel fluctuation and depositional cycles, has made accurate recognition of subaerial versus submarine exposure surfaces of great concern (Rose 1970, Allan and Matthews 1977, Harrison 1977, Perkins 1977, Klappa and James 1979, Videtich and Matthews 1980, Robbin 1981, Wright 1982, Longman et al. 1983, Doyle et al. 1985) (Lohmann, this volume).

Though paleokarstic surfaces are acknowledged as valuable paleoenvironmental, diagenetic, and stratigraphic horizons, relatively little attention has focused on the sequential changes in preservation and alteration of these surfaces during a sealevel flooding event. A notable exception is the work of Longman and Brownlee (1980), which suggests that coastal marine physical and biological erosion in the Philippines results in the destruction of spectacular tower karst developed there.

On the Bahama Banks the combination of broad, low banktops, episodic sealevel fluctuation, and intense coastal bioerosion should also yield a substantially altered record of subaerial exposure events and surfaces throughout the Quaternary (Neumann 1966, Neumann and Moore 1975). The present work describes the various unconformable facies relationships which exist beneath the broad, dish-shaped Bight of Abaco lagoon as a result of continued Holocene transgressive flooding. High-resolution seismic profiles reveal the regional geomorphic nature of the buried Pleistocene bedrock surfaces. Cores of overlying Holocene sediment provide information on the evolution of the lagoonal burial environment. Rock samples of exposed and buried substrate indicate the manner in which each rock surface was preserved, modified, or destroyed. It appears that the genesis of this unconformity was a product of the Pleistocene sealevel fall which exposed it, but its preservation and final expression is ultimately controlled by the conditions accompanying the Holocene sealevel rise which later flooded it.

Study Area

The Bight of Abaco is a shallow carbonate lagoon on Little Bahama Bank, approximately 200 km from the Florida coast. It is semienclosed, bounded by the Abaco Islands to the north and east, and by Grand Bahama Island to the west. Open communication with Northwest Providence Channel (NWPC) is maintained to the south, over a buried Pleistocene sill approximately 2 m deep (Fig. 6.1). Temperature and salinity vary seasonally from 20° to 30° C, and from 34 to 39 ‰, respectively. Tidal pumping and wind-driven vertical mixing maintain a stable, tropical lagoon environment (Neumann and Land 1975). Skeletal aragonite and Mg-calcite sands and muds are the predominant sediments, and are apparently the products of in situ skeletal deposition and breakdown (Neumann and Land 1975, Boardman 1976).

Figure 6.1 demonstrates the close correspondence of depth to Pleistocene basement and bathymetric contours. Seismic and sediment probe surveys have shown that the Pleistocene bedrock topography beneath the Bight of Abaco forms a broad, dish-shaped structure, rimmed by a submerged sill between Grand Bahama Island and Mores Island. The northern Bight of Abaco depocenter is the thickest accumulation of Holocene sediment (about 3 m), grading to a thin veneer at Black Point (BP) to the north. To the southwest, bedrock is exposed as a 5-km-wide coastal hardground. It extends along the eastern shore of Grand Bahama Island near Big Harbor Cay (BHC). The northern Bight of Abaco was chosen for this study because of its pronounced dished topography and the paleoenvironmental data which exist on the thick Holocene sequence deposited there (Boardman 1976). Five sample sites are located in the northern lagoon along a seismic profile extending between Big Harbor Cay (BHC) and Black Point (BP) (Fig. 6.2).

The Holocene depositional history of the Bight of Abaco reflects progressive sealevel flooding mediated by a peripheral bedrock sill (Boardman 1976, Boardman and Neumann 1977). A three-part depositional history of Holocene facies development has been recon-

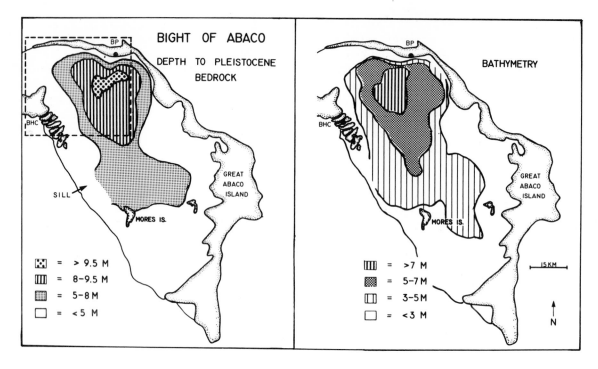

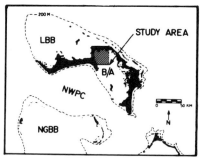

Figure 6.1. Bight of Abaco depth to Pleistocene bedrock and bathymetric contours, modified from Boardman (1976). Bedrock contours in meters below sealevel. Inset shows northern Bight of Abaco study area (boxed) in the northern Bahama Banks. Karstified Pleistocene bedrock formed a silled, dish-shaped antecedent topography over which Holocene sealevel rose. LBB = Little Bahama Bank; B/A = Bight of Abaco; NGBB = northern Great Bahama Bank; NWPC = Northwest Providence Channel; BP = Black Point; BHC = Big Harbor Cay.

structed by textural, chemical, and paleontological analyses of dated sediment cores (Fig. 6.3).

Phase A

The initial phase is characterized by formation of paleosols on, and locally filling dolines in a subaerially exposed karst field. This subaerial exposure phase began by the late Pleis-

tocene and ended during subsequent Holocene wetting by a rising meteoric lens about 6100 yr B.P.

Phase B

Upon further sealevel rise, meteoric and marine waters mixed and produced a fluctuating brackish to hypersaline (schizohaline) setting. As precipitation and evaporation varied, a stressful, unstable environment resulted, prohibiting colonization by normal, open-marine organisms.

Phase C

By 3600 yr B.P. sealevel rose over the peripheral bedrock sill, and stable open communication with the sea was established. This resulted in the normal marine biological and sedimentary facies which continue to the present.

Using Walther's law the three vertical Holocene depositional phases of (A) subaerial exposure/paleosols, (B) transitional/schizohaline, and (C) marine inundation/deposition are transformed laterally into three consecutive

burial environments for the Pleistocene exposure surface beneath the modern Bight of Abaco. The Bight of Abaco "dish" thus affords the opportunity to examine the diagenetic/taphonomic characteristics of three consecutive Holocene burial environments as they transgress and variably overprint the Pleistocene unconformity at their base (Fig. 6.2).

Materials and Methods

Three scales of resolution have been applied to the analysis of the Pleistocene unconformity traced beneath the modern Bight of Abaco. First, high-resolution seismic profiles reveal the

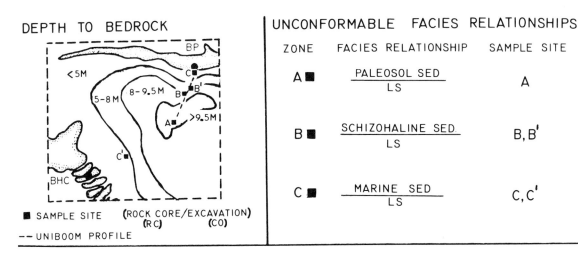

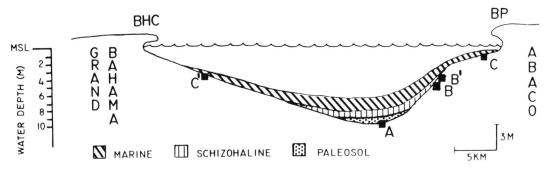

FIGURE 6.2. Northern Bight of Abaco sample locations and proposed unconformable facies relationships. Schematic cross-section illustrates three concentric seismic/geomorphic zones, and their respective sample sites: Central depresion Zone A (site A), Transitional slope Zone B (sites B, B'), and Marginal terrace Zone C (sites C, C'). Interpreted Holocene burial sediments are explained in Figure 6.3. BP = Black Point; BHC = Big Harbor Cay.

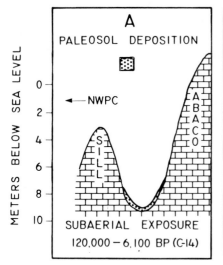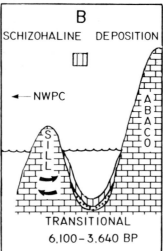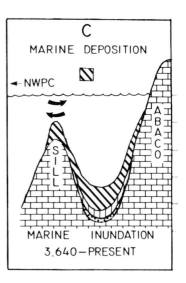

FIGURE 6.3. Bight of Abaco sealevel history and Holocene facies relationships, from Boardman and Neumann (1977). Holocene sedimentary evolution is based on textural, mineralogic, and faunal constituent changes in dated sediment cores. The peripheral sill at the bank margin (Fig. 6.1, left) results in at least three sequential depositional phases: paleosol, schizohaline, and normal, open marine. NWPC = Northwest Providence Channel.

Pleistocene and Holocene seismic sequences which bound the unconformable interface as well as describe the geomorphology of the upper bedrock surface. Second, bedrock samples from different seismic/geomorphic zones along the unconformity were recovered and reveal macroscopic differences in surface expression. Lastly, standard petrographic and SEM techniques were used to microscopically characterize the origin and preservation of the bedrock surface.

Approximately 50 km of "Uniboom" seismic lines were run in the study area. Figure 6.2 traces the path of one typical profile which illustrates characteristic seismic and geomorphic changes across the basin. Along this profile, four rock (RC) core tops and two excavations to bedrock (CO) reveal buried rock surfaces. These sites were chosen to sample seismically expressed variations in Pleistocene bedrock from the central, transitional, and marginal lagoon zones. Rock cores were recovered beneath up to 2.5 m of Holocene sediment using the hydraulic drill and casing system described by MacIntyre (1975). Larger areas of the bedrock surface were excavated at marginal sites C,C'

where burial by Holocene sediment is less than 0.5 m. Excavation methods employed a suction dredge driven by a 5HP waterpump modified from Brett (1964). Once a large area of bedrock surface was exhumed, a hydraulically powered underwater cutoff saw was used to sample the exposed surface.

All fieldwork conducted aboard the ORV *Calanus* of the University of Miami on cruise C8501, in February/March 1985. Laboratory analyses were performed at the Curriculum in Marine Sciences of the University of North Carolina at Chapel Hill.

Results

The analysis of seismic profiles, Holocene sediments, and underlying Pleistocene bedrock surface samples suggests that at least three different Holocene/Pleistocene unconformable facies relationships are expressed within the Bight of Abaco basin. We have designated three different burial zones which radiate from the northern lagoon depocenter: Central depression (Zone A), Transitional slope (Zone B), and

Marginal terrace (Zone C). Sample locations which targeted each zone are illustrated in Figure 6.2, and carry the same letter designation as the seismic/geomorphic zone they represent.

Zone A: Central Depression

A characteristic seismic expression from central lagoon Zone A is shown in Figure 6.4. Pleistocene bedrock reflectors are parallel, continuous, and locally wavy. The upper bedrock reflector is interrupted by frequent microkarstic depressions (dolines) approximately 0.6 to 1.0 m deep and at least 80 to 100 m wide. Analyses of additional intersecting seismic lines suggest dolines are common in only the central lagoon depression, and thus designate the area of Zone A. The overlying Holocene sedimentary reflectors are parallel and continuous, exhibiting onlap-fill in the localized dolines, consistent with the authigenic paleosols of depositional Phase A (Fig. 6.2, and Boardman 1976). More elevated surfaces between dolines are overlain by a second reflector, which continues laterally across the basin onlapping bedrock upslope.

Bedrock samples have ubiquitous vadose solution pipes 0.5 to 1.5 cm in diameter which penetrate the irregular, blackened bedrock surface (Fig. 6.5A). Intersecting solution pipes create delicate internal channels and bridgelike microkarstic surface features within the original lagoonal pelletoidal packstone. The black stain is a surface phenomenon extending only 0.2 mm into the lagoonal rock matrix, which otherwise maintains a whitish-gray hue throughout. Framboidal pyrite and plant residue is concentrated in microkarstic surface irregularities and leached voids. Smooth subaerial calcrete rinds or hardpans of Klappa (1980) and Esteban and Klappa (1983) are poorly developed, and occur only in a localized fashion around root molds and solution pipes. Instead, an irregular surface expression of leached pellet- and root-moldic voids is delicately preserved over the blackened karstic surface (Fig. 6.5B).

Zone B: Transitional Slope

The transitional lagoon Zone B is seismically defined as the broad, gradual bedrock slope between the area of central lagoon dolines (Zone A) and the marginal lagoon terrace (Zone C). Zone B terminates at the slope break which begins Zone C. The transitional zone lacks dolines, instead exhibiting continuous, parallel bedrock reflectors without upper surface truncations (Fig. 6.6). Overlying Holocene sediment

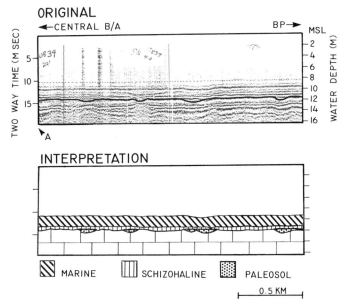

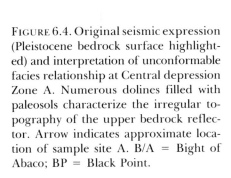

FIGURE 6.4. Original seismic expression (Pleistocene bedrock surface highlighted) and interpretation of unconformable facies relationship at Central depression Zone A. Numerous dolines filled with paleosols characterize the irregular topography of the upper bedrock reflector. Arrow indicates approximate location of sample site A. B/A = Bight of Abaco; BP = Black Point.

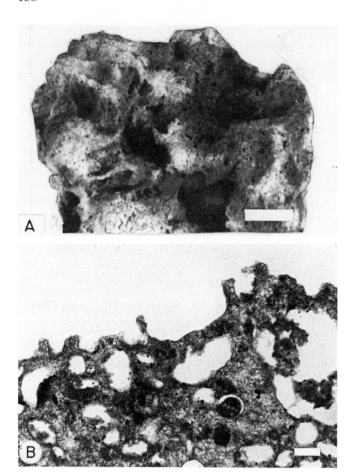

FIGURE 6.5. Lithologic expressions of the irregular, karstic bedrock surface at site A, Central depression Zone A of Figure 6.4. *A.* Macroscopic expression of delicate solution pipes and microkarstic surface irregularities, which are lined with a black, pyrite-rich stain. Scale bar = 1.0 cm. *B.* Photomicrograph showing the surficial expression of abundant leached porosity (root- and pelmolds). Pyrite accumulations (opaques) are associated with the surficial black stain and molds. Delicate microkarst at the bedrock surface is preserved intact, without development of an extensive caliche crust. PPL; scale bar = 0.1 mm.

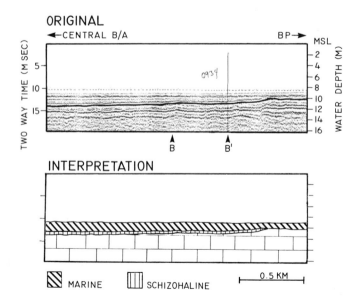

FIGURE 6.6. Original seismic expression (Pleistocene bedrock surface highlighted) and interpretation of the unconformable facies relationship at Transitional slope Zone B. An erosional bedrock slope break terminates this smooth Pleistocene surface, with basin onlap of a postpaleosol Holocene reflector (schizohaline?) at that point. Arrows indicate sites B and B'. B/A = Bight of Abaco; BP = Black Point.

reflectors are parallel, continuous, and onlap against the slope break at the start of Zone C. Additional seismic profiles radiating from the basin center suggest that the slope break may mark the beginning of a continuous erosional feature encircling the northern Bight of Abaco depression.

Pleistocene bedrock surfaces at Zone B have

FIGURE 6.7. Lithologic expressions of the smooth bedrock surface caliche at the Transitional slope Zone B of Figure 6.6. *A*. Macroscopic expression of a 0.2 to 1.0-cm-thick caliche crust which lines the upper surface, as well as solution pipes which penetrate deep within the rock. Scale bar = 1.0 cm. *B*. Photomicrograph of the vaguely clotted caliche crust which characterizes Zone B bedrock. Caliche preservation is excellent, lacking marine bioerosion or encrustation. PPL; scale bar = 0.5 mm.

been sampled at two rock core sites (B, B') placed approximately 0.8 km and 0.3 km downslope from the slope break. Figure 6.7A illustrates the rock surface sampled at site B. Bedrock from sites B and B' are composed of the same lagoonal pelletoidal packstone facies as at Zone A, but here they are *pervasively* altered by pedogenic calcretization. A smooth caliche crust up to 1.0 cm thick gives a vaguely clotted, micritic aspect to the unconformable bedrock surface (Fig. 6.7B). Vadose solution pipes similar to those of Zone A are smoothed and internally lined by the caliche rind. Upon final burial, the solution pipes were filled by a brown, organic-rich Holocene sediment. This infill contains abundant caliche and wood fragments, as well as freshwater gastropods in a friable matrix. It also contains a skeletal assemblage rich in interior platform foraminifera such as *Quinqueloculina agglutinans*, and *Q. lamarckiana* (Rose and Lidz 1977). A lack of biological modification such as boring and encrustation at the upper caliche surface, and the restricted nature of the biofacies within the solution pipes suggest that Zone B bedrock was buried and preserved beneath a schizohaline environment.

Zone C: Marginal Terrace

Marginal Zone C occupies the broad, flat terrace around the periphery of the northern Bight of Abaco, beginning at the bedrock slope break and ending at the island coasts. Figure 6.8 illustrates a characteristic seismic section from Zone C, located near the Black Point (BP) terminus at the basin margin. Pleistocene bedrock reflectors are subparallel and irregularly wavy, suggesting that a Pleistocene beach/dune complex has been flooded at this locality. Holocene sediment reflectors and probe depths indicate that less than 40 cm of sediment veneers the unconformity and pinches out at the Black Point coastal notch. The notch presently truncates and is eroding the exposed Pleistocene dune complex at the southern coast of Little Abaco Island.

Bedrock surfaces from Zone C have been excavated and sampled from beneath Holocene marine sediments at Black Point site C, as well as Big Harbor Cay site C' on the southwest

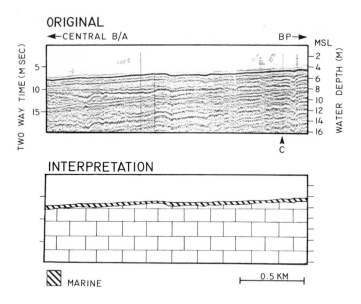

FIGURE 6.8. Original seismic expression (Pleistocene bedrock surface highlighted) and interpretation of the unconformable facies relationship at Marginal terrace Zone C. Pleistocene beach/dune construction has created irregular upper bedrock reflectors, which are overlain by < 40 cm of open marine sediments and truncated by marine erosion. Arrow indicates sample site C. B/A = Bight of Abaco; BP = Black Point.

margin (Fig. 6.2). Pleistocene bedrock facies at site C are oolitically coated pelletal grainstones of beach/dune origin, whereas pelletoidal packestones at site C′ denote subtidal lagoonal origin, similar to those at central and transitional sites. Despite differences in original Pleistocene depositional facies at sites C and C′, their subaerially exposed bedrock surfaces both exhibit abundant evidence of bioerosive truncation and colonization during sealevel rise. Figure 6.9A shows an eolian core top from site C with 1-2-cm-wide vadose solution pipes, which are variably bioeroded and encrusted. Each pipe has been bioeroded by endolithic sponges and worms which have removed all but scattered traces of caliche along the enlarged inner wall. The bioerosive community has extended well below the upper bedrock surface. The SEM photomicrograph of Figure 6.9B shows the fine epilithic scalloping effect of the sponge *Cliona*, which was actively removing the caliche lining of solution pipe walls prior to final burial beneath normal marine sediments. Scattered primary interparticle, leached, or rhizolithic voids have been bioerosively enlarged, preserving a record of marine hard substrate colonization and excavation deep within the rock.

A bedrock sample taken from the Big Harbor Cay site C′ illustrates the downward penetration of caliche via solution pipes, as well as the combined bioerosive effect of endolithic bivalves and sponges (Fig. 6.10A). Solution pipes here were lined or filled by extensive caliche crusts, and have been subsequently truncated by complete bioerosion of the upper caliche surface. The bioeroded surface is commonly "pocked" with the relict anterior impressions of bivalve borings, suggesting continued subtidal exhumation through ongoing bioerosion. Upper bedrock surfaces at both sites in Zone C preserve the evidence of dense clionid sponge infestation, as well as patchy populations of endolithic bivalves, including *Botula fusca* (Fig. 6.10B), *Gastrochaena stimpsonii*, and *Lithophaga nigra*. Dense spatfalls of monotypic boring molluscs (primarily *B. fusca*) contribute to upper surface erosion only, while boring sponges and polychaete worms extend their influence deep into solution pipes.

In addition to the endolithic community, the once subaerially exposed but now marine inundated bedrock surfaces in marginal zones contain evidence of nonboring chasmolithic (nestling) and encrusting taxa. The nestling opportunistic bivalve *Diplodonta semiaspera* commonly reinhabits vacant bivalve or sponge borings. Examples of multiple "nested" diplodonts, each showing predation by drilling gastropods are extremely common in rel-

FIGURE 6.9. Lithologic expressions of the bioeroded bedrock surface at site C of Marginal terrace Zone C. *A*. Macroscopic expression of bioeroded beach/dune facies, which are permeated by bioerosively enlarged vadose solution pipes. Caliche remnants and enhanced vadose cementation around the solution pipes create a discolored outline. Scale bar = 1.0 cm. *B*. SEM photomicrograph of *Cliona* excavations which have nearly removed all the caliche lining solution pipes at site C. Root-hair molds permeate the caliche and are subsequently truncated by epilithic scallops, yielding a pocked effect. Scale bar = 24.9 microns.

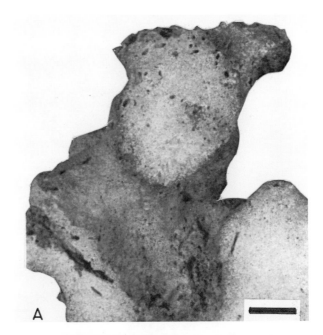

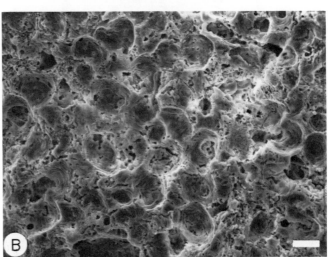

ict (and often themselves drilled) *B. fusca* shells. Gorgonian holdfasts (Fig. 6.11) and small stony corals are present on upper bedrock surfaces and are commonly bored at their bases.

Open vadose solution pipes and cavities created sheltered habitats for colonization by cryptobiontic, or cavity-dwelling marine biotas prior to final burial (Jackson et al. 1971). Common cryptobionts include serpulid worms which were often penetrated by clionids or overgrown by bryozoans and demosponges (Fig.

6.12). Additional taxa include tube-forming calcareous algae (see Jones and Goodbody 1984, Fig. 7), encrusting forams (*Homotrema rubrum*), and bivalves (*Chama macerophylla*). Examples of biotic interactions, including overgrowth, upturned margins, and apertures, and shell drilling are common at Zone C hardgrounds. These skeletal and therefore geologically preservable traces are evidence that the upper exposed and internally cryptic surfaces formed highly competitive biological marine substrates (Jackson 1977 and 1981).

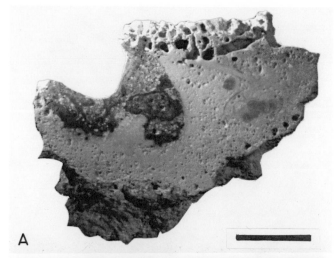

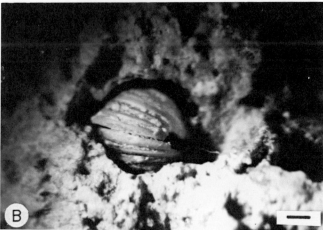

FIGURE 6.10. Macroscopic views of the upper bedrock surface at site C′ of Zone C, located near Big Harbor Cay. *A.* Cross-section illustrates the permeation of thick caliche through solution pipes at this site. Bioerosion of the upper surface demonstrates substrate truncation and removal by endolithic bivalves, polychaetes, and clionid sponges. Scale bar = 1.0 cm. *B.* Closeup of the most common endolithic bivalve *Botula,* which here maintains itself anomalously close to the surface, perhaps in response to potential overgrowth by encrusters. It is often drilled by predatory gastropods (naticids). The surrounding substrate shows surface expressions of internal clionid galleries and worm tubes. Scale bar = 1.0 mm.

Discussion

During Pleistocene subaerial exposure, karst processes created a dish-shaped, peripherally silled depression in bedrock below the modern Bight of Abaco. During Holocene transgression, the progression of freshwater, schizohaline, and open marine conditions of Boardman (1976) and Boardman and Neumann (1977) created three distinct burial environments for the once subaerially exposed Pleistocene basement. As a result, many of the surface features which would allow identification and lateral correlation of the subaerial exposure surface at the base of the modern lagoon are variably preserved, altered, or erased due to differing taphonomic characteristics of the successive burial environments. The resultant geologic products are at least three distinctly different, spatially related Holocene/Pleistocene unconformities preserved beneath the modern Bight of Abaco (Fig. 6.13). They are arranged concentrically from central to marginal lagoon.

Seismic and lithologic data from the central basin Zone A indicate that ubiquitous microkarstic depressions, or dolines, are filled with paleosols which blanket and preserve the original subaerial exposure surface. Microkarstic and pyrite-stained bedrock from this central zone exhibit preservation of delicate dissolution features. Such preservation suggests burial beneath paleosols and perhaps organic-rich, anoxic environments of stagnant freshwater ponds. This interpretation is consistent with Holocene data of Boardman (1976), in which terrestrial paleosols comprised the thin basal

FIGURE 6.11. Closeup of a relict Gorgonian holdfast attached to the bioeroded bedrock surface at site C'. The surface expression of a truncated caliche-filled solution pipe shows the crater-shaped anterior imprint of an exhumed boring bivalve (upper left). Corals, forams, and calcareous algae also encrust the upper bedrock surface.

sediments in the central lagoon depositional sequence.

Transitional slope Zone B is geomorphically and macroscopically smoother than Zone A, due to the absence of dolines and the existence of a well-developed pedogenic caliche crust. The onlap of a later, postpaleosol sediment reflector over bedrock Zone B extends to the distinct terminal slope break, suggesting a second type of burial environment for bedrock there. Wood and caliche fragments, freshwater gastropods, and restricted marine forams are found in organic-rich burial sediments there. Combined with the existence of an unmodified, well-preserved caliche crust below, these sedimentary characteristics are suggestive of burial beneath mixed salinity, perhaps schizohaline conditions. The transgressive removal of soil cover and final burial of Zone B caliche probably took place beneath a stressful, schizohaline marine depositional environment. This second burial environment resulted in the preservation of original bedrock surfaces by prohibiting the establishment of a marine bioerosive community within the basin.

Despite differences in the style of subaerial diagenesis (i.e., irregular karst vs. smooth caliche) found at bedrock Zones A and B, both exhibit a preserved surficial record of subaerial exposure and unconformity. A generally wetter central basin depression, collecting freshwater and vegetation-rich paleosols, contributes to both the production and ultimate burial preservation of karstic irregularity within that zone. In contrast, the smoother slope surface of Zone B developed beneath more well-drained, alternately wet and dry soil cover, creating the well-developed pedogenic caliche crust which is found there (Esteban and Klappa 1983, Warren 1983, Spencer et al. 1984). The restricted characteristics of the final schizohaline burial environment here preserved an intact surface of subaerial exposure. Though local variability in diagenetic paleokarst development is expected to complicate this model of unconformable facies relationships, central basin depressions would typically express karstic surfaces preserved beneath paleosols in the stratal record. Tracing the same unconformity laterally into transitional zones can lead to a quite different unconformable facies relationship, one characterized by schizohaline deposits over a smoother caliche surface.

The preservational characteristics of the Holocene burial environment change dramatically with increasingly open marine depositional conditions in the Bight of Abaco. The onset of normal marine deposition and hard substrate colonization is recorded at the ero-

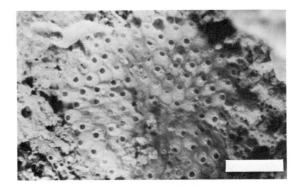

FIGURE 6.12. Closeup of a cryptic undersurface created by vadose solution, and subsequently colonized by marine hard substrate organisms at bedrock site C'. Overgrowth and cross-cutting relationships illustrate that cheilostome bryozoan colonies (*Celleporaria* sp.?) overgrew serpulid worms and were later overgrown and penetrated by clionid sponges. Scale bar = 0.2 cm.

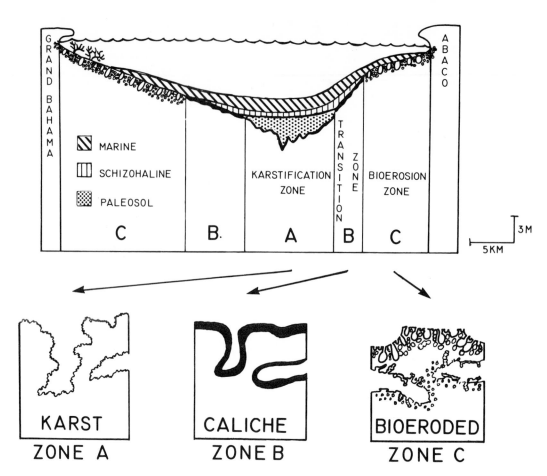

FIGURE 6.13. Idealized cross-section illustrating a model for variable expression of the Holocene/Pleistocene unconformity identified beneath the dish-shaped Bight of Abaco. Central and transitional zones preserve the original subaerial exposure surface beneath a protective blanket of paleosol, or fluctuating salinity (schizohaline) sediments. Open marine deposits at marginal zones bury highly altered (bioeroded and encrusted) subaerial exposure surfaces. Bedrock surfaces recovered from the three zones (bottom) illustrate marked genetic and preservational differences in the Holocene/Pleistocene unconformity expressed in the basin.

sional break in slope which marks the beginning of marginal terrace Zone C (see Figs. 6.2 and 6.6). Beginning at this point, bioerosive organisms severely truncate the record of subaerial unconformity and continue their geomorphic influence to the modern, actively eroding coastal notch. Aggressive intertidal and continued subtidal bioerosion substantially removes many of the diagnostic subaerial exposure criteria of Esteban and Klappa (1983). This erosive stage is accompanied by extensive surficial and cryptic community encrustation and nestling. Final burial beneath biogenic lagoonal sands and muds leaves behind an unconformable surface distinctly different from the intact karst and caliche found in Zones A and B. This process of extensive marine modification of unconformities at lagoon *margins* would undoubtedly inhibit lateral correlation with well-preserved central basin expressions of the same exposure event. Thus, biological overprint by marine endolithic, chasmolithic, and epilithic organisms at lagoon margins substantially removes the record of subaerial exposure and leaves in its place a biological surface of perhaps confused origin (Rose 1970, Klappa and James 1979, Wilkinson et al. 1982, Longman et al. 1983). The three unconformable facies relationships produced by differing production and preservation processes along a single lagoonal unconformity are illustrated in Figure 6.13.

In a positive sense, these lateral changes in the preservation of subaerial exposure surfaces are predictable, and taphonomic analysis may provide valuable information on paleoenvironmental gradients associated with sealevel rise. A correlated change from karstic to bioeroded surfaces suggests establishment of stable, open marine conditions within an internally depressed lagoon. In contrast to the silled, dish-shaped banktop lagoon in the Bight of Abaco, a gradually shallowing ramped margin could be flooded by open marine waters continuously. In this case, initial substrate alteration should progress landward as a linear front of physical and biological erosion associated with more open marine environments. Paleoecologic and taphonomic analysis applied to the record of lagoonal unconformities can help define ancient lagoons as silled or ramped, and, if silled, these can potentially indicate the position of the substrate within the basin and the paleoenvironments established by the waters which flooded it.

We suspect that purely paleokarstic unconformities may be most commonly preserved at the bottom and center of platform depressions where they are blanketed by peats or paleosols, or conversely at the tops of elevated islands where they are beyond the reach of marine modification. The potent bioerosive community concentrated in the coastal notch and adjacent subtidal terrace which is associated with open marine limestone coasts can render the preservation of subaerial exposure surfaces between depressed and elevated points extremely unlikely. The geologic record of limestone platform exposure could more often resemble marine hardgrounds buried by open marine deposits after the passage of an active front of coastal bioerosion, which behaves as a "biological buzzsaw" held against the topographic highs of the previous lagoon.

Summary

Differing preservation, and consequently variable expression of paleokarstic unconformities in lagoonal sequences can be attributed to at least three general parameters: antecedent platform topography, sealevel flooding history, and position of substrate within the flooded basin. The final record of subaerial exposure is limited by the paleoceanographic and preservational characteristics of the sealevel rise which preserves, alters, or erodes it. In the case of silled lagoons with internal depressions, central paleokarstic or caliche horizons are preserved intact below freshwater or schizohaline deposits; conversely, they are often modified or destroyed by marine bioerosion and buried beneath normal marine sediments at the margins.

In ramped lagoons facing open margins, continuously potent marine physical and biological erosion should alter relict paleokarst beginning at the onset of bank margin flooding. In this case, subaerial exposure surfaces could be lost throughout by bioerosive truncation, leaving a surface of biological overprint along the entire unconformity. A bored and encrusted surface of misleading marine hardground aspect would be preserved beneath open marine deposits—just as at dished lagoon margins.

The above scenario suggests that subaerial unconformities are most commonly preserved within interior basin depressions, or on elevated island tops not yet touched by the bioerosive edge of a rising sea. The areally extensive unconformity between buried banktop depressions and elevated exposed banktop islands has become a zone of biological truncation and overprint perhaps since the Lower Paleozoic (Kobluk et al. 1978) (Desrochers and James, this volume).

The historical onset of aggressive coastal bioerosion in lagoons can render subaerial exposure surfaces ephemeral and discontinuous, susceptible to miscorrelation, misinterpretation, or complete loss. In a more positive sense, taphonomic analysis applied to lateral changes from paleokarstic to bioeroded unconformities can yield significant paleoenvironmental information. This includes the reconstruction of antecedent basin geometry (dished, ramped), the geographic location of outcrop (central, marginal), flooding history (extent, timing), and the paleoecology of flooding environments (fresh, through open marine) developed during sea-level rise.

Acknowledgments. The authors wish to express their hearty thanks to the crew and scientists aboard the ORV *Calanus* during cruise C8501: Capt. Daniel Schwartz, First Mate Paul Tisevich, Mark Boardman, Matt Eaton, Glenn Safrit, Steve Snyder, and especially Bill Trumbull. Ian MacIntyre generously lent us the use of his rock-coring equipment. This project benefited from valuable field excursions in Italy with Gabrielle and Lucia Carannante, B. Abate, and G. LoCicero, as well as discussions with Mark Boardman, David Bush, and Bill Trumbull. Earlier drafts benefited from review by Joseph Carter, Albert Hine, and Ian MacIntyre—thanks to all. Alison M. Rasmussen is thanked for drafting the figures, as is Shirley S. Kilpatrick for typing the manuscript. William Dewar is thanked for his critical eye, and encouragement during the early stages of this project.

This work is part of a Ph.D. dissertation by Kenneth A. Rasmussen in progress at the Curriculum in Marine Sciences, University of North Carolina at Chapel Hill.

Research funding was provided by an NSF grant to A.C. Neumann, M.R. Boardman, and P. Baker (OCE 83-15203). Additional funding was provided by the Graduate School and the Curriculum in Marine Sciences at the University of North Carolina at Chapel Hill to K. Rasmussen.

References

Allan, J.R., and Matthews, R.K., 1977, Carbon and oxygen isotopes as diagenetic and stratigraphic tools; surface and subsurface data, Barbados, West Indies: Geology, v. 5, p. 16–20.

Beach, D.K., and Ginsburg, R.N., Facies succession of Pliocene-Pleistocene carbonates, northwestern Great Bahama Bank: American Association of Petroleum Geologists Bulletin, v. 64, No. 10, p. 1634–1642.

Boardman, M.R., 1976, Lime mud deposition in a tropical island lagoon, Bight of Abaco, Bahamas:

Unpubl. M.S. thesis, University of North Carolina at Chapel Hill, Chapel Hill, NC, p. 121.

Boardman, M.R., and Neumann, A.C., 1977, Lime mud deposition in an enclosed lagoon, Bight of Abaco, Bahamas: American Association of Petroleum Geologists—Society of Economic Paleontologists and Mineralogists Annual Meeting, Washington, DC, Program and Abstracts, p. 40.

Brett, C.E., 1964, A portable hydraulic diver-operated dredge-seive for sampling subtidal macrofauna: Journal of Marine Research, v. 22, p. 205–209.

Carannante, G., 1971, Ricerche sedimentologiche sulla successione ciclotemica dell' Infralias del Passo dell' Annunziata Lunga (Monti di Venafro): Bolletino Societa Naturalisti in Napoli, v. 80, 1971, p. 389–412.

Catalano, R., D'Argenio, B., and Lo Cicero, G., 1974a, Ritmi deposizionali e processi diagenetici nella successione triassica di piattaforma carbonatica dei Monti di Palermo: Nota di Societa Geologisti Italiana nell' assemblea del 12-7-1974, Palermo, p. 3–18.

Catalano, R., D'Argenio, B., and Lo Cicero, G., 1974b, I ciclotemi Triassici di Capo Rama (Monti di Palermo): Geologica Romana, v. 13, p. 125–145.

Coniglio, M., and Harrison, R.S., 1983, Holocene and Pleistocene caliche from Big Pine Key, Florida: Bulletin of Canadian Petroleum Geology, v. 31, p. 3–13.

Desrochers, A., and James, N.P., this volume, Early Paleozoic surface and subsurface paleokarst: Middle Ordovician carbonates, Mingan Islands, Quebec.

Doyle, L.J., Brooks, G., and Hebert, J., 1985, Submarine erosion and karstification on the West Florida continental margin: disparate environments yield similar features: Geological Society of America Annual Meeting, Orlando, Florida, Abstracts, p. 565.

Esteban, M., and Klappa, C.F., 1983, Subaerial exposure environment, *in* Scholle, P.A., Bebout, D.G., and Moore, C.H., eds., Carbonate depositional environments: American Association of Petroleum Geologists Memoir 33, p. 1–54.

Fischer, A.G., 1964, The lofer cyclothems of the alpine Triassic: Kansas Geological Survey Bulletin, v. 169, p. 107–149.

Garrett, P., and Gould, S.J., 1984, Geology of New Providence Island, Bahamas: Geological Society of America Bulletin, v. 95, p. 209–220.

Harrison, R.S., 1977, Subaerial versus submarine discontinuity surfaces in a Pleistocene reef complex, Barbados, W.I.: Proceedings of the Third International Coral Reef Symposium, Rosenstiel School of Marine and Atmospheric Science, University of Miami, Miami, Florida, p. 143–147.

Harrison, R.S., and Steinen, R.P., 1978, Subaerial crusts, caliche profiles, and breccia horizons: comparison of some Holocene and Mississippian exposure surfaces, Barbados and Kentucky: Geological Society of America Bulletin, v. 89, p. 385–396.

Isphording, W.C., 1976, Geomorphic evolution of tropical karst terranes, *in* Tolson, J.S., and Doyle, F.L., eds., Proceedings of the Twelfth International Congress on Karst Hydrogeology: International Association of Hydrogeologists Memoirs, v. 12, p. 115–129.

Jackson, J.B.C., 1977, Competition on marine hard substrata: the adaptive significance of solitary and colonial strategies: American Naturalist, v. 111, p. 743–769.

Jackson, J.B.C., 1981, Competitive interactions between bryozoans and other organisms: Geological Society of America Annual Meeting, Lophophorates Short Course Notes, Cinncinnati, OH, p. 22–36.

Jackson, J.B.C., Goreau, T.F., and Hartman, W.D., 1971, Recent brachiopod-coralline sponge communities and their paleoecological significance: Science, v. 173, p. 623–625.

James, N.P., 1972, Holocene and Pleistocene calcareous crust (caliche) profiles; criteria for subaerial exposure: Journal of Sedimentary Petrology, v. 42, p. 817–836.

James, N.P., and Choquette, P.W., 1984, Diagenesis 9. Limestones—The meteoric diagenetic environment: Geoscience Canada, v. 11, p. 161–194.

Jones, B., and Goodbody, Q.H., 1984, Biological alteration of beachrock on Grand Cayman Island, British West Indies: Bulletin of Canadian Petroleum Geology, v. 32(2), p. 201–215.

Klappa, C.F., 1979, Calcified filaments in Quaternary calcretes; organomineral interactions in the subaerial vadose environment: Journal of Sedimentary Petrology, v. 49, p. 955–968.

Klappa, C.F., 1980, Rhizoliths in terrestrial carbonates: classification, recognition, genesis and significance: Sedimentology, v. 27, p. 613–629.

Klappa, C.F., and James, N.P., 1979, Biologically induced diagenesis at submarine and subaerial carbonate discontinuity surfaces, *in* McIlreath, I.A., ed., Recent advances in carbonate sedimentology in Canada: Canadian Society of Petroleum Geologists Symposium, September 20–21, Calgary, Alberta, Canada, Abstracts, p. 16–17.

Kobluk, D.R., James, N.P., and Pemberton, S.G., 1978, Initial diversification of macroboring ichnofossils and exploitation of the macroboring niche in the lower Paleozoic: Paleobiology, v. 4(2), p. 163–170.

Lohmann, K.C, this volume, Geochemical patterns of meteoric diagenetic systems.

Longman, M.W., and Brownlee, D.N., 1980, Characteristics of karst topography, Palawan, Philippines: Zeitschrift fur Geomorphologie N.F. B.D. 24, Heft 3, p. 299–317.

Longman, M.W., Fertal, T.G., Glennie, J.S., Krazan, C.G., Suek, D.H., Toler, W.G., and Wiman, S.K., 1983, Description of a paraconformity between carbonate grainstones, Isla Cancun, Mexico: Journal of Sedimentary Petrology, v. 53, p. 533–542.

MacIntyre, I.G., 1975, A diver-operated hydraulic drill for coring submerged substrates: Atoll Research Bulletin, no. 185, p. 21–26.

Multer, H.G., and Hoffmeister, J.E., 1968, Subaerial laminated crusts of the Florida Keys: Geological Society of America Bulletin, v. 79, p. 183–192.

Mylroie, J.E., 1982, Karst geology and Pleistocene history of San Salvador Island, Bahamas: Proceedings of the First Symposium on the Geology of the Bahamas, College Center of the Finger Lakes, Bahamian Field Station, San Salvador, Bahamas, March 23–25, p. 6–11.

Neumann, A.C., 1966, Observations on coastal erosion in Bermuda and measurements of the boring rate of the sponge *Cliona lampa*: Limnology and Oceanography, v. 11, p. 92–108.

Neumann, A.C., and Land, L.S., 1975, Lime mud deposition and calcareous algae in the Bight of Abaco, Bahamas: A budget: Journal of Sedimentary Petrology, v. 45(4), p. 763–786.

Neumann, A.C., and Moore, W.S., 1975, Sea-level events and Pleistocene coral ages in the northern Bahamas: Quaternary Research, v. 5, p. 215–224.

Pierson, B.J., 1983, Cyclic sedimentation, limestone diagenesis and dolomitization in Upper Cenozoic carbonates of the southeastern Bahamas: Unpubl. Ph.D. dissertation, University of Miami, Coral Gables, FL, 275 p.

Perkins, R.D., 1977, Depositional framework of Pleistocene rocks in south Florida: Geological Society of America Memoir 147, pt. 2, p. 131–198.

Read, J.F., and Grover, G., Jr., 1977, Scalloped and planar erosion surfaces, Middle Ordovician limestones, Virginia: analogs of Holocene exposed karst or tidal rock platforms: Journal of Sedimentary Petrology, v. 47, p. 956–972.

Robbin, D.M., 1981, Subaerial $CaCO_3$ crust: a tool for timing reef initiation and defining sea-level changes: Proceedings of the Fourth International Coral Reef Symposium, Manila, v. 1, p. 575–579.

Rose, P.R., 1970, Stratigraphic interpretation of submarine versus subaerial discontinuity surfaces: an example from the Cretaceous of Texas: Geological Society of America Bulletin, v. 81, p. 2787–2798.

Rose, P.R., and Lidz, B., 1977, Diagnostic foraminiferal asemblages of shallow-water modern environments: south Florida and the Bahamas: Sedimenta VI, Comparative Sedimentology Laboratory, University of Miami, Miami, FL, 55 p.

Semeniuk, V., and Searle, D.J., 1985, Distribution of calcrete in Holocene coastal sands in relationship to climate, southwestern Australia: Journal of Sedimentary Petrology, v. 55, p. 86–95.

Spencer, T., Woodroffe, C.D., and Stoddart, D.R., 1984, Calcareous crusts and contemporary weathering of raised reef limestones, Grand Cayman Island, West Indies: Advances in Reef Science, Joint Meeting of the Atlantic Reef Committee, Rosenstiel School of Marine and Atmospheric Science, and the International Society for Reef Studies, Miami, FL, Oct. 26–28, Abstracts of Papers, p. 117.

Sweeting, M.M., 1979, Present problems in karst geomorphology: *in* Sweeting, M.M., and Pfeffer, K.H. eds., Karst processes: Berlin, Gebruder Borntraeger Publishers, p. 1–5.

Videtich, P.E., and Matthews, R.K., 1980, Origin of discontinuity surfaces in limestones: isotopic and petrographic data, Pleistocene of Barbados, West Indies: Journal of Sedimentary Petrology, v. 50, p. 971–980.

Warren, J.K., 1983, Pedogenic calcrete as it occurs in Quaternary calcareous dunes in coastal south Australia: Journal of Sedimentary Petrology, v. 53(3), p. 787–796.

Wilber, R.J., 1981, Late Quaternary evolution of a leeward carbonate bank margin, western Little Bahama Bank; a chronostratigraphic approach: Unpubl. Ph.D. dissertation, University of North Carolina at Chapel Hill, Chapel Hill, NC, 277 p.

Wilkinson, B.H., Janecke, S.U., and Brett, C.E., 1982, Low magnesian calcite marine cement in Middle Ordovician hardgrounds from Kirkfield, Ontario, Journal of Sedimentary Petrology, v. 52(1), p. 47–57.

Wright, V.P., 1982, The recognition and interpretation of paleokarsts: two examples from the Lower Carboniferous of South Wales: Journal of Sedimentary Petrology, v. 52(1), p. 83–94.

7
Neptunian Dikes and Fissure Fills: An Overview and Account of Some Modern Examples

PETER L. SMART, R.J. PALMER, F. WHITAKER, and V. PAUL WRIGHT

Abstract

Neptunian dikes (marine) and fissure infills (terrestrial) are deposits filling voids in older rocks and are widely reported from carbonate terrains, but sediment infills from brackish and freshwater phreatic environments (termed *cavern infills* here) have been less widely recognized. The initiation, development, and sedimentary filling of voids can take place in a variety of environments, which may be recognized from distinctive features including morphology, distribution, fauna, and sedimentology. These environments may change significantly during and between each stage; thus a detailed and complex record may be established by careful interpretation of both void and sediments. The initial voids may be depositional, synsedimentary, or tectonic in origin, while their development is controlled largely by the rapidity of erosion (slow in marine, fast in vadose terrestrial and in the mixing zone between freshwater and saline water). While the contemporaneous rate of infill is rapid, erosive development of the void may be limited. Sediment character is dependent on external environment and on the ratio of allochthonous to autochthonous sediment. Autochthonous sedimentation may be rapid in poorly lithified rock under strong erosional attack, particularly if large voids are present. These ideas are illustrated by an account of the morphology, distribution, fauna, and sediments of "blue holes" (underwater caves) from the Bahama Banks, which we recognize as modern examples of neptunian dikes.

Introduction

Neptunian dikes are bodies of younger sediment infilling fissures in rocks exposed on the sea floor (Bates and Jackson 1980, Visser 1980) and are the marine equivalents of terrestrial (often karstic) fissure fills, with which they may be genetically related. Surprisingly, there have been no previous general reviews of their genesis, morphology, and development. They are a minor but widely recognized sedimentary feature, particularly associated with carbonate buildups. This association is probably due to three factors. First, carbonates become lithified much more rapidly than siliciclastic rocks and thus respond in a brittle manner to stress, developing fractures even at an early stage. Second, carbonate slopes are steeper than siliciclastic slopes and tend to steepen upward (Schlager and Camber 1986). They therefore tend to develop fractures due to unloading and mass movement along the platform margins. Finally, because carbonates are soluble, circulation of undersaturated waters can form substantial voids.

In this chapter we discuss some of the factors significant in controlling the initiation, development, and sedimentation of neptunian dikes and cavern and fissure infills, and discuss their importance in paleoenvironmental reconstruction and interpretation. We also compare them to underwater caves ("blue holes") in the Bahamas, which we consider to be modern examples in which the active sedimentation processes can be studied. Finally, we discuss their importance in paleoenvironmental reconstruction and interpretation.

Nature and Formation of Void Infill Deposits

It is apparent from the numerous published accounts of neptunian dikes and fissure infills that there is great variety in both the morphology of initial voids and the sedimentary infill of voids. This reflects the different processes and geological environments in which void initiation, development, and infill can occur (Fig. 7.1). Furthermore, because these development stages can span considerable time periods, the geological environment may change completely between them. Thus a period of subaerial cavity development may be succeeded by subaerial, subaqueous, or submarine infill. Only the latter forms a neptunian dike *sensu strictu*, although the void itself may have been generated under subaerial conditions, because the conditions of sediment deposition have conventionally been used to define this term. Similarly, fissure fills are subaerial in origin. We suggest that the term *cavern fill* be restricted to subaqueous fills deposited in brackish or phreatic freshwater environments, although this does not imply that neptunian and fissure fills cannot be emplaced into roofed voids (cavities) as well as those directly open to the surface. In the following sections we will consider the three developmental stages in detail and suggest criteria for their recognition in the geological record.

Void Initiation

The morphology and particularly distribution of host voids is controlled primarily by the processes which initiate them. These provide the openings which permit the effective operation of later developmental processes, but in some cases sedimentation will rapidly follow and little modification will occur. Three types of initial voids may be recognized: primary depositional cavities, fractures arising from essentially synsedimentary processes, and fractures due to regional tectonism. Many neptunian dikes are associated with regional tectonism (Richter 1966, Wendt 1971, Szulcsewski 1973, Schlager 1969, Misik 1979, Stanton 1981, Wood 1981, Vera et al. 1984, Wendt et al. 1984, Blendinger 1986). They are characterized by strongly preferred orientations which can be related to the regional tectonic framework. For instance, in the Middle Triassic carbonate platforms of the Dolomites, Blendinger (1986) suggested that the east–west dikes resulted from extension as a result of the northwest–southeast-oriented sinistral shear. In some cases, different generations of dikes can be ascribed to specific tec-

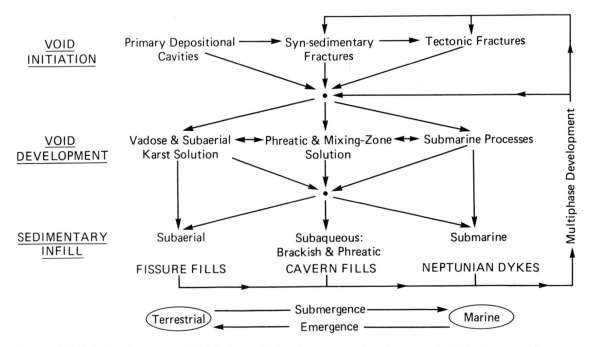

FIGURE 7.1. Relations between void initiation, void development, and sedimentary infill for fissure and cavern fills, and neptunian dikes.

tonic phases or events. This is well illustrated in the Subbetic ranges of southern Spain, where Vera et al.(1984) demonstrate that the orientation of the second phase of neptunian dikes reflects the commencement in the Jurassic of transverse movement of the African plate in relation to Europe after a previous phase of compression. Where the regional tectonic framework is more constant and the bedrock well lithified, reactivation of fractures after infill may occur. In the Mendip Hills of southwest England this reactivation of Variscan fractures has resulted in zoned neptunian dikes, with filling sediments of Keuper to Upper Jurassic age (Stanton 1981, Robinson 1956). Brecciation of the 'previous sedimentary infill has also occurred, and the fragments have been incorporated into the new infill.

The term *synsedimentary* is used here to describe fractures developing as a result of sedimentation and erosion processes, including compaction, unloading, and mass movement. In the Devonian Canning Basin reef complexes of Western Australia, Playford (1984) demonstrated that the majority of neptunian dikes developed parallel or perpendicular to the platform margin (those parallel tending to be wider), and were not directly related to fault activity. He suggested that slippage along marginal slope bedding and compaction of the basin sediments were the major causes. These fractures may be associated with considerable extension and also show reactivation. In the general model of Bernoulli and Jenkyns (1974) for the breakup of the Mediterranean Tethyan platform, although active block faulting is contemporaneous with development of neptunian dikes, joint development in the newly unconfined margins of the upstanding blocks may be the most important mechanism of development, again giving fractures paralleling the platform margin. Thus unless there is also lateral displacement, giving fractures oblique to the margin as in the example described by Blendinger (1986), it may often be difficult to differentiate between synsedimentary and tectonic fractures based on fracture orientation.

Wendt (1971) distinguishes between *Q fissures*, which cut the bedding, and *S fissures*, which parallel it. More generally, these would be called neptunian dikes and sills, respectively, although such an extension in terminology is probably not justifiable. Opening of bedding planes occurs both as a result of tectonic and synsedimentary induced fracturing. The relative frequency of the S and Q types is therefore probably not a generally useful technique for distinguishing these origins, although Playford (1984) suggests that there may be significant differences in this ratio between marginal slope and reef deposits. Few studies have examined these distributions in detail, although Vera et al. (1984) do observe significant spatial variations in the frequency of different types.

Primary depositional voids are only of real significance in reef and forereef breccia deposits. In the former, substantial cavities may be present between the main reef-forming elements, such as coral heads. These voids are largely infilled at an early stage by sediment derived from erosion of the reef itself (Land and Goreau 1970). However, some may become closed and, thus isolated from the source of sediment, persist, with slower authochtonous processes dominating the infill. Examples of this type of infill are figured by Misik (1979), Tucker (1973), and Playford (1984). Although these would not generally be described as neptunian dikes, they comprise a subset of such features, the only distinguishing characteristic being the relatively small difference in age between the host and infilling The irregular outline of such voids and their clear relation to the primary framework of the deposit readily distinguish these types from those developed by fracturing.

Void Development

Many of the neptunian dikes described in the literature show little evidence of void development after the initial opening (Robinson 1956, Richter 1966, Wendt 1971, Szulcsewski 1973, Playford 1984, Schlager 1969, Blendinger 1986). Conversely, others show evidence of both dissolutional and mechanical enlargement prior to sediment infill, particularly those that are associated with periods of emergence (Wood 1981, Wendt et al. 1984, Vera et al. 1984, Whiteside and Robinson 1983, Scholl and Wendt 1971, Halstead and Nicoll 1971, Gonzalez-Donoso et al. 1983).

Three factors are significant in determining

the extent of void development prior to infill: first, the relative rapidity of erosion in the particular environment in which the void is exposed; second, the resistance of the bedrock to that erosion; and third, the time available between initiation of the void and infill. Of the three sets of processes recognized in Figure 7.1, submarine processes are probably least effective in void development. Although solutional forms are reported on the northeastern shelf of Florida (Popenoe et al. 1984), these are thought to be inherited from earlier low stands of sealevel. Only Misik (1979) and Melendez et al. (1983) have suggested that submarine dissolution was significant in the development of some neptunian dikes. Other authors, after careful examination of the evidence, favor dissolution during emergence (Gonzalez-Donoso et al. 1984). Furthermore, relatively rapid rates of sedimentation would be expected in voids opening on the sea floor, and their lifetime would be sufficiently short that even the very efficaceous biological erosion by grazers and borers would have little effect. These types of neptunian dikes are therefore characterized by relatively planar walls with no erosional forms in evidence.

The uplift and flexing of the platform which causes development of the initial voids may also result in emergence. Thus subaerial karst processes may operate, giving irregular enlargement of the void from the initial fracture line and generating a distinctive suite of surface dissolution features including karren, phytokarst, and root molds. Because insoluble sediment is often lacking on carbonate platforms, true paleosols may not be present, but iron-rich and calcrete crusts may be formed (James and Choquette 1984). In general, void infill occurs much less rapidly in the terrestrial than in the marine environment, and there is thus potential for the development of large karst cavities. For instance, in the Cretaceous of the Sierra de Cabra (southern Spain), Vera et al. (1984) describe one karst cavity over 25 m wide and 30 m deep, although typically the cavities did not penetrate more than 8 m below the contemporaneous land surface. This decrease in the frequency of dissolution voids with depth is typical of vadose karst development and is due to progressive downward saturation of recharge water.

Surprisingly, development of voids under water-filled phreatic conditions has not found favor in the literature on neptunian dikes. Tubes and elliptical voids guided by joints and bedding planes are reported (Wendt et al. 1984, S. Wright pers. comm.), sometimes with dissolution-scalloped surfaces (Robinson 1956, Wendt et al. 1984), and are characteristic of this environment. They may also occur over a considerable depth range, but we are not aware of any observations from the fossil record on this point. Mixing corrosion (Bögli 1971) at both the water table and the base of the freshwater lens may permit relatively rapid dissolution (Wigley and Plummer 1976, Back et al. 1979, Back et al. 1986), and will be of some significance in emergent platforms. Changes in sealevel may also permit the mixing zone to affect many horizons in the platform. Furthermore, preexisting fracture systems allow the circulation of brackish waters, particularly along the bank margins, greatly increasing the area in which mixing zone solution can occur (Smart et al. in prep.). However, unless groundwater flow velocities are relatively high, distinctive scalloped surfaces will not be developed. Differential dissolution forms due to the varying stability of the component carbonate minerals should be apparent, both to the eye and under the microscope. These are reported for fissures in the Cambro-Ordovician of northern Scotland by S. Wright (pers. comm.), where differential dissolution of burrowed limestones along the fissure walls is apparent, and in thin sections figured by Misik (1979) and Molina et al. (1985). We believe that this type of void development is much more common than has previously been recognized and is of considerable significance in carbonate diagenesis, a suggestion supported by Playford (1984).

Sedimentation

The sediments and cements filling neptunian dikes and the cavern and fissure infills include all those commonly found in the geological record. They may represent surficial or cave terrestrial environments (Halstead and Nicoll 1971), through littoral (Liss and Wojcik 1960) and shallow marine (Gonzalez-Donoso et al. 1983) to pelagic deposits (Bernoulli and Jenkyns 1971). Because they are found predominantly in carbonate environments which frequently lack sources of siliceous sediments, the infills are also commonly carbonate. These include cements formed in a variety of environments, which produce distinctive textures (James and Choquette 1983 and 1984). Early marine and marine-burial cements are particularly copious in the neptunian dikes of the Canning Basin, for instance (Kerans et al. 1986). In some areas, up to ten separate phases of neptunian dike sedimentation can be recognized by differences in the character of the sediments (Scholl and Wendt 1971). As the analysis and interpretation of such sediments are in the general realm of the sedimentary geologist, we will concentrate here on the more general factors controlling sedimentation in large voids.

The change from void development to infill is usually caused by some change in the external environment, controlling the rate of sediment supply. A particularly good example occurs in the Subbetic platforms studied by Gonzalez-Donoso et al. (1983). Under emergent conditions void development predominated, but on submergence, sedimentation of the voids commenced. As the submergence was gradual there is a clear zoning, with the deeper voids hosting the earlier fills. This provides a sensitive sea-level indicator for the area, and together with interpretation of the nature of the sediments made possible a remarkably detailed account of the changing environment.

The nature of the sedimentary deposits preserved in voids is controlled primarily by the relative rates of autochthonous and allochthonous sedimentation. Autochthonous sediments include biogenic and chemical cements and material derived from the walls. Allochthonous sediments must be transported into the void by gravitational or fluid transport. Thus, in addition to the dependence of the allochthonous sedimentation rate on the external environment noted above, the rate is also dependent on the degree of interconnection between void and environment. This may be indicated by evidence of water movement such as cross-bedding and other similar sedimentary structures (for example, Misik 1979). Some cavities may be totally isolated and are decoupled from external sediment supply, while adjacent well-connected ones receive a copious supply (Kerans et al. 1986). Wendt (1971) noted that S fissures (parallel to the bedding) were characterized by much lower rates of sedimentation than Q fissures, whose greater vertical extent increased the probability of a direct connection to the surface. Similarly, fractures developed near the surface, for instance by vadose karstification, will be associated with relatively high sedimentation rates compared with deep cavities such as those formed at the base of the freshwater lens. There is therefore some interdependence between the environment of void development and the nature of the sediment infill.

Although relatively rapid biological processes such as deposition by encrusting organisms have been reported for some neptunian dikes (Gonzalez-Donoso et al. 1983, Playford 1984), these examples appear to be rare and are associated with shallow bank and reef environments. In a modern case described by MacIntyre (1984), rates of deposition by serpulid worms were very low. Autochthonous sedimentation is thus dominated by the rate of supply of material derived from the walls, as chemical precipitation is rather slow. Three factors are of importance here: the nature of the wallrock, the erosional environment in the void, and the size of the void. In general, small voids are mechanically more stable than larger ones, thus chemical sedimentation giving sparry calcite and other spelean infills may predominate in this type, as described by Wendt (1971) and Assereto and Folk (1980). In larger voids breakdown of the wallrock is dependent on its relative resistance and the strength of the erosive forces. Most voids only develop in lithified sediments, but in some cases cementation and conversion of metastable minerals will have proceeded further than in others, and the bedrock will therefore be both mechanically stronger and chemically more stable. The erosional environment is also important in controlling both the amount and nature of sediment supply; thus chemical dissolution in the mixing zone may cause disaggregation of the rock, liberating component grains such as oolites, while mechanical erosion will give large fragments of unaltered bedrock. Such contrasts are apparent in the literature, but generally such detailed sedimentological considerations have been less important than information pertaining to the age of the deposits.

One particular process which has been suggested for the development of neptunian dikes is injection of unconsolidated sediment from the sea floor following the sudden opening of a fissure. Robinson (1956) and Duff et al. (1986) believe this process is important for the neptunian dikes in the Carboniferous limestone of southwest England. They suggest that it is indicated by parallel zonation of the fissure fill on both walls, but fail to recognize that this may also be achieved by multiphase movements causing a central suture to develop in the preexisting sediment (and also causing some brecciation), which is infilled by the later sediment. Other authors have also noted that in vertical fractures sediments appear to show weak lamination parallel to the wall (Misik 1979), but this appears to be commonly observed in present-day caves, where sediment-bearing waters enter from cracks in the roof and walls (Bull 1981). There is some evidence of plastic deformation, such as flame structures and disoriented primary sedimentary structures in the Mendip examples, but this may simply be due to slumping into the fissure. Sediment injection as a cause of neptunian dikes appears at most to be of minor importance.

Blue Holes of the Bahamas

Given numerous studies of carbonate sedimentation conducted on the Bahama Bank, it is surprising that the parallels between sedimentation in the underwater caves (blue holes) found both on the land and offshore on the shallow banks and some types of neptunian dikes have not been noted. The existence of these features is mentioned in the classic study of Newell and Rigby (1957), and an indication of their potential geological significance is given in a paper by Dill (1977). We are presently involved in an extensive study of the development, geochemistry, wallrock diagenesis, and sedimentation of blue holes and present here only a brief description of some of the sites we have surveyed.

The Bahama Bank (Fig. 7.2) is a tectonically stable platform whose major boundaries are defined at great depth by tectonic fractures (Mullins and Lynts 1977). The great relief of the banks and steepness of their margins is related both to depositional processes and to erosion of the intervening channels by turbidity currents (Hooke and Schlager 1980). Geophysical, sedimentological, and underwater surveys using submersibles demonstrate that the banks have erosional margins, and that at

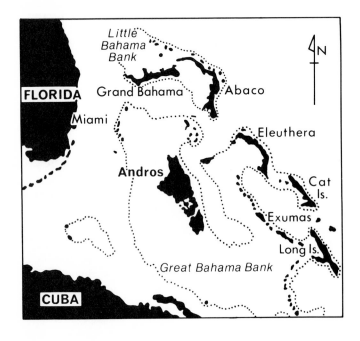

Figure 7.2. Location of the Bahama Banks, Andros Island, and Grand Bahama (-500 m contour dotted).

depth spalling and mass movement are active processes (Freeman-Lynde et al. 1981, Mullins and Neumann 1979). Lateral unloading parallel to the bank margins has produced a series of fractures. These are often multiple and complex, sometimes curvilinear, and near vertical, but show no displacement. Many can be traced laterally for tens of kilometers, both on the submarine banks and across the land (Fig. 7.3). Major perpendicular joint sets are also present. The geometry of these fractures is thus very similar to that described for the Canning Basin by Playford (1984, Fig. 24). On land, both sets of fractures are enlarged by surface dissolution processes giving straight but irregular rifts up to half a meter wide, which taper downward to a depth of several meters (Fig. 7.4). The rock surfaces are extensively fretted, but karren are generally absent. Infill is limited to detached bedrock blocks and coarse organic material (leaves, branches, etc.), including living tree roots which penetrate to considerable depth. These rifts are characteristic of the typical karstic voids widely recognized in the geological record (Vera et al. 1984). Some narrow fractures have been infilled with pink and brown micrite banded parallel to the walls, and in places these infills have been brecciated. No

detailed studies have yet been conducted but they appear similar to those described by Perkins (1977) from south Florida.

Some of the fractures paralleling the bank margin have been greatly enlarged, and direct access may be gained to extensive cave systems below present sealevel, using specialized cave diving techniques. A survey of several related fracture-controlled blue holes is presented in Figure 7.5 (Palmer 1985). The passages average 2 m to 5 m wide (but reach 30 m in Lothlorien), and are in excess of 90 m deep, the floor being beyond the limits of air diving. They are laterally continuous, and can be followed from the submarine entrances under the land, where a brackish water lens may be developed (Smart et al. in prep.). The bedrock walls are rough but planar and show evidence of both solution and spallation (Fig. 7.6). Upwards, the passages terminate in bedding-plane ceilings or collapsed boulders jammed in the narrowing but continuing fissure. These voids are identical to some of the bigger neptunian dikes and fissures reported in the literature, for instance Playford 1984, Palmer et al. 1980, Molina et al. 1985. Fallen boulders of wallrock up to several meters in diameter form a jumbled mass on the floor and in places jam as bridges across the passage.

FIGURE 7.3. Aerial view of the north shore of South Bight, Andros Island, looking northwest. A major fracture (arrowed) can be seen with blue holes (dark zones) developed in the floor of a shallow bay.

Figure 7.4. Enlargement of fracture paralleling bank margin by vadose solution, South Andros Island.

The autochthonous sediment is not limited to this coarse material, however, but sand and silt have also accumulated from disintegration of the poorly lithified and generally friable wall-rock. These caves are thus representative of neptunian dikes in which internal sediment sources predominate, and the infill is primarilly of the host rock.

Wave action and currents on the banks, combined with the tidally induced semidiurnal inflow and outflow from the submarine bank entrances, transport allochthonous sediments into some of the caves. These are predominantly coralgal and shell sands and finer lime muds derived from the surface of the surrounding banks (Fig. 7.7). Within the blue holes the tidally induced currents may exceed 5 knots near entrances and at restricted passage cross-sections (Warner and Moore 1984), and allochthonous sediment may be in transport over 600 m into some smaller passages. Size sorting

along the passage and ripple and dune structures including cross-bedding are also observed, as reported in the fossil record by Misik (1979). These currents would be sufficient to mobilize the spar balls reported by Playford (1984), but no comparable sediments have yet been observed in the blue holes. At some marine sites, the entrance slopes are colonized by corals forming an upstanding patch reef (Fig. 7.8). Within, the cave the walls are encrusted with a marine fauna including sponges, wall-boring bivalves, polychaetes, serpulid worms, and ahermatypic corals (Trott and Warner 1986). Bioturbation of the floor sediments by *Holothuria* and by crustacea is also observed. Similar faunas are reported for neptunian dikes by Wendt (1971).

Inland where a freshwater lens is present, comparable faunal activity is absent but a specialized microcrustacean food chain based on the input of organic matter entering through the roof is observed (Cunliffe 1985). Bacterial activity is vigorous (cf. Playford 1984), particularly at the saline interface, and conditions are frequently reducing in this zone. This fauna, which is limited to deep saline groundwaters, may be the equivalent of the ostracod faunas observed by Wendt (1971) and Misik (1979) in neptunian dikes. Thus, the presence of these apparently marine faunas in neptunian dikes does not necessarily preclude contemporaneous emergence. Indeed, the remarkable juxtaposition of Triassic terrestrial mammals and reptiles with brackish water diatoms in the cavern fills of the Bristol region described by Marshall and Whiteside (1980) may be explained by just such a brackish lens situation. The contemporaneous fissure faunas described by Robinson (1956) and Halstead and Nicoll (1971) for the same area also find their equivalents in the distinctive terrestrial fauna reported from the infill of a modern solution pit near Nassau by Brodkorb (1959).

Although the blue holes have originated on synsedimentary fractures, their present size and extent are due to dissolutional activity in the mixing zone at the base of the freshwater lens (Back et al. 1979, Rudnicki (1980), Palmer and Williams 1984, Smart et al. in prep.). Here mixing of fresh and saline waters occurs in a fairly narrow zone (about 10 m maximum), causing undersaturation of the water with re-

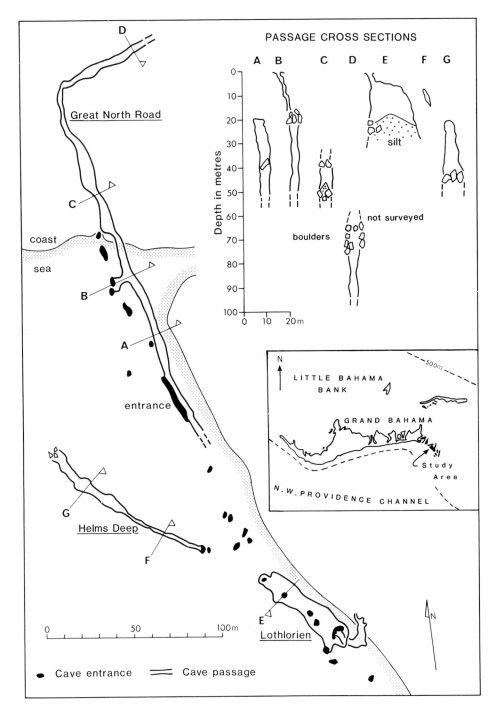

FIGURE 7.5. Plan and passage cross-sections of fracture controlled blue holes in the Big Creek area of eastern Grand Bahama. Inset: Location map.

spect to calcium carbonate (Wigley and Plummer 1976, Back et al. 1979). Preferential enlargement of the fractures therefore occurs in this zone, and tubular and elliptical passages may also develop along bedding planes. These can be seen in Aquarius Cavern (Fig. 7.9), which has developed at a previous lens base 15 m to 20 m below present sealevel (Palmer 1985). This mixing zone level is associated with the much more extensive freshwater lens which existed

FIGURE 7.6. Stargate blue hole, South Andros Island. The diver is swimming along a 10-m-wide, dissolutional-enlarged cave passage developed on a fracture paralleling the bank margin. Speleothems deposited subaerially during Pleistocene low sealevel stands and now submerged mask the entire left wall, while on the right dissolutional etching produces a much rougher surface.

FIGURE 7.7. Sand waves advancing across the shallow submerged bank toward the 15-m-wide entrance of Manta blue hole (Zodiac inflatable for scale arrowed). Sediment from the bank is transported by tidal currents through a 10-m-high, 2-m-wide passage developed on a fracture leading inland and orthogonal to the coast (the latter is marked by double arrows). The entrance is some 100 m south of the area shown in Figure 7.5. The much smaller entrances to Lothlorian (L) and Helms Deep (H) can also be seen, as can the trace of the fractures which guide them.

FIGURE 7.8. Patch reef developed around entrance to marine blue hole (diver center for scale). Gorgonian corals are developed on the exposed front face of the patch reef (right), while hard-bodied corals are developed along the left and rear walls.

existed in the area during the glacio-eustatic emergence of the Bahama Bank. Dissolution in a static or slowly moving mixing zone appears to be characterized by complex "Swiss cheese" macroporous channelling (Fig. 7.10), but while this is reported from oil field boreholes (Lapre pers. comm.), we are not aware of any instances reported from neptunian dikes or cavern infills. During the periods of emergence, extensive subaerial speleothems were deposited in the blue holes (Gascoyne et al. 1979). Similar banded spelean infills are also observed in much smaller voids intersected by cored boreholes in the area. No other extensive cement has been observed in either terrestrial or marine blue holes.

While our observations do not extend to the very small cavities which preclude direct exploration, but are well represented in the geological record, it is clear that blue holes provide a very good present-day example of active neptunian dike, fissure, and cavern infill. Our

initial work also confirms that the clear-cut separations between these three may not be easy to demonstrate in the geological record, where periods of brief emergence and submergence are not necessarily recorded.

Significance of Neptunian Dike, Cavern, and Fissure Infills

Void-infill deposits are of great importance, because once formed they are more protected from subsequent erosion than the contemporaneous surface sediments. They may thus preserve sediments and faunal remains for periods from which all other records are lacking. When the void is poorly supplied with sediment from the surface, sedimentation rates will be low, and a particularly complete sequence of deposits spanning a substantial time period may be emplaced before the cavity becomes full to capacity. This is well illustrated by the remarkable

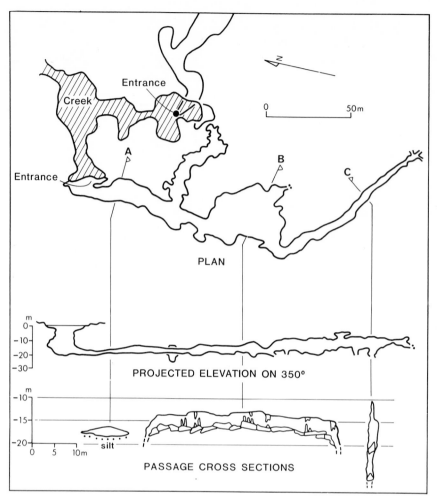

Figure 7.9. Plan, elevation, and passage cross-sections of Aquarius Cavern, eastern Grand Bahama, a blue hole in which development of bedding-plane tubes and elliptical passages has occurred at a previous position of the mixing zone. Stoping of the roof has subsequently occurred, and much of the bedrock floor is now obscured by breakdown blocks.

condensed sequences described by Wendt (1971) from the Jurassic of Sicily.

The sediments may also allow a very detailed account of the changing environment to be built up, while the processes of void initiation and development provide information on periods of emergence not represented in the sediment record. The work of Gonzalez-Donoso et al. (1983) is exemplary in this respect. The environmental interpretations may also be placed into a firm stratigraphic context because faunal remains are trapped by the voids. These may be washed in from outside the fissure and simply trapped, like the Triassic lizard and mammal faunas from the Mendip Hills (Robinson 1956), or actually inhabit the void, as is the case with the ostracods described by Misik (1979).

The development of fractures indicates the degree of sediment lithification, while the orientation and age of these may yield significant detailed information on the tectonic history of an area not available from regional studies of larger rock units. Some care is needed however, as nontectonic synsedimentary processes acting on bank margins are clearly very effective in developing fractures. While the often substantial voids thus developed (Playford (1984) describes an example 6 km long, 80 m deep, and up to 20 m wide) are filled on burial, the enhanced groundwater circulation which can oc-

FIGURE 7.10. A blue hole developed at the interface between fresh (above) and saline waters (below). The sharp interface (arrow) is being disrupted by movement of the diver, causing mixing seen in the hazy refracted zone (bottom right). Note the extensive dissolutional modification of the wallrock giving a "Swiss cheese" appearance which may be typical of mixing zone dissolution.

cur while they are open can cause significant dissolution of the surrounding rock (Smart et al. in prep.). Such sites may therefore be particularly favorable for the accumulation and extraction of hydrocarbons and metalliferous deposits.

Acknowledgments. Our recent fieldwork in the Bahamas was financed by Shell Exploration and Production Limited and the Royal Society of London. Commercial sponsorship was also provided by many firms, in particular Virgin Airlines Limited and Bahamas Air, while Lester Cole allowed us to use his house as a field base.

References

Assereto, R., and Folk, R.L., 1980, Diagenetic fabrics of aragonite, calcite and dolomite in an ancient peri-tidal-spelean environment: Triassic Calcare Rosso, Lombardia, Italy: Journal of Sedimentary Petrology, v. 50, p. 371–394.

Back, W., Hanshaw, B.B., Herman, J.S., and Van Driel, J.N., 1986, Differential dissolution of a Pleistocene reef in the groundwater mixing zone of coastal Yukatan, Mexico: Geology, v. 14, p. 137–140.

Back, W., Hanshaw, B.B., Pyle, T.E., Plummer, L.N., and Weidie, A.E., 1979, Geochemical significance of groundwater discharge and carbonate solution to the formation of Caleta Xel Ha, Quintana Roo, Mexico: Water Resources Research, v. 15, p. 1521–1535.

Bates, R.L., and Jackson, J.A., 1980, Glossary of geology: American Geological Institute,

Bernoulli, D., and Jenkyns, H.C., 1974, Alpine Mediterranean and Central Atlantic Mesozoic facies in relation to the early evolution of the Tethys: Society of Economic Paleontologists and Mineralogists, Special Publication 19, p. 129–160.

Blendinger, W., 1986, Isolated stationary carbonate platforms: the Middle Triassic (Landinian) of the Marmolada area, Dolomites, Italy: Sedimentology, v. 33, p. 159–183.

Bögli, A., 1971, Corrosion by mixing of karst water: Transactions of the Cave Research Group of Great Britain, v. 13, p. 109–114.

Brodkorb, P., 1959, Pleistocene birds from New Providence Island: Bulletin of the Florida State Museum Biological Science, v. 4, p. 339–371.

Bull, P.A., 1981, Some fine-grained sedimentation phenomena in caves: Earth Surface Processes, v. 6, p. 11–22.

Cunliffe, S., 1985, The flora of Sagittarius, an anchaline cave and lake in Grand Bahama: Transactions of the British Cave Research Association, v. 12, p. 103–109.

Dill, R.F., 1977, The blue holes—geologically significant submerged sinkholes and caves of British Honduras and Andros, Bahama Islands: Proceedings of the Third International Coral Reef Symposium, p. 237–242.

Duff, K.L., McKirdy, A.P., and Harley, M.J., 1986, New sites for old: a student's guide to the geology of the East Mendips: Nature Conservancy Council, p. 191.

Freeman-Lynde, R.P., Cita, M.B., Jadoul, F., Miller, E.L., and Ryan, W.B.F., 1981, Marine geology of the Bahama Escarpment: Marine Geology, v. 44, p. 119–156.

Gascoyne, M., Benjamin, G.J., Shwarcz, H.P., and Ford, D.C., 1979, Sea-level lowering during the Illinosian glaciation: evidence from a Bahama blue hole: Science, v. 205, p. 806–808.

Gonzalez-Donoso, J.M., Linares, D., Martin-Algarra, A., Rebollo, M., Serrano, F., and Vera, J.A., 1983, Discontinuidades estratigraficas durante el Cretacico en el Penibetico (Cordillera Betica): Estudios Geologica, v. 39, p. 71–116.

Halstead, B.L., and Nicoll, P.G., 1971, Fossilized caves of Mendip: Studies in Speleology, v. 2, p. 93–102.

Hooke, R.B., and Schlager, W., 1980, Geomorphic evolution of the Tongue of the Ocean and the Providence Channels, Bahamas: Marine Geology, v. 35, p. 343–366.

James, N.P., and Choquette, P.W., 1983, Diagenesis 6; Limestones—the sea floor diagenetic environment: Geoscience Canada, v. 10, p. 162–179.

James, N.P., and Choquette, P.W., 1984, Diagenesis 9; Limestones—the meteoric environment: Geoscience Canada, v. 11, p. 161–194.

Kerans, C., Hurley, N.F., and Playford, P.E., 1986, Marine diagenesis in Devonian reef complexes of the Canning Basin, Western Australia, in Schroeder, J.H., and Purser, B.H., eds., Reef diagenesis: Berlin, Springer-Verlag, p. 357–380.

Land, L.S., and Goreau, T.F., 1970, Submarine lithification of Jamaican reefs: Journal of Sedimentary Petrology, v. 40, p. 457–462.

Lis, J., and Wojcik, Z., 1960, Triassic bone breccia and karst forms in Stare Gliny Quarry near Olkusz (Cracow Region): Kwartalnik Geologiczny, v. 1, p. 55–74.

MacIntyre, I.G., 1984, Extensive submarine lithification in a cave in the Belize barrier reef platform: Journal of Sedimentary Petrology, v. 54, p. 221–235.

Marshall, J.E.A., and Whiteside, D.I., 1980, Marine influence in the Triassic "uplands": Nature, v. 287, p. 627–628.

Melendez, G., Sequeiros, L., and Brochwicz-Lewinski, W., 1983, Lower Oxfordian in the Iberian Chain, Spain; Part 1—Biostratigraphy and nature of gaps: Bulletin Academic Polonaire des Sciences, v. 30, p. 159–172.

Misik, M., 1979, Sedimentologicke a mikrofacialne studium jury bradla vrsateckeho hradu (neptunicke dajky, bioherny vyvoj oxfordu): Zapadne Karpaty Seria Geologica, v. 5, p. 7–56.

Molina, J.M., Ruiz-Ortiz, P.A., and Vera, J.A., 1985, Sedimentacion marina somera entre sedimentos pelagicos en el Dogger del Subbetico extero (Sierras de Cabra y de Puente Geril, Prov. de Cordoba): Trabajos de Geologica de University de Oviedo, v. 15, p. 127–146.

Mullins, H.T., and Lynts, G.W., 1977, Origin of the northwestern Bahama Platform; review and reinterpretation: Geological Society of America Bulletin, v. 88, p. 1447–1461.

Mullins, H.T., and Neumann, A.C., 1979, Deep carbonate bank margin structure and sedimentation in the northern Bahamas: Society of Economic Paleontologists and Mineralogists, Special Publication 27, p. 165–192.

Newell, N.D., and Rigby, J.K., 1957, Geological studies on the Great Bahama Bank: Society of Economic Paleontologists and Mineralogists, Special Publication 5, p. 15–72.

Palmer, R.J., 1985, The blue holes of eastern Grand Bahama: Transactions of the British Cave Research Association, v. 12, p. 85–92.

Palmer, R.J., McKerrow, W.S., and Cowie, J.W., 1980, Sedimentological evidence for a stratigraphical break in the Durness Group: Nature, v. 287, p. 720–722.

Palmer, R.J., and Williams, D.W., 1984, Cave development under Andros Island, Bahamas: Transactions of the British Cave Research Association, v. 11, p. 50–52.

Perkins, R.D., 1977, Depositional framework of Pleistocene rocks in south Florida, in Enos, P., and Perkins, R.D., eds., Geological Society of America Memoir 147, p. 131–198.

Playford, P.E., 1984, Platform-margin and marginal-slope relationships in Devonian reef complexes of the Canning Basin: The Canning Basin Western Australia: Proceedings of the Geological Society of Australia and Petrological Exploration Society of Australia Symposium, Perth, Western Australia, p. 189–214.

Popenoe, P., Kohout, F.A., and Manheim, F.T., 1984, Seismic-reflection studies of sinkholes and limestone dissolution features on the northeastern Florida shelf, in Beck, B.F., ed., Sinkholes: their geology, engineering and environmental impact: Rotterdam, Balkema, p. 43–57.

Richter, D., 1966, On the New Red Sandstone neptunian dykes of the Tor Bay area (Devonshire): Proceedings of the Geological Association, v. 77, p. 173–186.

Robinson, P.L., 1956, The Mesozoic fissures of the Bristol Channel area and their vertebrate faunas: Journal of the Linnean Society, Zoology, v. 43, p. 260–286.

Rudnicki, J., 1980, Kras wybrzezy morskich: Studia Geologica Polonica, v. 65, p. 54–70.

Schlager, W., 1969, Das Zusammenwirken von Sedimentation und Bruchtektonik in den triadischen Hallstätterkalken der Ostalpen: Geologische Rundschau, v. 59, p. 289–308.

Schlager, W., and Camber, O., 1986, Submarine slope angles, drowning unconformities and self-erosion of limestone escarpments: Geology, v. 14, p. 762–765.

Scholl, W.U., and Wendt, J., 1971, Obertriadische und jurassische Spaltenfülungen in Steineven Meer (Nördliche Kalkalpen): Neues Jahrbuch für Geologie und Paläontologie Abhandlungen, v. 139, p. 82–98.

Smart, P.L., Whitaker, F., and Palmer, R.J., in prep, Enhanced groundwater flow along a bank marginal fracture, South Andros Island, Bahamas.

Stanton, W.I., 1981, Further field evidence of the age and origin of the lead-zinc-silica mineralization of the Mendip region: Proceedings of the Bristol Naturalists Society, v. 41, p. 25–34.

Szulczewski, M., 1973, Famennian-Tournasian neptunian dikes and their conodont fauna from Dalnia in the Holy Cross Mountains: Acta Geologica Polonica, v. 23, p. 15–59.

Trott, R.J., and Warner, G.F., 1986, The biota in the marine blue holes of Andros Island: Transactions of the British Cave Research Association, v. 13, p. 13–19.

Tucker, M.E., 1973, Sedimentology and diagenesis of Devonian pelagic limestones (Cephalopodenkalk) and associated sediments of the Rhenohercynian Geosyncline, West Germany: Neues Jahrbuch für Geologie und Paläontologie Abhandlungen, v. 142, p. 320–350.

Vera, J.A., Molina, J.M., and Ruiz-Ortiz, P.A., 1984, Discontinuidades estratigraficos diques neptunicos y brechas sinsedimentarias en la Sierra de Cabra: Publicaciones de Geologica, Universidad Autonoma de Barcelona, no. 20, p. 141–162.

Visser, W.A., 1980, Geological nomenclature: Royal Geological and Mining Society of the Netherlands, The Hague, Martinus Nijoss, 540 pp.

Warner, G.W., and Moore, C.A.M., 1984, Ecological studies of the marine blue holes of Andros Island, Bahamas: Transactions of the British Cave Research Association, v. 11, p. 30–40.

Wendt, J., 1971, Genese und Fauna submariner sedimentärer Spaltenfüllungen im mediterranen Jura: Palaeontographica Abhandlungen A, v. 136, p. 122–192.

Wendt, J., Aigner, T., and Neugebauer, J., 1984, Cephalopod limestone deposition on a shallow pelagic ridge: the Tafilait Platform (Upper Devonian), eastern Anti-Atlas, Morocco: Sedimentology, v. 31, p. 601–625.

Whiteside, D.L., and Robinson, D., 1983, A glauconitic clay-mineral from a speleological deposit of Rhaetian age: Palaeogeogeography, Palaeoclimatology and Paelaeoecology, v. 41, p. 81–85.

Wigley, T.M.L., and Plummer, N.L., 1976, Mixing of carbonate waters: Geochimica et Cosmochimica Acta, v. 40, p. 985–989.

Wood, W.A., 1981, Extensional tectonics and the birth of the Lagonegro Basin (southern Italian Apennines): Neues Jahrbuch für Geologie und Palaeontologie Abhandlungen, v. 161, p. 93–131.

PART II   Examples of Paleokarst Terranes

8
Proterozoic Paleokarst Profile, Dismal Lakes Group, N.W.T., Canada[1]

CHARLES KERANS and J. ALAN DONALDSON

Abstract

The Middle Proterozoic Dismal Lakes Group provides a rare example of Precambrian paleokarst which contains a remarkably complete spectrum of solutional, precipitative, and depositional features. The Dismal Lakes Group is a 1.5-km siliciclastic to carbonate succession that records dominantly stable peritidal carbonate sedimentation along the northwestern margin of the Canadian Shield. Karst features are widespread along a regionally extensive subaerial exposure surface associated with regression near the end of Dismal Lakes deposition, and are especially abundant near a depositional high developed earlier near September Lake.

An east-to-west transect down depositional dip from the crest of the profile to the flanking shoreline displays the following karst features: 30 m-thick collapse breccias with preserved roof pendants and cross-stratified chert granulestone cave-floor sediments, grike systems filled by transgressive marine sand, fibrous flowstone, cave popcorn, regolith, and finally, a sharp erosional surface overlain by mixed siliciclastic-carbonate facies of the succeeding marine transgressive phase.

Evolution of the karst profile occurred in three stages: (1) initial subaerial exposure and phreatic dissolution leading to formation of subsurface karst, (2) continued regression and vadose infill by autochthonous and allochthonous clastic sediments and flowstone precipitation, and (3) cave collapse prior to marine transgression.

The common development of grikes and the concomitant paucity of small- or medium-scale karren indicate either a temperate climate during subaerial exposure, or karst dissolution beneath a soil mantle now only locally preserved. The eustatic versus local tectonic nature of the regressive event leading to karst formation is speculative; both factors appear to have had an influence.

Introduction

The Middle Proterozoic paleokarst profile described here is remarkable not only in terms of its antiquity but also in terms of the variety of karst features preserved and their arrangement along a preserved paleotopographic profile. The purpose of this report is to outline the stratigraphic and structural setting of the paleokarst profile and to present an initial description of the main karst features. The relationship of this exposure event to evolution of the Dismal Lakes carbonate platform, and to regional geologic events, is reviewed in light of existing incomplete knowledge of this area of the Canadian Shield. However, treatment of such pertinent issues as paleoclimatic implications (Wright 1982, James and Choquette 1984), the relative importance of phreatic versus mixing corrosion in cave excavation (Bögli 1980), or the relative position of the water table during initial stages of subsurface karst development must be left to future research. Considering the economic importance of paleokarst breccias as hosts for base metal deposits (Kyle 1983) (Sangster, this volume) and petroleum (e.g., Wilson 1985), it seems likely that this previously little-known paleokarst terrane will warrant more detailed study.

[1]Publication authorized by the Director, Bureau of Economic Geology, The University of Texas at Austin.

Regional Geologic Setting

Paleokarst deposits of the Dismal Lakes Group occur along the northwestern margin of the Canadian Shield. The Dismal Lakes Group is one component of an unmetamorphosed and virtually undeformed sequence of Early to Late Proterozoic sandstones, dolostones, and minor volcanic rocks known as the Coppermine Homocline. The Hornby Bay Group and Dismal Lakes Group which comprise the lower half of the succession are two major siliciclastic to carbonate cycles which record progressive stabilization of the underlying Early Proterozic orogenic suite of Wopmay Orogen (Baragar and Donaldson 1973, Hoffman 1981, Kerans et al. 1981) (Figs. 8.1 and 8.2). This stabilization is marked by a change in depositional style from an early phase of terrigenous braided stream deposition in fault-bounded basins (lower Hornby Bay Group) to a phase of widespread carbonate-dominated peritidal platform sedimentation, and by an increased lateral conti-

nuity of facies. Continuity of facies is demonstrated by individual units that can be traced throughout the middle and upper Dismal Lakes Group and into the adjacent Elu basin, 300 km to the east (Kerans et al. 1981).

The age of the Dismal Lakes Group is poorly constrained by Phanerozoic standards. Volcanic rocks of the uppermost Hornby Bay Group which conformably underlie basal Dismal Lakes Group sandstones have been dated by U-Pb zircon methods at 1663 +/− 7 m.y. (Bowring and Ross 1985). Rb/Sr dates on the conformably overlying Coppermine River Group basalts are variable, but give an average age of 1200 m.y. (Smith et al. 1967, Fahrig and Jones 1969). The implication is that the 1.5-km-thick Dismal Lakes Group, dominated by peritidal and shallow-water shelf deposits and containing only one recognized unconformity, was deposited during a time span of 460 m.y., roughly equivalent to the interval from the base of the Cambrian into the Cretaceous. Part of this apparent discrepancy must reflect the presence of ad-

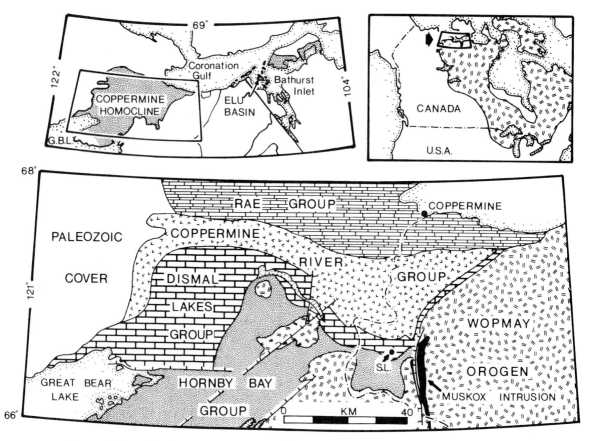

FIGURE 8.1. Regional geologic setting of Dismal Lakes Group. S.L. is September Lake.

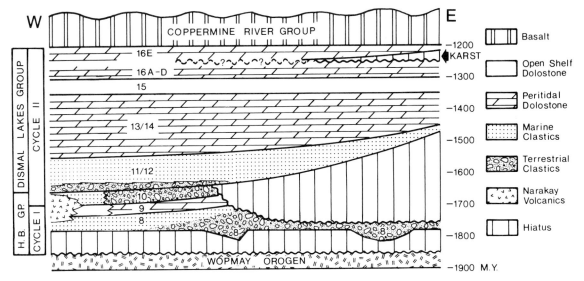

FIGURE 8.2. Chronostratigraphy of the lower Coppermine Homocline showing age relationships, unconformities, and general facies relationships. Position
of paleokarst interval is indicated. Cycles I and II
refer to the two main clastic-to-carbonate cycles in
the Dismal Lakes–Hornby Bay Group succession. H.
B. GP. = Hornby Bay Group.

ditional unconformities not yet recognized due
to the regional scale of mapping and absence
of biostratigraphic control. Regardless, it is clear
that the paleokarst unconformity discussed
here may be a record of significant hiatus. Alternatively, Bowring and Ross (1985) have suggested that existing Rb/Sr and K/Ar dates for
the Coppermine River Group may be in significant error.

Strata of the Dismal Lakes Group comprise
a lower fining-upward siliciclastic succession of
fluvial conglomerates and sandstones through
marginal marine sabkha mudstones (units 11
to 13 of Baragar and Donaldson 1973) (Fig.
8.3), followed by an upper series of peritidal
and open-shelf carbonates (now all dolostones)
with minor shales and sandstones (units 14 to
16 of Baragar and Donaldson 1973) (Fig. 8.3).
The paleokarst deposits discussed here occur
in the upper portion of this upper carbonate
sequence, referred to as unit 16 (Kerans et al.
1981). The upper carbonate-dominated portion
of the Dismal Lakes Group is characterized by
marked facies continuity, recording the development of a stabilized, regionally extensive
north- and west-facing carbonate shelf which
rimmed the northern margin of the Canadian
Shield during the Middle Proterozoic.

Stratigraphy of the Karst Interval (Unit 16)

Unit 16 (Baragar and Donaldson 1973, Kerans
et al. 1981) has been subdivided into five informal members comprising two main onlap-
offlap packages (members A to D and E: Kerans 1982) (Fig. 8.4). These packages (approximately equivalent to seismic stratigraphic
sequences) are separated by a significant unconformity ascribed to a relative sealevel fall
during which time erosion and karstification
occurred (Fig. 8.4). Dolostones in and upon
which the paleokarst developed are uniform 20
to 40 cm bedded, fine-to-medium crystalline
peritidal deposits displaying small-scale stromatolites, cryptalgal-laminites, intraclastic
breccias, and mudstone. Thin discontinuous
layers and nodules of black and white chert occur throughout this dolostone and are particularly common as replacements of mm-lami-
nated cryptalgal mats. Lateral and vertical
lithologic variability within the paleokarst zone
is minimal.

The most extensive exposure of paleokarst
development occurs in the eastern part of the

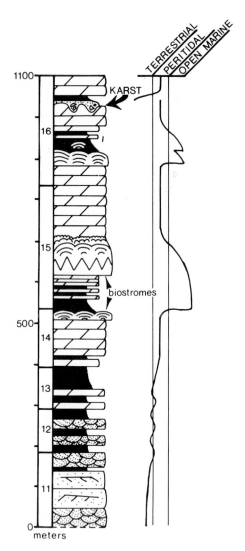

FIGURE 8.3. Stratigraphic column of Dismal Lakes Group, showing unit subdivisions and generalized depositional environments. Lithologies shown in black are shale or mudstone.

Dismal Lakes outcrop belt near September Lake (Fig. 8.1), which corresponds to an area of depositional thinning throughout deposition of the group (Kerans et al. 1981) (Figs. 8.1 and 8.4). The unconformity surface becomes increasingly subtle as it is traced westward into a recessively weathering shaly carbonate succession where its disconformable relationship cannot be demonstrated on the basis of outcrop features.

Immediately overlying the paleokarst surface, and modifying it in part, is a transgressive

mixed siliciclastic/carbonate facies which contains a basal lag rich in carbonate boulders reworked from the underlying karstified carbonates. The siliciclastic content of this lithofacies fines and becomes less abundant upward, eventually passing upward into peritidal dolostones which persist to the end of Dismal Lakes deposition (Fig. 8.5).

Sedimentary structures in this fining-upward mixed sequence include low-angle planar cross-stratification and early-cemented intraformational blocks (beachrock?), flaser and lenticular bedding, intraclastic conglomerates, and well-developed mudcracks. The depositional environment for this sequence is interpreted to be a transgressive beach/tidal flat complex. Terrigenous sediments in this facies cannot have been derived locally, because all underlying exposed carbonates are sand-free; transport by longshore processes is believed to be responsible for input of allochthonous sand. The shoreline environment of this onlap succession is significant, for its lowest occurrence along the disconformity surface gives a minimum estimate of the extent of subaerial exposure and downward shift in coastal onlap (cf. Vail et al. 1976).

A distinctive marker bed containing branching digitate stromatolites marks the approximate upper limit of the mixed siliciclastic/carbonate transgressive facies throughout the area of karst development. This marker bed is used as a datum for measured sections in the paleokarst zone, allowing accurate reconstruction of paleorelief and extent of erosion in this zone (Fig. 8.5). Where this marker bed is not present, sections are correlated relative to the Coppermine River Group basal contact. Both of these markers are assumed to approximate time lines with no significant depositional slope. An important feature of this reconstructed disconformity surface is the absence of localized subsidence over areas of major collapse, demonstrating the pre-onlap timing of all karst processes from solution to collapse. This is consistent with the lack of internal disruption in the the transgressive facies (member E), which would be expected had collapse occurred after onlap.

The easternmost area illustrated in Figures 8.5 and 8.6 is directly above the Muskox intrusion, an ultramafic layered intrusion emplaced

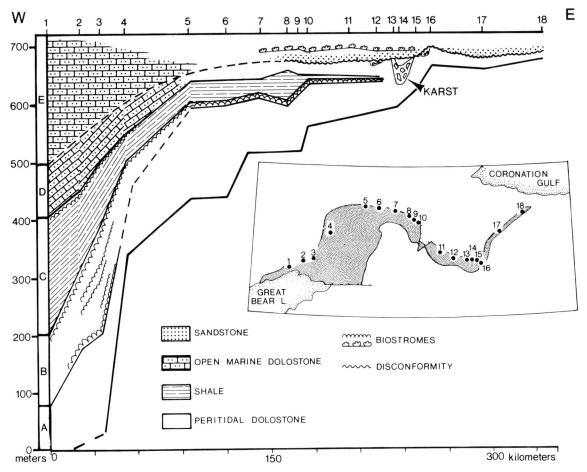

FIGURE 8.4. Stratigraphic framework of unit 16, upper Dismal Lakes Group, showing informal member subdivision and position of paleokarst interval. Location of measured sections is given in inset.

into Dismal Lakes strata near the end of deposition. The Muskox intrusion is a component of the Mackenzie igneous events, which mark a major episode of magmatism and local doming, and include the Coppermine River Group basalts and Mackenzie diabase dikes (Fahrig and Jones 1969). This area above the roof of the intrusion displays maximum paleorelief and contains faults showing minor syndepositional displacement which occurred during exposure of the platform (Kerans 1983).

The key feature of the stratigraphic framework in terms of paleokarst development is the west-sloping nature of the disconformity, which displays a minimum of 60 m of relief above sealevel across its 90-km length, thus providing

a paleogeomorphic framework for the karst features described below (Fig. 8.6).

Karst Features

Karst features recognized at the unit 16 disconformity include both surface and subsurface types, the latter by far the more abundant. The types and extent of karst features developed during exposure vary along the length of the measured profile, reflecting the extent of weathering and relative time of exposure. Seven localities are described sequentially from the west to east along the profile, illustrating a range from least to most extensive karst development. This discussion is followed by a

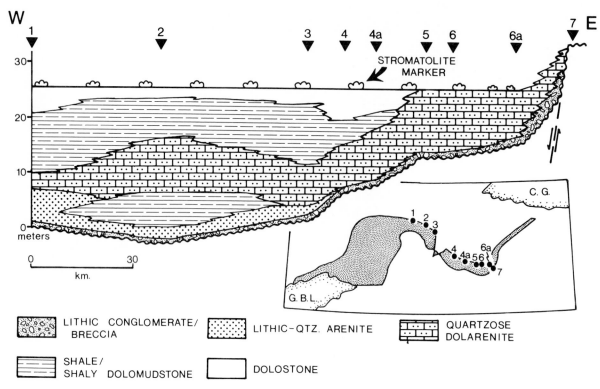

FIGURE 8.5. Lithofacies cross-section of onlap succession which overlies the karst interval. The stromatolite marker used as a datum is interpreted to approximate a time line.

synthesis of these features with respect to the karst profile as a whole, and finally by presentation of an evolutionary model for karst development.

Sharp Erosional Contact

Localities 1 and 2 exhibit sharp erosional truncation surfaces beneath the transgressive boulder lag of overlying member E (Fig. 8.6). Local relief of up to 20 cm is common, but the surface morphology is distinctively smooth and devoid of sharp scalloped surfaces or solution basins (*kamenitzas*). *Rillenkarren, rundkarren,* and other typical surface karst features are also absent. Signs of subaerial exposure in the underlying peritidal dolostones, such as solution-widened fractures (grikes), vugs, vadose silt, pendular cements, aggrading neomorphic fabrics, dissolution-related microporosity, or other indicators of exposure are also absent. Dolostone clasts in the basal conglomerate lack signs of dissolution or recrystallization and are typically angular to subrounded.

Regolith Development

A regolith is present at locality 3, situated 6 km east of locality 2 and 60 km west of the main karst collapse zone (Fig. 8.6). The upper 1 m below the disconformity surface displays a system of vertical and horizontal fractures filled with chalky dolostone. These chalky fracture zones outline a series of slightly tilted and rotated blocks of more resistant host dolostone (here an ooid packstone), which range up to 50 cm long and 20 cm thick. Minor green argillite and quartz sand occur within the chalky dolostone and apparently have filtered down from the overlying disconformity surface. Features of mature caliche profiles such as hardpan, platy structure, or pisolites (James 1972, Walls et al. 1975, Esteban and Klappa 1983) were not observed. Reworking of the regolith by the overlying conglomerate appears minimal, despite local truncation of up to 20 cm.

The weathered dolostone at locality 3 is an ooid packstone. Ooids throughout the Dismal Lakes Group typically display well-defined

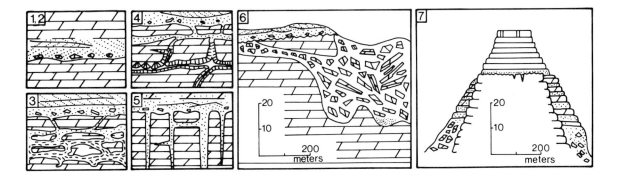

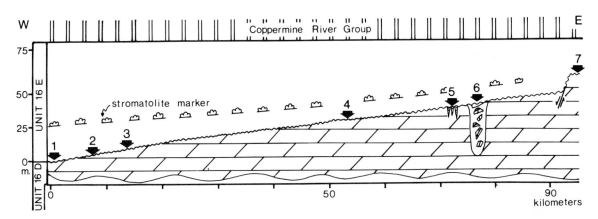

FIGURE 8.6. Schematic outcrop relationships at each of the 7 localities discussed in the text, showing their relative position on the reconstructed paleorelief profile of the unconformity surface. 1, 2: Sharp disconformity surface with a basal lag of dolostone clasts contained in cross-stratified sandstones of the overlying transgressive deposits. 3: Regolith development with chalky dolostone (dashed lines) as matrix to slightly tilted blocks of host dolostone. 4: Fibrous dolomite flowstone lining grike system for 60 cm below the disconformity. 5: Grikes filled with sand from the overlying transgressive sand sheet. 6: Major zone of collapse brecciation, showing a large intact block interpreted as a roof pendant capping the western portion of the deposit. 7: Paleohorst structure (unpatterned core) onlapped by breccias, quartzose dolostones, and finally Coppermine River Group basalt (uppermost bed). The outer limit of present exposure mimicks that of the paleohorst. Vertical dimensions for sketches 1–5 are 1 m. Location of sections is given in inset on Figure 8.5.

concentric lamination in submicrospar size mosaic dolomite, with oomoldic and leach porosity being extremely rare. This excellent fabric retention, despite replacement by dolomite, is a phenomenon common to many Proterozoic dolostones (Ricketts 1983, Tucker 1983). In contrast, samples from the regolith at locality 3 show nearly total destruction of ooid microstructure, the development of intercrystalline porosity, recrystallization to a microspar-pseudospar crystal mosaic, and collapse of some ooids (Fig. 8.7).

Grikes

The main features at localities 4 and 5 are grikes or kluftkarren (Fig. 8.6). Grikes are solution-widened joints which may or may not have formed under a soil cover (Sweeting 1972, Bögli 1980). At locality 4 a distinctive shallow grike system is filled by a combination of speleothemic flowstone and sandstone, and at locality 5 abundant grikes are infilled by siliciclastic sand of the overlying transgressive sequence.

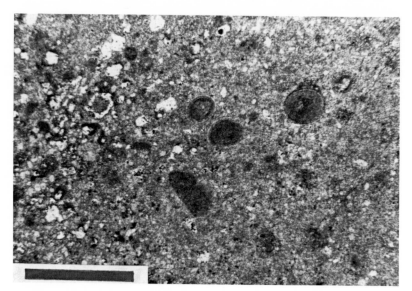

FIGURE 8.7. Coarse recrystallization texture and dissolution porosity in ooid packstone of regolith zone, locality 3. Ooids show a range of preservation, from those with rare concentric laminae (right side of photo) to relict grains which are now a combination of microspar and pore space (left side of photo). Photomicrograph in plane-polarized light; scale bar is 1 mm.

Sand-filled Grikes. The most common grike systems in the Dismal Lakes profile are those filled with siliciclastic sand from the overlying transgressive sequence. At locality 5, grike openings are up to 10 cm across and extend up to 4 m into the underlying dolostone. In plan view these grikes display a diamond pattern, with predominant joint orientations averaging 308° and 25° (Figs. 8.6 and 8.10). Grike systems are generally restricted to the upper 10 m of the karst profile, but at locality 7, chert-rich insoluble residues and quartz sand occur in a solution-widened fracture system 90 m below the surface. This grike is coincident with a syndepositionally active fault (Kerans 1983), which may explain this particularly deep penetration of dissolving waters. Sweeting (1972) points out that fault zones are commonly zones of selective dissolution in karst terranes.

Flowstone-Filled Grikes. At locality 4, a system of vertically oriented grikes and associated horizontal cavities with openings up to 15 cm across extends 2 m into the medium-bedded dolostone below the disconformity (Fig. 8.6). The upper 1 m of this system of jointed dolostone shows rotation of blocks and infill of grikes by siliciclastic sand from the overlying transgressive sequence.

Grikes in the lower meter are filled by banded fibrous dolomite, or flowstone, which encrusts roofs of host dolostone in a fairly even manner (Fig. 8.8). The crusts developed on cavity roofs are up to 12 cm thick, and those on floors are generally less than 2 to 4 cm. In one horizontal cavity a 10-cm crust of fibrous dolostone extends from the roof, whereas the floor is covered by a 2-cm layer of 2 to 5-mm-diameter pisolites of similar fibrous dolostone (cave pearls). In thin section these crusts display a distinctive millimeter-scale parallel banding with mamillary outline. The flowstone crusts consist of 20 × 100 micron bladed crystals with c-axes perpendicular to the grike walls (Fig. 8.9). This crystal fabric is similar to that described from modern speleothems (Folk and Assereto 1976, Kendall and Broughton 1978). Distinctive micritic structures with crude cellular structure and dendritic habit encrust the roofs of flowstone-filled cavities and occur along some growth surfaces of the fibrous crusts (Fig. 8.9). These structures may be equivalent to cave popcorn, a common modern vadose cave precipitate formed by evaporation-driven precipitation (Sweeting 1972).

Collapse Breccias and Internal Sediments

Breccia deposits containing blocks up to 5 m across occur from locality 5 in the west to areas north and east of the illustrated section, generally weathering out as massive knobs along the contact. The thickness of these breccias is variable, ranging from a few meters at locality

FIGURE 8.8. Grike in gray host dolostone (H) of locality 4, which is filled by lighter gray laminated dolostone (L). Laminated fill encrusts both the roof and floor of the cavity system (arrows). See Figure 8.9 for photomicrograph of laminated material.

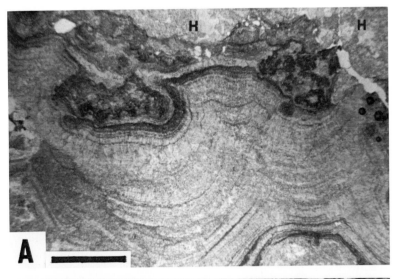

FIGURE 8.9. Laminated flowstone from grike at locality 4. *A.* Photomicrograph in plane light of laminar-fibrous dolostone crust growing downward from the contact with microsparitic host dolostone roof (H). The contact is partly plucked. Dark clotted areas at crust-host roof contact and lining growth surfaces within the crust may be cave popcorn. Scale bar is 2 mm. *B.* View in cross-polarized light of *A,* showing that the preferred orientation of c-axes in the fibrous crust is perpendicular to the host dolostone substrate. Scale is same as for 8.9 *A.*

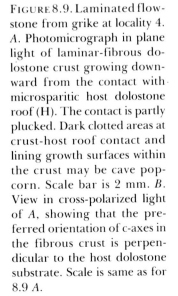

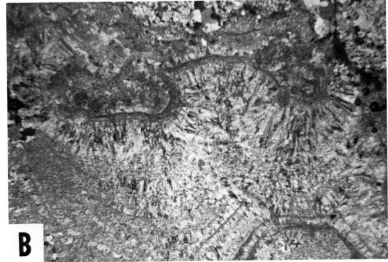

FIGURE 8.10. Sand-filled grikes (S) exposed at locality 5, approximately 1 m below the unconformity surface. The white notebook resting on bedding is 16 cm long.

5 to more than 30 m in the vicinity of locality 6 (Fig. 8.6). They consistently occur directly below the disconformity surface; no examples of collapse breccias entirely within the underlying dolostone are known.

Breccia fabrics differ according to clast size, orientation, and matrix content. In most breccias the clasts are tabular 5-to-40 cm angular fragments which display random orientation. The matrix of most breccias is a siliciclastic sand-rich dolomicrite, but some breccias contain clay-rich matrix and others consist almost entirely of clean quartz sand. Angular chert fragments of coarse sand to pebble size are an ubiquitous minor component. These breccias commonly rest sharply on unweathered dolostone, although small sand-filled grikes occur locally.

At locality 6 a spectacular breccia containing blocks up to 5 m long is developed to a maximum thickness of 30 m (Figs. 8.6 and 8.11). The basal contact of this breccia is sharp but displays significant local relief over the 250 m lateral extent of the exposure (Fig. 8.6). This breccia thins both to the east and west, from a maximum of 30 m to 2 m in a lateral distance of a few hundred meters. A large (6 m thick by 40 m wide) slab of intact dolostone caps the breccia deposit near its western margin. This block is surrounded by breccias both below and to either side, but displays bedding concordant with unbrecciated host dolostone (Fig. 8.6). This remnant apparently represents a preserved roof pendant. Orientation of blocks in the main collapse zone is subparallel, with dips generally less than 45°, and the base of the breccia is defined by zones of intact or only slightly tilted blocks (Fig. 8.11). Blocks within the breccia are highly angular and display a range of sizes from a few centimeters to several meters.

The matrix sediment of this megabreccia is a claystone, containing chert granules and pebbles, which displays contorted lamination wrapping around blocks. Grikes filled with quartzose sand from the overlying transgressive facies are a prominent feature of this breccia deposit. At the base of the breccia zone at locality 6 is a meter-thick pocket of chert-pebble conglomerate which displays low-angle trough cross-stratification and parallel lamination, representing a cave-floor stream deposit (Figs. 8.12 and 8.13). This chert-pebble conglomerate rests on unbrecciated dolostone; blocks of dolostone breccia depress and contort bedding in the upper portion. Chert pebbles are tabular and display fine mm-scale laminae, closely resembling silicified cryptalgal-laminite of the unbrecciated host dolostone (Fig. 8.13). Coarse subarkosic sand comprises approximately 30% of this deposit, which is cemented by chalcedony, euhedral quartz prisms, and poikilitic baroque dolomite, in that order of deposition.

FIGURE 8.11. Collapse breccia deposit of locality 6. Large blocks up to 5 m across display a crude sub-parallel orientation. Areas marked "I" are intact or only slightly disturbed and probably represent remnants of cave-floor pillars. Person for scale is 1.6 m tall (arrow).

Marine Onlap of Erosional Remnant

The crest of the paleotopographic profile occurs at locality 7, directly above the roof of the Muskox intrusion. Structural relationships discussed by Kerans (1983) suggest that this area is a fault-bounded horst formed immediately before karst development. Syndepositional fault offset is minimal on the order of a few tens of meters. The trace of the disconformity surface in this area defines an erosional remnant with 30 m of local topographic relief, bounded by east- and west-facing walls with steep (>45°) slopes (Fig. 8.6). Cherty claystone

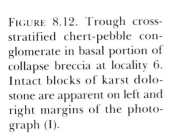

FIGURE 8.12. Trough cross-stratified chert-pebble conglomerate in basal portion of collapse breccia at locality 6. Intact blocks of karst dolostone are apparent on left and right margins of the photograph (I).

FIGURE 8.13. Photomicrograph in cross-polarized light of chert-pebble conglomerate shown in Figure 8.12. Submillimeter scale lamination in rectangular chert clasts is interpreted as silicified cryptalgal mat fabric typical of chert in the host dolostone. Scale bar is 5 mm.

fracture-fill sediments occur up to 90 m below the disconformity surface along a fracture plane coincident with a syndepositional fault that bounds the horst (Kerans 1983).

Breccias stacked against the horst remnant have a mixed dolomudstone–siliciclastic sandstone matrix, a clast size range of 0.05 to 5 m, and a random clast orientation. These breccias grade upward into dolostone breccias with clasts of 5 to 10 cm size and a matrix of quartz arenite. At the top of the erosional remnant, only a thin veneer of quartz sand and centimeter-scale sand-filled grikes mark the contact between dolostones of the erosional bluff and similar peritidal dolostones of the onlap succession (Figs. 8.5 and 8.6). On the flank of the bluff, a few meters below its crest, eroded dolostones are encrusted by laminar-to-pustular fibrous dolomite which has been partially silicified. These crusts appear to be analogous to marine-marginal aragonitic crusts of the Persian Gulf, called coniatolites by Purser and Loreau (1973), or marine caliche by Scholle and Kinsman (1974).

Reconstruction of the Karst Profile

The various features described from beneath the unit 16 disconformity surface display a progression of typical karst-related deposits which define a variably exposed and weathered karst terrane. Pertinent aspects of the various deposits have been described individually and are now considered in the context of a composite profile.

The sharp erosional contacts at localities 1 and 2 are interpreted to define the approximate western limit of prolonged subaerial exposure along the profile. The absence of caliche textures, scalloped surfaces, vuggy porosity, or grikes in the eroded dolostone could be a result of erosional stripping. The preservation of chalky regolith at locality 3, below the same transgressive sequence only 6 km farther east, and the seaward position of these exposures along the profile suggest minimal exposure.

The regolith developed at locality 3, located

6 km west of the position of inferred maximum regression (locality 2), is similar to chalky caliche described from Tertiary profiles (Esteban and Klappa 1983). Recrystallization of ooids to microspar, development of micro- and macroporosity, and incipient brecciation of host carbonate are all typical of lower portions of caliche profiles (James 1972, Walls et al. 1975, Esteban and Klappa 1983). Lack of rhizocretions is consistent with the Precambrian age of the deposit. This deposit, being directly overlain by marine transgressive deposits with only minimal reworking, appears to be a product of a single exposure event.

Flowstone was noted only at locality 4 in the profile. Such precipitative fabrics typically are formed in the vadose portions of caves where CO_2 degassing can occur, such as near openings of caves (James and Choquette 1984). The position of this flowstone deposit in the most seaward preserved cavity system may indicate proximity to the point of cave emergence.

The fracture systems or grikes filled with flowstone and sandstone which are developed from locality 4 eastward are characteristic features of the vadose portion of a karst terrane, being formed by the dissolving action of downward percolating rainwater (Sweeting 1972, Bögli 1980). Grikes extending downward to 2 m at locality 4 and 4 m at locality 5 give a minimum estimate of the extent of the zone of percolation in these areas. Coupled with features at localities 1 through 3, the eastward increase in depth of grike development appears to demonstrate a westward shallowing of the piezometric surface, which presumably merged with the marine base-level.

The main breccia zone at locality 6 must represent the formation and subsequent collapse of a major cavern system. That these breccias formed by collapse, rather than by tectonic processes or as talus, is demonstrated by the presence of abundant internal sediment which predated collapse. Both the clay-rich matrix and the chert-pebble conglomerate represent insoluble residues formed by the dissolution process which formed the cave system. The subarkosic sand in the chert-pebble conglomerate may represent insoluble residue or may be of allochthonous origin; the rarity of siliciclastic sand in the host dolostone favors an al-

lochthonous origin. Blocks of collapsed material which deform upper portions of the cross-stratified conglomerate further illustrate that these stream deposits formed in an enclosed cavity. The characteristically contorted bedding of the claystone matrix throughout the main breccia also reflects post-depositional collapse.

Collapse of the breccia must have occurred prior to marine transgression, because the onlap succession shows no signs of collapse, sag, or preferential thickening over the main breccia zone. The infill of interparticle pore space in the upper parts of the main breccia (and in several other related breccias) is also an indication that brecciation and collapse must have predated the transgression. Thus the karst landscape must have evolved from juvenile through senile stages (cf. Sweeting 1972, p. 169) during the period of regression.

The crest of the paleokarst landscape is unique in that it is situated directly atop the Muskox intrusion, where it is associated with local minor syndepositional faulting. Syndepositional faulting may explain the development and preservation of the talus-like deposits stacked against the erosional remnant, as opposed to the typical collapse breccias. The upper sand-rich breccias at this locality may represent marine reworking of the talus during the ensuing transgression. The presence of coniatolite crusts, also unique to this area, is consistent with their present-day environment of formation in a marginal-marine position receiving abundant sea spray (Purser and Loreau 1973, Scholle and Kinsman 1974).

The karst features described above occur along a disconformity surface which can be shown by independent stratigraphic reconstruction to have been uplifted at least 60 m above contemporaneous sealevel (Figs. 8.5 and 8.6). Although it cannot be proved that all of the karst features formed simultaneously, their distribution is consistent with longer exposure and weathering in the eastern, more uplifted portions of the profile.

Kerans (1983) suggested that doming during initial stages of emplacement of the intrusion may have controlled localization of the karsting. Stratigraphic reconstruction shows that 60 m of topographic relief occurs between the crest of the profile and the inferred coastline at locality 1, a distance of some 90 km (Fig. 8.6). Although this uplift is not great, its effect on karst development is clearly significant, as is seen by variability across the profile. In assessing the importance of local tectonic versus regional (eustatic) controls on karst development, it is critical to recognize that karst collapse breccias occur at an identical stratigraphic position in the Elu Basin of Bathurst Inlet, 300 km to the east (Campbell 1979, Kerans et al. 1981). More detailed work will be required to distinguish regional from local effects.

Summary of Stages of Karst Development

The evolution of the Dismal Lakes paleokarst profile can be summarized in five main evolutionary stages: (1) Initial exposure of the broad carbonate platform probably reflects the combined and possibly related effects of eustatic sealevel drop and local uplift related to emplacement of the Muskox intrusion. (2) Following uplift, juvenile karst development involved dissolution in both vadose and phreatic zones, resulting in the formation of a regolith, grikes, and phreatic caves. (3) During the mature stage of karst development the main cave system was largely within the vadose zone, while through-going streams reworked insoluble residues (chert fragments and clay) and brought in allocthonous sand from the flanking Early Proterozoic terrane. Flowstones also formed at this time near the seaward (western) entrance to this cave system. (4) Wholesale collapse of the cave and grike system during the senile stage occurred before transgression and marine onlap, as demonstrated by the conformable and undisturbed nature of the onlap succession. (5) Infill by siliciclastic sand of uncollapsed cavity spaces such as grikes and some breccia-related porosity, minor erosion of regolith, reworking of talus breccias, and precipitation of coniatolite crusts all resulted from marine transgression and coastal onlap.

Discussion

So far it has been assumed that the karst features described are related to subaerial exposure prior to deposition of the overlying mixed clastic-carbonate interval. Wright (1982) has discussed features which aid in the recognition of subaerial as opposed to intrastratal karst (karst developed at the interface between more permeable cover rocks and less permeable underlying carbonates). Such intrastratal karst may form at any time after deposition of the more permeable cover sequence, and thus has a different connotation than subaerial paleokarst (Wright 1982). Features which demonstrate the subaerial nature of the Dismal Lakes paleokarst include regolith deposits and peritidal transgressive deposits capping the karst surface, and flowstone and cave-floor stream deposits within the subsurface karst system. Furthermore, both grikes and the major collapse breccias must have been subaerially exposed, because they remained open until sand from the onlapping marine shoreline facies covered them.

Surface karst landforms have been used by several authors to deduce conditions of paleoclimate (Wright 1982) (Desrochers and James, this volume). The only surface karst landforms observed along the Dismal Lakes disconformity surface are grikes; rillenkarren, rinnenkarren, solution pools (kamenitza), solution pipes, and dolinas are all absent. The absence of these solution features may imply a temperate climate during subaerial exposure. Equally likely is the possibility that karstification occurred beneath a more extensive regolith or caliche horizon such as the one preserved only at locality 3.

The nature of the regressive event which led to development of this major paleokarst profile has significant implications for interpretation of regional geologic evolution of this portion of the Canadian Shield. Major paleokarst deposits throughout the Phanerozoic are commonly linked to tectono-eustatic events of regional or global scale. An example of such regionally extensive karst is that of Lower Ordovician age (Knox equivalent) which is recorded throughout the Appalachians, Arbuckles, the Permian Basin, and the mid-Continent (Walters 1946, Barnes et al. 1959) (Desrochers and James, this volume; Mussman et al., this volume). This paleokarst is in part contemporaneous with both a global eustatic lowstand and early phases of closure of Iapetus Ocean (Desrochers and James, this volume; Mussman et al., this volume).

Regression leading to paleokarst development in the Dismal Lakes Group appears to be related at least in part to local tectonism associated with early phases of emplacement of the Muskox intrusion (Kerans 1983), but it also can be related to a synchronous regression in the Bathurst Inlet area 300 km to the east. Other occurrences of correlative paleokarst are not known to the authors, but if present on other continents, they may serve as a start in developing a much-needed eustatic framework for the Proterozoic.

Conclusions

Subaerial exposure of landward portions of the upper Dismal Lakes Group peritidal carbonate platform (unit 16 of Baragar and Donaldson 1973, Kerans et al. 1981) led to the development of a spectrum of surface and subsurface karst features along a 90-km dip-parallel profile. The approximate position of maximum regression is marked by a sharp erosional disconformity lacking signs of subaerial exposure (locality 1). Progressively more extensive karst development, from regolith in the west (locality 3) through a zone of grikes (localities 4 and 5), into an area of major cave-collapse breccias (locality 6 and surrounding areas), reflects differential exposure associated with the 60 m of paleorelief.

The initial or juvenile stage of karstification includes formation of subsurface karst features such as grikes and caves (presumably by phreatic and mixing corrosion: Bögli 1980, James and Choquette 1984), and the development of minor regolith. The intermediate or mature stage records continued regression and vadose modification of subsurface karst, including deposition of autochthonous and allochthonous cave-floor sediments by free-flowing streams and the precipitation of flowstone near cave entrances. The final or senile stage of karst development is recorded by cave collapse. Evolution of the karst profile from juvenile through senile stages occurred prior to onlap by mixed siliciclastic-carbonate deposits of the ensuing transgression. The youngest features of the paleokarst surface are reworked talus deposits and coniatolite-like crusts which formed during onlap of the remnant horst block summit of the profile at locality 7.

The relative abundance of grikes and paucity of karren may indicate either that the climate was temperate during karst formation, or that deposits similar to the regolith at locality 3 covered the disconformity surface during exposure, and were stripped away during the ensuing transgression. The regional significance of this major episode of subaerial exposure and karst development must be proven by careful documentation of correlative karst features such as that in the Elu Basin of Bathurst Inlet (Campbell 1979). By analogy with some of the major paleokarst deposits of the Phanerozoic, the Dismal Lakes profile may record a major tectono-eustatic event during the Middle Proterozoic.

Acknowledgments. This report is based on data collected during regional field mapping of the Hornby Bay and Dismal Lakes Groups. Excellent logistical support in the field was supplied by the Department of Indian and Northern Affairs, and additional financial support for field and laboratory portions of this work were supplied by grant A5536 to Donaldson from the National Science and Engineering Research Council of Canada. The support of these organizations is gratefully acknowledged. G.M. Ross helped during mapping of the paleokarst terrane, and F.H.A. Campbell shared his knowledge of the correlative Elu Basin stratigraphy and paleokarst.

References

Baragar, W.R.A., and Donaldson, J.A., 1973, Coppermine and Dismal Lakes map areas: Geological Survey of Canada, paper, 71-39, 20 p.

Barnes, V.E., Cloud, P.E. Jr., Dixon, L.P., Folk, R.L., Jonas, E.C., Palmer, A.R., and Tynan, E.J., 1959, Stratigraphy of the pre-Simpson Paleozoic subsurface rocks of Texas and southeast New Mexico: Bureau of Economic Geology, publication no. 5924, 294 p.

Bögli, J., 1980, Karst hydrology and physical speleology: Berlin and Heidelberg, Springer-Verlag, 285 p.

Bowring, S.A., and Ross, G.M., 1985, Geochronology of the Narakay volcanic complex: implications for the age of the Coppermine Homocline and Mackenzie igneous events: Canadian Journal of Earth Sciences, v. 22, p. 774–781.

Campbell, F.H.A., 1979, Stratigraphy and sedimentation in the Helikian Elu Basin and Hiukitak Platform, Bathurst Inlet-Melville Sound, Northwest Territories: Geological Survey of Canada, paper 79-8, 18 p.

Esteban, M., and Klappa, C.F., 1983, Subaerial exposure environment, in Scholle, P.A., Bebout, D.G., and Moore, C.H., eds., Carbonate depositional environments: American Association of Petroleum Geologists, Memoir 33, p. 1–54.

Fahrig, W.F., and Jones, D.L., 1969, Paleomagnetic evidence for the extent of the Mackenzie igneous events. Canadian Journal of Earth Sciences, v. 6, p. 679–688.

Folk, R.L., and Assereto, R., 1976, Comparative fabrics of length-slow and length-fast calcite and calcitized aragonite in a Holocene speleothem, Carlsbad Caverns, New Mexico: Journal of Sedimentary Petrology, v. 46, p. 486–496.

Hoffman, P.F., 1981, Wopmay Orogen: a Wilson cycle of Early Proterozoic age in the northwest of the Canadian Shield, in Strangway, D.F., ed., Geological Association of Canada, special paper 20, p. 523–552.

James, N.P., 1972, Holocene and Pleistocene calcareous crust (caliche) profiles; criteria for subaerial exposure: Journal of Sedimentary Petrology, v. 42, p. 817–836.

James, N.P., and Choquette, P.W., 1984, Diagenesis 9—limestones—the meteoric diagenetic environment: Geoscience Canada, v. 11, p. 161–194.

Kendall, A.C., and Broughton, P.L., 1978, Origin of fabrics in speleothems composed of columnar calcite crystals: Journal of Sedimentary Petrology, v. 48, p. 519–538.

Kerans, C., 1982, Sedimentology and stratigraphy of the Dismal Lakes Group, Proterozoic, Northwest Territories: unpubl. PhD. thesis, Carleton University, Ottawa, 304 p.

Kerans, C., 1983, Timing of emplacement of the Muskox intrusion: constraints from Coppermine homocline cover strata: Canadian Journal of Earth Sciences, v. 20, p. 673–683.

Kerans, C., Ross, G.M., Donaldson, J.A., and Geldsetzer, H.J., 1981, Tectonism and depositional history of the Helikian Hornby Bay and Dismal Lakes Groups, District of Mackenzie, in Campbell, F.H.A., ed., Proterozoic basins of Canada: Geological Survey of Canada, paper 81-10, p. 157–182.

Kyle, J.R., 1983, Economic aspects of subaerial carbonates, in Scholle, P.A., Bebout, D.G., and Moore,

C.H., eds., Carbonate depositional environments, American Association of Petroleum Geologists, Memoir 33, p. 73–92.

Purser, B.H., and Loreau, J.P., 1973, Aragonitic, supratidal encrustation on the Trucial Coast, Persian Gulf, in Purser, B.H., ed., The Persian Gulf, New York, Springer-Verlag, p. 343–376.

Read, J.F., and Grover, G.A., Jr., 1977, Scalloped and planar erosion surfaces, Middle Ordovician limestone, Virginia: analog of Holocene exposed karst on tidal rock platforms: Journal of Sedimentary Petrology, v. 47, p. 956–972.

Ricketts, B.D., 1983, The evolution of a Middle Precambrian dolostone sequence—a spectrum of dolomitization regimes: Journal of Sedimentary Petrology, v. 53, p. 565–586.

Scholle, P.A., and Kinsman, D.J.J., 1974, Aragonitic and high-magnesium calcite caliche from the Persian Gulf—a modern analog for the Permian of Texas and New Mexico: Journal of Sedimentary Petrology, v. 44, p. 904–916.

Smith, C.H., Irvine, T.N., and Findlay, D.C., 1967, Muskox intrusion: Geological Survey of Canada, map 1213A.

Sweeting, M.M., 1972, Karst landforms: London, MacMillan, 362 p.

Tucker, M.E., 1983, Diagenesis, geochemistry, and origin of a Precambrian dolomite: the Beck Spring dolomite of eastern California: Journal of Sedimentary Petrology, v. 53, p. 1097–1120.

Vail, P.R., Mitchum, R.M., Jr., and Thompson, S., 1976, Seismic stratigraphy and global changes of sea level, part 3: relative changes of sea level from coastal onlap; in Payton, C.E., ed., Seismic stratigraphy—applications to hydrocarbon exploration, American Association of Petroleum Geologists, Memoir 26, p. 63–82.

Walls, R.A., Harris, W.B., and Nunan, W.E., 1975, Calcareous crust (caliche) profiles and early subaerial exposure of Carboniferous carbonates, northeastern Kentucky: Sedimentology, v. 22, p. 417–440.

Walters, R.F., 1946, Buried Precambrian hills in northeastern Barton County, central Kansas: American Association of Petroleum Geologists Bulletin, v. 30, p. 660–710.

Wilson, J.L., 1985, Petroleum reservoirs and karst: Society of Economic Geologists and Mineralogists, Annual Midyear Meeting Abstracts, v. 2, p. 97–98.

Wright, V.P., 1982, The recognition and interpretation of paleokarsts: two examples from the lower Carboniferous of south Wales: Journal of Sedimentary Petrology, v. 52, p. 83–94.

9
Early Paleozoic Surface and Subsurface Paleokarst: Middle Ordovician Carbonates, Mingan Islands, Québec

André Desrochers and Noel P. James

Abstract

Middle Ordovician platform carbonates on the Mingan Islands are characterized by two major paleokarst unconformities and numerous local paleokarst surfaces which are today exposed on extensive bedding planes.

The lower paleokarst unconformity was developed on Lower Ordovician dolomites (Canadian to earliest Whiterockian) of the Romaine Formation. It corresponds to the Knox or Beekmantown unconformity present elsewhere in eastern North America. The post-Romaine unconformity represents an extensive karst plain that was veneered by eolian sediments and exhibits a variety of surface solution features ranging from karren to small dolines. Subsurface features are local in the form of small caves with collapse breccias.

The upper paleokarst unconformity, within the Middle Ordovician limestones (Chazyan) of the Mingan Formation, displays substantial karst relief which was modified by intertidal erosion preceding renewed submergence. The intra-Mingan unconformity, at a smaller scale, is sculpted into a variety of sharp-crested karren and basin features.

These two unconformities developed during uplift caused by Taconic Orogenesis some 500 km to the east. Eustatic sealevel changes also occurred but were probably less important in the formation of these unconformities.

Local paleokarst horizons also separate meter-scale shallowing-upward calcarenite cycles. The short-lived exposure events resulted in lithification followed by karst erosion forming sharp karren. If exposure was prolonged, the karren progressively widened and eventually developed into extensive planar surfaces. The origin of these cycles is explained by shoaling with the paleokarst surfaces, capping the cycles, formed during periods of relative sealevel fall.

Introduction

The Mingan Islands, located along the Québec North Shore of the Gulf of St. Lawrence (Fig. 9.1), consist of superb coastal exposures and are among the few localities where a carbonate sequence may be studied at various scales in three dimensions. Of particular importance is the presence of paleokarst exposed today on extensive bedding planes.

Two different styles of paleokarst are recognized in the sequence: (1) major paleokarst unconformities of regional extent, and (2) local paleokarst horizons associated with shallowing-upward calcarenite cycles. Paleosols and calcrete are entirely missing from these karst surfaces, and the majority of their features have been produced essentially by surface water runoff. The purposes of this chapter are to document the different types of paleokarst and to explain their development.

Geological Setting

The Lower and Middle Ordovician sequence exposed in the Mingan Islands is about 120 m thick and consists mainly of shallow marine carbonates of late Canadian to Chazyan age (Nowlan 1981). Strata are essentially undisturbed with an east–west strike and very gentle (usually 1° to 2°) dip to the south (Fig. 9.2). The general stratigraphy, previously recognized by Twenhofel (1938), comprises dolostones of the Romaine Formation and overlying limestones of the Mingan Formation (Fig. 9.3). A thin sil-

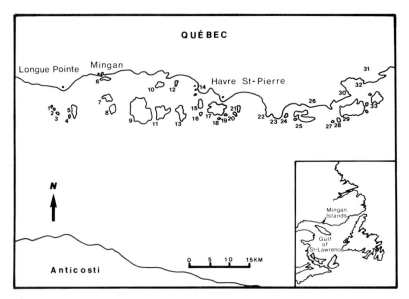

FIGURE 9.1. Location map of the Mingan Islands, Québec. Locality names follow the Commission de Toponymie du Québec (Gauthier-Larouche 1981). 1: Ile aux Perroquets. 2: Ile de la Maison. 3: Ile du Wreck. 4: L'Ilot. 5: Ile Nue de Mingan. 6: Ile du Havre de Mingan. 7: Ile à Bouleaux de Terre. 8: Ile à Bouleaux du Large. 9: Grande Ile. 10: Ile de la Grosse Romaine. 11: Ile Quarry. 12: Ile de la Petite Romaine. 13: Ile Niapiskau. 14: Pointe aux Morts. 15: Ile du Fantôme. 16: Ile à Firmin. 17: Ile du Havre. 18: Ile à Calculot. 19: Ile aux Goélands. 20: Petite Ile au Marteau. 21: Grosse Ile au Marteau. 22: Grande Pointe. 23: Pointe Enragée. 24: Ile de la Fausse Passe. 25: Ile Saint-Charles. 26: Baie Puffin. 27: Ile à Calculot des Betchouanes. 28: Ile Innu. 29: Ile à la Chasse. 30: Pointe de la Perdrix. 31: Mont Sainte-Geneviève. 32: Pointe du Sauvage. 33: Ile Sainte-Geneviève.

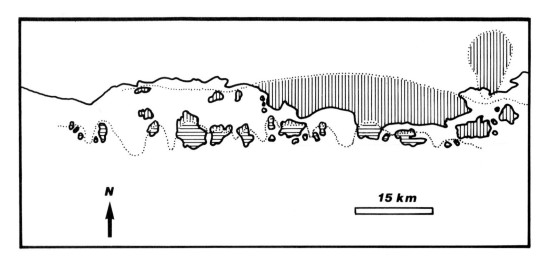

 Mingan Formation   Romaine Formation

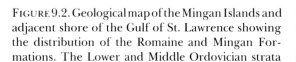

FIGURE 9.2. Geological map of the Mingan Islands and adjacent shore of the Gulf of St. Lawrence showing the distribution of the Romaine and Mingan Formations. The Lower and Middle Ordovician strata rest unconformably on Precambrian basement with an east–west regional strike and gentle dip, usually 1° to 2°, toward the south. The dotted lines indicate geological contacts.

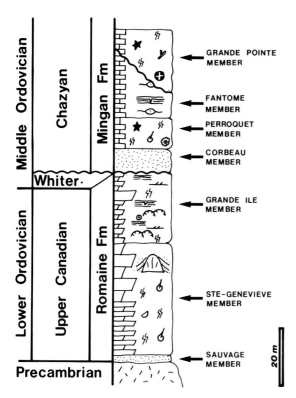

FIGURE 9.3. Generalized stratigraphic section of the Romaine and Mingan Formations showing ages, lithologies, and provisional member names. "Whiter." stands for Whiterockian.

iciclastic unit, however, is present at the base of both formations.

The Romaine Formation represents a "classic" shallowing-upward sequence deposited during early to mid Ordovician (late Canadian to earliest Whiterockian) time. The sequence is composed of three facies associations (Fig. 9.3): (1) a basal assemblage of mature sandstones deposited during initial flooding on Precambrian basement (Sauvage Member), (2) a middle, "noncyclic" assemblage of open shelf limestones (now dolomitized) deposited as extensive bioturbated mudbanks and thrombolite patch reefs (Sainte-Geneviève Member), and (3) an upper, "cyclic" assemblage of sabkha-type tidal flat dolostones (Grande Ile Member).

An extensive paleokarst unconformity occurs on top of the Romaine Formation and is usually overlain by basal sandstones of the Mingan Formation. Most of Whiterockian time, estimated to be 10 m.y. in length (Ross et al. 1982), is missing and represented by this unconformity in the study area.

The Mingan Formation defines an overall transgressive sequence but with numerous second-order superimposed transgressions and regressions. The formation is composed of two major sedimentary packages separated by a marked paleokarst unconformity (Fig. 9.3). Mingan strata beneath the intra-Mingan unconformity consist of two lithofacies assemblages: (1) lower tidal flat and nearshore siliciclastics (Corbeau Member), and (2) upper Bahamas-type tidal flat and open shelf limestones (Fantôme and Perroquet Members), the latter geographically restricted to westernmost localities.

The intra-Mingan unconformity developed on exposed peritidal limestones and resulted in the formation of a karst topography with substantial relief (ca 20 m). Biostratigraphic data indicate that this paleokarst unconformity represents a relatively short depositional break during Chazyan time (Nowlan 1981).

In contrast to the peritidal siliciclastics and limestones of the lower Mingan Formation, the limestones (Grande Pointe Member) above the unconformity comprise a wide spectrum of shallow, subtidal facies, including sediments deposited in semirestricted lagoons, beaches, small patch reefs, sand shoals, and low-energy, open shelf settings. Subtidal sedimentation was locally controlled by karst erosion topography with sand shoals forming tidal deltas, 1 to 10 km wide, in the large depressions. Three distinct calcarenite cycles, each capped by a paleokarst surface, make up these sand shoal deposits. Details concerning the stratigraphy and depositional environments can be found in Desrochers (1985).

Paleokarst Unconformities: Post-Romaine Paleokarst

The post-Romaine paleokarst is well exposed at more than 20 localities and is usually overlain by basal sandstone of the Mingan Formation. These sandstones are characterized by a marine fauna (mainly brachiopods) and trace fossils and by quartz particles with a mature, bimodal texture commonly found with deflationary sands in modern eolian environments (Folk 1968). The basal sandstones were reworked from a blanket of residual eolian sediments covering the exposed Romaine sequence, and redeposited as strandline sands with renewed submergence during Chazyan time. The post-Romaine paleokarst is characterized by: (1) a regionally widespread mappable unconformity, (2) various solution surface features ranging from small dolines to karren, (3) solution subsurface features (collapse breccias), and (4) associated features such as borings, and pebble lags.

Regional Unconformity

On a regional scale, the post-Romaine unconformity in cross-section appears to be a planar surface with little or no relief (Fig. 9.4). This is also reflected by the uniform thickness (averaging 4–5 m) of the overlying Corbeau Member at the base of the Mingan Formation which is usually capped by a marker bed of silty dololaminite. At some localities however, variations in thickness of the member may result from local relief of a few meters superimposed on the regional unconformity. The uncon-

formity truncates older beds toward the east, where the Mingan Formation directly overlies the Sainte-Geneviève Member of the Romaine Formation (Fig. 9.4). As much as 30 m of Romaine strata are missing in the easternmost localities, indicating that the Romaine platform was tilted and eroded during the development of this paleokarst.

Solution Surface Features

Various solution features are present on top of Romaine Formation dolostones and include: dolines forming first-order relief, rundkarren, kamenitza, and planar surfaces with small-scale karren with two-end member forms. The reader is referred to James and Choquette (1984, p. 171–173) for a general summary on surface karst.

Dolines. Small depressions are present throughout the area, especially in the eastern Mingan Islands, where they are common (Fig. 9.5A). These bowl-shaped basins range from 1

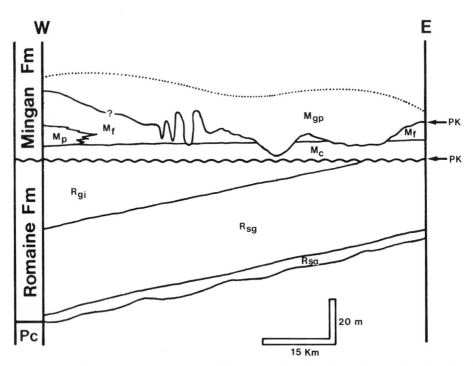

Figure 9.4. Generalized cross-section parallel to regional strike illustrating the distribution of stratigraphic units within the Romaine and Mingan Formations. Letters correspond to the different members. R_{sa}: Sauvage Member. R_{sg}: Sainte-Gene-viève Member. R_{gi}: Grande Ile Member. M_c: Corbeau Member. M_f: Fantôme Member. M_p: Perroquet Member. M_{gp}: Grande Pointe Member. Note the position of paleokarst unconformities (PK).

to 3 m in depth and 5 to 20 m in diameter. Intraformational breccias (see below for description) are sometimes associated with the depressions but their exact relationship is difficult to evaluate because pervasive dolomitization obscures lithological fabrics. In some cases, breccia masses characterized by flat floors, steep, irregular walls, and roofs with a distinct V-shape are present beneath the dolines.

Dolines on or near the top of the Romaine Formation are thought to result from the subsurface development of discrete bodies of carbonate breccia which formed by removal of the underlying carbonates and subsequent foundering of the cave walls and roof.

Rundkarren. Rundkarren, or rounded solution runnels, form dendritic to subparallel grooves on the paleokarst unconformity (Fig. 9.5B, C). The grooves range from 10 to 40 cm in width and 5 to 30 cm in depth. They are round to flat in cross-section, closely spaced, and separated by rounded-crest ridges. Their length varies from a few meters to over 15 m in some parallel forms. These features are common on extensive, homogeneous bedding surfaces of sucrosic dolomite, especially in the western Mingan Islands (i.e., Grande Ile, Ile Quarry, Ile Niapiskau), where a thin unit of these porous dolomites is present at the top of the Romaine Formation.

Modern rundkarren are solution sculptures forming beneath a soil cover where only smooth and rounded forms are created and an enriched soil CO_2 atmospheric source is present (Sweeting 1973, Bögli 1980). The orientation of these karren was clearly related to local drainage conditions which depended primarily on the slope and dip of the cover/rock interface. Although rounded, the Romaine rundkarren lack the depth and some other characteristics of what most karst geomorphologists would consider to be rundkarren and may also be regarded as a subtype of rinnenkarren (D. Ford, pers. comm., 1985). This type of karren may develop beneath unconsolidated material where the flow of meteoric water which carves them is concentrated into point discharges, rather than being in sheet flow, by some controls operating at and upstream of the head of the karren channels themselves. In this case, an atmospheric source of CO_2 may be sufficient to

develop these features without the contribution of soil CO_2.

Against this background it is suggested that the post-Romaine rundkarren formed beneath a sediment cover and on subhorizontal surfaces for the short and dendritic forms, whereas longer and parallel forms developed on more inclined surfaces.

Kamenitzas. Kamenitzas are solution basins with circular to oval outlines, flat bottoms, and rounded edges (Fig. 9.5D). These small basins range from 20 to 80 cm in diameter and from 5 to 30 cm in depth. They are commonly interconnected with dendritic rundkarren and rarely found alone.

Modern kamenitzas are usually developed on horizontal to slightly inclined substrates where solution by stagnant water forms pools and basins (Sweeting 1973). On the other hand, kamenitzas also form today in the intertidal/supratidal zone through the destruction of limestone coasts by bioerosion and mixing corrosion, but they are characterized by sharp-crested forms and by a highly pitted surface (Schneider 1976).

The post-Romaine kamenitzas are interpreted to be subaerial solution sculptures developed beneath a sediment cover as evidenced by their smooth and rounded edges and their association with rundkarren.

Planar Surfaces. The paleokarst unconformity is commonly featureless and exhibits a planar surface ornamented only by rare cm-deep pits. Rundkarren and kamenitzas are rare or poorly developed, if present. These surfaces are generally found where finely crystalline dolostones of the Grande Ile Member are present at the top of the Romaine Formation.

Of particular importance to the interpretation of this surface is the work of Pluhar and Ford (1970), who studied dissolution of fresh dolomite blocks with dilute hydrochloric acid in the laboratory. On bare, sloping surfaces (5° to 30°), small sharp-crested karren were produced, whereas dolomite karsting was limited to micropitting on sloping surfaces covered with an artificial silica sand.

The formation of karren on the post-Romaine planar surfaces was probably similarly inhibited by the presence of a cover of residual eolian sand (Desrochers 1985), which caused

Figure 9.5 A–D. Post-Romaine paleokarst (solution surface features). *A*. Field photograph (bedding plane view) of a doline forming a bowl-shaped basin on top of the Romaine Formation. Basin dimensions are 3.0 m in depth and 20.0 m in diameter. Baie Puffin. *B*. Field exposure of the contact between dolostones of the Romaine Formation and overlying cross-bedded sandstones of the basal Mingan Formation. Note the presence of subparallel rundkarren (arrows) on top of the Romaine Formation. Grande Ile E. *C*. In closer view, rundkarren consist of rounded grooves that are separated by rounded-crest ridges. Field notebook is 20 cm. *D*. Field photograph (bedding plane view) of a kamenitza on top of the Romaine Formation. Kamenitza consist of solution basins with circular outlines, flat bottoms, and rounded edges. Note the presence of abundant *Trypanites* borings on the kamenitza surface (detail in Figure 9.7A). Ile Niapiskau W. Scale in cm.

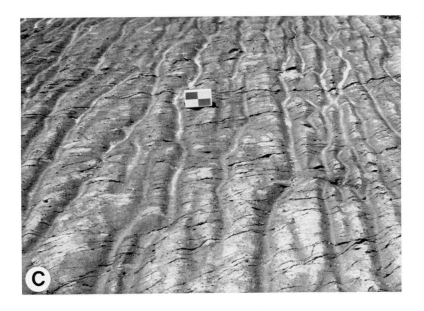

FIGURE 9.5

spreading and diffusion of meteoric water over the surface. Furthermore, the fine-grained nature of dolomite may have formed a nonporous substrate influencing the rate of karren formation. Pluhar and Ford (1970) describe similar relationships on modern karren developed in Silurian dolomites of southern Ontario.

Subsurface Solution Features

Collapse Breccias. Small, local intraformational breccias are ubiquitous, and especially common in the eastern Mingan Islands, where they occur 10 to 30 cm beneath the post-Romaine unconformity (Fig. 9.6A). These breccias form irregular masses that range from a few meters to 20 m in size. Breccia margins, at least in one locality, are parallel oriented and delimited by sharp, vertical fractures. Joints or fractures, however, have not been observed elsewhere. The breccias are clast-supported with a matrix of fine detrital dolomite grains or coarse, white dolomite cement. The clasts are angular to subrounded, and range from 1 to 30 cm in size (Fig. 9.6B). Large fragments exhibit fitted fabrics. They are poorly sorted, oligomictic in

FIGURE 9.6 A,B. Post-Romaine paleokarst (subsurface solution features). *A*. Field photograph (vertical cliff exposure) of intraformational breccia (b). Note that the adjacent strata(s) display no evidence of disruption. Baie Puffin. *B*. Field photograph (cross-section view) of intraformational breccia composed of chaotic, commonly vertically oriented, clasts. Baie Puffin. Hammer is 30 cm.

composition, and clearly derived from the same stratigraphic horizon.

A tectonic origin is unlikely because the breccias are stratigraphically controlled, and adjacent strata display no evidence of disruption. These intraformational breccias most likely represent collapse features due to partial removal of soluble material. Fractures may have controlled the development and distribution of these features and facilitated circulation of meteoric water in the subsurface. On the other hand, the formation of these breccias may also be due to dissolution in a mixing zone concomitant with pervasive dolomitization. There is growing evidence that dissolution is important in the mixing zone; for instance, the northeastern Yucatan coast exhibits scalloped mor-

phology and related collapse features which apparently developed in the coastal mixing zone (Back et al. 1979 and 1984). The second hypothesis, if correct, would mean that the breccias are relict, intrastratal features rather than breccias generated contemporaneously with the subaerial exposure during Whiterockian time.

Associated Features

Borings. Abundant borings (up to 40 tubes/cm^2) are present on the paleokarst unconformity but are patchy in distribution and occur on both finely crystalline and sucrosic dolomites (Fig. 9.7A, B). These borings cut through sucrosic dolomite crystals. They are simple, unbranched

FIGURE 9.7 A–C. Post-Romaine paleokarst (associated features). *A.* Field photograph (bedding plane view) showing abundant *Trypanites* borings (dark spots) on a planar surface at the top of the Romaine Formation. Boring aperture is 1 mm to 2 mm in diameter. Ile Niapiskau W. *B.* Field photograph (cross-section view) of *Trypanites* borings at the top of the Romaine Formation. Borings are 3 mm to 5 mm long. Ile Niapiskau W. *C.* Field photograph (bedding plane view) of chert and dolomite lithoclasts in an argillaceous sandstone matrix overlying the Romaine Formation. Grosse Ile au Marteau. Lens cap is 50 mm in diameter.

tubes with a circular aperture oriented normal to the surface and belong to the ichnogenus *Trypanites* (Pemberton et al. 1980). Their size averages 1–2 mm in width and 3–5 mm in length. They are filled with terrigeneous material of the basal Mingan Formation.

Trypanites borings were formed after the karst surface developed on well-lithified dolomite substrates and were subsequently filled with sediments. It is not clear what organisms produced these borings but marine siphunculid worms are most likely, because of the overall morphology of the borings (Pemberton et al. 1980).

Pebble Lags. Pebble lags 10 to 30 cm thick are found on local depressions above the post-Romaine paleokarst (Fig. 9.7C). They consist of chert and dolomite lithoclasts in an argillaceous sandstone matrix in which the quartz particles have a mature, bimodal texture. Lithoclasts range from 1 to 20 cm in size, clast- to matrix-supported in texture, and are poorly sorted. Chert lithoclasts are subspherical in shape and have pitted surfaces (bored?). Many of the chert pebbles are identical to chert nodules found in the upper Romaine lithofacies, especially dololaminites, and so suggest that they were eroded out and concentrated at the unconformity surface. Dolomite lithoclasts are angular to subrounded with flat, discoid shapes and are sometimes bored by *Trypanites*. These borings, however, are usually absent on the unconformity beneath the pebble lags.

Pebble lags are probably a regolith (i.e., residual material) on top of the Romaine Formation that was slightly reworked but not completely removed by waves and currents with the renewed marine transgression. This is suggested by their stratigraphic position above the unconformity and their local derivation.

Interpretation

On the basis of the preceding evidence, the sequential development of post-Romaine paleokarst is envisaged to have taken place in four stages, summarized in Figure 9.8. It began with a marine regression that exposed the Romaine sequence during early Whiterockian time. Romaine lithofacies were pervasively dolomitized prior to this time by contemporaneous dolomitization of peritidal limestones in sabkha-

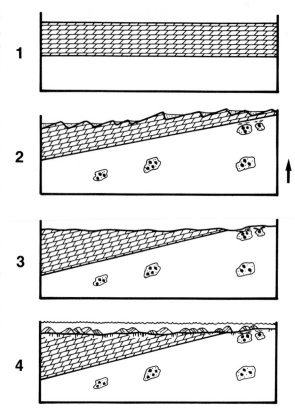

FIGURE 9.8. Schematic cross-sections illustrating the sequential development of the paleokarst unconformity capping the Romaine Formation. Stage 1: Romaine Formation prior to exposure consisted of a lower unit of sucrosic dolostones and an upper unit of dolomicrites (dolomite pattern). Stage 2: epeirogenic movements caused tilting of the platform and differential weathering of the Romaine dolomites. Stage 3: progressive weathering resulted in an extensive karst plain with little relief on which a variety of solution features developed. Stage 4: Chazyan transgression reworked an eolian sand blanket overlying the post-Romaine paleokarst, which was eventually exposed, at least locally, on the sea floor, bored by endolithic organisms, and finally covered by sandy material. Note the presence of the intraformational breccias of possible collapse origin beneath the paleokarst surface (clastlike objects in stages 2, 3, and 4). See text for further details.

like environments, whereas sucrosic dolomites probably formed by replacement of subtidal limestones in shallow mixing zones (Desrochers 1985).

During initial exposure (Fig. 9.8, stage 2), tectonic movements caused tilting of the sequence, and irregular topographic features formed by differential weathering of the Romaine dolomites. Scarce dolomite clasts in the postunconformity sandstones suggest that dissolution, rather than physical weathering, was the main process.

During stage 3, positive topographic elements were progressively eroded and eventually peneplained to produce an extensive karst plain with little relief (Fig. 9.8). Stratigraphic relief along the unconformity, however, increases in the study area from west to the eastern part where approximately 30 m of upper Romaine strata are missing.

A variety of small karst features formed on top of the unconformity, controlled by: (1) the presence of a sediment cover, (2) the substrate nature (i.e., finely crystalline or sucrosic dolomites), (3) the local drainage conditions. A sediment cover above the paleokarst surface was responsible for the development of smooth karren forms rather than sculpted forms. Well developed solution features were formed only over areas with substrates of sucrosic dolomite. This type of dolomite is relatively porous and permeable due to high intercrystalline porosity, and so was probably more susceptible to dissolution than the impermeable dolomicrites. The style of the solution features, ranging from subparallel karren to kamenitzas, indicates that local drainage conditions were present and controlled by the dip slope of the substrate. It should be noted that highly inclined substrates are not necessary to form subparallel solution forms, and substrates with less than 5° may develop such features (Sweeting 1973). Local drainage conditions are probably relict from the earlier tectonic tilting of the platform (i.e., stage 2) because most rundkarren forms are oriented in a persistent east–west direction.

Subsurface solution features consist of intraformational breccias that are stratigraphically restricted to the lower unit of sucrosic dolomite. It is not clear whether their origin was due to concomitant dissolution and dolomitization of preexisting subtidal limestones or to dissolution by meteoric water percolating downward from the paleokarst unconformity. The first hypothesis is more likely because permeability was probably insufficient to allow downward water movement, due to little or no primary porosity in the dolomicrites capping the formation, and also due to poor secondary porosity along joints

and fractures. Furthermore, minor topographic relief following stage 2 of the unconformity development may have provided insufficient hydraulic gradient to assure groundwater circulation in the system. Permeability and groundwater circulation are two fundamental requirements for the development of an active karst system regardless of adequate precipitation (Stringfield et al. 1979). Thus, surface solution processes were more likely favored over those operating in the subsurface. Nevertheless, large, bowl-shaped structures observed today on the top of the Romaine Formation are thought to result from the collapse of the rocks overlying the intraformational breccias. Whether these structures are true dolines or relict subsurface features exhumed by erosion at the unconformity surface remains unclear.

The final stage of the unconformity development occurred with renewed marine transgression during Chazyan time (Fig. 9.8, stage 4). The cover of residual eolian sand overlying the paleokarst unconformity was completely reworked during the transgression, but local accumulations of chert–rubble conglomerate remained in more protected sites. It is possible that the sediment cover and possibly shoreline deposits (beaches, barrier-island coasts) were destroyed by the transgression and reworked by marine currents in a process similar to the Holocene transgressive barrier of the U.S. middle Atlantic coast (Swift 1975). The proposed mechanism consists of shoreface erosion and offshore deposition under conditions of relatively slow and steady sealevel rise, and as a result a transgressive disconformity or "ravinement" underlies sediments deposited in shelf environments (Swift 1968). Deep shoreface erosion during the Chazyan transgression across the area, however, was limited by well lithified Romaine dolomites which were eventually exposed on the sea floor, bored by marine endolithic organisms, and finally covered by recycled sandy material deposited as subtidal sandbars near the strandline.

Intra-Mingan Paleokarst

Another paleokarst unconformity is present within limestones of the Mingan Formation separating the uppermost Grande Ile Member from the underlying members (Fig. 9.3). This unconformity exhibits substantial relief, which in turn controls present distribution of the underlying members due to erosion and appears to have controlled sedimentation in the uppermost member with renewed submergence. The intra-Mingan paleokarst unconformity is also characterized by (1) a regionally widespread mappable surface, (2) minor solution surface features exhibiting sharp-crested forms, and (3) associated features such as sediment-filled fissures, paleonotches, and encrusting organisms.

Regional Unconformity

The paleokarst unconformity exhibits irregular topographic features of landforms which are smooth on a regional scale. The unconformity in the eastern Mingan Islands is characterized by two regional depressions (6–10 km wide), one centered on Grande Pointe and one on Ile Saint-Charles (Fig. 9.4). In contrast, the unconformity in the western islands is more irregular with closely spaced depressions and ridges or pinnacles, usually less than 2 km to 3 km apart. Landforms (ridges, depressions, pinnacles) demonstrate up to 20 m of local relief as measured either by correlation of key beds between closely spaced sections or by mapping along extensive seacliff exposures (Fig. 9.9). The topographic gradient associated with the unconformity (Fig. 9.10A, B, C) is usually 1° to 5°, but reaches 20° in some areas (e.g., Ile du Fantôme, Ile du Havre SW).

Surface Solution Features

Small superficial karren are superimposed on the paleokarst unconformity and consist of sculpted runnels (rinnenkarren) and basins (kamenitza). Their surface is in sharp contact with underlying beds. In thin section, both sedimentary particles and cements are truncated along this surface.

Rinnenkarren. Rinnenkarren or crested solution runnels are similar in size and in shape to rundkarren except that their steep to vertical sides are characterized by sharp edges (Fig. 9.11A, B). These karren have rounded to scalloped bottoms. They are commonly sinuous, randomly oriented, and 20 cm to 50 cm in

length. In some areas however, they are longer (up to 5 m) and oriented parallel with respect to the slope of the unconformity surface.

Modern rinnenkarren form only on bare surfaces where meteoric water flows unhindered (Sweeting 1973, Bögli 1980). Their length and parallelism usually increase with increasing slope. A similar origin is envisaged for rinnenkarren present at the top of the intra-Mingan paleokarst unconformity.

Kamenitza. Kamenitzas are similar to those described from the paleokarst unconformity capping the Romaine Formation but have steep sides with irregular and sharp edges that in places overhang. Kamenitzas are commonly associated with short and sinuous rinnenkarren and grade laterally into subparallel rinnenkarren within tens of meters.

These "fossil kamenitzas" formed in either subaerial or coastal settings. Kamenitzas with similar sculpted edges are observed today on slightly inclined to horizontal surfaces lacking cover where running water is more stagnant and forms pools and basins (Sweeting 1973). They are also present in the tidal zone of modern tropical limestone coasts (Scheinder 1976).

Associated Features

Sediment-Filled Fissures. Fissures with sedimentary fillings are locally present beneath the paleokarst unconformity. These fissures form a complex network and link together more porous horizons with abundant biomoldic and vuggy pores (Fig. 9.12A). Vertical fissures are planar to slightly curved and commonly oriented east–west. Most of them are less than 100 m long and 8 cm wide. They are occasionally cut by minor secondary fissures which show no preferred orientation. They are partly filled by marine internal fibrous cement and/or internal sediment similar to those found in adjacent moldic pores. The most common internal sediment, however, is red geopetal carbonate mud containing abundant ostracods and some gastropod fragments (Fig. 9.12B). Angular, cm-sized clasts of host material (fenestral mudstones) are sometimes present in the fissures. In some cases, several generations of internal sediment are laterally superimposed in these fissures, suggesting that additional material was injected as the fissures were opening.

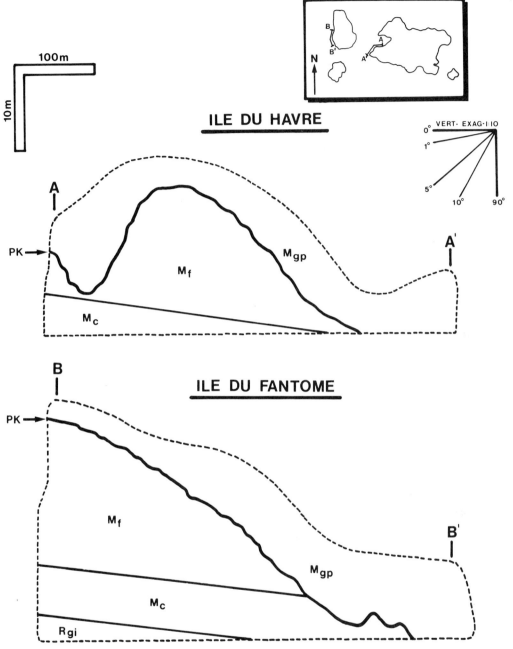

FIGURE 9.9. Outcrop sketches showing the intra-Mingan paleokarst unconformity in cross-section. Profile A–A¹ located at Grande Anse on Ile du Havre (SW). Profile B–B¹ located at Anse à Michel on Ile du Fantôme (west). Note the vertical exaggeration (1:10). M_c: Corbeau Member. M_p: Perroquet Member. M_f: Fantôme Member. M_{gp}: Grande Pointe Member. R_{gi}: Grande Ile Member of the Romaine Formation. Note the presence of paleokarst surface (PK).

Fissuring of the rock is most likely a product of fracturing and solution widening associated with subaerial exposure of the Mingan sequence. Fissures probably acted as conduits bringing internal sediment into porous sub-surface units. Subsequent filling of these features indicates that the sequence was, at least periodically, in contact with circulating marine waters during incipient exposure as indicated by the fossiliferous marine infilling. Unlike

Figure 9.10 A–C. Intra-Mingan paleokarst (regional unconformity). *A.* Field photograph (vertical cliff exposure) of the intra-Mingan paleokarst unconformity (black line). The unconformity is irregular and directly overlain by a cluster of small sponge-bryozoan bioherms. Ile du Fantôme W. Hammer for scale. *B.* Field photograph (cliff exposure) of the intra-Mingan paleokarst unconformity (black line). The unconformity is characterized by a regular topographic gradient from the right to the left where 3.5 m of relief is present. Note the presence of numerous coral bioherms forming small massive units above the unconformity. Field assistant is 2.0 m high (circle) for scale. Ile de la Fausse Passe E. *C.* Field photograph (cliff exposure of the intra-Mingan paleokarst unconformity (arrows). The unconformity is characterized by a regular topographic gradient from the left to the right where 4.0 m of relief is present. Note the slight angular relationship between strata on both sides of the unconformity. Ile du Fantôme S.

FIGURE 9.11 A,B. Intra-Mingan paleokarst (solution surface features). *A.* Field photograph (cross-section view) of small superficial karren or kamenitzas overlain by skeletal limestones of the Grande Pointe Member. Note the steep to vertical sides of sculpted karren characterized by sharp edges. Ile de la Fausse Passe E. Hammer is 30 cm. *B.* Field photograph showing (arrows) small oriented karren both in cross-section and on bedding plane. Ile de la Fausse Passe E. Hammer is 30 cm.

modern ostracods that are ubiquitous in all aquatic environments, ostracods during the Early Paleozoic are found only in marine environments (Horowitz and Potter 1971, Sohn 1985). Red sediment, often referred to as *terra rossa*, is also a common fill in cracks of emergent rocks in such places as the Bahamas (Roehl 1967). Terra rossa is considered as the by-product of pedogenic processes and subaerial solution of limestones and consists of red non-calcareous mudstone (Pye 1983). On that basis, the red sediment filling the fissures beneath the intra-Mingan paleokarst cannot have been derived from a true residual soil, although it is

possible that minor insoluble residues (oxidized clays) were washed into the fissures and mixed with internal marine sediment, staining the latter a characteristic red color. Furthermore, the occurrence of karren with sharp-crested forms argues against a significant development of soil above the paleokarst surface.

Paleonotch. The paleokarst surface is locally characterized by steep, vertical walls (Fig. 9.12C) which are 50 to 150 cm high and morphologically similar to intertidal–subtidal notches observed along tropical carbonate coasts (Neumann 1966, Semeniuk and Johnson

1985). The Mingan paleonotches generally exhibit smooth surfaces, and borings such as *Trypanites* are absent.

The ecological zonation of modern carbonate shorelines is expressed in both epilithic and endolithic animals and plants, which play an important role in their destruction (Scheinder 1976). The origin of these Ordovician features is unclear, as we know little about bioerosion of carbonate shorelines during the Early Paleozoic. The absence of unquestionable borings suggests that either physical destruction due to wave action or dissolution at the shoreline (i.e., mixing) was responsible. Physical erosion is discounted because erosional blocks at, or near, the unconformity contact are never observed. Nevertheless, there is growing evidence for the presence of endolithic sponges in the geological record as early as Middle Ordovician (Kobluk 1981, Pickerill and Harland 1984). Endolithic sponges are important in the bioerosion of modern limestone coasts (Neumann 1966).

Encrusting Organisms. Rare encrusting organisms are found in situ on the paleokarst unconformity. The most common fossils are laminar trepostome bryozoans. Crinoid holdfasts with simple discoid forms are present but are noticeably less common (Fig. 9.12D). In addition to these organisms, lithistid sponges with various growth forms (globular, cylindrical, saucer) are commonly present above the paleokarst surface but display some evidence of slight reworking and transport. These marine organisms colonized or lived attached to the paleokarst surface, which provided a hard substrate for their growth during marine transgression.

Interpretation

The sequential development of the intra-Mingan paleokarst unconformity is also envisaged in four stages and illustrated in Figure 9.13. Prior to subaerial exposure, the Mingan sequence was composed of a lower siliciclastic unit (shales, sandstones) and an upper limestone unit (Fig. 9.13, stage 1). Depositional environments indicate that peritidal conditions prevailed across the study area, and so any significant sealevel fall would have simultaneously affected the entire area.

The lower Mingan sequence, affected by tec-

tonic movements (see below), was gradually exposed causing extensive fissuring of the lithified early Mingan limestones (Fig. 9.13, stage 2). Subaerial exposure was also accompanied by dissolution in the upper limestone unit as evidenced by common biomoldic and vuggy porosity. Dissolution, however, apparently did not affect the underlying Romaine sequence, possibly because shales in the upper part of the siliciclastic unit acted as an aquiclude, inhibiting

FIGURE 9.12 A–D. Intra-Mingan paleokarst (associated features). *A.* Field photograph (cross-section view) of vertical sediment-filled fissures (arrows) occurring in fenestral mudstones of the Fantôme Member beneath the intra-Mingan paleokarst. Ile du Fantôme S. Lens cap for scale. *B.* Polished slab (cross-section view) of a sediment-filled fissure. Note the geopetal nature of the sediment filling (F), carbonate mud with abundant ostracods (oval particles in the upper half of the fissure). *C.* Field photograph (cross-section) of a notch along the intra-Mingan paleokarst. Note the steep vertical wall of the paleonotch (0.6 m high). Ile du Fantôme E. Hammer is 30 cm. *D.* Field photograph (bedding plane view) of crinoid holdfasts (arrows) encrusting the paleokarst surface. Ile Quarry N.

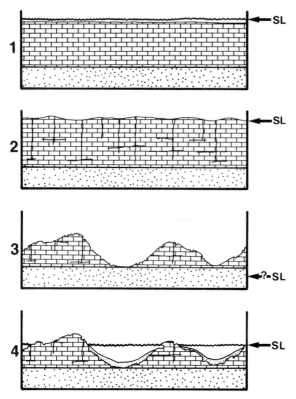

FIGURE 9.13. Schematic cross-sections illustrating the sequential development of the intra-Mingan paleokarst unconformity. Stage 1: Deposition of the Mingan Formation, prior to subaerial exposure, consisting of a lower unit of peritidal siliciclastics and an upper unit of peritidal limestones. Stage 2: epeirogenic movements caused extensive fissuring of the early lithified Mingan limestones. Stage 3: karst landforms developed with further sealevel fall, and their surface was sculpted into a variety of small karren forms. Stage 4: restricted, subtidal carbonate sediments with renewed submergence were deposited first in the lowest depressions but eventually in a wide spectrum of normal marine subenvironments as marine circulation increased. See text for further details.

A period of 200,000 to 400,000 years may have been necessary to produce the karst landforms, based on known rates (50–100 mm/1000 years) of limestone corrosion in different areas (Sweeting 1973, p. 42). During this stage (Fig. 9.13, stage 3), it is possible that nonfilled or re-opened fissures and fractures were progressively enlarged by dissolution and eventually developed karst landforms with topographic ridges and pinnacles. The absence of erosional blocks indicates that chemical weathering was also important here; however, physical weathering, if present, left no record. The downward development of karst landforms was generally limited by the presence of the underlying siliciclastic unit. In a few areas (Grande Pointe, Ile du Fantôme) where the unconformity occurs within the siliciclastic unit, erosion was more likely caused by physical processes.

Small, sharp-crested solution sculptures, superimposed on the larger-scale karst landforms, developed on bare substrate. The morphological style of the sculptures appears to be controlled by the local topography of karst landforms. Today, in areas where weathering predominates, soils are commonly formed (Klappa 1983); this contrasts with their absence above the intra-Mingan paleokarst surface. It is possible that chemical weathering (mainly dissolution) was not conducive to the breakdown of rock into unconsolidated sediments and that erosion by runoff waters was sufficient to remove all weathered detritus or insoluble material above the paleokarst surface. Esteban and Klappa (1983), however, point out that weathering and soil-forming processes were probably different without the influence of land plant vegetation during Early Paleozoic time.

The final stage in the development of the intra-Mingan unconformity occurred with renewed transgression over the study area (Fig. 9.13, stage 4). With initial submergence, sediments in the lowest topographic depressions were deposited in more restricted marine environments. Continued submergence enhanced marine circulation, and sediment deposition occurred in a variety of open, normal marine environments. The submerged karst surface provided a hard substrate for the growth of various encrusting organisms and also acted as foundation for the development of reefs. Paleoshorelines were characterized by beaches

downward movement of meteoric fluids. During initial exposure, however, the sequence was periodically affected by marine waters as the strandline was still fluctuating. Marine waters circulating in the fissure network were saturated with lime mud that eventually filled them and connected porous horizons.

Karst landforms developed with the eventual sealevel fall which reached at least 20 m, as indicated by erosional relief of the Mingan sequence observed on Ile du Fantôme (Fig. 9.9).

and rocky coasts. Rocky shorelines modified slightly the pre-existing karst surface and exhibited distinctive intertidal notches and kamenitzas but their origin (dissolution vs biological erosion) remains uncertain.

The Cause of These Regionally Extensive Paleokarst Unconformities

The formation of regional paleokarst unconformities requires either tectonic uplift or eustatic sealevel fall, or both. The paleokarst unconformities in the Mingan Islands were formed during Middle Ordovician time (Whiterockian–Chazyan) when both global marine regressions (Fortey 1984, Barnes 1984) and significant changes in the Appalachian tectonic regimes (Rodgers 1971, Williams 1979) are well documented.

The factors responsible for the formation of the post-Romaine unconformity are in part eustatic, because this unconformity is also recognized elsewhere in eastern North America. In fact, a pre-Chazyan unconformity is recorded along the entire Appalachian system, above the Beekmanton-Knox strata, which are equivalent in age with the Romaine Formation (Ross et al. 1982) (Mussman et al., this volume). Furthermore, the presence of major regressive phases in both shelf and off-shelf localities from widely separated cratonic areas during Whiterockian time indicates a time of global sealevel lowstand (Fortey 1980 and 1984, Barnes 1984). The post-Romaine unconformity, however, cannot be explained by sealevel changes alone but also requires tectonic causes to explain tilting and erosion (up to 30 m of stratigraphic relief) of the Romaine sequence during Whiterockian time.

The period of this unconformity and its southern equivalents also marks the transition from passive margin to convergent margin sedimentation, which left an extensive record in rocks from Newfoundland to Tennessee. Attempts to explain these unconformities in terms of modern tectonic environments have been presented recently for the Canadian Appalachians (Jacobi 1981), New England Appalachians (Rowley and Kidd 1981), Central

Appalachians (Shanmugam and Lash 1982), and Southern Appalachians (Shanmugam and Walker 1980, Mussman 1982). All these studies suggest that uplift and erosion of the carbonate bank resulted from a peripheral bulge related to rapid subsidence of the bank margin. In general, subsidence was diachronous towards the west, and an adjacent peripheral bulge also migrated simultaneously toward the craton to form an east–west diachronous unconformity. It is unclear whether subsidence was due to loading of the lithosphere by emplacement of thrust sheets (Hiscott et al. 1983, Quinlan and Beaumont 1984) or to downwarping of the shelf margin as it approached the subduction zone (Shanmugam and Lash 1982). Nevertheless, a peripheral bulge, based on rheological models, must form adjacent to any flexural downwarp of the lithosphere regardless of its cause (Walcott 1970). A peripheral bulge model showing the development of the post-Romaine unconformity is illustrated in Figure 9.14.

The origin of the intra-Mingan paleokarst unconformity, however, is more equivocal. In spite of local regressive-transgressive sequences reported by Fortey (1984) from the Llanvirnian-Llandelian (i.e., Chazyan) of Britain, there is not yet evidence of sealevel fall during Chazyan time. In fact, evidence for such depositional trends with karst development is unknown in the Chazy Group of New York and Vermont, where small on-shelf reefs grew in shallow water above fair-weather wave base for most of Chazyan time (Pitcher 1964, Kapp 1975). This argues strongly in favor of local tectonic movements as having been responsible for the formation of this paleokarst unconformity. It is possible that the passage of a second peripheral bulge affected the Mingan sequence. This bulge could have been generated by the collapse of the platform margin and emplacement of the Canadian Taconic allochthon in nearby Québec (Gaspé Peninsula), which took place later than in western Newfoundland.

Paleokarst Horizons

In addition to major paleokarst unconformities, numerous local paleokarst horizons are also present, associated with meter-scale, shallowing-upward calcarenite cycles in the Mingan For-

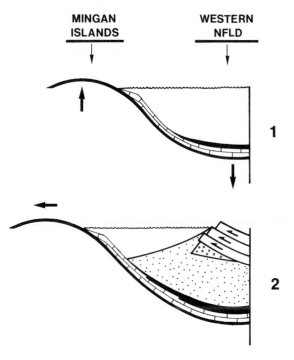

MINGAN
ISLANDS

WESTERN
NFLD

1

2

FIGURE 9.14. Schematic cross-sections illustrating development of the paleokarst unconformities and overlying transgressive sequences. The formation of the paleokarst unconformity capping the Romaine Formation is shown here in two stages. Stage 1: an upward flexure or peripheral bulge formed in response to the shelf collapse to basinal depths characterized by the deposition of black graptolitic shales of the Table Head Group in western Newfoundland. Stage 2: progressive development of a pericratonic foreland basin in western Newfoundland forced the peripheral bulge to migrate toward the cratonic interior and the deposition of the onlap facies of the Mingan limestones and siliciclastics. A similar model but operating in a diachronous fashion is also used to explain the intra-Mingan paleokarst unconformity. See text for further details.

mation. The paleokarst surfaces record subaerial emergence, lithification of newly exposed carbonate sediments, and karst erosion forming sharp karren. If exposure was prolonged, karren progressively widened and eventually developed into extensive planar surfaces.

Calcarenite Cycles

Calcarenites cycles occur only in the Perroquet and Grande Pointe Members of the Mingan Formation (Fig. 9.3). At least three superim-

posed cycles, all capped by paleokarst surfaces, are present in the Grande Pointe Member overlying the irregular intra-Mingan paleokarst unconformity. Individual cycles range from 2 to 6 m in thickness and can be traced laterally over from 1 to 10 km. The lithofacies associations within an "ideal" calcarenite cycle are summarized in Figure 9.15. Calcarenite cycles in the Perroquet Member, exposed only in the westernmost islands, were not studied in detail because of limited exposure, but paleokarst surfaces are present there as well.

The upper part of each calcarenite cycle represents sand shoal sediments composed of skeletal-ooid grainstones (Fig. 9.16A). The upper grainstone units form lens-shaped bodies and grade laterally into muddy skeletal deposits. Unlike the sand-shoal deposits, these sediments were never subaerially exposed, as indicated by the absence of paleokarst surfaces or other diagnostic features of meteoric diagenesis (Fig. 9.17). Instead, they became the locales of more restricted deposition caused by arrested water circulation behind the exposed sand shoal. This indicates that the sand shoals had a synoptic relief of several meters over the intershoals areas.

Paleokarst Surfaces

The paleokarst surfaces, when present, always occur on top of calcarenite cycles. These surfaces, some of which can be traced over several kilometers, are laterally discontinuous and stop where grainstones grade laterally into skeletal muddy sediments that were deposited in less turbulent, intershoal areas (Fig. 9.15). The surfaces are irregular, scalloped to planar in vertical section, with less than 30 cm of relief. The same surfaces, if exposed on bedding planes, exhibit a variety of solution sculptures which include: (1) rinnenkarren and kamenitza, (2) trittkarren, and (3) planar surfaces.

Rinnenkarren and Kamenitzas. These features are similar in size and in shape to those present on the intra-Mingan paleokarst and are characterized by sharp crested edges (Fig. 9.16C). Rinnenkarren are locally well developed but display no consistent orientation along the same paleokarst surfaces. Small forms are sometimes superimposed on larger depressions, which

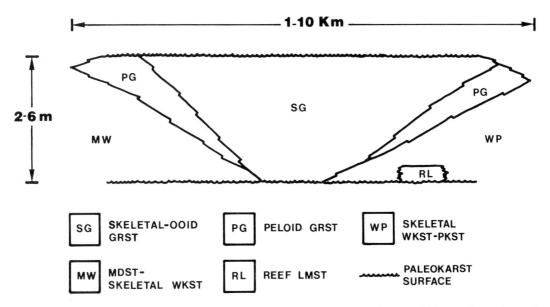

FIGURE 9.15. Schematic diagram illustrating the lateral lithofacies variation within a calcarenite cycle.

may form channels up to 1 m wide and several tens of meters long (Fig. 9.16B).

Comparison with modern karst surfaces (Sweeting 1973) indicates that these sharp-crested sculptures developed on bare limestone substrates exposed to rainwater. The randomly oriented rinnenkarren and their common association with kamenitzas probably reflect slightly inclined substrates disturbed only by minor topographic gradients which locally controlled the water flow.

Trittkarren. Trittkarren, which are best described as resembling the imprint of a heel, are asymmetrical in vertical section with a flat tread (20–40 cm long) and a vertical riser (5–15 cm high). They are semicircular and laterally coalescent on the bedding plane (Figs. 9.18A, B). Furthermore, they are commonly organized in steps and oriented perpendicular relative to other karren forms, such as rinnenkarren.

Modern trittkarren are present only on slightly inclined substrates directly exposed to rainwater. They are thought to result from a two-water-layer process; the upper layer is more "aggressive" on the vertical back wall, the lower layer is highly alkaline on the flat tread (Sweeting 1973). Solution by water flowing perpendicular to their orientation causes preferential solution of the back wall.

Planar Surfaces. The paleokarst surfaces are also characterized by extensive planar surfaces which can be traced for several hundreds of meters, if not kilometers, with little or no relief (Fig. 9.18C, D). The microrelief on these planes consists only of oriented ridges (5–10 cm high, 20–150 cm long) and isolated pinnacles (5–15 cm high). In some areas they pass laterally, over a few tens of meters, into surfaces with well-developed, sharp crested karren (Fig. 9.16C, D, E, F). Centimeter-sized pits are ubiquitous features that are superimposed on the planar surfaces. These pits are circular to irregular in plan view and have a smooth relief that is filled by limestones associated with the overlying calcarenite cycle. Furthermore, large *Chondrites* burrows (Filion and Pickerill 1984) with branches up to 80 cm long are truncated just at the base of their mastershaft, where they generally branch to form a horizontal dendritic network (Fig. 9.18E). The planar surfaces are clearly erosional, because both grains and cements are truncated along the surfaces (Fig. 9.18F).

Karren forms in modern karst terrains converge toward flat surfaces at the lower end of the topographic gradient where flowing water becomes more stagnant and sideways enlargement is promoted. We believe that the Ordovician planar surfaces were similarly formed by the progressive enlargement of karren and eventual development of extensive planar surfaces which are textured by relict features (small ridges and pinnacles). The truncated master-

FIGURE 9.16 A–C. Paleokarst horizons. *A.* Field photograph (vertical cliff exposure) showing the upper part of a calcarenite cycle composed of skeletal-ooid grainstones characterized by herringbone cross-bedding. Grande Pointe. Hammer is 30 cm. *B.* Field photograph (bedding plane view) of sculpted karren on top of a calcarenite cycle. Karren are separated by well-oriented depressions or channels filled by overlying limestones. Grande Pointe. Hammer for scale. *C.* Field photograph (bedding plane view) of a paleokarst surface illustrating well-developed and oriented karren on top of a calcarenite cycle. Ile Nue de Mingan N. Hammer is 30 cm.

FIGURE 9.16 D–F. *D*. Field photograph (bedding plane view) of a paleokarst surface showing karren with widened forms but still recognizable oriented ridges. Ile Nue de Mingan N. Scale bar is 2 cm. *E*. Field photograph (bedding plane view) of a paleokarst surface showing more flattened karren with discontinuous ridges. Ile Nue de Mingan N. Scale bar is 2 cm. *F*. Field photograph (bedding plane view) of a paleokarst surface characterized by a planar surface with only relict features such as small pinnacles. Note the paleokarst surface disappearing beneath overlying limestones (top left). Ile du Havre SE. Scale bar is 2 cm. Photographs C, D, E, and F represent a temporal evolution in the style of paleokarst surfaces capping the calcarenite cycles.

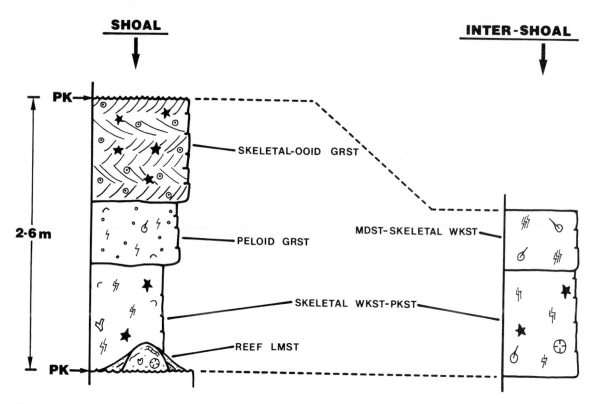

FIGURE 9.17. Description of an "ideal" calcarenite cycle developed on sand shoal deposits and capped by a paleokarst surface (PK). Sediments were deposited in the inter-shoal areas, and muddy lithofacies deposited under more restricted conditions when the sand shoals were exposed.

shafts of *Chondrites* burrows suggest that at least several tens of centimeters were removed at the top of calcarenite cycles by karst erosion.

The origin of the small pits is more equivocal. Sharp circular pitting on modern exposed surfaces, sometimes considered as small kamenitzas, are thought to be a direct result of precipitation, whereas pits with smoother and irregular form are most likely the result of biological activity (Corbel 1963). The problem with the second hypothesis is the absence of higher plants during early Paleozoic time. Lichens and algae, present at this time, however, are known to produce strong organic acids (Moore and Bellamy 1974). Cowell (1976) interpreted pitting forming today as a product of biological solution by lichens and algae growing directly on the carbonate bedrock. This offers a possible explanation for small pits observed on the planar surfaces in this study.

Interpretation

The origin of the calcarenite cycle may be explained by shoaling with the paleokarst surfaces, capping these cycles, formed during periods of relative sealevel fall. These surfaces are erosional, and removed the upper part of each calcarenite cycle by karst erosion as indicated by: (1) paleokarst surfaces superimposed directly upon subtidal lithofacies, and (2) crossbedding in subtidal lithofacies truncated by paleokarst surfaces. Either or both, eustatic or tectonic causes may be responsible for the formation of the paleokarst surfaces. The amount of sealevel fall, however, is clearly on the scale of meters because inter-shoal areas, located in relatively deeper water, are not affected by subaerial exposure.

In summary, the paleokarst surfaces represent exposure surfaces which formed contem-

FIGURE 9.18 A–C. Paleokarst horizons. *A*. Field photograph (bedding plane view) of trittkarren on top of a calcarenite cycle. Trittkarren are asymmetrical in vertical section with a flat tread and a vertical riser and are laterally coalescent on the bedding plane. Grande Pointe. Hammer is 30 cm. *B*. Field photograph of trittkarren showing different steps on the bedding plane. Ile de Fantôme W. Hammer is 30 cm. *C*. Field photograph showing sharp contact at the top of the calcarenite cycle. Note the presence of sculpted karren, here shown in cross-section. Ile Nue de Mingan N. Lens is 50 mm (*Continued*).

FIGURE 9.18 D–F. *D*. Field photograph (vertical cliff exposure) of a planar erosion surface (arrow) capping a calcarenite cycle and separating grainstones from overlying skeletal muddy limestones (L). Ile du Havre SE. *E*. Field photograph (bedding plane view) of the trace fossil, *Chondrites*, truncated by a planar surface capping a calcarenite cycle. Ile du Havre SE. Lens cap for scale. *F*. Field photograph (bedding plane view) of gastropod shell truncated by a planar surface capping a calcarenite cycle. Ile du Havre SE. Lens cap is 50 mm in diameter.

poraneously with sedimentation and developed only on sand shoal deposits in response to minor fluctuations in sealevel. The style of solution sculpture (karren) was controlled by the absence of soil or sediment cover and by the local drainage conditions at the surface. The profile of these sculptures changed through time and developed into peneplaned surfaces. With subsequent sealevel rise, such a surface would have been rapidly submerged and most likely preserved without significant modification.

Conclusions

Three different types of paleokarst surfaces are recognized in Middle Ordovician carbonates exposed in the Mingan Islands:

1. An extensive karst plain developed on top of the Romaine Formation.
2. An extensive irregular karst surface developed within the limestones of the Mingan Formation.

These two major paleokarst unconformities resulted from tectonic movements associated with the Taconic Orogeny occurring several hundreds of km to the east. Eustatic sealevel changes also occurred but were probably less important in the formation of these unconformities.

3. Local paleokarst contemporaneous with sedimentation developed on sand shoal deposits in the Mingan Formation in response to minor fluctuations in sealevel.

Acknowledgments. This study represents part of a Ph.D. dissertation by A.D. completed while the authors were affiliated with the Department of Earth Sciences, Memorial University of Newfoundland, St. John's, Nfld., Canada. Early stages of fieldwork were supported by the Ministère de l'Energie et des Ressources du Québec. Subsequent research was financed by Natural Science and Engineering Research Council of Canada Grant A-9159 to N.P.J. We thank Parks Canada for permission to sample in the Mingan Islands National Park.

References

Back, W., Hanshaw, B.B., Plummer, L.N., and Weidie, A.E., 1979, Geochemical significance of groundwater discharge and carbonate solution in the formation of Caleta Xel Ha, Quintana Roo, Mexico: Water Resources Research, v. 15, p. 1531–1535.

Back, W., Hanshaw, B.B., and Van Driel, J.N., 1984, Role of groundwater in shaping the eastern coastline of the Yucatan Peninsuala, Mexico, *in* Lafleur, R.G., ed., Groundwater as a geomorphic agent: Winchester, Allen and Unwin, p. 157–172.

Barnes, C.R., 1984, Early Ordovician eustatic events in Canada, *in* Bruton, D.L., ed., Aspects of the Ordovician System: Paleont. Contributions from the Univ. of Oslo (295), p. 51–64.

Bögli, J., 1980, Karst hydrology and physical speleology: Berlin-Heidelberg, Springer-Verlag, 285 p.

Corbel, Jean, 1963, Marmites de géants et microformes karstiques: Norois Ann., v. 10, (10), p. 121–132.

Cowell, D.W., 1976, Karst geomorphology of the Bruce Peninsula, Ontario: Unpubl. M.S. thesis, McMaster University, Hamilton, Ont., Canada.

Desrochers, A., 1985, The Lower and Middle Ordovician platform carbonates of the Mingan Islands, Québec: Stratigraphy, sedimentology, paleokarst, and limestone diagenesis: Unpubl. Ph.D. thesis, Memorial University of Newfoundland, St. John's, Nfld., Canada, 454 p.

Esteban, M., and Klappa, C.F., 1983, Subaerial exposure, *in* Scholle, P.A., Bebout, D.G., and Moore, C.H., eds., Carbonate depositional environments: Amer. Assoc. Petroleum Geologists Memoir 33, p. 1–54.

Filion, D., and Pickerill, R.K., 1984, Systematic ichnology of the Middle Ordovician Trenton Group, St. Lawrence Lowland, eastern Canada: Maritime Sediments and Atlantic Geology, v. 20, p. 1–41.

Folk, R.L., 1968, Bimodal supermature sandstone: product of the desert floor: Proc. of the 23rd Intern. Geol. Congress, Section 8, Prague, p. 8–32.

Fortey, R.A., 1980, The Ordovican of Spitzbergen, and its relevance to the base of the Middle Ordovician in North America, *in* Wones, D.R., ed., The Caledonides in the USA: Virginia Polytechnic Institute and State University, Dept. of Geol. Sciences, Memoir 2, p. 33–40.

Fortey, R.A., 1984, Global earlier Ordovician transgressions and regressions and their biological implications, *in* Bruton, D.L., ed., Aspects of the Ordovician system: Paleontological Contributions from the Univ. of Oslo (295), p. 37–50.

Gauthier-Larouche, G., 1981, Origine et toponymie de l'archipel de Mingan: Commission de Toponymie du Québec, Etudes et recherches toponymiques (1), 165 p.

Hiscott, R.N., Quinlan, G.M., and Stevens, R.K., 1983, Comments on "Analogous tectonic evolution of the Ordovician foredeeps, soutern and central Appalachians": Geology, v. 10, p. 502–566.

Horowitz, A. S., and Potter, P.E., 1971, Introductory petrogrpahy of fossils: Berlin-Heidelberg-New York, Springer-Verlag, 302 p.

Jacobi, R.D., 1981, Peripheral bulge—a causal mechanism for the Lower/Middle Ordovician unconformity along the western margin of the northern Appalachians: Earth and Planetary Science Letters, v. 56, p. 245–251.

James, N.P., and Choquette, P.W., 1984, Diagenesis 9. Limestones—the meteoric diagenetic environment: Geoscience Canada, v. 11, p. 161–194.

Kapp, U.S., 1975, Paleoecology of Middle Ordovician stromatoporoid mounds in Vermont: Lethaia, v. 8, p. 195–207.

Klappa, C.F., 1983, A process-response model for the formation of pedogenic calcretes, in Wilson, R.C., ed., Residual deposits: Jour. Geol. Soc., London, Spec. Pub. 11, p. 211–220.

Kobluk, D.R., 1981, Middle Ordovician (Chazy Group) cavity-dwelling boring sponges: Can. Jour. Earth Sciences, v. 18, p. 1101–1108.

Moore, P.D., and Bellamy, D.J., 1974, Peatlands: Berlin-Heidelberg-New York, Springer-Verlag, 221 p.

Mussman, W.J., 1982, The Middle Ordovician Knox unconformity, Virginia Appalachians: transition from passive to convergent margin; Unpubl. M.S. thesis, Virginia Polytechnic Institute, Blacksburg, VA.

Neumann, A.C., 1966, Observations on coastal erosion in Bermuda and measurements of the boring rate of the sponge Cliona lampa: Limnol. Oceanogr., v. 11, p. 92–108.

Nowlan, G.S., 1981, Stratigraphy and conodont faunas of the Lower and Middle Ordovician Romaine and Mingan Formations, Mingan Islands, Quebec (abstr.): Maritime Sediments and Atlantic Geology, v. 17, p. 67.

Pemberton, S.G., Kobluk, D.R., Yeo, R.K., and Risk, M.J., 1980, The boring Trypanites at the Silurian-Devonian disconformity in southern Ontario: Jour. Paleontology, v. 58, p. 885–891.

Pickerill, R.K., and Harland, T.L., 1984, Middle Ordovician micro-borings of probable sponge origin from eastern Canada and eastern Norway: Jour. Paleontology, v. 58, p. 885–891.

Pitcher, M., 1964, Evolution of Chazyan (Ordovician) reefs of eastern United States and Canada: Can. Petroleum Geol. Bull., v. 12, p. 632–691.

Pluhar, A., and Ford, D.C., 1970, Dolomite karren of the Niagara Escarpment, Ontario, Canada: Zeitschrift für Geomorphologie, v. 14, p. 392–410.

Pye, Kenneth, 1983, Red beds, in Goudie, A.S., ed., Chemical sediments and geomorphology: precipitates and residua in the near-surface environment: London, Academic Press, p. 227–263.

Quinlan, G.M., and Beaumont, C., 1984, Appalachian thrusting, lithopheric flexure, and the Paleozoic stratigraphy of the eastern interior of North America: Can. Jour. Earth Sciences, v. 21, p. 973–996.

Rodgers, John, 1971, The Tactonic orogeny: Geol. Soc. America Bull., v. 82, p. 1141–1177.

Roehl, P.O., 1967, Stony Mountain (Ordovician) and Interlake (Silurian) facies analogs of recent low-energy marine and subaerial carbonates, Bahamas: Amer. Assoc. Petroleum Geol. Bull., v. 51, p. 1979–2032.

Ross, R.J., Jr., et al., 1982, The Ordovician System in the United States: Int. Union Geol. Sciences, Pub. 12, 73 p.

Rowley, D.B., and Kidd, W.S., 1981, Stratigraphic relationships and detrital composition of the Middle Ordovician flysch of western New England: implications for the tectonic evolution of the Taconic orogeny: Jour. Geol., v. 89, p. 199–218.

Scheinder, J.H., 1976, Biological and inorganic factors in the destruction of limestone coasts: Stuttgart, Schweizerbart, Contributions to Sedimentology, v. 6, 112 p.

Semeniuk, V., and Johnson, D.P., 1985, Modern and Pleistocene rocky shore sequences along carbonate coastlines, southwestern Australia: Sedimentary Geology, v. 44, (314), p. 225–285.

Shanmugam, G., and Lash, G.G., 1982, Analogous tectonic evolution of the Ordovician foredeeps, southern and central Appalachians: Geology, v. 10, p. 562–566.

Shanmugan, G., and Walker, K.R., 1980, Sedimentation, subsidence, and evolution of a foredeep basin in the Middle Ordovician, southern Appalachians: Amer. Jour. Science, v. 280, p. 479–496.

Sohn, I.G., 1985, Latest Mississippian (Namurian A) nonmarine ostracods from west Virginia and Virginia: Jour. Paleontology, v. 59, 446–460.

Stringfield, V.T., Rapp, J.R., and Andes, R.B., 1979, Effects of karst and geologic structure on the circulation of water and permeability in carbonate aquifers: Jour. Hydrology, v. 43, 313–332,

Sweeting, M.M., 1973, Karst landforms: London, MacMillan, 362 p.

Swift, D.J., 1968, Shoreface erosion and transgressive stratigraphy: Jour. Geol., v. 76, p. 444–456.

Seift, D.J., 1975, Barrier-island genesis: evidence from the Atlantic shelf, eastern USA: Sed. Geol., v. 14, p. 1–43.

Twenhofel, W.H., 1938, Geology and paleontology of the Mingan Islands: Geol. Soc. America Special Paper 11, 132 p.

Walcott, R.I., 1970, Isostatic response to loading of the crust in Canada: Can. Jour. Earth Sciences, v. 7, p. 2–13.

Williams, H., 1979, Appalachian orogen in Canada: Can. Jour. Earth Sciences, v. 16, p. 792–807.

10
Ordovician Knox Paleokarst Unconformity, Appalachians

WILLIAM J. MUSSMAN, ISABEL P. MONTANEZ, and J. FRED READ

Abstract

The Ordovician Knox unconformity in the Appalachians developed in less than 10 m.y. during a time of initial collision of the passive margin and of eustatic sealevel lowering. It formed on cyclic limestones and dolomites of the 200- to 1200-m-thick Upper Knox–Beekmantown Group, and provides an example of the effects of long-term exposure on a carbonate shelf and the subsequent diagenesis related to karsting followed by deep burial. Erosional relief is over 100 m in the south. It increases over syndepositional structures and bevels down to Upper Cambrian rocks on the craton. The disconformity is virtually absent in the Pennsylvania depocenter.

Paleokarst features include paleotopographic highs, sinkholes and caves that extend to over 65 m below the unconformity, and intrastratal breccias down to 300 m. Near-vertical sinkholes are filled with carbonate breccia and gravels with fine dolomite matrix. Caves are filled with breccia and laminated dolomite. Intrastratal breccias up to 35 m thick and over 200 m long contain dolomite clasts commonly with fitted fabrics, in a fine dolomite matrix.

Nonluminescent calcite cements fill leached grains and intergranular spaces down to 200 m below the unconformity, and formed from slowly moving, oxidizing meteoric waters undergoing diffuse flow in an unconfined aquifer. Rapidly moving, conduit flow cave waters caused extensive dissolution of limestone beds locally forming intrastral breccias; some of these also may have resulted from dissolution in the mixing zone. During burial, compaction further fractured breccia beds, and in the Late Paleozoic, warm saline basinal brines (80°C to 200°C) caused further dissolution, dolomitization fronts surrounding breccia zones, and precipitation of saddle dolomite and associated sulfides, within permeable horizons.

Introduction

The Knox unconformity is the major stratigraphic break in the Paleozoic sequence in the U.S. Appalachians. It occurs between Lower to Middle Ordovician Knox/Beekmantown carbonates and overlying Middle Ordovician limestones. The unconformity is of economic interest because of associated base metal deposits (Harris 1971, Collins and Smith 1975), and potential hydrocarbon reservoirs. The unconformity developed on passive margin carbonates (Rodgers 1968, Bird and Dewey, 1970) and is overlain by foreland basin deposits (Shanmugan and Walker 1980, Read 1980). Actively subsiding depocenters in the Appalachians (Colton 1970, Read 1980) influenced unconformity development (Fig. 10.1).

Paleokarstic features such as erosional highs, caves, sinkholes, and intrastratal breccias are described in Mussman and Read (1986). They are discussed briefly here, along with diagenetic phenomena associated with aquifer development and later deep burial diagenesis (described in detail in Montanez, in prep.).

Similar unconformities elsewhere may control the distribution of lead–zinc sulphide and perhaps hydrocarbon deposits associated with permeable breccia horizons beneath the unconformity. These horizons could have acted as conduits for warm basinal brines which were expelled from shales following deep burial beneath synorogenic deposits.

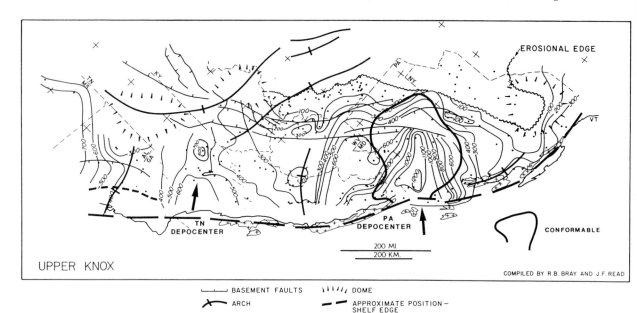

FIGURE 10.1. Isopach map of Upper Knox/Beekmantown carbonates in the United States Appalachians, plotted on a palinspastic base (from Read, in press). Thickest sections occur in Pennsylvania and Tennessee depocenters (sites of maximum subsidence), whereas thinnest sections occur over arches. Sections in the Pennsylvania depocenter tend to be relatively conformable, reflecting subsidence approximately equaling sealevel fall rate here, during development of the unconformity.

Setting

The unconformity is exposed within imbricate thrust sheets that moved from southeast to northwest, with displacements of up to tens of kilometers. Overthrust Precambrian and Lower Cambrian igneous and metasedimentary rocks of the Blue Ridge occur to the southeast, and nearly flat-lying Late Paleozoic sediments of the Appalachian Plateau lie to the northwest.

Ordovician carbonates below the unconformity are referred to as Upper Knox Group (200 to 700 m thick) in Virginia–Tennessee and Beekmantown Group (up to 1200 m thick) in northern Virginia to New York (Fig. 10.2). Middle Ordovician foreland basin carbonates and shales overlie the unconformity. The unconformity passes into conformable sequences in Pennsylvania, the site of an actively subsiding depocenter (Fig. 10.1).

The widespread, global distribution of the unconformity suggests eustatic sealevel fall (cf. Sloss 1963, Mussman 1982, Mussman and Read 1986). However, tectonics also influenced its development which was synchronous with conversion of the passive margin to a convergent margin (Shanmugan and Walker, 1980 and Mussman and Read 1986). The unconformity may have formed when the shelf passed over a peripheral bulge with a calculated uplift of up to 180 m, and which formed during eastward subduction (Jacobi 1981, Quinlan and Beaumont 1984) of the North American continental margin beneath a magmatic arc (Hatcher 1972, Slaymaker and Watkins 1978, Shanmugam and Walker 1980).

Uppermost Knox carbonates in southwest Virginia are Lower Ordovician, Canadian age, and are overlain by Chazyan limestones (Fig. 10.2). The unconformity here probably spanned about 10 million years (duration of the Whiterockian; Ross et al. 1982). In northern Virginia, sections are more conformable and Beekmantown carbonates are early Middle Ordovician (Whiterockian) age, and are overlain by latest Whiterockian/earliest Chazyan limestone (Sutter and Tillman 1973, Tillman 1976, A. Harris pers. comm.). Subaerial emergence, even though of short duration (few tens to hundreds of thousands of years), formed distinctive karstic features here.

The Knox Group (southwest Virginia) is

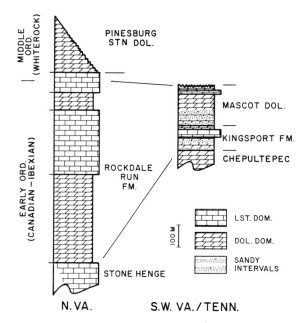

FIGURE 10.2. Simplified stratigraphic columns representative of Virginia–Tennessee and N. Virginia/Maryland/Pennsylvania, for the Lower and early Middle Ordovician.

mainly cyclic dolomite, with abundant bedded and nodular chert and quartz sand stringers, whereas the cyclic Beekmantown Group (northern Virginia) has many limestone interbeds (locally up to 50%) (Fig. 10.2). Cycles are 2 m to 9 m thick, and consist of: cryptalgal laminated dolomite and fenestral limestone (0 to 5.5 m thick), which overlies massive and burrowed, thin-bedded dolomite or limestone (0.5 to 3 m thick), that overlies coarsely crystalline dolomite and thrombolitic limestone (0.5 to 4 m thick).

These beds are cyclic, upward-shallowing sequences that formed in a semiarid, peritidal setting.

Paleokarst Features on the Knox Unconformity

Although mainly a disconformity, in some areas the unconformity is angular with discordance of up to 12°. The unconformity increases in magnitude across strike from southeast to northwest. Along strike, erosional relief decreases from 140 m in southwest Virginia (Webb 1959) to a few meters in northern Vir-

ginia. The unconformity has paleokarstic highs and lithoclastic breccias, conglomerates, and redbeds, and below the unconformity, filled caves, sinkholes, and intraformational breccias (Fig. 10.3).

Paleotopographic Highs

Paleohighs are a few meters to 30 m in relief. Postunconformity beds thin or pinch out onto the highs. Some unconformity highs form "islands" of Knox dolomite surrounded by peritidal beds (Webb 1959). Paleohighs are veneered by thin (up to 30 cm) beds of mud-supported, chert-dolomite breccia and fine detrital dolomite. Away from the highs, the unconformity commonly is covered by up to 2 m of discontinuous breccia (Fig. 10.4A). Locally, these were reworked in fluvial channels to form chert-rich conglomerates, or they were reworked during transgression to form fossiliferous breccias or conglomerates, locally with a fenestral carbonate matrix. Locally, unconformity lows are filled with up to 70 m of red or gray detrital carbonate, and with quartz sand/silt and shale interbeds (Fig. 10.4A). Commonly these sequences are chert-dolomite lithoclast layers (up to 15 cm thick) of graded clast- or mud-supported breccia grading up into detrital dolomite sand with dolomite mud caps. These mainly are subaerial debris flows and sheet flood deposits shed from paleohighs.

Sinkholes or Dolines

Sinkholes on the unconformity are narrow (3 to 35 m wide, up to 65 m deep) breccia-filled depressions (Fig. 10.4B). Many are vertical to bedding, but some are horizontal at depth. Sides are subparallel and contacts between host beds and fills are sharp and irregular. They are filled with mud-supported and lesser, clast-supported lithoclast breccia with rare pods and lenses of granule conglomerate in detrital dolomite matrix. There are angular, rotated carbonate blocks (up to 2 m diameter, and locally showing in-place brecciation), and gravel- to cobble-size, poorly rounded, corroded carbonate and angular chert clasts. Matrix within breccias is quartzose, fine to medium crystalline dolomite.

Many of these sinkholes probably formed by

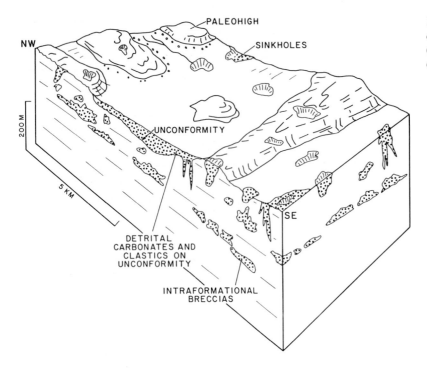

FIGURE 10.3. Block diagram of karstic surface on Knox Group during early Middle Ordovician.

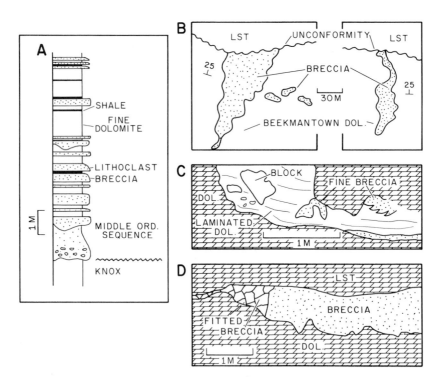

FIGURE 10.4. A–D. A. Measured section of detrital carbonates of Blackford Formation veneering unconformity at East River Mountain, Virginia. B. Map view of breccia-filled sinkholes on unconformity near Fincastle, Virginia (remapped after Campbell 1975). C. Cross-section through portion of cave-fill within Beekmantown Group, shown on Figure 10.3. Note symbols on Figures 10.3 and 10.4 are different. D. Cross-section of intraformational breccia near Leaksville, Virginia. Breccia fills cavity after dissolution of limestone, relicts of which occur at base of fill. Breccia grades laterally into rocks with fitted-fabric into unaltered host dolomite. Note symbols in Figures 10.3 and 10.4 are different.

dissolution of host limestone along joints or bedding planes, whereas some formed by collapse of caves. They were filled by material shed from roofs, or washed in from the unconformity, and only rarely by marine sediments deposited during transgression.

Caves

These are thin, sheet- to podlike bodies (up to 12 m long, and 2 m wide) of detrital dolomite that extend to 35 m below the unconformity. They are mainly subvertical to randomly oriented to bedding. Cave walls are sharp and irregular, and fills are fine dolomite, with local breccias near the unconformity (Fig. 10.4C). They are brown, massive, or rarely laminated. Layering may be inclined toward the center of cavities and may contain small slumps and scours filled with laminated dolomite mud or carbonate conglomerate which also forms rare sheets at bases of laminated beds (Fig. 10.4C). Conglomerates are clast- and mud-supported. Breccias are poorly sorted and clast-supported

and have sandy argillaceous fine dolomite matrix.

Intraformational Breccias

Intraformational breccia bodies (Fig. 10.4D) commonly occur between 200 m and 250 m below the unconformity surface (Misra et al. 1983, Kyle 1983). They are lenticular to stratiform with irregular lateral and upper contacts with host beds and sharp basal contacts. They are a few meters to over 400 m in diameter and up to 30 m high (McCormick and Rasnick 1983), and form a reticulate pattern (Fig. 10.5). Breccias grade upward and laterally into unaltered host rock through incipiently brecciated (fitted fabric) dolomite. Basal contacts have remnants of limestone beds or solution-scalloped limestone clasts. Laterally, many contacts with unaltered limestone beds are corrosional. The breccias are composed predominantly of clasts of dolomite and minor limestone and chert, are poorly sorted, and many show compaction-brecciation. They have interstitial medium- to

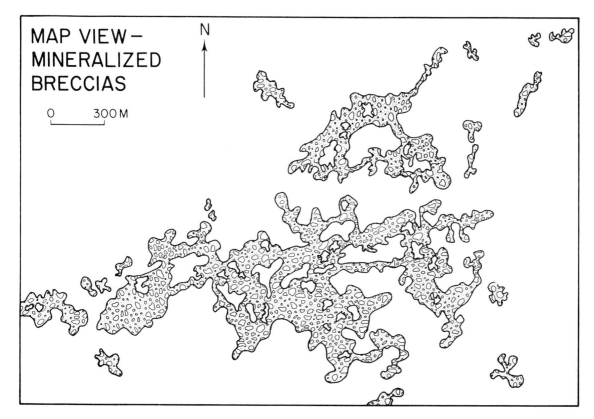

FIGURE 10.5. Map view of intraformational breccia in subsurface of Young Mine, Tennessee. Note cavern-like distribution of mineralized breccias. Modified from mine map provided by ASARCO.

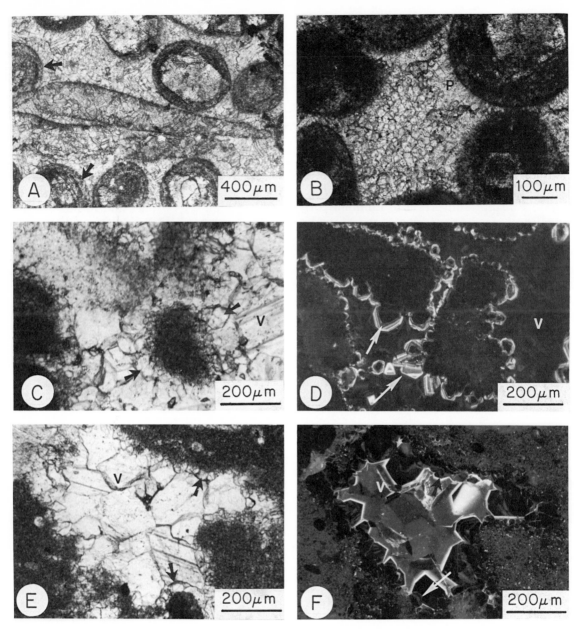

FIGURE 10.6. A–F. Early dissolution fabrics and early nonferroan calcite cements. *A.* Leached molluscs with only micritic envelopes remaining. Clear nonferroan druse and blocky spar cement fill leached molluscs and intergranular voids. Leached ooids (arrows) are filled with nonferroan calcite microspar. *B.* Intergranular, fine equant cement in primary void between ooids. Clear prismatic crystals (P) form thin isopachous rims on ooids. *C, D.* Plane light and cathodoluminescent paired photographs of nonluminescent to bright/dull to nonluminescent zonation in early nonferroan calcite cement. Calcite druse (arrows) rimming grains shows a nonluminescent to bright–dull (subzoned) luminescence. Void-filling cement (V) is nonluminescent. *E, F.* Plane light and cathodoluminescent paired photographs of nonluminescent to bright-to-dull zonation in early nonferroan calcite cement. Calcite druse (arrows) lining primary void is primarily nonluminescent with slightly luminescent subzones (Fe and Mn concentrations indistinguishable from those values typical of nonluminescent calcite cement). Void-filling cement (V) shows a nonluminescent to bright-to-dull zonation (subzoned).

coarse-grained dolomite or cements of coarse calcite, saddle dolomite, and sulfides.

Early Dissolution Fabrics in Host Carbonates

Besides large-scale dissolution features, there is abundant evidence of precompaction leaching of individual grains within host carbonates. Former aragonitic gastropods are leached and filled with clear calcite druse and blocky spar cement (Fig. 10.6A). Partially leached ooids occur in discrete horizons but are relatively rare (Fig. 10.6A). Pellets are wholly to partly leached and are filled with fine calcite cement. Leaching of echinoderms is rare; most inverted to calcite and microdolomite (up to 20%).

Dissolution within the Knox Paleoaquifer: Discussion

Unconfined modern coastal aquifers (Fig. 10.7) lack confining beds, thus fresh groundwaters form a wedge that slopes landward over seawater and tends to pinch out at the coastline (Back and Hanshaw 1970). Such an unconfined aquifer would have developed within Knox carbonates during unconformity formation. Initially, the aquifer would have been characterized by diffuse flow (Fig. 10.8) typical of immature, recently exposed carbonate sediments with homogeneous porosity and permeability, a well-defined water table, and a random arrangement of poorly integrated solution cavities (White 1969, James and Choquette 1984).

Water movement under conditions of diffuse flow obeys Darcy's Law and is relatively deep.

Abundant leached and spar-filled molluscan grains indicate that initial meteoric waters were undersaturated with respect to aragonite. Most echinoderm grains are neomorphosed and contain abundant microdolomite. This suggests they inverted to low-Mg calcite in waters saturated or only slightly undersaturated with respect to high-Mg calcite (cf. Dorobek 1984, Niemann 1984) or in diagenetic systems with low water/rock ratios (Chafetz et al. 1985). Leached ooids and pellets may have been high-Mg calcite, but leaching of former high magnesian calcite echinoderms is rare, suggesting that ooids and pellets were aragonite or more Mg-rich than echinoderms.

With continued dissolution, the Knox aquifer became increasingly characterized by conduit or open flow (Fig. 10.8). Recharge to such an aquifer probably was from surface runoff draining into sinkholes (Harris 1971). Large sediment loads carried into the shallow caves by conduit flow resulted in local sediment-choked cavities (White 1969). Transmissivity of such conduit flow aquifers is high, and groundwater flow paths are localized by a well-integrated system of caves which typically extend down to 100 m below the water table (White 1969). Conduit flow commonly reaches velocities of several meters per second (turbulent regime). In the Knox aquifer, the rapidly flowing meteoric waters moving by conduit flow probably were CO_2 charged and undersaturated with respect to calcite and aragonite (Back and Hanshaw 1970) and possibly dolomite; this may have caused extensive dissolution of host

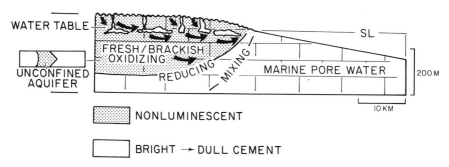

FIGURE 10.7. Schematic cross-section illustrating unconfined aquifer recharged from unconformity on Knox carbonates. Rectangle at left illustrates cement sequence likely to result from deposition of nonluminescent cement from shallow oxidizing pore fluids, followed by stagnation and deposition of bright or dull cements.

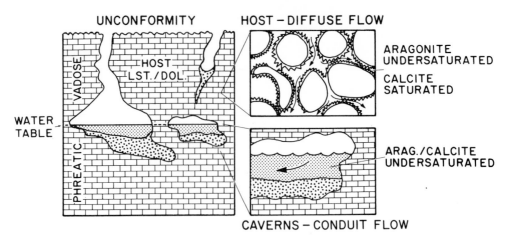

Figure 10.8. Schematic diagram illustrating types of flow associated with evolving aquifer in Knox carbonates. Initially flow would be diffuse, but as caverns develop, these local areas become sites of conduit flow.

limestone, collapse of cavity walls and roofs, and subsequent formation of intraformational breccias (interstratal karst of Quinlan 1972, Sweeting 1972, p. 298). Diffuse flow conditions were maintained within the adjacent host sediment, where shallow phreatic cementation diminished porosity and permeability.

Recharge in northwestern belts was by surface drainage from the unconformity, suggested by close association of unconformity surface sinkholes with intraformational breccias (Harris 1971). Also, post-unconformity Middle Ordovician Blackford detrital carbonates and siliciclastics occur in collapse breccias down to 280 m below the unconformity (Harris 1969 and 1971, Kyle 1976). There is little correspondence between paleokarst features and collapse breccias in southeastern belts. Harris (1971) suggests that dissolution here was caused by movement of meteoric water through a paleoaquifer (Kingsport Formation of Knox Group) with recharge from the northwest. Grover and Read (1983) suggest that during the Middle Ordovician orogeny, Knox carbonates also were exposed in tectonic highlands to the southeast. These carbonates may have provided recharge areas for meteoric waters that moved downdip toward the northwest (cratonward) in a confined aquifer (Fig. 10.9).

The stratigraphic distribution of breccias is coincident with the Kingsport-Mascot contact and may be due to increase in limestone from Mascot Dolomite down into the Kingsport Formation. This may have been coupled with the presence of an Ordovician deep water table or mixing zone. Localization of the breccias deep below the unconformity and their lesser development high in the section suggest that intense dissolution of limestone and concomitant brecciation of dolomite interbeds also may have been controlled by a mixing zone located at the base of the freshwater lens, where the solution was undersaturated with respect to calcite (Thrailkill 1968, Plummer 1975). Dissolution in a mixing zone occurs in the "boulder zone" of the deeper, Tertiary Florida aquifer (Back and Hanshaw 1970) and in the Yucatan aquifer (Back et al. 1979, Back and Hanshaw 1970). Extensive dissolution of subtidal limestone beds in the mixing zone of the Knox/Beekmantown aquifer may have resulted in the development of solution channels. Concentration of flow along these conduits might have formed intraformational breccias.

Cementation

Early Cements

Microspar and earliest sparry calcite cements occur in Knox packstone and grainstone beds. Microspar (0.005–0.020 mm) primarily occurs as intergranular, fine equant cement (Fig. 10.6B), but may be present in leached cavities as cement and rarely as crystal silt. Microspar which occurs as neomorphosed intergranular and host micrite is nonluminescent to bright/

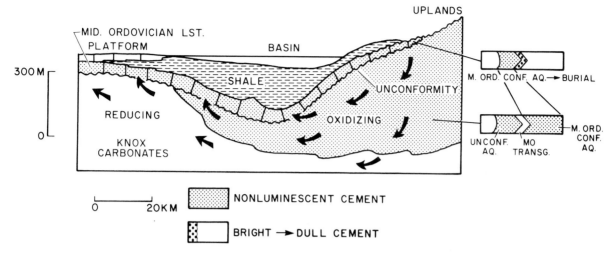

FIGURE 10.9. Schematic cross-section showing distribution of cements from upland sourced confined aquifer system, as in the Middle Ordovician. Updip areas characterized by Fe- and Mn-poor, nonluminescent cements which become slightly more Fe- and Mn-rich (bright or dully luminescent) downdip, as waters become more reducing. Small rectangles at right illustrate cement sequences observed in Knox carbonates versus cements in the Middle Ordovician. Note that second nonluminescent cement in the Knox is probably equivalent to the first nonluminescent cement in the Middle Ordovician.

dull or dull with nonluminescent to bright/dull overgrowths. Micrite (< 4μm) commonly is dull with local areas of chalky microporosity filled with clear nonluminescent microspar. Clear cements occur as fine to coarse prismatic, scalenohedral and equant crystals and syntaxial overgrowths. Clear prismatic crystals (0.02–0.08 mm) form thin isopachous rims on pellets and ooids (Fig. 10.6B). Scalenohedral crystals (0.005–0.10 mm) rim grains, line primary and dissolution voids (Fig. 10.6C, E), and rarely occur as pendant cement in shelter voids and on grains.

Some early cements show a zonation from nonluminescent to bright/dull to nonluminescent (Fig. 10.6D). Younger-generation nonluminescent cement is irregularly distributed and commonly is less abundant than first-generation nonluminescent cement. More commonly, early cements show a nonluminescent to bright to dull zonation (Fig. 10.6F). or are uniformly nonluminescent or rarely, in southeastern sections, uniformly dull. These cements predated compaction, as suggested by sheared nonluminescent to bright/dull cement and by spalling of these cements from ooids. Both generations of nonluminescent cements have low Fe and Mn concentrations, with most analyses indicating Fe concentrations below limits of detection (400 ppm). Limited analyses of bright luminescent cement indicate low Fe (below 400 ppm) and slightly elevated Mn (up to 1000 ppm) concentrations. Dull cement has elevated Fe (up to 1100 ppm) and Mn (up to 750 ppm) concentrations.

Interpretation. With the exception of rare pendant and meniscus cements that precipitated in the vadose zone, most meteoric diagenesis and cementation in Knox carbonates is shallow phreatic. Shallow phreatic calcite cements in the Appalachians have Fe < 1200 ppm and Mn < 1000 ppm, according to studies of various Paleozoic units (Grover and Read 1983, Dorobek 1984, Niemann 1984). Knox early cements plot within these limits for Fe and Mn, suggesting that they precipitated from shallow, oxidizing to moderately reducing fluids. Such conditions would have existed in the shallow phreatic zone within the early Middle Ordovician unconfined aquifer and possibly in the subsequent confined aquifer.

Dissolution of host aragonite and possibly high-Mg calcite by shallow phreatic meteoric waters resulted in local supersaturation of pore fluids with respect to calcite. Consequently, nonluminescent calcite cement with low Fe and Mn precipitated from oxidizing meteoric waters concomitantly with aragonite leaching in waters undergoing diffuse flow (Meyers 1978). Fe^{2+} concentrations of waters were below 1 ppm, using a $K_D = 4$ (Ichikuni 1983). Some nonluminescent cement that is discontinuously distributed may have precipitated from shallow freshwater lenses that developed beneath tidal flats during emergence associated with cyclic deposition (pre-unconformity). However, correlation between widespread distribution of earliest nonluminescent cement and areas of intense development of the unconformity and karstic features suggests that this cement precipitated within an unconfined meteoric aquifer system (Fig. 10.7) that developed with the Knox unconformity under humid conditions (Reinhardt and Hardie 1976, Read and Grover 1977). Early dull/bright luminescent cement, which contains only slightly elevated Fe and Mn concentrations relative to nonluminescent cements, precipitated from reducing fluids possibly during stagnation of the aquifer (Carpenter and Oglesby 1976, Champ et al. 1979). Stagnation may have occurred during incipient arc-continent collision in the Middle Ordovician, when the platform foundered and marine transgression occurred. This caused cratonward recession of the freshwater lens, shut off direct recharge from the unconformity surface, and caused stagnation of the unconfined aquifer.

The second nonluminescent cement present locally in Knox carbonates may have formed in the Middle Ordovician during uplift of tectonic highlands. These uplands recharged a confined paleoaquifer that may have developed within Upper Knox and Middle Ordovician carbonates which were blanketed by Ordovician shale (Fig. 10.9) (Harris 1971, Grover and Read 1983). Post-Knox Middle Ordovician carbonates have a single generation of nonluminescent cement. This suggests that oxidizing fluids were reintroduced locally into Knox/Beekmantown carbonates following collision of the passive margin and deposition of Middle Ordovician carbonates. Oxidizing fluids would have moved along permeable beds, possibly to great depths under high hydrostatic head (White 1969, Hitchon 1969). The oxidizing fluids were saturated with respect to calcite and precipitated a second generation of nonluminescent calcite cement. Dissolution of limestone beds and continued development of intraformational breccias may have occurred in the mixing zone at the base of the confined aquifer.

Stable Isotopes. Early Ordovician marine calcites (Fig. 10.10) probably had $\delta^{18}O$ values of -5 to

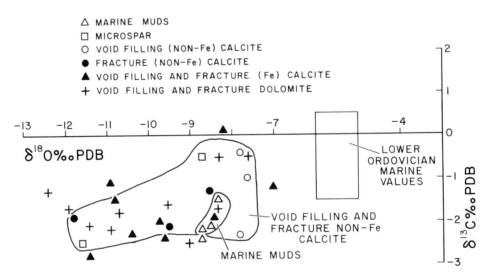

FIGURE 10.10. Stable isotope diagram for major components of Knox carbonates. Marine muds appear to be diagenetically modified from the Ordovician marine values (rectangle) by meteoric and/or burial fluids. Nonferroan (below 1100 ppm) calcites are the early nonluminescent to early dull cements related to the Knox aquifer, while the Fe calcites are typically burial calcites, along with the dolomite cements.

−6‰ PDB (K.C Lohmann pers. comm. 1985). $\delta^{18}O$ values for marine muds (−8.3 to −8.8‰ PDB) and microspar (−8.8 to −11.6‰ PDB) in the Knox are lighter than these Ordovician marine values. Microspar commonly is non-luminescent to bright/dull or has nonlumines-cent microcement overgrowths; marine muds (micrite) have patchy chalky microporosity filled with clear nonluminescent to bright/dull microspar. These suggest recrystallization and cementation of lime muds by meteoric waters. Some marine micrite and microspar are uni-formly dull, possibly indicating cementation by later warm basinal brines or depleted mixed waters. Precompaction, nonferroan, early non-luminescent and bright/dull cements are de-pleted in ^{18}O ($\delta^{18}O$ −7.6 to −7.8‰ PDB) rel-ative to Ordovician marine values, which is compatible with a meteoric fluid origin.

Early Ordovician marine calcite $\delta^{13}C$ values are likely to have been 0.5 to −1.5‰ (Lohmann pers. comm. 1985). $\delta^{13}C$ values (−0.4 to −2.5‰) for nonferroan cements in the Knox overlap this range or show slight depletion in ^{13}C. Significant depletion in Ordovician aqui-fers would not be expected, given the absence of terrestrial vegetation, although some light C could have been contributed from oxidation of organic matter in the sediments, especially in algal laminates.

Burial Moldic Porosity

Much dissolution of host grains and matrix af-ter early cementation resulted in moldic po-rosity in the burial environment. Coarser-grained layers, rare mud layers, and burrows locally were leached and subsequently cement-ed with saddle dolomite and Fe-rich calcite, which stains mauve to blue with Dickson's so-lution (Dickson 1965) (Fig. 10.11A). Solution-enlarged fractures are filled with mauve- to purple-staining calcite (Fig. 10.11B), which has abundant microdolomite inclusions (Fig. 10.11C). Small-scale fabric-selective dissolution of nonskeletal carbonate grains and muds re-sulted in chalky microporosity. Fabric-selective dissolution of previously cemented ooids re-sulted in microporous ooids lacking collapsed cores typical of early dissolution (Fig. 10.11D). Leached pellets are cemented with blocky Fe-rich calcite spar and saddle dolomite (Fig. 10.11A,B; 10.12A,B). Remaining porosity and

large leached cavities are filled with mauve- or blue-staining ferroan calcite and/or saddle do-lomite (Figs. 10.11A and 10.12A).

Interpretation. A major burial dissolution event(s) that predates ferroan calcite and saddle dolomite cements was responsible for much moldic porosity in the Knox carbonates. Dis-solution may have been associated with organic maturation (Carothers and Kharaka 1978 and 1980, Kharaka et al. 1983). In the subsurface, organic acids (mainly acetic acid) are produced between 80°C and 140°C by thermal degra-dation of kerogen and are abundant in many oil field waters; decarboxylation of acetate to form CO_2 continues at temperature up to 200°C (Carothers and Kharaka 1978). Either of these reactions might have caused leaching of un-stable grains.

Late Spar Cements and Dolomite

Late Calcite Cements. Bright and/or dull to very dull luminescent cement commonly postdates earlier nonluminescent cements and commonly is subzoned (Fig. 10.6F). This locally occludes intergranular primary and secondary porosity, and fills compaction-induced fractures. Earliest generations stain pink (low Fe) with Dickson's solution (Dickson 1965) and have moderately elevated Fe (up to 1800 ppm) concentrations with slightly elevated Mn (up to 900 ppm) con-centrations. Ferroan cements stain mauve to purple or blue, are inclusion-rich with up to 20% (visual estimate) microdolomite inclusions (Fig. 10.11C) and are enriched in Fe (up to 3500 ppm), although Mn concentrations rarely exceed 900 ppm. Nonferroan fracture-filling cements have $\delta^{18}O$ values of −8.5 to −11.8‰ PDB and $\delta^{13}C$ values of −1.3 to −2.1‰ PDB. Ferroan cements have similar values with $\delta^{18}O$ values of −7 to −11.4‰ PDB and $\delta^{13}C$ values of 0.1 to −2.8‰ (Fig. 10.10). Distribution of ferroan cement appears to be limited geo-graphically to southwestern Virginia and Ten-nessee. Ferroan cement always postdates non-ferroan cements within a sample, but may be coeval with some burial nonferroan cements in northern sections where Fe-rich cements are rare.

Saddle Dolomite. In east Tennessee, saddle do-lomite, lesser quartz, and late-stage calcite occur

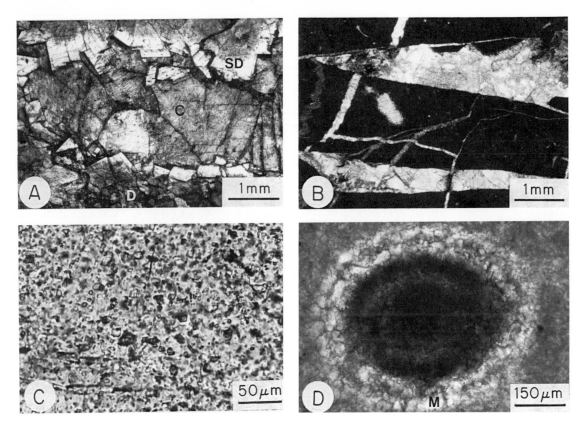

FIGURE 10.11. A–D. Burial moldic porosity and late calcite and dolomite cement. *A.* Leached coarse-grained layer subsequently filled with saddle dolomite (SD) and later ferroan blocky calcite cement (C). Saddle dolomite nucleated on finer-grained replacement dolomite mosaic of host (D). *B.* Solution-enlarged fractures in mud host are filled with ferroan calcite cement. *C.* Microdolomite inclusions in ferroan calcite from fracture-filling cement of B. Inclusions are crystallographically oriented within the calcite spar. *D.* Small-scale fabric selective dissolution of outer layers of an ooid. Chalky microporosity is cemented by ferroan calcite microspar (M).

as pods or lenses associated with intraformational breccias or as replacement mosaics that form haloes within host rocks surrounding ore zones. In general, saddle dolomite occurs both as a cement filling leached cavities (leached interhead fills within thrombolotes, coarse-grained layers, and burrows) in subtidal facies and in fractures, and as a replacement of host microspar, grains, and cements. Saddle dolomite cement is subhedral to euhedral with individual crystals ranging from 30 μm to several mm in diameter. Crystals have curved faces, are milky to brownish, and are inclusion-rich. Cathodoluminescent zonation can be well developed and is regionally mappable (Ebers and Kopp 1979, Gorody 1980). Saddle dolomite replacing host microspar (calcite) is inclusion-rich and exhibits pseudocorrosional growth fronts where it is in contact with the host microspar or grains. Replacement saddle dolomite has splotchy luminescence with relict calcite inclusions preserving early cement (nonluminescent to bright/dull) luminescence. Complete replacement of subtidal facies results in a subhedral to euhedral dolomite mosaic which displays well-developed cathodoluminescence (Fig. 10.12C, D). Individual saddle dolomite crystals grow into secondary voids which rarely show correlation to original fabric. Late-stage pink- to blue-staining calcite and lesser quartz fill remaining porosity.

These dolomites are Ca rich (up to 68 mole %) and have high Fe (up to 12,000 ppm) and moderate Mn (up to 3000 ppm) concentrations. $\delta^{18}O$ values range from values similar to host rocks (-7.6 to -8.7‰ PDB) to values depleted

FIGURE 10.12. A–D. Burial moldic porosity and late calcite and dolomite cement. *A.* Leached pellets cemented with ferroan microspar and saddle dolomite (arrows). Intergranular primary voids are lined with nonferroan calcite druse and microspar (D). Ferroan, blocky spar (V) fills remaining void space. *B.* Leached pellet filled with ferroan fine-grained and blocky calcite. *C, D.* Plane light and cathodoluminescent paired photographs of saddle dolomite mosaic. Rhombs that fill a secondary void have well-developed cathodoluminescent zonation. Late-stage ferroan calcite fills remaining porosity (C).

in ^{18}O (-9 to $-12.5\permil$ PDB). δ^{13}C values range from -0.5 to $-2.5\permil$ PDB (Fig. 10.10).

Fluid Inclusion Data. Fluid inclusion data for Pb–Zn sulfides and associated carbonate cements within intraformational breccias from the East Tennessee mining district indicate homogenization temperatures of 80°C to 170°C (Roedder 1971). Combined fluid inclusion data from main-stage sphalerite suggest homogenization temperatures ranging from 100°C to 200°C (Taylor et al. 1983a and 1983b, Caless 1983). Reported temperatures are not pressure corrected, which suggests that true temperatures could have been 25°C to 50°C higher (Caless 1983). Fluid inclusion freezing temperatures range from $-30°$C (Taylor et al. 1983a) for main-stage sphalerite to $-9°$C (Roedder 1971, Caless 1983) for postore, vug-filling fluorite, indicating NaCl brines. A few data have anomalously low freezing temperatures ($-37.0°$C to $-36.4°$C), suggesting the presence of $CaCl_2$ and $MgCl_2$ salts in solution (Crawford 1981).

Interpretation. Elevated Fe concentrations relative to earlier cements suggest that most later dull (both pink and purple/blue staining) cements precipitated from reducing pore waters under burial conditions. These burial fluids were enriched in Fe^{2+} (possibly up to 50 ppm; calculated using $K_D = 4$; Ichikuni 1983). That these dull cements postdate the Knox aquifer is suggested by their occurrence in post-Knox Middle Ordovician carbonates, where they overlie the last nonluminescent cements, which in turn are cut by compaction fractures filled by dull cement (Grover and Read 1983). Most dull cements precipitated from basinal fluids at elevated temperatures. Abundant micro-dolomite inclusions in these cements may indicate elevated temperatures which favored either precipitation of a Mg-rich calcite (Fücht-bauer and Hardie 1976) that exsolved to calcite and microdolomite, or coprecipitation of dolomite and calcite (Rosenberg and Holland 1964). Light $\delta^{18}O$ values for the calcites (Fig. 10.10) are compatible with precipitation from warm basinal brines or mixed fluids with depleted $\delta^{18}O$ values. Overlap of $\delta^{18}O$ and $\delta^{13}C$ values with those of the host rocks and early nonferroan calcite cements suggests that pressure solution was a major source of carbonate.

Reducing pore-water conditions would have developed following erosion of Ordovician tectonic highlands (decreased hydraulic head) and burial beneath thick foreland basin clastics. With renewed uplift of these highlands in the Devonian and Carboniferous, there probably was increased and deep penetration of slow-moving, reducing meteoric fluids recharged from the highlands (Hitchon 1969, Kyle 1976). These fluids would have been reducing because they would have had to migrate through thick sequences of organic-rich Middle to Late Paleozoic shales and sands. Ferroan calcite cement (mauve to purple/blue staining) appears to be geographically limited to areas peripheral to the Tennessee depocenter, where as much as 2000 m of Ordovician shales (a major source of Fe) accumulated. Furthermore, these Middle to Late Paleozoic upland-sourced fluids (Bethke 1986) would have been added to fluids being expelled by compaction dewatering of basinal shales (Anderson 1983, Cathles and Smith 1983), and waters released by smectite-illite diagenesis as the shales passed through the oil window (Fig. 10.13) (Burst 1969, Perry and Hower 1970, Johns and Shimoyama 1972). Optimum conditions thus developed during the Carboniferous for geopressuring of shales and fluid expulsion. Such conditions accompanied increased subsidence which resulted from deposition of synorogenic clastics and emplacement of thrust sheets. The combination of the high relief Late Paleozoic highlands, plus dewatering of geopressured basinal shales sug-

gests that expulsion of warm saline brines most likely occurred during the Carboniferous/Permian. Previous to and during expulsion of these brines, much compaction and fracturing of intraformational breccias enhanced permeability of Ordovician karstic conduits. This facilitated channelization and rapid migration of brines along preexisting breccias. Apparently, basinal brines were undersaturated with respect to calcite and commonly dolomite, resulting in further extensive dissolution of preexisting breccias and host carbonate in east Tennessee (Hoagland et al. 1965, Kyle 1976, Churnet et al. 1982, Churnet and Misra 1981). Pb–Zn sulfides and associated gangue minerals (saddle dolomite, calcite, quartz, and fluorite) precipitated from these brines, plugging porosity of intraformational breccias and fracture systems.

Conclusions

The Ordovician Knox unconformity developed on top of passive-margin carbonates during arc-continent collision, and later was buried beneath a thick foredeep sequence. Paleohighs up to 30 m high occur on the unconformity and are flanked by accumulations of coarse and fine detrital carbonates and clastics that thicken into topographic lows. Paleokarst features include filled sinkholes, caves, and intraformational breccias. These features were formed by dissolution associated with conduit flow in the Knox paleoaquifer, whose recharge area was the vast expanse of exposed shelf. Waters moving by diffuse flow through host carbonates of the paleoaquifer caused initial leaching of unstable grains, followed by cementation, which may have continued into the shallow burial environment when waters became recharged from a tectonic highland along the uplifted basin margin.

During burial, widespread leaching of host grains and matrix formed a chalky microporosity, possibly during organic maturation of kerogen. Subsequently, remaining pore-spaces in intergranular voids, dissolution pores, and compaction- or tectonically induced fractures were filled by later sparry calcite cements (commonly Fe-rich), along with saddle dolomites and local sphalerite. These were depos-

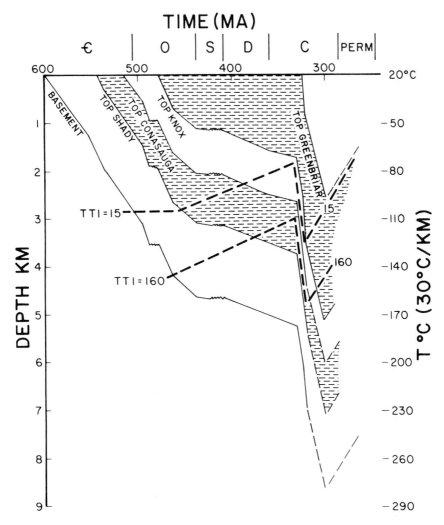

FIGURE 10.13. Burial history plot (not decompacted) of Paleozoic sequence in southwest Virginia. Time–temperature index (TTI) is calculated from Waples (1980) and defines the oil window. Most of the shales in the sequence pass into the oil window in the Late Paleozoic; this in conjunction with high relief uplands and rapid subsidence related compaction would favor fluid expulsion at this time. This time coincides with the major period of overthrusting.

ited from warm basinal brines that tended to be channeled by the intraformational breccias. Major fluid expulsion appears to have been related to compaction and shale diagenesis associated with passage of the shale sequence through the oil window, along with elevated hydraulic head due to uplift of Late Paleozoic mountains during overthrusting.

The karstic features appear to have provided a major plumbing system that controlled subsequent diagenesis and base metal and possible hydrocarbon emplacement in the Appalachian Knox carbonates.

Acknowledgments. Thanks are extended to Dr. W.D. Lowry, Dr. C.G. Tillman (deceased), Fred Webb Jr., George Grover Jr., John A. Bova, William A. Koerschner, and Hank Ross. Technical assistance was provided by R. Harris, C. Ross, S. Walker, and M. Ostrand, T. Cooney, Brent Bray, S. Haythornthwaite, S. Chiang, M.

Eiss, J. Webb, and Melody Wayne (drafting), and C. Zauner and Llyn Sharp (photography). Ada Simmons typed final drafts. Financial assistance was provided largely by grants EAR 7911213, EAR 8108577, and EAR 8305878 from the National Science Foundation, Earth Sciences Section, to J.F. Read, and a grant-in-aid from Sigma Xi, the Research Society of North America.

References

Anderson, G.M., 1983, Some geochemical aspects of sulphide precipitation in carbonate rocks, in Kisvarsanyi, G., Grant, S.K., Pratt, W.P., and Koenig, J.W., eds., Proceedings of the International Conference on Mississippi Valley-type Lead–Zinc Deposits: Univ. of Missouri–Rolla, MO, p. 61–76.

Back, W., and Hanshaw, B.B., 1970, Comparison of chemical hydrology of the carbonate peninsulas of Florida and Yucatan: Jour. Hydrology, v. 10, p. 330–368.

Back, W., Hanshaw, B.B., Pyle, T.E., Plummer, L.N., and Weidie, A.E., 1979, Geochemical significance of groundwater discharge and carbonate solution to the formation of Caleta Xel Ha, Quintana Roo, Mexico: Water Resources Research, v. 15, p. 1531–1535.

Bethke, C.M., 1986, Hydrologic constraints on the genesis of the Upper Mississippi Valley mineral district from Illinois basin brines: Econ. Geology, v. 81(2), p. 233–249.

Bird, J.M., and Dewey, J.F., 1970, Lithosphere plate-continental margin tectonics and the evolution of the Appalachian orogen: Geol. Soc. Amer. Bull., v. 81, p. 1031–1060.

Burst, J.F., 1969, Diagenesis of Gulf Coast clayey sediments and its possible relation to petroleum migration: Am. Assoc. Petroleum Geologists Bull., v. 53, p. 73–93.

Caless, J.R., 1983, Geology, paragenesis, and geochemistry of sphalerite mineralization at the Young Mine, Mascot-Jefferson City zinc district, east Tennessee: M.S. thesis, Virginia Polytechinic Institute and State University, Blacksburg, VA, 237 p.

Carothers, W.W., and Kharaka, Y.K., 1978, Aliphatic acid anions in oil-field waters—Implications for origin of natural gas: Amer. Assoc. Petroleum Geologists Bull., v. 62(12), p. 2441–2453.

Carothers, W.W., and Kharaka, Y.K., 1980, Stable carbon isotopes of HCO₃ in oil-field waters—Implications for the origin of CO₂: Geochim. Cosmochim. Acta, v. 44, p. 323–332.

Carpenter, A.B., and Oglesby, T.W., 1976, A model for the formation of luminescently zoned calcite cements and its implications: Geol. Soc. America Abstracts with Programs, v. 8, p. 469–470.

Cathles, L.M., and Smith, A.T., 1983, Thermal constraints on the formation of Mississippi Valley-type lead–zinc deposits and their implications for episodic basin dewatering and deposit genesis: Econ. Geology, v. 78, p. 983–1002.

Chafetz, H.S., Wilkinson, B.H., and Love, K.M., 1985, Morphology and composition of nonmarine carbonate cements in near-surface settings, in Schneidermann, N., and Harris, P.M., eds., Carbonate cements: Soc. Econ. Paleontologists & Mineralogists Spec. Publ. 36, p. 337–369.

Champ, D.R., Gulens, J., and Jackson, R.E., 1979, Oxidation–reduction sequences in groundwater flow systems: Canadian Jour. Earth Sci., v. 16, p. 12–23.

Churnet, H.G., and Misra, K.C., 1981, Genetic implications of the trace element distribution pattern in the Upper Knox carbonate rocks, Copper Ridge district, east Tennessee: Sed. Geol., v. 30, p. 173–194.

Churnet, H.G., Misra, K.C., and Walker, K.R., 1982, Deposition and dolomitization of Upper Knox carbonate sediments, Copper Ridge district, east Tennessee: Geol. Soc. America Bull., v. 93, p. 76–86.

Collins, J.A., and Smith, L., 1975, Zinc deposits related to diagenesis and intrakarstic sedimentation in the Lower Ordovician St. George Formation, western Newfoundland: Bull. Canadian Petroleum Geologists, v. 23, p. 393–427.

Colton, G.W., 1970, The Appalachian Basin—its depositional sequence and their geologic relationships, in Fisher, G.W., et al., eds., Studies of Appalachian geology—central and southern: New York, Interscience, p. 5–47.

Crawford, M.L., 1981, Phase equilibria in aqueous fluid inclusions, in Hollister, L.S., and Crawford, M.L., eds., Short course in fluid inclusions—applications to petrology: Mineralogical Assoc. Canada Short Course Handbook, 6, p. 75–100.

Dickson, J.A.D., 1965, A modified staining technique for carbonates in thin section: Nature, v. 205, p. 587.

Dorobek, S.L., 1984, Stratigraphy, sedimentology, and diagenetic history of the Siluro-Devonian Helderberg Group, central Appalachians: M.S. thesis, Virginia Polytechnic Institute and State University, Blacksburg, VA, 237 p.

Dorobek, S.L., 1987, Petrography, geochemistry, and origin of burial diagenetic facies, Siluro-Devonian Helderberg Group (carbonate rocks), Central Appalachians; A.A.P.G. Bull., v. 7(5), p. 492–514.

Ebers, M.L., and Kopp, O.C., 1979, Cathodoluminescent microstratigraphy in gangue dolomite, the Jefferson City district, Tennessee: Econ. Geology, v. 74, p. 908–918.

Füchtbauer, H., and Hardie, L.A., 1976, Experimentally determined homogeneous distribution coefficients for precipitated magnesian calcites—application to marine carbonate cements: Geol. Soc. America Abstracts with Programs, v. 8, p. 887.

Gorody, A.W., 1980, The Lower Ordovician Mascot Formation, Upper Knox Group, in north central Tennessee, Part I: Paleoenvironmental history; Part II: Dolomitization and paleohydraulic history: Ph.D. thesis, Rice University, Houston, TX, 181 p.

Grover, G., Jr., and Read, J.F., 1983, Paleoaquifer and deep burial related cements defined by cathodoluminescent patterns, Middle Ordovician carbonates, Virginia: Amer. Assoc. Petrologists Bull,, v. 7(8), p. 1275–1303.

Harris, L.D., 1969, Kingsport Formation and Mascot Dolomite (Lower Ordovician) of east Tennessee: Tennessee Division of Geology Report of Investigations 23, 139 p.

Harris, L.D., 1971, A Lower Paleozoic paleoaquifer—the Kingsport Formation and Mascot Dolomite of Tennessee and southwest Virginia: Econ. Geology, v. 66, p. 735–743.

Hatcher, R.D., Jr., 1972, Developmental model for the southern Appalachians: Geol. Soc. America Bull., v. 83, p. 2735–2760.

Hitchon, B., 1969, Fluid flow in the Western Canada sedimentary basin, 1. Effect of topography: Water Resources Research, v. 5(1), p. 186–195.

Hoagland, A.D., Hill, W.T., and Fulweiler, R.E., 1965, Genesis of the Ordovician zinc deposits in east Tennessee: Econ. Geology, v. 60, p. 693–714.

Ichikuni, M., 1983, Anionic substitution in calcium carbonate, in Augustithis, S.S., ed., The significance of trace elements in solving petrographic problems and controversies: Athens, Greece, Theophrastus Publications, S.A., p. 81–94.

Jacobi, R.D., 1981, Peripheral bulge—a causal mechanism for the Lower/Middle Ordovician unconformity along the western margin of the Northern Appalachians: Earth and Planetary Sci. Letters, v. 56, p. 245–251.

James, N.P., and Choquette, P.W., 1984, Diagenesis 9—limestones—the meteoric diagenetic environment: Geosci. Canada, v. 11(4), p. 161–194.

Johns, W.D., and Shimoyama, A., 1972, Clay minerals and petroleum-forming reactions during burial and diagenesis: Amer. Assoc. Petroleum Geologists Bull., v. 56, p. 2160–2167.

Kharaka, Y.K., Carothers, W.W., and Rosenbauer, R.J., 1983, Thermal decarboxylation of acetic acid—implications for origin of natural gas: Geochim. Cosmochim. Acta, v. 47, p. 397–402.

Kyle, J.R., 1976, Brecciation, alteration and mineralization in the central Tennessee zinc district: Econ. Geology, v. 71, p. 892–903.

Kyle, J.R., 1983, Economic aspects of subaerial carbonates, in Scholle, P.A., Bebout, D.G., and Moore, C.L., eds., Carbonate depositional environments: Amer. Assoc. Petroleum Geologists, Tulsa, OK, p. 73–92.

McCormick, J.E., and Rasnick, F.D., 1983, Mine geology of ASARCO's Young mine, in Tennessee zinc deposits field trip guide book, March 9–11, 1983: Virginia Tech Dept. Geol. Sci. Guide Book No. 9, p. 39–44.

Meyers, W.J., 1978, Carbonate cements: their regional distribution and interpretation in Mississippian limestones of southwestern New Mexico: Sedimentology, v. 25, p. 371–400.

Misra, K.C., Churnet, H.G., and Walker, K.R., 1983, Carbonate-hosted zinc deposits of east Tennessee, in Tennessee zinc deposits field trip guide book, March 9–11, 1983: Virginia Tech Dept. Geol. Sci. Guide Book No. 9, p. 2–20.

Montanez, I.P. (in prep.), Diagenesis and dolomitization of Upper Knox carbonates, southern Appalachians: Ph.D. thesis, Virginia Polytechnic Institute and State University, Blacksburg, VA.

Mussman, W.J., and Read, J.F., 1986, Sedimentology and development of a passive- to convergent-margin unconformity—Middle Ordovician Knox unconformity, Virginia Appalachians: Geol. Soc. America Bull., v. 97, p. 282–295.

Niemann, J.C., 1984, Regional cementation associated with unconformity-sourced aquifers and burial fluids, Mississippian Newman Limestone, Kentucky: M.S. thesis, Virginia Polytechnic Institute and State University, Blacksburg, VA, 200 p.

Perry, E.A., and Hower, J., 1970, Burial diagenesis in Gulf Coast pelitic sediments: Clays and Clay Minerals, v. 18, p. 165–177.

Plummer, L.N., 1975, Mixing of seawater with calcium carbonate groundwater: Geol. Soc. America Memoir, v. 142, p. 219–236.

Quinlan, G.M., and Beaumont, C., 1984, Appalachian thrusting, lithospheric flexure, and the Paleozoic stratigraphy of the eastern interior of North America: Canadian Jour. Earth Sci., v. 21, p. 973–996.

Quinlan, J.F., 1972, Karst-related mineral deposits and possible criteria for the recognition of paleokarsts—a review of preservable characteristics of Holocene and older karst terranes: 24th Internat. Geological Congress, v. 6, p. 156–167.

Read, J.F., 1980, Carbonate ramp-to-basin transitions and foreland basin evolution, Middle Ordovician, Virginia Appalachians: Amer. Assoc. Petroleum Geologists Bull., v. 64, p. 1575–1612.

Read, J.F., (in press), Evolution of Cambro-Ordovician passive margin, U.S. Appalachians: Decade of North American Geological Synthesis, Appalachians-Ouachitas Volume, Geological Society of America.

Read, J.F., and Grover, G.A., Jr., 1977, Scalloped and planar erosion surfaces, Middle Ordovician limestones, Virginia—analogues of Holocene exposed karst or tidal rock platforms: Jour. Sed. Petrology, v. 47, p. 956–972.

Reinhardt, J., and Hardie, L.A., 1976, Selected examples of carbonate sedimentation, lower Paleozoic of Maryland: Maryland Geol. Survey Guidebook 5, 53 p.

Rodgers, J., 1968, The eastern edge of the North American continent during the Cambrian and Early Ordovician, in Zen, E-an, et al., eds., Studies of Appalachian geology—northern and maritime: New York, Interscience, p. 141–149.

Roedder, E., 1971, Fluid inclusion evidence on the environment of formation of mineral deposits of the southern Appalachian Valley: Econ. Geology, v. 6, p. 777–791.

Rosenberg, P.E., and Holland, H.D., 1964, Calcite-dolomite-magnesite stability relations in solutions at elevated temperatures: Science, v. 145, p. 700–701.

Ross, R.J., Jr., et al., 1982, The Ordovician system in the United States, correlation chart and explanatory notes: International Union of Geological Sciences, Publ. No. 12, 73 p.

Shanmugam, G., and Walker, K.R., 1980, Sedimentation, subsidence, and evolution of a foredeep basin in the Middle Ordovician, southern Appalachians: Amer. Jour. Science, v. 280, p. 479–496.

Slaymaker, S.L., and Watkins, J.S., 1978, A plate tectonics model of the southern Appalachians suggested by gravity data (abstr.): Geol. Soc. America Abstracts with Programs, v. 10, p. 198.

Sloss, L.L., 1963, Sequences in the cratonic interior of North America, Geol. Soc. America Bull., v. 74, p. 93–114.

Suter, D.R., and Tillman, C.G., 1973, The conodont genus Multioistodus from supratidal limestones in the Beekmantown Formation of the Appalachians of west-central Virginia (abstr.): Geol. Soc. America Abstracts with Programs, v. 5(5), p. 441–442.

Sweeting, M.M., 1972, Karst landforms: London, MacMillan, 362 p.

Taylor, M., Kelly, W.C., Kessler, S.E., McCormick, J.E., Rasnick, F.D., and Mellon, M.V., 1983a, Relationship of zinc mineralization in east Tennessee to Appalachian orogenic events: in Kisvarsanyi, G., Grant, S.K., Pratt, W.P., and Koenig, J.W., eds., Proceedings of the International Conference on Mississippi Valley-type Lead–Zinc Deposits: Univ. of Missouri–Rolla, MO. p. 271–278.

Taylor, M., Kessler, S.E., Cloke, P.L., and Kelly, W.C., 1983b, Fluid inclusion evidence for fluid mixing, Mascot–Jefferson City zinc district, Tennessee: Econ. Geology, v. 78(7), p. 1425–1439.

Thrailkill, J., 1968, Chemical and hydrologic factors in the excavation of limestone caves: Geol. Soc. America Bull., v. 79, p. 19–46.

Tillman, C.G., 1976, A Prioniodus apparatus from beds of Whiterock age (Ordovician), Harrisonburg, Virginia (abstr.): Geol. Soc. America Abstracts with Programs, v. 8(4), p. 513.

Waples, D.W., 1980, Time and temperature in petroleum formations—application of Lopatin's method to petroleum exploration: Amer. Assoc. Petroleum Geologists Bull., v. 64, p. 916–926.

Webb, F., Jr., 1959, Geology of the Middle Ordovician limestones in the Rich Valley area, Smyth County, Virginia: M.S. thesis, Virginia Polytechnic and State University, Blacksburg, VA, 96 p.

White, W.B., 1969, Conceptual models for carbonate aquifers: Groundwater, v. 7, p. 15–21.

11
Surface and Subsurface Paleokarst, Silurian Lockport, and Peebles Dolomites, Western Ohio

CHARLES F. KAHLE

Abstract

The Lockport Dolomite in northwestern and west-central Ohio and the stratigraphically equivalent Peebles Dolomite in southwestern Ohio collectively display a variety of previously unrecognized subsurface and surface paleokarst features. The former include molds, vugs, in situ breccia with corroded clasts, collapsed strata, solution-enlarged joints, internal sediment derived partly from soil, boxwork, and caves. Surface paleokarst features at the top of these units are represented by locally developed paleosol and sinks. Erosional relief at the top of the Lockport/Peebles ranges from fractions of an inch (typical) up to 9 feet (2.7 m) (rare). The contact between the Lockport/Peebles and the overlying Greenfield Dolomite is typically a paraconformity in the form of a paleokarst planar erosion surface. As such, the contact can be difficult to recognize, but it is not a facies contact nor is it gradational.

Surface karstification and erosion of the Lockport and Peebles dolomites occurred prior to formation of the overlying Greenfield Dolomite. Subsurface karstification of Lockport/Peebles strata is interpreted to have occurred concomitantly with surface karstification of these units. Climatic conditions during karstification were probably semi-arid.

A Gley paleosol with hydromorphic properties occurs locally at the top of the Lockport/Peebles; it is remarkably similar to modern Gley soils. This indicates that conditions of soil formation have been similar from the Silurian to the present in spite of the dominance of land plants in post-Silurian rocks.

Formation of marine strata in the Lockport Dolomite in portions of northwestern Ohio was interrupted by (a) syndepositional faulting that led to (b) subaerial exposure of at least some fault blocks. Evidence of (a) is recorded by at least one fault, terra rossa, and scalloped and planar erosion surfaces within the Lockport, and by apparently laterally restricted subaerial unconformities within this unit; (b) is recorded by concentrations of caves and/or vugs and molds in rocks immediately below the unconformities. Thus, some paleokarst features within the Lockport Dolomite at some localities originated earlier than did the great majority of paleokarst features within this unit.

Introduction

To what extent, if any, Silurian rocks in the Michigan Basin region contain evidence of subaerial exposure in the form of calcrete and/or paleokarst facies (Esteban and Klappa 1983) is of fundamental importance in deciphering the history of these rocks (Shaver 1977). Conflicting models have been proposed to account for the history of these rocks; an excellent review is by Shaver (1977). Much of the conflict is founded on whether or not different workers recognized evidence of subaerial exposure in Silurian rocks they studied, especially in a rock unit known as the Lockport Group or Lockport Dolomite (e.g., Mesolella et al. 1974, Janssens and Stieglitz 1974, Norris 1975, Droste and Shaver 1976, Gill 1977. Huh et al. 1977, Janssens 1977, Kahle 1978, Sears and Lucia 1979, Petta 1980, McGovney et al. 1982). In southwestern Ohio the Lockport is termed the Peebles Dolomite (Bowman 1961, Cook 1975, Shaver et al. 1983, Fig. 5).

The main purpose of this paper is to document the nature of surface and subsurface paleokarst features in portions of the Lockport Dolomite and the Peebles Dolomite that crop out in quarries along the lengths of the Findlay arch in western Ohio (Fig. 11.1, Table 11.1). The Lockport Dolomite was studied at 13 quarries and the Peebles Dolomite at one quarry (Fig. 11.1, Table 11.1). No such regionally comprehensive study has been published previously. An additional purpose of this paper is to consider the timing, and the significance, of surface and subsurface paleokarst features in the Lockport Dolomite and the Peebles Dolomite.

Stratigraphy

Detailed treatment of the evolution of stratigraphic nomenclature, and existing stratigraphic nomenclature, for Silurian strata in

FIGURE 11.1. Location of 14 quarries (solid squares) where information for this report was obtained. Quarries are named after the town in which they occur or after the closest town. Actual quarry names and exact quarry locations can be obtained from the author. Quarry numbers are keyed to Table 11.1. Location of Findlay arch is from Green (1957) and Coogan and Parker (1984, Fig. 2).

western Ohio are beyond the scope of this report. Readers are referred to Summerson (1963), Ulteig (1964), Cook (1975), Droste and Shaver (1976), Janssens (1977), Shaver (1977), and Shaver et al. (1983) for details concerning these topics.

The focal point of the present paper is a regionally extensive mass of dolomite in northwestern and west-central Ohio (Cook 1975) which has been referred to as the Guelph, Lockport, Lockport Formation, Lockport Dolomite, Lockport Group, and Guelph–Lockport (Alling and Briggs 1961, Table 1; Summerson 1963, Fig. 9; Ulteig 1964, Fig. 6; Cook 1975; Norris 1975, p. 225; Janssens 1977, Fig. 1 and p. 15), and is referred to herein (Fig. 11.2), following Norris (1975), as the Lockport Dolomite. The present author applies this name in northwestern and west-central Ohio to dolomite whose characteristics include the following features: (1) The Lockport Dolomite in western Ohio displays exceptional purity relative to subjacent and superjacent units. Insoluble residue is always less than 1.0% and averages about 0.5%. According to Ulteig (1964, p. 18) the most easily recognized characteristic associated with what he terms the Lockport Group is the relative purity of the dolomite. Likewise, Norris (1975, p. 227) considers exceptional purity to be a major characteristic of what he terms the Lockport Dolomite. (2) Following Norris (1975, p. 227) and Ulteig (1964), the Lockport Dolomite is white to gray to brown, extremely massive to bedded, commonly sucrosic, and typically coarsely crystalline. (3) The Lockport typically contains a variety of marine fossils including corals, stromatoporoids, bivalves, gastropods, cephalopods, brachiopods, bryozoans, crinoids, echinoids, and trilobites (e.g., Summerson 1963). (4) A major characteristic of the Lockport Dolomite is reefs (Norris 1975, Janssens 1977). Reef cores are characterized by massive bedding; reef flank rocks are well to indistinctly bedded (e.g., Winzler 1974; Shaver et al. 1978, Figs. 13A, 21B; Kahle 1978). Detailed information about the Lockport Dolomite as seen in quarries in western Ohio occurs in Winzler (1974), Requarth (1978), Lanz (1979), Lehle (1980), and Tomassetti (1981).

Janssens (1977), as a part of a subsurface study of Silurian rocks in northwestern Ohio, recognized (p. 15) what he termed the Lockport

TABLE 11.1. Summary of evidence and incidence of evidence for surface and subsurface paleokarst in the Lockport Dolomite and the Peebles Dolomite at 14 quarries in western Ohio.

	1	2	3	4	5	6	7	8	9	10	11	12	13	14
	MAUMEE	LIME CITY	CLAY CENTER	FREMONT	MAPLE GROVE	WEST MILGROVE	CAREY	BUCKLAND	GENOA	WOODVILLE N	WOODVILLE SE	GIBSONBURG	MILLERSVILLE	PEEBLES
Lockport present	Y	Y	Y	Y	Y	Y	Y	Y	Y	Y	Y	Y	Y	
(thickness in ft)	100	60	100	30	80	85	140	45	110	110	105	90	40	
Peebles present														Y
(thickness in ft)														45
Reef rocks	+	+	+	+	+	+	+	+	+	+	+	+	0	?
Interreef rocks	+	+	+	+	+	+	+	+	+	+	+	+	+	?
Vugs and molds	C	C	C	C	C	C	C	C	C	C	C	C	R	C
Sponge rock	R	R	C	R	R	C	C	R	R	C	R	R	R	R
Insitu breccia	R	R	C	R	R	R	R	R	C	R	R	R	R	R
Collapsed strata	R	R	C	R	0	0	C	0	R	R	0	0	R	0
Solution-enlarged joints	R	R	C	C	0	R	C	R	C	C	R	R	R	R
Mud and/or shale in joints	R	R	C	C	0	0	C	0	R	0	0	0	0	0
Red/orange discoloration of rocks	R	R	C	R	0	0	R	0	0	R	R	0	0	0
Internal sediment	R	R	R	R	0	R	R	0	R	R	R	0	R	0
Channel breccia	R	0	0	R	0	0	C	0	0	0	0	0	0	0
Boxwork	R	0	0	R	0	0	C	0	0	0	0	0	0	0
Caves	C	R	C	0	0	0	C	0	0	R	0	0	0	0
Pocket breccia	R	R	R	R	R	R	R	0	R	R	R	0	0	0
Unconformity within unit	?	0	+	0	0	0	0	0	+	0	0	0	0	0
Scalloped and planar erosion surfaces within unit	R	0	0	0	0	0	0	0	0	0	0	0	0	0
Possible terra rossa within unit	R	0	0	0	0	0	0	0	0	0	0	0	0	0
Top of unit exposed	Y	Y	Y	Y	Y	Y	Y	Y	N	N	N	N	N	Y
Flat bottomed closed depressions at top of unit	0	R	R	C	0	0	0	0						C
U- or V-shaped closed depressions at top of unit	?	C	C	C	0	0	0	0						C
Paleosol at top of unit	0	R	R	R	R	0	0	R						C
Erosional truncation at top of unit	C	C	C	C	R	R	R	R						C
Calcrete in top 1 ft of unit	C	0	0	0	0	0	0	R						0
Greenfield present	Y	Y	Y	Y	Y	Y	Y	Y						Y
(thickness in ft)	67	30	15	22	25	21	14	37						28
Onlap of Greenfield strata onto underlying Lockport or Peebles	R	R	R	?	R	R	R	R						?

Note: Y = Yes. N = No. + = Present. 0 = Not observed and presumed absent. ? = Questionable. C = Common. R = Rare. Quarry locations are shown in Figure 11.1.

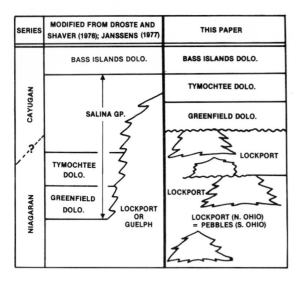

FIGURE 11.2. Two different viewpoints concerning stratigraphic relationships for some Silurian rock units in western Ohio.

Group and Lockport Dolomite. The present author was unable to recognize the formations that Janssens recognized in the subsurface in portions of the Lockport that crop out in quarries in western Ohio, and, accordingly, the Lockport is referred to herein as the Lockport Dolomite and is not treated as a group. Janssens (1977, Fig. 16) further noted that the Lockport may be up to about 450 feet (137 m) thick in parts of the subsurface in northwestern Ohio. Maximum thickness of the Lockport Dolomite in quarry exposures in western Ohio known to the author is about 140 feet (43 m) at Carey (Fig. 11.1, Table 11.1).

In southwestern Ohio the Lockport Dolomite (or Group) is termed the Peebles Dolomite (Bowman 1961; Horvath 1969, Fig. 6; Cook 1975; Shaver et al. 1983, Fig. 5). The characteristics of the Peebles (Bowman 1961, Horvath 1969) are similar or identical to those of the Lockport (Ulteig 1964, Horvath 1969, Janssens 1977). The maximum thickness of the Peebles Dolomite in southwestern Ohio appears to be about 125 feet (38 m) (Horvath 1969). For purposes of the present report the Peebles was studied at one quarry in southwestern Ohio (Fig. 11.1, quarry 14) where the exposed thickness of this unit is 45 feet (14 m) (Bowman 1961, Fig. 19). Details about the Peebles at this quarry occur in Bowman (1961).

Rocks termed Greenfield in western Ohio (Cook 1975) have been referred to as the Greenfield Formation (Bowman 1961, p. 264) and the Greenfield Dolomite (Summerson 1963, p. 51); the terminology of Summerson is followed herein. The Greenfield Dolomite (Fig. 11.2) is a tidal flat unit characterized by a finely crystalline brown to light gray dolomite, laminations, a variety of algal stromatolites, mud cracks, flat pebble conglomerates, and evaporite molds (Summerson 1963, Textoris and Carozzi 1966, Kahle 1974 and 1978; Requarth 1978, Lehle 1980). Insoluble residue content averages about 4.0%. Cross-stratification occurs locally and may be bipolar. Fauna is nonexistent to sparse and includes brachiopods, corals, molluscs, ostracods (Summerson 1963), crinoids (Winzler 1974), and extremely rare stromatoporoids. The Greenfield Dolomite may be about 100 feet (30 m) thick, but its precise thickness is unknown (Summerson 1963).

Silurian strata in western Ohio present a fundamental problem in terms of the stratigraphic relationship of the Lockport Dolomite and the Greenfield Dolomite (Fig. 11.2). Do these units have a facies relationship (Janssens and Stieglitz 1974, Fig. 10; Droste and Shaver 1976, Fig. 2; Janssens 1977, Fig. 1; Shaver 1977, Fig. 7) or does the Greenfield superpositionally overlie the Lockport (Summerson 1963, Cook 1975)? If the latter, is the contact conformable or unconformable (Fig. 11.2)? Resolution of these problems, and of the stratigraphic relationship between the Peebles Dolomite and the Greenfield Dolomite in southwestern Ohio (Fig. 11.2), has a direct bearing on the validity of any possible scenario concerning the Silurian history of the Michigan Basin region (e.g., Alling and Briggs 1961, Gill 1977, Shaver 1977).

Terminology

Paleokarst refers to karst features formed by solution associated with a landscape of the past (Wright 1982, p. 83). Types of paleokarst are surface and subsurface (Wright 1982, p. 84). The former represents solution which took place at the atmospheric–rock interface to produce features such as sinks, karren, and scalloped and planar erosion surfaces (Maslyn 1977, Read and Grover 1977, Kobluk 1984).

Subsurface paleokarst (caves, etc.) results from solution caused by underground drainage. Terms for pore types are from Choquette and Pray (1970). Unless otherwise specified, terminology for karst and paleokarst follows Monroe (1970). Pedological terms are mainly from Freytet and Plaziat (1982).

Types, Abundance, and Occurrence of Paleokarst Features

Features inferred to represent paleokarst (Fig. 11.3) are (1) molds, (2) vugs, (3) in situ breccia, (4) collapsed strata, (5) solution-widened joints which may be open, filled with (a) clayey mud, (b) shale, or any combination of (a) and (b), (6) red-orange discoloration of rocks, (7) internal sediment, (8) breccia-filled channel pores, (9) boxwork, (10) caves, (11) scalloped and planar erosion surface, (12) breccias interpreted to represent terra rossa, (13) closed depressions interpreted to be sinks, and (14) dolomitic shale considered by the author to be a paleosol. Features 1 to 14 are associated with the Lockport Dolomite; only features 1, 2, 13 and 14 are associated with the Peebles Dolomite (Fig. 11.4). The relative abundance of each feature in the Lockport Dolomite or Peebles Dolomite at each of the 14 quarries studied (Fig. 11.1) is shown in Table 11.1, and the vertical stratigraphic oc-

currence of the features is shown in Figure 11.4.

Evidence of Subsurface Paleokarst

Molds and Vugs

Description. Fossil molds (Fig. 11.5A) and solution-enlarged fossil molds (Fig. 11.5B) occur throughout the Lockport/Peebles (Table 11.1). Fossils responsible for molds and solution-enlarged molds include stromatoporoids (Fig. 11.5A), corals, bivalves, and gastropods. Megascopically, the most common type of fossil mold porosity is due to dissolution of hemispherical stromatoporoids (Fig. 11.5A), although not all hemispherical forms display any evidence of leaching. Laminar stromatoporoids typically show no or little evidence of dissolution. Corals (Fig. 11.5A) may show signs of dissolution but rarely occur as molds.

Partially to totally leached hemispherical stromatoporoids (Fig. 11.5A) occur throughout the Peebles Dolomite at Peebles but are concentrated in the upper 10 to 20 feet (3 to 6 m) of this unit. Similarly, leached hemispherical stromatoporoids are common in reef core rocks, rare in reef flank rocks, and very rare in interreef rocks in the Lockport Dolomite. As seen locally in quarry walls at Clay Center, Genoa, Fremont, West Milgrove, and Carey,

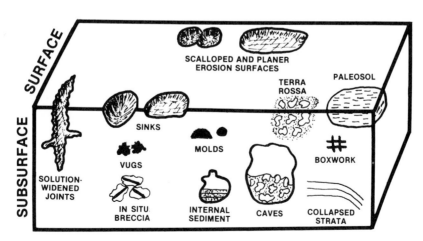

FIGURE 11.3. A sketch showing the assortment of surface and subsurface paleokarst features associated with the Lockport/Peebles Dolomite in western Ohio. No scale intended. See text for details.

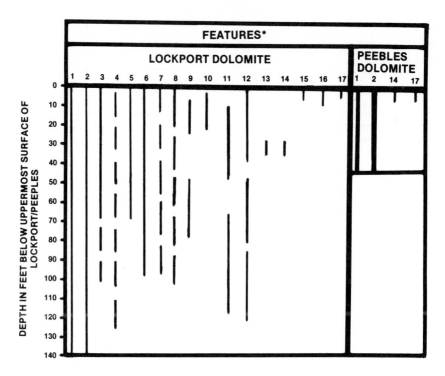

*Key to paleokarst features. 1 = vugs, 2 = molds, 3 = sponge rock (see text), 4 = insitu breccia, 5 = collapsed strata, 6 = solution-enlarged joints, 7 = mud and/or shale in solution-enlarged joints, 8 = red and orange discoloration of rock, 9 = internal sediment, 10 = channel breccia, 11 = boxwork, 12 = caves, 13 = scalloped and planar erosion surfaces, 14 = possible terra rossa, 15 = flat-bottomed closed depressions, 16 = V- or U-shaped closed depressions, and 17 = shale (paleosol).

FIGURE 11.4. Vertical stratigraphic occurrence (vertical lines below numbers keyed to Figure 11.1 and Table 11.1) of paleokarst features associated with the Lockport Dolomite and the Peebles Dolomite in western Ohio. Data for the Lockport is based on study of this unit at 13 quarries; the range in thickness of the portion of this unit exposed at these quarries is 30 to 140 feet (9 to 43 m). The thickness of the Peebles exposed at the only quarry where this unit was studied (Fig. 11.1, Table 11.1) is 45 feet (14 m).

evidence of such leached fossils in the Lockport is greatest in the uppermost 30 to 40 feet (9 to 12 m) of this unit.

Vugs (Fig. 11.5C, D, E, F) are extremely common in the Lockport Dolomite and the Peebles Dolomite, and occur throughout both units (Fig. 11.3, Table 11.1). The maximum diameter of vugs ranges from fractions of an inch to about 10 inches (25 cm). Vugs and passageways in the Lockport Dolomite at Clay Center, West Milgrove, and Carey, (Fig. 11.1, Table 11.1) display a statistical preferred orientation versus depth below the stratigraphic top of this unit. At these quarries vugs and passageways are predominantly vertical in the top 40 to 60 feet (12 to 18 m) of the Lockport (Fig. 11.5C), and predominantly horizontal at greater depths.

Moreover, the walls of vugs in the upper 40 to 60 feet (12 to 20 m) of the Lockport at Clay Center, West Milgrove, and Carey tend to display highly irregular reentrants and projections of wallrock (Fig. 11.5C), whereas vug walls at greater depths tend to be more smooth (Fig. 11.5E).

A distinctive rock type in the Lockport Dolomite formed by high concentrations of vugs is herein termed *sponge rock* (Fig. 11.6A) for descriptive purposes. Sponge rock is characterized by: (1) vugular porosity of about 10 to 40%, (2) some degree of interconnection of vugs (revealed by serial sections), (3) rock surface that is much more irregular than surrounding rock (Fig. 11.6A), and especially, by (4) obliteration of bedding in areas where sponge rock occurs. Areas of bedding obliterated by sponge rock as

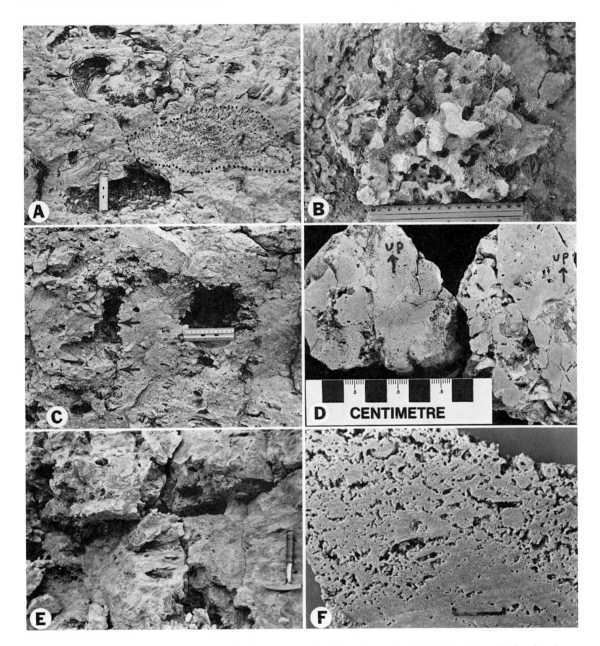

FIGURE 11.5. A–F. *A.* Molds of hemispherical stromatoporoids (arrows) in a reef core in the Lockport Dolomite at Clay Center. In contrast, note excellent preservation of coral (outlined by dotted line). Scale is about 5.5 inches (14 cm) long. *B.* Fossil molds and solution-enlarged fossil molds in Lockport Dolomite at Woodville North. *C.* Vugs, about 50 feet (15 m) below the stratigraphic top of the Lockport Dolomite, in the east wall at Clay Center. Note highly irregular walls of vugs and vertical elongation of vug indicated by arrows. Scale is about 5.5 inches (14 cm) long. *D.* View of vugs and passageways that are predominently vertical in the Lockport Dolomite. Specimen on left is from about 40 feet (12 m) below the top of the Lockport at Clay Center; specimen on right is from about 50 feet (15 m) below the top of the Lockport at Carey. *E.* View of predominatly horizontal vugs and passageways about 100 feet (30.5 m) below the top of the Lockport at Carey. Hammer, lower right, provides scale. *F.* Predominantly horizontally oriented vugs in specimen from about 80 feet (24 m) below the top of the Lockport Dolomite at Clay Center. Scale bar is 1 inch (2.5 cm) long.

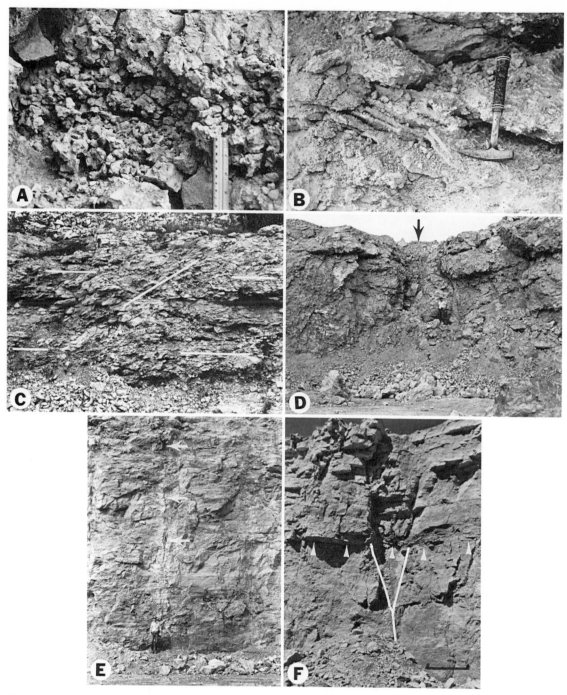

FIGURE 11.6. A–F. Lockport Dolomite. A. Sponge rock at Gibsonburg; scale 4 in (10.2 cm). B. Solution collapse about 90 ft (29 m) below top at Clay Center. C. Solution collapse zone, 12 ft (4 m) thick, about 5 ft (1.5 m) below top; north wall, C-3 pit at Carey; bedding attitude shown by white lines. D. Solution collapse zone about 90 ft (29 m) below top, SW part of Clay Center, view to east. Rocks at center fractured and brecciated; opposing dips on either side; vugs decrease in abundance away from collapse zone. E. Solution enlarged joint (arrows); south wall, C-3 pit at Carey. F. Contact (arrows) with overlying Greenfield Dolomite, north wall, Fremont. Solution enlarged joint (V shaped part of Y) partly filled with green-gray dolomitic shale; no solution in joint below V; scale bar 5 ft (1.5).

seen in quarry walls range from a few ft^2 (Fig. 11.6A) to about 10 square yards (8.4 m^2). Locally, sponge rock is chalky and discolored in subdued shades of red and yellow. The occurrence of sponge rock is relatable to the development of bedding; the better the development of bedding the greater the likelihood of sponge rock. Sponge rock occurs in the Lockport Dolomite at nearly all quarries where this unit was studied (Fig. 11.1, Table 11.1).

The abundance of molds, solution-enlarged molds, and vugs in the Lockport/Peebles may be random in the sense that abundance cannot be related to any other feature(s). Alternatively, the abundance of molds and vugs is relatable to other features, as follows. (1) Molds and vugs are commonly most abundant in rocks on either side of enlarged joints and decrease in abundance away from joints. (2) Some molds and vugs occur preferentially parallel to bedding surfaces. (3) Vugs typically decrease in abundance away from solution collapse zones. (4) The abundance of fossil molds, solution-enlarged fossil molds, and vugs in the Lockport Dolomite and Peebles Dolomite is locally at a maximum in the uppermost 10 to 40 feet (3 to 12 m) of these units and decrease in abundance at greater depths. (5) Vugs typically are more abundant in rock debris in caves and in rocks immediately adjacent to caves than in rocks farther away from caves.

Interpretation. Vugs and molds (Fig. 11.5) provide evidence that rocks containing these features have been subjected to dissolution in the vadose and/or phreatic environments (Read and Grover 1977, p. 961; Longman 1980, p. 461; Esteban and Klappa 1983). Formation of such features in a marine phreatic environment (Longman 1980) does not seem plausible because dissolution effects in this environment appear to result only in microscopic etching of carbonate substrates (Alexandersson 1972). Formation of molds, solution-enlarged molds, and vugs could occur in the deep burial (mesogenetic) environment, as experimental work has shown (Dawson and Carozzi 1985, p. 248). This idea fails, however, to explain why molds, solution-enlarged molds, and vugs in the Lockport/Peebles typically occur preferentially in rocks alongside of enlarged joints, in rocks in collapse zones, in rocks in and immediately

surrounding caves, and, especially, in rocks in the uppermost portions of the Lockport/Peebles. It is concluded that molds, solution-enlarged molds, and vugs in the Lockport/Peebles are subsurface paleokarst features.

As noted above, vugs in the upper 40 to 60 feet (12 to 18 m) of the Lockport at Clay Center, West Milgrove, and Carey are predominantly vertical, whereas vugs at depths greater than 40 to 60 feet below the top of this unit at these quarries are predominantly horizontal. Moreover, highly irregular vug walls and vertically oriented passageways predominate with the former (Fig. 11.5C, D), and relatively smooth vug walls and horizontally oriented passageways predominate with the latter (Fig. 11.5E, F). These differences probably record a change from a vadose zone characterized by predominantly vertical flow of acidic waters and/or high flow rates to a phreatic zone characterized by predominantly horizontal flow of relatively less acidic waters and/or lower flow rates during a time of karstification of the Lockport at these quarries.

Sponge rock (Fig. 11.6A) is interpreted to represent a paleokarst feature formed as a result of intense leaching of rock by groundwater. This interpretation is strengthened by the occurrence of some sponge rock in collapse zones and in rocks in and immediately surrounding caves. These relationships strongly suggest that dissolution responsible for caves, solution collapse zones, and sponge rock in the Lockport Dolomite occurred at or about the same time, and are an expression of the same process, namely, karstic dissolution. In areas where sponge rock occurs in otherwise bedded dolomite as seen in quarry walls, sponge rock is always restricted in extent vertically and laterally. This is interpreted to reflect confined flow of groundwater undersaturated relative to CaCO$_3$ along a group of adjacent bedding surfaces that, for whatever reason, were more permeable than subjacent, superjacent, and laterally adjacent bedding surfaces.

In Situ Breccia

Description. Such breccia is always clast-supported, displays small passageways, rotated and displaced clasts, corroded clasts, and shows vuggy corrosion (Fig. 11.5D). As seen in quarry walls, areas of in situ breccia never occupy more than a few square feet. In situ breccia is relatively rare (Table 11.1).

Interpretation. In situ breccia shows no signs that it results from collapse of cave ceilings. The breccia is interpreted to result from less than wholesale collapse brought about by solution-widening of minute joints and creation of vugs associated with karstification.

Collapse

Description. Collapse in the Lockport Dolomite (Table 11.1) takes several forms. First, some beds display abrupt changes in dip (Fig. 11.6B). Second, bedding in some masses of rock dips up to 60°, and such rock masses are structurally discordant with adjacent flat-bedded rocks (Fig. 11.6C). Third, some otherwise horizontal beds have opposing dips into a central zone of highly fractured, brecciated rock containing rotated and displaced clasts (Fig. 11.6D). The location of collapsed rocks is not stratigraphically controlled. Maximum depth of collapsed rocks below the top of the Lockport (Fig. 11.4) is 70 feet (21 m) at Clay Center (Fig. 11.6B). Minimum depth of collapsed rocks below the top of the Lockport is 10 feet (3 m) at Carey (Fig. 11.6C). Evidence of solution is pervasive in areas of collapsed rocks and always involves dolomite. Neither evaporite minerals nor molds of such minerals occur in the Lockport. Solution-widening of bedding surfaces (Fig. 11.6B) and joints is common in collapse zones. Sponge rock (Fig. 11.6A) occurs in many collapsed rocks and is absent in adjacent rocks that have not collapsed. Collapsed rocks may show little evidence of breakage (Fig. 11.6C) or may be highly brecciated (Fig. 11.6D).

Interpretation. Collapsed rocks are among the most characteristic features resulting from dissolution associated with karst (Esteban and Klappa 1983, Fig. 19) and paleokarst (Roberts 1966; Sando 1974, p. 133). Collapsed rocks associated with unquestioned surface and subsurface paleokarst in the Mississippian Madison Limestone are frequently brecciated, display solution-embayed rock-surfaces, show dips in

one direction or in opposing directions at marked variance with the dip of adjacent rocks, and contain solution-widened joints (Roberts 1966, Sando 1974). These features are identical to those associated with solution collapse zones in the Lockport Dolomite (Fig. 11.6B, C, D).

Joints

Description. Most joints are nonsystematic. Joints in the Lockport Dolomite (Table 11.1) may be closed, enlarged and open (Fig. 11.6E), or enlarged and partially to completely filled with tan to brown to green dolomitic mud, shale, or both (Figs. 11.6F, 11.7A). Exceptions occur at Genoa (Lanz 1979) and Clay Center, where joints in the Lockport Dolomite may contain celestite. Enlarged joints are rare to common (Table 11.1), and typically range in width from several inches to several feet (Fig. 11.6E). Most joints display evidence of solution-widening in the form of erratic width along joint length, corroded joint walls, lack of correspondence in the shape of opposing joint walls, or any combination of these attributes (Fig. 11.7A). Some joints which extend downward from the top of the Lockport Dolomite are solution-widened up to 4 feet (1.2 m) at the top of the joint, and taper downward within vertical distances of the 3 to 5 feet (1 to 1.5 m) continuations of the same joint that are not solution-widened (Fig. 11.6F). At Clay Center (Fig. 11.1, Table 11.1) shale-filled joints in the upper 50 to 70 feet (15 to 21 m) of the Lockport tend to be predominantly vertical, whereas continuations of the same joints at depths below 50 to 70 feet tend to be predominantly horizontal (Fig. 11.7A).

Rarely, joints in the Lockport can be traced upward into the overlying Greenfield. No solution-widened joints in the Lockport were found, however, to be traceable upward into the Greenfield.

Dolomitic shale occurs in all or portions of many enlarged joints regardless of the attitude of the joints (Figs. 11.6F, 11.7A). Clay minerals in the shale are kaolinite and illite. The same clay minerals occur in shale at the top of the Lockport Dolomite. The only clay mineral in insoluble residues of the Lockport Dolomite is

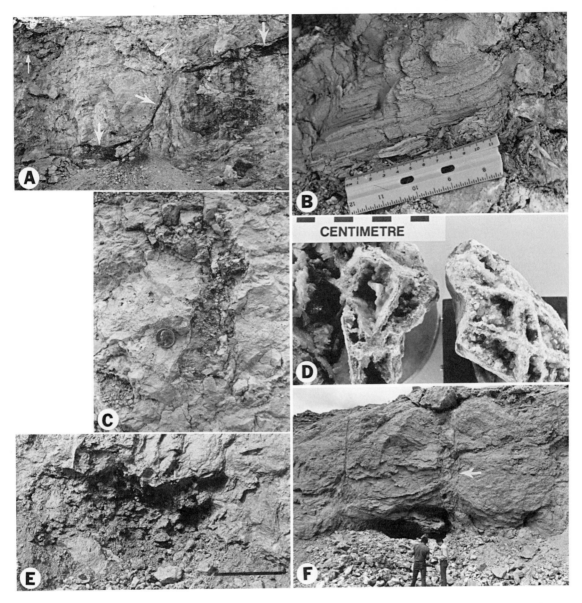

FIGURE 11.7. A–F. *A*. Solution-widened joint (large arrows) filled with dolomitic shale and breccia about 95 feet (29 m) below the top of the Lockport Dolomite in the east wall at Clay Center. Note additional solution-widened joints indicated by small arrows. Height of rock shown is about 15 feet (5 m). *B*. Laminated internal sediment (dolomitic mud containing illite and kaolinite) about 80 feet (24 m) below the top of the Lockport Dolomite in the north wall at Clay Center. *C*. Channel breccia (see text) about 5 feet (1.5 m) below the top of the Lockport Dolomite in the south wall at West Milgrove. Coin (left of center) is about 1 inch (2.5 cm) in diameter. *D*. Paleokarst boxwork. Specimen on the left is from near the top of the Mississippian Leadville Limestone about 5 miles (9.7 km) south of Silverton, Colorado. Specimen on the right is from about 70 feet (21 m) below the top of the Silurian Lockport Dolomite at Carey, Ohio. *E*. Cave filled with rock debris about 115 feet (35 m) below the top of the Lockport Dolomite in the south wall of C-3 pit at Carey. Scale bar is 5 feet (1.5 m) long. *F*. Cave about 90 feet (27 m) below the top of the Lockport Dolomite at Clay Center. Note solution-enlarged joints (indicated by an arrow) above and to the right of the cave.

illite. Some enlarged joints containing dolomitic shale in the Lockport Dolomite can be traced downward from the top of this unit for distances of about 90 feet (27 m).

Interpretation. The most logical process that serves to explain the combined evidence of joint enlargement by solution and mud and/or shale in such joints in the Lockport Dolomite (Figs. 11.6F, 11.7A) is the process of water circulation that occurs in a karst environment (Jennings 1971, Sweeting 1973). Such circulation accounts for (1) solution-enlargement of joints, (2) creation of an illitic clay resulting from dissolution of Lockport rocks and accumulation of some of this clay along joints, and (3) filling of all or portions of some joints that had continuity with soil on top of the Lockport Dolomite by downward infiltration of some of the soil into the underlying joints.

Joints, many of which are enlarged, are a well-known feature in Silurian pinnacle reefs in the Michigan basin (e.g., Petta 1980). It is possible that these enlarged joints originated at the same time as the enlarged joints in the Lockport Dolomite in western Ohio.

Color

Description. The color of an estimated 99% of all Lockport rocks is some shade of gray to buff to white. A fraction of a percentage of these rocks is discolored by way of a surficial (fractions of an inch) "rind" in shades of red and orange. A key point is that these colors do not occur randomly but are restricted in occurrence to some rocks in caves, channel breccias (see below), joint systems, internal sediment, and collapse zones. Localized red to orange discoloration of Lockport/Peebles rocks below the top of these units ranges from 1 foot (0.3 m) to about 100 feet (30 m).

Interpretation. Discoloration of rocks in shades of red and orange typically results from oxidation due to subaerial exposure (Wilson 1975, p. 80, 86; Kobluk 1984, p. 398; Prather 1985, p. 21), and is an expected consequence of karstification (Esteban and Klappa, 1983). As noted above, reddish to orange discoloration occurs up to 100 ft below the top of the Lockport Dolomite (Fig. 11.4). Assuming the discolora-

tion is due to oxidation that occurred entirely in the vadose environment it is possible that water tables in the Lockport Dolomite, at least locally and possibly at different times locally or regionally, were as much as 100 ft (31 m) below the top of this unit during the Silurian.

Internal Sediment

Description. Such sediment (Dunham 1969, p. 139) is geopetal, occurring in the bottom of some vugs (Fig. 11.7B) and caves. The sediment may be mud or weakly lithified dolomitic shale containing dolomite, organic matter, illite, and kaolinite, and is structureless to laminated (Fig. 11.7B). Where laminations occur they are horizontal or dip up to 15°. Colors are typically gray to buff to black and, rarely, red to orange.

Interpretation. Internal sediment has been noted previously in some Silurian rocks in the Michigan basin region (Gill 1977, p. 84; Cercone and Budai 1985, p. 17). Formation of internal sediment can involve different environments. (1) Internal sediment can result from dissolution of evaporites containing inclusions of carbonate in a mesogenetic environment (Cercone and Budai 1985), but the Lockport Dolomite contains no evidence of evaporites. (2) Internal sediment can be marine in origin (Hsu 1983). Evidence would be marine fossils in the internal sediment; no such fossils occur in internal sediment in the Lockport Dolomite. (3) Another idea is that internal sediment is indicative of dissolution of carbonate rocks in the vadose zone to create an insoluble residue consisting entirely or primarily of silt-size sediment followed by transport by vadose water(s), and subsequent accumulation of the residue in secondary solution voids and small fractures (Dunham 1969, Hsü 1983, Cercone and Budai 1985). Illite is the only clay mineral in the Lockport Dolomite, and illite and kaolinite are the only clay minerals in a paleosol (see below) that occurs at the top of this unit. Illite and kaolinite are the only clay minerals in internal sediment in the Lockport Dolomite. Thus, it is reasonable to conclude that (a) some illite in internal sediment in the Lockport Dolomite represents a portion of insoluble residue that resulted from vadose dissolution of this unit, and (b) some illite and all kaolinite in internal sediment in the Lockport Dolomite were derived by way of downward infiltration of soil that once existed at the top of this unit into underlying vugs and caves. In terms of the idea that karstification in ancient rocks is indicated by evidence of downward infiltration of sediment and its eventual accumulation in cavities, along with local reddish discoloration of the sediment due to oxidation, the Redwall Limestone in Arizona (McKee and Gutschick 1969, p. 80 and 84) provides a perfect analog for the Lockport Dolomite (Fig. 11.7B).

Channel Breccias

Description. Channel pores are markedly elongate pores which typically originate by indiscriminate solution along one or more fractures or by lateral coalescence through enlargement of other types of pores (Choquette and Pray 1970, p. 245). Breccias within such pores are herein termed *channel breccias* (Fig. 11.7C); they are rare (Table 11.1). They may be clast- or matrix-supported; clasts are angular to rounded, and matrix consists of illite, kaolinite, black organic material, and dolomite. Some channel breccias are stained in shades of red (Tomassetti 1981, Fig. 15).

Interpretation. Channel breccias formed by solution enlargement of joints and concomitant or subsequent filling of the joints by breccia during karstification following formation of the Lockport Dolomite and prior to formation of the overlying Greenfield Dolomite. Breccia matrix originated in the same way as internal sediment, noted above.

Boxwork

Description. Boxwork, also termed *honeycomb* (Sweeting 1973, p. 142), consists of a reticulated network of thin blades projecting out from rock surfaces wherein the blades enclose irregularly shaped and sized spaces so that the overall appearance is that of a series of open, connected, odd-shaped boxes (Bretz 1942, p. 733). Examples of boxwork in Wind Cave, South Dakota (Bretz 1942, Fig. 27), and in some Devonian reef/bank rocks in western Canada (Beales and Oldershaw (1969, Fig. 1C), are identical for all practical purposes to examples of boxwork in

the Lockport Dolomite (Fig. 11.7D). At localities where boxwork occurs in the Lockport Dolomite (Table 11.1), it typically occurs at random. The only occurrences of boxwork observed in caves was at Carey and Maumee (Fig. 11.1, Table 11.1).

Interpretation. Boxwork may or may not originate in caves but it does originate in the meteoric environment (e.g., Bretz 1942, Beales and Oldershaw 1969, Warren 1982). The author (unpublished research) has observed boxwork, 99% of which does not occur in caves, in the upper part of the Mississippian Leadville Limestone in Colorado (Fig. 11.7D), a unit with documented paleokarst features (Maslyn 1977) (De Voto, this volume). This boxwork is identical in appearance to boxwork in the Lockport Dolomite (Fig. 11.7D). Analogy indicates that boxwork in the Lockport Dolomite is a paleokarst feature.

Caves

Description. Caves ranging from about 5 feet to 50 feet (1.5 to 15 m) in diameter are common in the Lockport Dolomite (Table 11.1), and are partially to completely filled with a variety of sizes of rock debris (Fig. 11.7E, F, Fig. 11.8A, B, C, D). Depth of caves relative to the stratigraphic top of the Lockport Dolomite is highly variable. Cave fill in a cave at Maumee is truncated by an unconformity at the top of the Lockport Dolomite (Fig. 11.8B). Maximum depth of a cave floor below the top of the Lockport Dolomite is approximately 120 ft (37 m) at Carey (Fig. 11.7E). Walls of caves range from smooth and planar, smooth and undulatory (but not scalloped), to highly irregular. Beds in the floor of some caves have undergone solution collapse.

Cave fill is entirely rock or a mixture of sediment and rock and can be partial (Fig. 11.7F) or complete (Fig. 11.8A). Cave sediment is uncemented; particles are clay- to boulder-size. Cave rocks are breccias (common), conglomerates (extremely rare), mixtures of breccias and conglomerates (rare to common), and dolomitic shale (rare to common). Breccias are mainly clast-supported (Fig. 11.8A, B). In general, the larger the cave the greater the tendency for the cave to contain breccia, the larger

FIGURE 11.8. A–F. Breccia-filled cave in the Lockport Dolomite in the north wall at Clay Center. Arrows point to the ceiling of the cave which is about 15 feet (5 m) below the top of the Lockport Dolomite. Width of field of view is about 22 feet (7 m). *B*. Breccia-filled cave (outlined by dotted lines) in the uppermost part of the Lockport Dolomite in the south wall of the west pit at Maumee. The top of the cave is truncated by an erosional unconformity (dashed line) between the Lockport Dolomite and the overlying Greenfield Dolomite. *C*. Corroded clasts in cave-fill in the Lockport Dolomite at Maumee. Scale bar is 1 inch (2.5 cm) long. *D*. pocket cave (see text) about 15 feet (5 m) below the top of the Lockport Dolomite in the east wall at Clay Center. Sledgehammer (left of center) is 14 inches (36 cm) long. *E*. Cave popcorn. Specimen on right is modern and is composed of calcite. Specimen on left is dolomite and is from a cave in the Lockport Dolomite at Maumee. Scale is in inches. *F*. Modern pocket cave in Tertiary Bluff Limestone, White Rock quarry, Grand Cayman Island. Scale is 6 inches long. Compare this figure with Figure 11.8D.

the breccia clasts, and the greater the tendency for the cave to be completely filled with breccia. Relatively small caves are herein termed *pocket caves*, and breccias in such caves are herein termed *pocket breccias* (Fig. 11.8D). Clasts in caves and pocket caves may be highly corroded (Fig. 11.8C). Conglomerate clasts range from moderately to very well rounded. Matrix in cave breccia and conglomerates consists of dolomitic shale (common) and dolomitic mud (rare). Dolomitic shale veneers the floor of some caves to a depth of 2 to 5 in (5 to 13 cm) and may be laminated or massive. Clay minerals in dolomitic mud and dolomitic shale in caves are illite and kaolinite.

Caves in the Lockport Dolomite may be intersected by joints (Fig. 11.7F), suggesting a genetic relationship between joints and caves (cf. Jennings 1971, Sweeting 1973). Alternatively, some caves apparently are not intersected by joints (Fig. 11.7E). Stalactites and cave popcorn (Fig. 11.8E) occur in caves at Maumee.

Interpretation. Caves are a primary manifestation of karst and paleokarst (Jennings 1971, Sweeting 1973, Esteban and Klappa 1983). A karst pocket cave (Fig. 11.8F) is strikingly similar to a paleokarst pocket cave in the Lockport Dolomite (Fig. 11.8D). Likewise, a modern example of cave popcorn (Fig. 11.8E) is identical, except mineralogically, to a Silurian counterpart (Fig. 11.8E).

The absence of horizontal layering and imbricated clasts in the fills of caves in the Lockport Dolomite, and the apparent absence of scalloped cave walls in this unit suggests that the caves were never subjected to free flow of water. The caves must have been partially to completely filled with debris almost from their inception regardless of whether they formed entirely in the phreatic zone (Bretz 1942) or entirely in the vadose zone (Wood 1985). Cave fill originated in different ways. Collapse of portions of cave walls and ceilings was the primary source of clasts in caves. Some clasts may represent rock fragments eroded from joint walls and subsequently deposited in caves. Collapse of cave ceilings and walls resulted in partial to complete fill of caves with clasts. The latter is possible because collapse can create clasts that accumulate with various degrees of po-

rosity. The volume of the clasts plus pores can be far greater than the volume of the original collapsed rock. Following deposition of angular clasts in caves some clasts may have been subjected to rounding in situ as a result of solution by groundwater (Bretz 1942, p. 736). Matrix in cave conglomerates and breccias represents (a) infiltration of soil into caves, and (b) accumulation of insoluble residue resulting from dissolution of Lockport strata.

Surface Paleokarst within the Lockport Dolomite

Evidence of surface paleokarst is confined to three quarries: Maumee, Genoa, and Clay Center (Table 11.1, Fig. 11.1).

Maumee

Description. The existence of surface paleokarst within the Lockport Dolomite at this quarry is indicated by scalloped and planar erosion surfaces (Fig. 11.9A), and by possible terra rossa, whose matrix ranges from dull red to pale brown (Fig. 11.9B) to grey (Fig. 11.9C) at the same stratigraphic level about 30 ft. (9 m) below the top of the Lockport. Although these features may be hundreds of feet apart laterally, they also occur laterally juxtaposed. None of the breccia occurs in caves.

Interpretation. Modern scalloped and planar erosion surfaces as well as examples of such surfaces in some Middle Ordovician limestones in Virginia (Read and Grover 1977, Figs. 2, 7) correspond almost exactly to such surfaces in the Lockport Dolomite at Maumee (Fig. 11.9A). The existence of scalloped and planar erosion surfaces about 30 ft. (9 m) below the top of the Lockport Dolomite at Maumee indicates that at least a portion of this unit at this locality was at the atmosphere–rock interface prior to the end of Lockport deposition. Whether the scalloped and planar erosion surfaces originated as a karst feature developed on limestone where soil cover is lacking or on limestone in the tidal zone along a coast (Read and Grover 1977, p. 964) is unknown. Breccias are commonly associated with such surfaces (Read and Grover

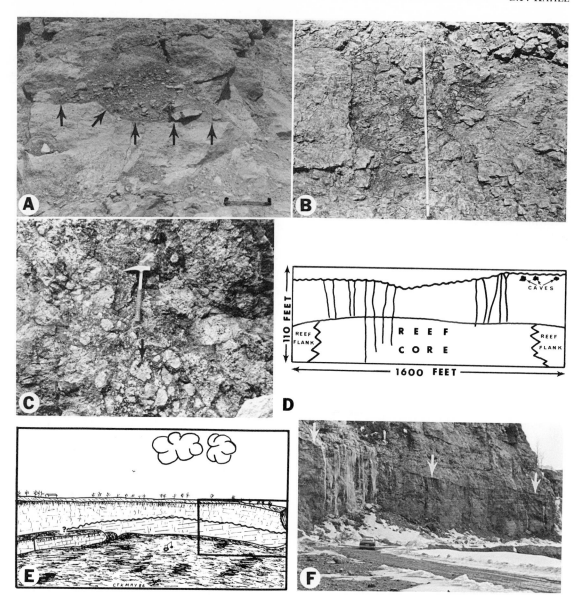

FIGURE 11.9. A–F. *A*. Scalloped and planar erosion surfaces (indicated by arrows) about 30 feet (9 m) below the top of the Lockport Dolomite in the west pit at Maumee. Scale bar is 2 feet long (0.6 m). *B*. Probable terra rossa (see text) about 30 feet below the top of the Lockport Dolomite in the north wall of the east pit at Maumee. Scale is 3 ft long (0.9 m). The matrix of the probable terra rossa is dull red to brown. *C*. Probable terra rossa about 30 ft below the top of Lockport Dolomite in the west pit at Maumee. The arrow indicates a corroded clast in the breccia. Compare this clast with a clast in terra rossa illustrated by Esteban and Klappa (1983, Fig. 34). *D*. A sketch (from photographs) of joints truncated by an erosional unconformity within the Lockport Dolomite in the southwest wall at Genoa. Vugs (not shown) and caves are concentrated in rocks just below the unconformity. *E*. A sketch (from a photograph) of the Lockport Dolomite exposed in the east wall of Clay Center. Note the fault indicated by the near vertical line in the left side of sketch, and the unconformity. Unconformity is about 60 feet (18 m) below the top of the Lockport Dolomite. Whether or not the unconformity truncates the fault is unknown. A closer view of the area outlined on the right side of the sketch is shown in 9F. *F*. A view of a part of 11.9 9E. The unconformity is indicated by arrows. Note truncation of Lockport strata below the unconformity.

1977). Similarly, some breccias in the Lockport Dolomite at Maumee (Fig. 11.9B, C) occur at the same stratigraphic level as do scalloped and planar erosion surfaces (Fig. 11.9A). These breccias are interpreted as terra rossa, a feature which is often associated with karst and which occurs at or typically within tens of feet of the earth's surface (Esteban and Klappa 1983). Modern terra rossa (personal observation by the author), as well as some examples of ancient terra rossa (Goldhammer and Elmore 1984, Park and Jones 1985) display sharp clast–matrix boundaries, fractures extending into clasts representing incomplete brecciation of parent rock, no signs of transport other than in-place rotation and slumping, corrosion of clasts, and red to pale brown matrix colors. Breccias at Maumee interpreted as terra rossa display the exact same characteristics (Fig. 11.9B, C).

Genoa

Description. A system of joints truncated by an unconformity occurs within the Lockport Dolomite at this quarry (Fig. 11.9D). The joints are filled by a combination of dolomite, celestite, and green dolomitic shale (Lanz 1979). No dolomitic shale is present on the unconformity. Vugs are more common in rocks 5 to 10 feet (1.5 m to 3 m) below the unconformity than in rocks farther below the unconformity (Lanz 1979). Caves filled with rock debris occur below the unconformity; the tops of the caves are within 3 to 5 feet (.9 to 1.5 m) of the unconformtiy (Fig. 11.9D).

Interpretation. Concentration of vugs and caves in rocks immediately below the unconformity suggests (Purdy 1974) that the unconformity is an expression of an ancient (but now buried) karst surface.

Clay Center

Description. An unconformity occurs within the Lockport Dolomite in the east wall about 60 feet (18 m) below the top of this unit (Fig. 11.9E, F). Strata below the unconformity are truncated by erosion (Fig. 11.9F). Strata above the unconformity are more fossiliferous than strata

below the unconformity and the former are reefal in character (Requarth 1978). A fault occurs just north of the unconformity, but whether the fault is truncated by the unconformity is unknown (Fig. 11.9E).

Interpretation. The unconformity (Fig. 11.9E) is interpreted as a surface-developed paleokarst surface within the Lockport because of red discoloration along the surface, evidence of pitting of the surface, and a high concentration of vugs in rocks immediately below the unconformity (Requarth 1978). Evidence of leaching immediately below buried karst surfaces and of reefs that formed on such surfaces have been summarized by Purdy (1974, Table 1). Comparison suggests that these features (Purdy 1974) have an ancient analog in the Lockport exposed in the east wall of Clay Center quarry (Fig. 11.9E) (Requarth 1978). Assuming the unconformity at Clay Center is a surface paleokarst feature and truncates the fault (Fig. 11.9E), then at least in the quarry area Lockport sedimentation was interrupted by an episode of karsting caused by uplift of a portion of the Lockport above sealevel along one or more faults. The possibility of such faulting is not out of the question, considering known and postulated faults in western Ohio (Fig. 11.10).

Surface Paleokarst at the Top of the Lockport/Peebles Dolomite

Closed Depressions

Description. Closed depressions occur at the top of the Lockport Dolomite at three out of eight locations where the top of this unit is exposed and at the top of the Peebles Dolomite at Peebles (Table 11.1, Fig. 11.11). Some of the depressions are V-shaped or U-shaped, with one side of the V or U being higher than the other side (Fig. 11.11A). Depth of these depressions, measured along the high side of the depression, ranges from about 4 to 8 feet (1.2 to 2.4 m) and width ranges from about 5 to 15 feet (1.5 to 4.5 m). V- and U-shaped depressions may be lined with dolomitic shale (Fig. 11.11A). Closed depressions with flat bottoms and relatively steep sides occur at the top

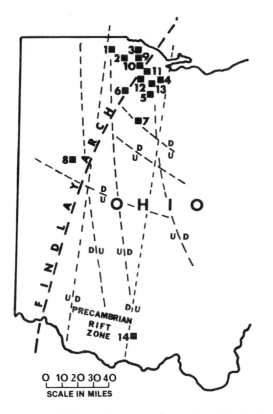

FIGURE 11.10. Some known and postulated faults in western Ohio. Compiled from Green (1957), Telljohann (1978), Keller et al. (1983), and Beardsley and Cable (1983).

of the Lockport Dolomite at Fremont (Fig. 11.11B, C). The depressions are filled with shale. Locally, the shale contains nodules and mud cracks (Fig. 11.11D). Similar flat-bottomed closed depressions filled with dolomitic shale containing nodules of dolomite occur at the top of the Peebles Dolomite (Fig. 11.11E) at Peebles (Fig. 11.1, Table 11.1). None of the closed depressions display any evidence of collapse, breccias, or both in rocks immediately below the depressions. A very rough estimate is that one closed depression occurs for each 500 feet (150 m) of contact between the Lockport/Peebles and the Greenfield.

Interpretation. Closed depressions at the top of the Lockport/Peebles are sinks (also termed *dolines*; see Monroe 1970, p. K7). The cross-

sectional appearance of known sinks (Sweeting 1973) matches those of sinks at the top of the Lockport/Peebles (Fig. 11.11A). Modern sinks are commonly lined or filled with soil and/or rock debris (Jennings 1971, Sweeting 1973). Sinks at the top of the Lockport/Peebles are lined with or filled by dolomitic shale that frequently contains rock debris (Fig. 11.11A, B, C, E). The absence of evidence of collapse, breccias, or both in rocks immediately below the depressions, and the presence of shale lining or filling the depressions strongly suggest (Sweeting 1973) that the sinks were caused by solution under different amounts of soil cover and did not form due to collapse.

Shale

Description. Dolomitic shale occurs at the top of the Lockport/Peebles at six out of the nine locations involved in this study where the top of the Lockport/Peebles is exposed (Table 11.1). Such shale at these quarries is typically restricted to closed depressions (Fig. 11.11A, B, C, E). It is relatively thin (1 to 6 inches) in V- or U-shaped depressions (Fig. 11.11A), and relatively thick (6 inches to 5 feet, or 15 cm to 1.5 m) in flat-bottomed depressions (Fig. 11.11B, C, E). Rarely, it forms a 1 to 3-inch-thick (2.5 to 7.6 cm) shale seam along the contact between the Lockport/Peebles and the Greenfield Dolomite. At Lime City, and Fremont (Fig. 11.11B, C) the shale is gray-green to greenish gray, fissile to massive, composed of dolomite, illite, and kaolinite, and displays rare mottles, slickensides (polished surfaces with faint striations), and mud cracks (Fig. 11.11D). Zonation is lacking. Shale at the top of the Lockport Dolomite at Clay Center and at the top of the Peebles Dolomite at Peebles (Fig. 11.12A) displays two zones: (1) an upper zone of dolomitic shale mottled in shades of red, green, and purple, and containing illite and kaolinite with some nodules of dolomite near the base (Figs. 11.12B, C), and (2) a lower zone of gray to bluish-gray shale containing illite and kaolinite and variable amounts of nodules of dolomite. Nodules are not soil peds but rounded to subrounded fragments of Lockport Dolomite (at Clay Center) and Peebles Dolomite (at Peebles) (Fig. 11.12B, C). Nodules in the upper zone typically are

FIGURE 11.11. A–E. *A.* A sink lined with greenish-gray shale at the top of the Lockport Dolomite in the south wall at Lime City. The hammer and hard hat (near center) provide scale. The contact between the Lockport Dolomite and the overlying Greenfield Dolomite is indicated by arrows. *B.* A view of the contact (indicated by arrows pointing down) between the Lockport Dolomite and the overlying Greenfield Dolomite in the east wall at Fremont. Two sinks filled with greenish-gray shale are indicated by arrows pointing up. *C.* Closeup view of shale-filled sink shown on the left side of 11.11B. Sledgehammer (to right of center) is 14 inches (36 cm) long. *D.* Mud cracks in greenish-gray shale at the top of the Lockport Dolomite at Fremont. Scale is 6 in long. *E.* A panorama of the contact (marked by arrows pointing down) between the Peebles Dolomite and the overlying Greenfield Dolomite. Location of rocks shown is the west side of Peebles quarry. View is looking north. Randy Court (in circle, far right) is 6 ft (1.9 m) tall. A shale-filled sink at the top of the Peebles is indicated by an arrow pointing up. Note that the contact to the left of the sink is a planar erosion surface.

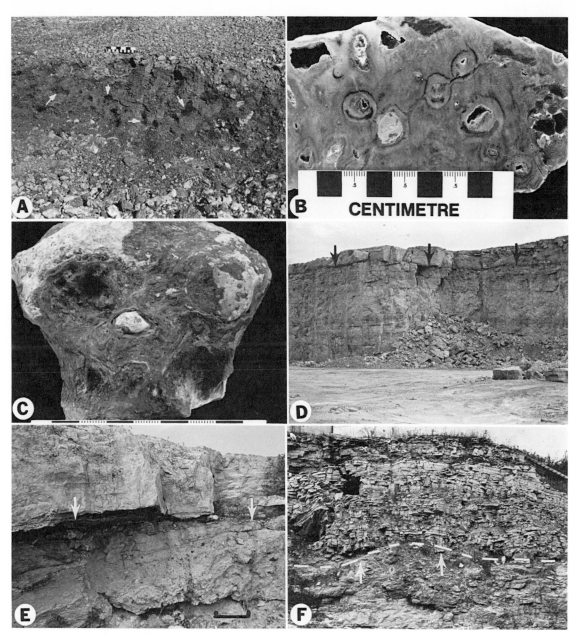

Figure 11.12 A–F. *A.* Portion of a Gley paleosol with hydromorphic properties at the top of the Peebles Dolomite in the south pit at Peebles. The scale is 6 inches (15 cm) long. The paleosol displays two zones, an upper zone displaying mottles (some of which are indicated by arrows) in shades of red, orange, and purple, and a lower unmottled zone the uppermost portion of which is barely visible in the lower right. Clay minerals in this paleosol are illite and kaolinite. *B.* Slabbed and polished surface of a mottled and vuggy nodule of Peebles dolomite from the upper mottled zone of the Gley paleosol at Peebles. *C.* Nodule from the upper mottled zone of the Gley paleosol at Peebles. The nodule has a partial crust (cutan?) of a yellow-brown mixture of clay minerals and iron oxide. Note pits in the surface of the nodule. Each black segment of the scale is equal to 1 cm. *D.* Erosional contact (indicated by arrows) between the Lockport Dolomite and the overlying Greenfield Dolomite in the northwest corner of Clay Center. View is to the northeast. The wall is about 40 feet (12 m) high. *E.* Closeup view of the erosional contact (indicated by arrows) between the Lockport Dolomite and the overlying Greenfield Dolomite in the north wall at Clay Center. Note erosional truncation of Lockport strata. Scale bar is 1 ft. (0.3 m) long. *F.* Erosional contact (indcated by dashed line) between the Lockport Dolomite and the overlying Greenfield Dolomite in the south end of the quarry at Carey. Sites of erosional truncation of Lockport strata are indicated by arrows. Height of rock shown is about 28 feet (8 m).

mottled (Fig. 11.12B), have 0.1 to 0.5 inch-thick crusts of red/yellow iron stained dolomitic shale (Fig. 11.12C), and contain pits 0.3 to 1.5 in (0.76 to 3.8 cm) in diameter (Fig. 11.12C).

Interpretation. Shale at the top of the Lockport/Peebles is a paleosol. Paleosols are a primary record of emergence, solution activity, and of surface-developed paleokarstic surfaces (Maslyn 1977, p. 312; Wright 1982, p. 84). Manifestations of some paleosols are mottles (Retallack 1983), slickensides (Ahmad 1983, Fig. 3.7; Meyer and Pena Dos Reis 1985), and solution-rounded nodules (Prather 1985, Fig. 3a). Dolomitic shale at the top of the Lockport Dolomite and the Peebles Dolomite (Fig. 11.12B, C) displays all of these characteristics.

Gley soils develop where the groundwater table is close to the surface and the water table fluctuates up and down (Buurman 1980, p. 593). This causes the soil to develop hydromorphic properties brought about by fluctuations in the water table (Buringh 1979, p. 61; Freytet and Plaziat 1982, Fig. 23). The part of the soil below the lowest point of fluctuation of the water table is always subjected to reducing conditions and develops a gray or bluish-gray color (Buringh 1979, p. 61). Soil within the zone of alternating oxidation and reduction brought about by fluctuations in the water table becomes mottled; the mottles can be in many colors including red, orange, yellow, gray, and bluish-gray (Buringh 1979, Buurman 1980). Nodules are common in the lower part of the upper mottled zone and less common in the underlying unmottled zone (Freytet and Plaziat 1982, Fig. 34). Dolomitic shales at Clay Center and Peebles have nearly identical properties (Fig. 11.12A); they are gley paleosols with hydromorphic properties. Known gley soils with hydromorphic properties contain large amounts of kaolinite (Buurman 1980); so does the paleosol at the top of the Lockport Dolomite at Clay Center, and at the top of the Peebles Dolomite at Peebles. Also, crusts of iron oxide on rock surfaces (cutans?) and pits in rocks as a result of karstification are known to occur in Ordovician rocks on Manitoulin Island in Ontario (Kobluk 1984). Similar features occur on and in nodules (Fig. 11.12B, C) in the Gley paleosol at the top of the Peebles Dolomite at Peebles.

Nature of Lockport/Peebles Dolomite and Greenfield Dolomite Contact

Evidence that this contact is a surface-developed paleokarst unconformity, and does not involve a facies relationship (Fig. 11.2), is the occurrence of sinks and a paleosol at the top of the Lockport Dolomite and the Peebles Dolomite, noted above. Several additional features document the subaerial, unconformable nature (Fig. 11.2, right-hand column) of this contact. The Greenfield Dolomite onlaps onto the Lockport Dolomite at seven out of the eight quarries involved in this study where both units crop out in the same quarry (Table 11.1). Onlap of strata of the Greenfield Dolomite onto the underlying Lockport Dolomite is especially evident at Maumee (Kahle 1974, Fig. 10A), Buckland (Shaver et al. 1978, Fig. 13A), and West Milgrove (Shaver et al. 1978, Fig. 21B; Lehle 1980). Conceivably, such onlap could have evolved entirely in a submarine setting. But this idea is invalidated by the occurrence of calcrete (Table 11.1) in the top 1 foot (0.3 m) of the Lockport at Maumee (Kahle 1978) and Buckland (Winzler 1974), documenting that the Lockport Dolomite was subaerially exposed prior to the formation of the overlying Greenfield Dolomite. That the Lockport/Peebles and Greenfield Dolomite contact is an erosional unconformity is demonstrated by erosional truncation of strata at the top of the Lockport and Peebles (Table 11.1). Evidence of such truncation ranges from striking (Fig. 11.12D, E) to subtle (Fig. 11.12F). Most parts of the contact between the Lockport Dolomite and the Greenfield Dolomite display no or only very minor (1 to 2 inch) erosional relief (Fig. 11.11B, E). Locally, however, as at Lime City (Fig. 11.11A), erosional relief at the top of Lockport Dolomite is up to 9 ft. (2.7 m). Erosional relief at the top of the Peebles Dolomite has a maximum value of about 5 ft. (1.5 m), as also pointed out by Bowman (1961).

Discussion

Surface-developed paleokarst features in the form of sinks and paleosol occur at the top of the Lockport/Peebles (Figs. 11.11, 11.12A).

Locally developed calcrete occurs in the top one foot (0.3 m) of the Lockport (Table 11.1). Erosional truncation of strata is fairly common at the top of the Lockport (Fig. 11.12C, D, E). Clearly, the sinks, paleosol, calcrete, and erosional truncation must have formed prior to the time of deposition of the overlying Greenfield Dolomite (Fig. 11.2, right-hand column).

It is probable that surface and subsurface karstification of the Lockport/Peebles occurred during the same period of time. First, as to the Lockport, if surface and subsurface karsting of this unit occurred concomitantly then, following Esteban and Klappa (1983, Fig. 5), it could be expected that maximum widening of many solution-widened joints that extend downward from the stratigraphic top of this unit would occur at the top of the joint, and the width of such joints would diminish progressively with depth. This is exactly what is seen in some joints that extend downward from the top of the Lockport (Fig. 11.6F). More important, if substrate karsting of the Lockport Dolomite occurred at some time during or after deposition of the Greenfield Dolomite which overlies the former unit (Fig. 11.2, right-hand column), then a complete karst profile (Esteban and Klappa 1983, Fig. 5) would not be expected to occur exclusively within the Lockport Dolomite. But by combining evidence from exposures of the Lockport Dolomite at Clay Center, Fremont, and Carey (Fig. 11.1), it is reasonably certain that a complete karst profile does occur exclusively within the Lockport Dolomite in northwestern Ohio. A paleosol occurs at the top of the Lockport at Clay Center and Fremont (Fig. 11.11B). Sinks occur at the top of the Lockport at Fremont (Fig. 11.11B). At Clay Center and at Carey, vugs and passageways in the Lockport are dominantly vertical (reflecting vadose conditions) in the uppermost 50 ± 10 feet (15 ± 3 m) of this unit and dominantly horizontal (reflecting phreatic conditions) at depths greater than 50 ± 10 ft. (15.5 m) below the top of this unit (Fig. 11.5A, D, F). Similarly, solution-widened joints, many of which are filled with shale, are dominantly vertical in the uppermost 60 ft. (18 m) of the Lockport at Clay Center, and dominantly horizontal (Fig. 11.7A) at depths beyond 60 ft. This probably records a transition from vadose conditions (above) to

phreatic conditions (below). The abundance of vugs and molds decreases with depths below the top of the Lockport Dolomite at Clay Center and Carey, and below the top of the Peebles Dolomite at Peebles. It is concluded that surface karstification which affected the top of the Lockport Dolomite and the Peebles Dolomite occurred concomitantly with subsurface karstification of these units, and therefore that both the surface and the subsurface karstification ended prior to the time that the Greenfield Dolomite was formed above the Lockport/Peebles (Fig. 11.2, right-hand column).

It is probable that the paleosol was once far more extensive at the top of the Lockport/Peebles than now appears. Most paleosol at the top of these units is restricted to closed depressions (Fig. 11.11A, B, E), and is absent on adjacent paleotopographic highs (Fig. 11.11E). The paleotopographic high areas at the top of the Lockport/Peebles are relatively flat, planar areas (Fig. 11.11B, E). Paleosol was probably eroded from paleotopographic highs but protected from erosion in paleotopographic lows.

Of extreme interest is the fact that modern Gley soils with hydromorphic properties (Buringh 1979, Plate 3) and Gley paleosols with hydromorphic properties in Cretaceous to Early Tertiary rocks in the Paris basin (Freytet and Plaziat 1982, Fig. 34) are remarkably similar to Gley paleosol with hydromorphic properties in Silurian rocks in western Ohio (Fig. 11.12A). Thus, in spite of the dominance of land plants in post-Silurian rocks (Esteban and Klappa 1983, Fig. 10), conditions of soil formation appear to have been remarkably similar from the Silurian to the present.

The Lockport/Peebles and Greenfield contact in western Ohio is a major, regional, subaerial unconformity even though as seen in quarry walls this contact never has over 10 feet (3 m) of erosional relief. Assuming that adequate subsurface evidence can eventually be brought to bear on the problem, it is predicted that this contact will be shown to have over 100 feet (30 m) of erosional relief both locally and regionally. One result of erosional relief of this magnitude as seen in cores or well cuttings of the Lockport and Greenfield from two or more wells spaced, say, one to five miles (1.6 to 8.0 km) apart would be that parts of the two units occur at the same subsurface elevation (e.g., Janssens 1977). From this it might be inferred that the Lockport/Peebles and Greenfield have a facies relationship (Janssens 1977) when, in fact, they do not (this report).

An identical characteristic of some surface-developed karst surfaces in modern carbonates in Cancun, Mexico (Longman et al. 1983) and in late Pleistocene carbonates in the Florida Keys (Harrison et al. 1984) is that most portions of the surfaces have only minute (fractions of an inch) erosional relief, and appear to be essentially planar over several square yards to 10 or more square yards (8.4 m^3), but locally the surfaces display relief measurable in feet. A similar surface-developed paleokarst surface is described by Desrochers and James (1985 and this volume) in Ordovician carbonates in the Mingan Islands of Quebec. The surface-developed karst and paleokarst surfaces described by the aforementioned authors are, in turn, similar to the surface-developed paleokarst surface at the top of the Lockport/Peebles (Figs. 11.11B, E). As seen in the walls of quarries this contact typically displays only minor relief (less than 1 in) for lateral distances of 100 to 1000 feet (30 to 300 m). Such portions of the contact are planar for all practical purposes (Fig. 11.11B, E). Locally, however, this otherwise planar contact displays variable relief measured in inches to about 9 feet (2.7 m). Thus, the contact between the Lockport/Peebles and the Greenfield is typically in the form of what Folk and McBride (1978, p. 1075) term a *planar erosion surface* and what Longman et al. (1983) term a *paraconformity,* which they define as an obscure unconformity in which the contact over large areas appears to be a bedding plane.

The contact between the Lockport Dolomite and the Greenfield Dolomite is not a facies contact as depicted by some workers (Janssens and Stieglitz 1974, Fig. 10; Droste and Shaver 1976, Fig. 2; Shaver 1977, Fig. 23; Janssens 1977, Fig. 1; Droste and Shaver 1982, p. 13), nor is it a gradational contact as portrayed by Shaver et al. (1978, Fig. 13A). This contact may not represent the contact between Niagaran age and Cayugan age rocks in the Great Lakes region, and it may well be that the idea of a regional unconformity separating Niagaran and Cayugan rocks in this region is a myth (Droste

et al. 1975, p. 749). Nevertheless, a regional surface-developed paleokarst unconformity does occur in Silurian rocks in all of western Ohio between reef-bearing rocks (Lockport/Peebles) and overlying tidal flat rocks (Greenfield Dolomite), as shown in the present report. The present author believes that evidence presented herein, combined with additional evidence (Stout and Lamey 1940, Summerson 1963, Horvath 1969, Summerson and Swann 1970, Mesolella et al. 1974, Gill 1977, Petta 1980, Wickstrom et al. 1985) for a regional subaerial unconformity at or near the top of reef-bearing Silurian strata in Michigan and Ohio, countermands the viewpoint which opposes, makes trivial and belittles the possibility of such an unconformity (Shaver 1977, p. 1422, Fig. 28; Shaver et al. 1978; Droste and Shaver 1982, p. 3).

On sedimentological and paleontological grounds, paleoclimatological conditions during Lockport/Peebles karstification were semiarid. Caves in the Lockport Dolomite contain no evidence of free flow of water, suggesting that water was not plentiful, possibly due to aridity. The Greenfield Dolomite is a tidal flat carbonate that contains evaporite molds (Textoris and Carozzi 1964, Kahle 1974, Lehle 1980). At Peebles the basal one inch (2.5 cm) of the Greenfield Dolomite contains anhydrite molds. Ostracods are common in the Greenfield Dolomite (Summerson 1963); they occur in the basal one inch of the Greenfield Dolomite at Carey and Lime City. Clearly, the onset of aridity indicated by the evaporite molds and the ostracods (Smosna and Warshauer 1981, Table 1) in the Greenfield Dolomite did not take place instantly at the beginning of deposition of this unit but must have been initiated some time earlier, probably at or near the end of marine conditions responsible for the reef-bearing Lockport/Peebles strata or during the subsequent interval of subaerial exposure of these units. Gley paleosol with hydromorphic properties at the top of the Lockport/Peebles (Fig. 11.12A) provides additional evidence that climatic conditions during karstification of these units were semiarid. Such soils can be indicative of formation of soils in depressions and a climate with a fairly high annual temperature accompanied by seasonal dryness (Curtis et al. 1976, p. 215; Buurman 1980, p. 605).

Conclusions

1. The Lockport/Peebles in western Ohio contains caves and sinks which are diagnostic, respectively, of subsurface and surface paleokarst. Features within the Lockport/Peebles indicative of subsurface paleokarst are vugs, molds, collapsed strata, in situ breccias, solution-enlarged joints, internal sediment, and boxwork; features at the top of the Lockport/Peebles indicative of surface paleokarst are paleosol and planar erosion surfaces.

2. At some quarries the Lockport/Peebles contains a karst profile which is expressed by either a downward decrease in the abundance of vugs and molds, a downward change from predominantly vertical vugs to predominantly horizontal vugs, or a downward change in the orientation of shale-filled, solution-widened joints from primarily vertical to primarily horizontal, or a combination of these changes. These downward changes reflect a transition from vadose to phreatic conditions.

3. The Lockport/Peebles and Greenfield contact in western Ohio is a regional surface-developed paleokarst unconformity which is typically expressed as a paleokarst planar erosion surface. The contact is not a facies contact nor is it gradational.

4. Gley paleosols with hydromorphic properties at the top of the Lockport Dolomite and the Peebles Dolomite are remarkably similar to modern Gley soils with hydromorphic properties; this indicates that conditions of soil formation have been similar from the Silurian to the present.

5. Syndepositional tectonics locally affected the formation of the Lockport Dolomite in northwestern Ohio. This caused the development of local subaerial unconformities with variable relief (which now appear within this unit), some subsurface karstification below such unconformities, and preferential reef growth on paleotopographically higher portions of the subsequently drowned subaerial surfaces.

6. Surface paleokarst at the top of the Lockport/Peebles, subsurface paleokarst within the Peebles Dolomite, and the vast majority of subsurface paleokarst within the Lockport Dolomite formed during the same time interval prior to the time interval represented by the Greenfield Dolomite which overlies the Lockport/Peebles.

Acknowledgments. My most sincere recognition for their work, enthusiasm, and ideas goes to Bowling Green State University Master's degree students who have worked with me on the Lockport/Peebles in western Ohio: Chris Anderson, Randy Court, Jack Floyd, Keith Dupont, Bob Lanz, Peter Lehle, Saadi Motameedi, Jeff Requarth, Ron Riley, and Ted Winzler. I am especially grateful to quarry owners for allowing me access to their property. Louis Briggs (deceased), University of Michigan, is fondly remembered for encouragement and inspiration, and for nurturing my passion for Silurian rocks in western Ohio. I thank Dan Gill, Geological Survey of Israel, for sharing his knowledge and ideas about Silurian rocks in the Michigan basin region; Robert Anderhalt (Philips Electronics), for help with X-ray analyses, and for clarifying some of my ideas; Brian Jones, University of Alberta, for helping me to learn more about karst; Mateu Esteban, University of Barcelona, for pointing out to me that shale at the top of the Peebles Dolomite at Peebles quarry has hydromorphic properties; Noel James and Philip Choquette for careful editing and thoughtful suggestions for improvement of the manuscript; Charlotte Parker and Connie Milliron for typing assistance; William Butcher for printing of photographs; and the Faculty Research Committee, Bowling Green State University, for funding of some of my research on the Lockport Dolomite. My most profound thanks go to my wife, Rosemary; without her innumerable sacrifices on my behalf, this study would not have been possible.

References

Ahmad, N., 1983, Vertisols, *in* Wilding, L.P., Smeck, N.E., and Hall, G.F., eds., Pedogenesis and soil taxonomy: developments in soil science 11B: Amsterdam, Elsevier, p. 91–124.

Alexandersson, T., 1972, Micritization of carbonate particles: processes of precipitation and dissolution in modern shallow-marine sediments: Bulletin of the Geological Institute of Upsala, v. 7, p. 201–236.

Alling, H.L., and Briggs, L.I., 1961, Stratigraphy of

Upper Silurian Cayugan evaporites: American Association of Petroleum Geologists Bulletin, v. 45, p. 515–547.

Beales, F.W., and Oldershaw, A.E., 1969, Evaporite-solution brecciation and Devonian carbonate reservoir porosity in western Canada: American Association of Petroleum Geologists Bulletin, v. 53, p. 503–512.

Beardsley, R.W., and Cable, M.S., 1983, Overview of the evolution of the Appalachian basin: Northeastern Geology, v. 5, p. 137–145.

Bowman, R.S., 1961, Silurian–Post-Medinan stratigraphy of southern Ohio (Highland and Adams Counties): Geological Society of America Guidebook for Field Trips for Cincinnati Meeting, 1961, p. 264–269, 285–288.

Bretz, J.H., 1942, Vadose and phreatic features of limestone caverns: Journal of Geology, v. 50, p. 675–911.

Buringh, P., 1979, Introduction to the study of soils in tropical and subtropical regions, 3rd ed.: Netherlands, Wageninger, 124 p.

Buurman, P., 1980, Paleosols in the Reading Beds (Paleocene) of Alum Bay, Isle of Wight, UK: Sedimentology, v. 27, p. 593–606.

Cercone, K.R., and Budai, J.M., 1985, Formation of geopetal diagenetic sediment by evaporite solution: evidence that not all crystal silt is vadose: Society of Economic Paleontologists and Mineralogists Annual Midyear Meeting Abstracts, v. 2, p. 17.

Choquette, P.W., and Pray, L.C., 1970, Geologic nomenclature and classification of porosity in sedimentary carbonates: American Association of Petroleum Geologists Bulletin, v. 54, p. 207–250.

Coogan, A.H., and Parker, M.M., 1984, Six potential trapping plays in Ordovician Trenton limestone, northwestern Ohio: Oil and Gas Journal, p. 121–126.

Cook, T.O., 1975, Stratigraphic atlas. North and Central America: Shell Oil Company, Houston, Texas.

Curtis, L.F., Courtney, F.M., and Trudgill, S.T., 1976, Soils in the British Isles: London, Longman Group Ltd., 364 p.

Dawson, W.C., and Carozzi, A.V., 1985, Experimental fabric-selective porosity in phylloid-algal limestones (abstr.): American Association of Petroleum Geologists Bulletin, v. 69, p. 248.

Desrochers, A., and James, N.P., Early paleosurface surface and subsurface paleokarst; middle Ordovician carbonates, Mingan Islands, Quebec: Society of Economic Paleontologists and Mineralogists Annual Midyear Meeting Abstracts, v. 2, p. 23.

Droste, J.B., Janssens, A., Liberty, B.A., and Shaver, R.H., 1975, The mythical Niagara-Cayugan unconformity, southern and eastern Michigan basin area: Geological Society of America, North Central Section, Annual Meeting Abstracts, p. 749.

Droste, J.B., and Shaver, R.H., 1976, The Limberlost Dolomite of Indiana—a key to the great Silurian facies in the southern Great Lakes area: Indiana Department of Natural Resources Geological Survery Occasional Paper 15, 21 p.

Droste, J.B., and Shaver, R.H., 1982, The Salina Group (Middle and Upper Silurian) of Indiana: Indiana Department Natural Resources, Geological Survey Special Report 24, 41 p.

Dunham, R.J., 1969, Early vadose silt in Townsend mound (reef), New Mexico, in Friedman, G.M., ed., Depositional environments in carbonate rocks: Society of Economic Paleontologists and Mineralogists Special Publication 14, p. 169–181.

Esteban, M., and Klappa, C.F., 1983, Subaerial exposure environment, in Scholle, P.A., Bebout, D.G., and Moore, C.H., eds., Carbonate depositional environments: American Association of Petroleum Geologists Memoir 33, Tulsa, OK, p. 1–54.

Folk, R.L., and McBride, E.F., 1978, Radiolarites and their relation to subjacent "oceanic crust" in Liguiria, Italy: Journal of Sedimentary Petrology, v. 48, p. 1069–1101.

Freytet, P., and Plaziat, J.C., 1982, Continental carbonate sedimentation and pedogenesis—Late Cretaceous and Early Tertiary of southern France: Contributions to Sedimentology 12, Stuttgart, E. Schweizerbart'sche Verlagsbuchhandlung, 213 p.

Gill, D., 1977, The Belle River mills gas field: productive Niagaran reefs encased by sabkha deposits, Michigan basin: Michigan Basin Geological Society Special Paper No. 2, 187 p.

Goldhammer, R.K., and Elmore, R.D., 1984, Paleosols capping regressive carbonate cycles in the Pennsylvanian Black Prince Limestone, Arizona: Journal of Sedimentary Petrology, v. 54, p. 1124–1137.

Green, D.A., 1957, Trenton structure in Ohio, Indiana, and northern Illinois: American Association of Petroleum Geologists Bulletin, v. 41, p. 627–642.

Harrison, R., Cooper, L.D., and Coniglio, M., 1984, Late Pleistocene carbonates of the Florida Keys, in Carbonates in subsurface and outcrop: Canadian Society of Petroleum Geologists Core Conference, Calgary, Alberta, Canada, p. 291–306.

Horvath, A.L., 1969, Relationships of Middle Silurian strata in Ohio and West Virginia: Ohio Journal of Science, v. 69, p. 321–342.

Hsu, K.J., 1983, Neptunic dikes and their relation to the hydrodynamic circulation of submarine hydrothermal systems: Geology, v. 11, p. 455–457.

Huh, J.M., Briggs, L.I., and Gill, D., 1977, Depositional environments of pinnacle reefs, Niagara and Salina Groups, northern shelf, Michigan Basin, *in* Fisher, J.H., ed., Reefs and evaporites: Studies in Geology No. 5, American Association of Petroleum Geologists Bulletin, p. 1–21.

Janssens, A., 1977, Silurian rocks in the subsurface of northwestern Ohio: Ohio Department of Natural Resources, Geological Survey Report Investigations No. 100, 96 p.

Janssens, A., and Steiglitz, R.D., 1974, The lower and middle Paleozoic history of the Findlay Arch in Ohio: Ontario Petroleum Institute Inc., 13th Annual Conf., p. 1–25.

Jennings, J.N., 1971, Karst: Cambridge, MA, MIT Press, 252 p.

Kahle, C.F., 1974, Nature and significance of Silurian rocks at Maumee quarry, Ohio, *in* Kesling, R.V., ed., Silurian reef-evaporite relationships: Michigan Basin Geological Society Field Conference Guidebook, p. 31–54.

Kahle, C.F., 1978, Patch reef development and effects of repeated subaerial exposure in Silurian shelf carbonates, Maumee, Ohio, *in* Kesling, R.V., ed., Geological Society of America, North Central Section Guidebook, p. 63–115.

Keller, G.R., Lidiak, E.G., Hinze, W.J., and Braile, L.W., 1983, The role of rifting in the tectonic development of the midcontinent, USA: Tectonophysics, v. 94, p. 391–412.

Kobluk, D.R., 1984, Coastal paleokarst near the Ordovician–Silurian boundary, Manitoulin Island, Ontario: Bulletin of Canadian Petroleum Geology, v. 32, p. 398–407.

Lanz, R.C., 1979, Development of Lockport (Niagaran) carbonate buildups and associated rocks at Genoa, Ohio: Unpubl. M.S. thesis, Bowling Green State University, OH, 118 p.

Lehle, P.F., 1980, Deposition and development of Lockport and Salina (Silurian) rocks at West Milgrove, Ohio: Unpubl. M.S. thesis, Bowling Green State University, OH, 105 p.

Longman, M.W., 1980, Carbonate diagenetic textures from near-surface diagenetic environments: American Association of Petroleum Geologists Bulletin, v. 64, p. 461–487.

Longman, M.W., Fertal, T.G., Glennie, J.S., Krazan, C.G., Suek, D.H., Toler, W.G., and Wiman, S.K., 1983, Description of a paraconformity between carbonate grainstones, Isla Cancun, Mexico: Journal of Sedimentary Petrology, v. 53, p. 533–542.

Maslyn, R.M., 1977, Recognition of fossil karst features in the ancient record: a discussion of several common fossil karst forms: Rocky Mountain Association of Geologists Symposium 1977, p. 311–319.

McGovney, J.E., Lehmann, P.J., and Sarg, J.F., 1982, Eustatic sea-level control of Silurian (Niagaran) reefs, Michigan basin: American Association of Petroleum Geologists, Abstracts with Program, p. 85.

McKee, E.D., and Gutschick, R.C., 1969, History of the Redwall Limestone of northern Arizona: Geological Society of America Memoir 114, 172 p.

Mesolella, K.J., Robinson, J.D., McCormick, L.M., and Ormiston, A.R., 1974, Cyclic deposition of Silurian carbonates and evaporites in Michigan Basin: American Association of Petroleum Geologists Bulletin, v. 58, p. 34–62.

Meyer, R., and Pena Dos Reis, R.B., 1985, Paleosols and alunite silcretes in continental Cenozoic of western Portugal: Journal of Sedimentary Petrology, v. 55, p. 76–85.

Monroe, W.H., 1970, A glossary of karst terminology: U.S. Geological Survey Water-Supply Paper 1899-K, 26 p.

Norris, S.E., 1975, Geologic structure of near-surface rocks in western Ohio: Ohio Journal of Science, v. 5, p. 225–228.

Park, D.G., and Jones, B., 1985, Nature and genesis of breccia bodies in Devonian strata, Peace Point area, Wood Buffalo Park, northeast Alberta: Bulletin of Canadian Petroleum Geology, v. 33, p. 275–294.

Petta, T.J., 1980, Silurian pinnacle reef diagenesis— northern Michigan: effects of evaporites on pore-space distribution, *in* Halley, R.B., and Loucks, G., eds., Carbonate reservoir rocks: Notes for Society of Economic Paleontologists and Mineralogists Core Workshop No. 1, p. 32–42.

Prather, B.E., 1985, An upper Pennsylvanian desert paleosol in the D-zone of the Lansing–Kansas City groups, Hitchcock County, Nebraska: Journal of Sedimentary Petrology, v. 55, p. 213–221.

Purdy, E.G., 1974, Reef configurations: cause and effect, *in* Laporte, L.F., ed., Reefs in time and space: Society of Economic Paleontologists and Mineralogists Special Publication No. 18, p. 9–76.

Read, J.F., and Grover, G.A., Jr., 1977, Scalloped and planar erosion surfaces, Middle Ordovician limestones, Virginia: analogues of Holocene exposed karst or tidal rock platforms: Journal of Sedimentary Petrology, v. 47, p. 956–972.

Requarth, J.S., 1978, The evolution of Guelph (Silurian) dolomite multistory reefs, White Rock Quarry, Clay Center, Ohio: Unpubl. M.S. thesis, Bowling Green State University, OH, 221 p.

Retallack, G.J., 1983, Late Eocene and Oligocene paleosols from Badlands National Park, South Dakota: Geological Society of America Special Paper 193, 82 p.

Roberts, A.E., 1966, Stratigraphy of Madison Group near Livingston, Montana, and discussion of karst and solution–breccia features: US Geological Survey Prof. Paper 526-B, p. B1–B23.

Sando, W.J., 1974, Ancient solution phenomena in the Madison Limestone (Mississippian) of north-central Wyoming: Journal of Research, US Geological Survey, v. 2, p. 133–141.

Sears, S.O., and Lucia, F.J., 1979, Reef-growth model for Silurian pinnacle reefs, northern Michigan reef trend: Geology, v. 7, p. 299–302.

Shaver, R.H., 1977, Silurian reef geometry—new dimensions to explore: Journal of Sedimentary Petrology, v. 47, p. 1409–1424.

Shaver, R.H., Ault, C.H., Ausich, W.I., Droste, J.B., Horowitz, A.S., James, W.C., Okla, S.M., Rexroad, C.B., Suchomel, D.M., and Welch, J.R., 1978, The search for a Silurian reef model: Great Lakes area: Indiana Department of Natural Resources, Geological Survey Special Report 15, 36 p.

Shaver, R.H., Sunderman, J.A., Mikulic, D.G., Kluessendorf, J., McGovney, J.E.E., and Pray, L.C., 1983, Silurian reef and interreef strata as responses to a cyclical succession of environments, southern Great Lakes area: Geological Society of America Field Trip 12, p. 141–196.

Smosna, R., and Warshauer, S.M., 1981, Rank exposure index on a Silurian carbonate tidal flat: Sedimentology, v. 28, p. 723–731.

Stout, W., and Lamey, C.A., 1940, Paleozoic and Precambrian rocks of the Vance well, Delaware County, Ohio: American Association of Petroleum Geologists Bulletin, v. 24, p. 672–692.

Summerson, C.H., 1963, A resume of the Silurian stratigraphy of Ohio, in Stratigraphy of the Silurian rocks in western Ohio: Michigan Basin Geological Society Guidebook, p. 47–58.

Summerson, C.H., and Swann, D.H., 1970, Patterns of Devonian sand on the North American craton and their interpretation: Geological Society of America Bulletin, v. 81, p. 469–490.

Sweeting, M.M., 1973, Karst landforms: New York, Columbia University Press, 362 p.

Telljohann, M., 1978, Bowling Green fault as an ap-plication of geophysical modeling from gravity data: Unpubl. M.S. thesis, Bowling Green State University, OH, 78 p.

Textoris, D.A., and Carozzi, A.V., 1966, Petrography of a Cayugan (Silurian) stromatolite mound and associated facies, Ohio: American Association of Petroleum Geologists Bulletin, v. 50, p. 1375–1388.

Tomassetti, J.A., 1981, Geology of Lockport (Silurian) rocks at the Ohio Lime Company quarry, Woodville, Ohio: Unpubl. M.S. thesis, Bowling Green State University, Bowling Green, OH, 72 p.

Ulteig, J.R., 1964, Upper Niagaran and Cayugan stratigraphy of northeastern Ohio and adjacent areas: Department of Natural Resources, Ohio Geological Survey Report Investigations 51, 48 p.

Warren, J.K., 1982, The hydrological significance of Holocene tepees, stromatolites, and boxwork limestones in coastal salinas in south Australia: Journal of Sedimentary Petrology, v. 52, p. 1171–1201.

Wickstrom, L.H., Botoman, G., and Stith, D.A., 1985, Report on a continuously cored hole drilled into the Precambrian in Seneca County, northwestern Ohio: Ohio Geological Survey Information Circular No. 51.

Wilson, J.L., 1975, Carbonate facies in geologic history: New York, Springer-Verlag, 471 p.

Winzler, T.J., 1974, Petrology and evolution of Silurian reef and associated rocks, Buckland quarry, Ohio: Unpubl. M.S. thesis, Bowling Green State University, OH, 104 p.

Wood, W.W., 1985, Origin of caves and other solution openings in the unsaturated (vadose) zone of carbonate rocks: a model for CO_2 generation: Geology, v. 13, p. 822–824.

Wright, V.P., 1982, The recognition and interpretation of paleokarst: two examples from the Lower Carboniferous of South Wales: Journal of Sedimentary Petrology, v. 52, p. 83–94.

12
Madison Limestone (Mississippian) Paleokarst: A Geologic Synthesis

WILLIAM J. SANDO

Abstract

Widespread karst was developed in the upper 400 feet (120 m) of the Madison Limestone (late Osagean to early Meramecian) in Wyoming and adjacent states during an estimated 34 m.y. period spanning Late Mississippian (early Meramecian to Chesterian) and Early Pennsylvanian (Morrowan) time. The karst was then covered by deposits of an eastward-transgressing sea (Amsden Formation and Big Snowy Group) during the Late Mississippian and Early Pennsylvanian. Paleokarst features can be distinguished from Tertiary and Holocene solution overprints by fillings of terrigenous sediments related to the overlying transgressive deposits. Madison paleokarst features include enlarged joints (mainly vadose zones), sinkholes (vadose and phreatic), caves (mainly phreatic), and two solution–breccia zones (phreatic). A composite reconstruction of the Madison karst topography in Wyoming reveals a mature landscape having a maximum relief of about 200 feet (60 m) and dominated by three major river systems and low, rounded hills. A highland area in southeast Wyoming was a source for much of the terrigenous detritus deposited on the karst surface. Analysis of the Madison karst bedrock suggests a complex system of block faults that probably represent Precambrian shear zones reactivated by prekarst uplift. The inferred Madison paleoaquifer system was dominated by flow along bedding planes in directions radial from the uplift in southeast Wyoming and through a conduit created by leaching of the lower of the two evaporitic intervals. Because the bedrock was broken complexly by block faults, the distribution of paleosolution features was controlled by local structural and topographic factors. Madison paleokarst offers excellent opportunities for further research on the karst process and for further exploration for mineral, energy, and water resources localized by the karst.

Introduction

The interior of the North American paleocontinent was occupied by a broad platform that received subtidal to supratidal carbonate and evaporite sediments during middle Mississippian time (Fig. 12.1). The platform can be divided into shelves corresponding to the principal Mississippian carbonate formations (Fig. 12.1).

The entire interior platform was subaerially exposed, with consequent intensive solution of the carbonate bedrock, during Late Mississippian and Early Pennsylvanian time, between about 345 and 312 m.y. BP. Some effects of this karst episode are described in other papers in this volume by De Voto on the Leadville shelf and by Meyers on the Lake Valley shelf.

Karstification of the Madison shelf in Wyoming and Montana was described and discussed in earlier papers by the writer (Sando 1974, Sando et al. 1975). This study presents new data on and new interpretations of Madison paleokarst, focused primarily on evidence from Wyoming.

History of Madison Shelf and Karst Plain

Madison Shelf and Karst Plain

Mississippian to Early Pennsylvanian history of the Madison shelf comprises the Madison depositional episode (Kinderhookian to early Meramecian), followed by the Madison paleokarst episode (early Meramecian to Morrowan), then

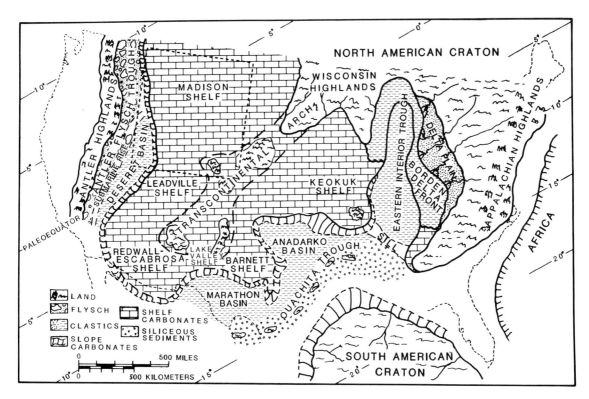

FIGURE 12.1. Paleogeography of conterminous United States during middle Mississippian time (late Osagean and early Meramecian, about 345–353 m.y. BP). Area of Madison shelf outlined with dashed lines is area shown on paleogeographic maps in Figure 12.3. Modified from Gutschick and Sandberg (1983, Fig. 5).

succeeded by the Amsden–Big Snowy depositional episode (middle Meramecian to Desmoinesian) (Sando et al. 1975, Sando 1976, Rose 1976, Gutschick et al. 1980). A chronometric lithofacies profile from a point in southeast Idaho just west of the western edge of the Madison shelf across Wyoming to the Transcontinental Arch in Nebraska (Fig. 12.2) illustrates the geometry of the Mississippian to Early Pennsylvanian sequence. The time scale is a composite biometric scale based on foraminifers, conodonts, and corals and is calibrated in millions of years BP by correlation with the latest radiometric scale (Sando 1986). The profile synthesizes data useful for an estimation of the duration of the Madison paleokarst episode.

The Madison depositional episode terminated about 345 m.y. BP (early Meramecian, composite biozone 14) (Fig. 12.2), when the Madison shelf became emergent, to form a broad

karst plain exposed to subaerial weathering. The shoreline at this time was in western Wyoming (Fig. 12.2, inset).

Amsden–Big Snowy Transgression

The basal member of the Amsden transgressive sequence is the Darwin Sandstone Member of the Amsden, Casper, and Hartville Formations, which represents beach and offshore bar deposits (Fig. 12.2). The Darwin rests disconformably on the Bull Ridge and Little Tongue Members of the Madison Limestone. Dating of the Humbug Formation, an offshore equivalent of the Darwin, was based on conodonts and foraminifers recovered from the Mt. Darby section south of the line of profile (DeJarnett 1984). The Darwin was dated mainly by biostratigraphic constraints placed on it by conodonts, foraminifers, and brachiopods recovered from the overlying Horseshoe Shale Member,

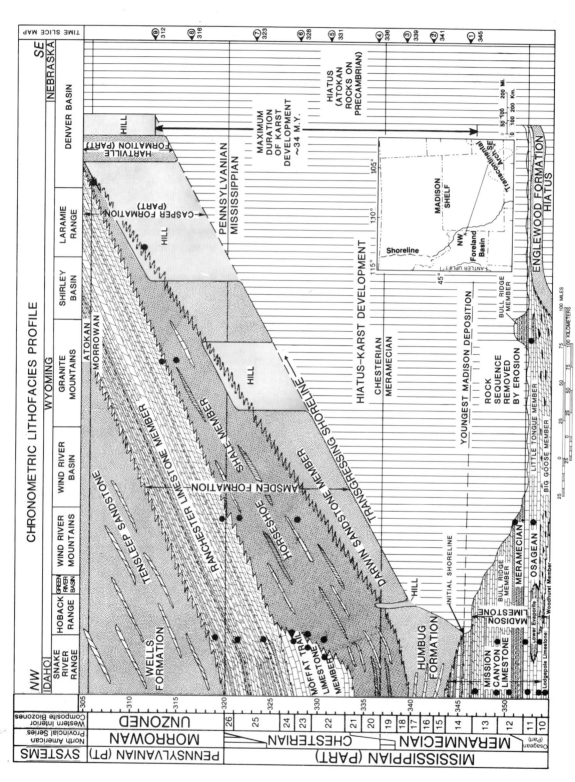

FIGURE 12.2. Chronometric lithofacies profile of the Mississippian to Lower Pennsylvanian sequence in Wyoming and adjacent states. Inset shows location of profile. Dots indicate critical faunal control points along profile; ages of parts of the sequence not directly dated by fossils were determined by regional correlation with other fossiliferous sequences and lithostratigraphy. Composite biozonation and radiometric calibration from Sando (1985). Time-slices at right margin refer to maps shown in Figure 12.3.

which represents tidal flat and lagoonal facies (Sando et al. 1975).

Because the Amsden transgression began at the western edge of the Madison shelf and proceeded eastward, the western part of the emergent limestone plain was exposed to erosion for the shortest period of time, and eastern parts of the plain were exposed for progressively longer periods of time.

Radiometrically calibrated biostratigraphic limits on the transgression permit an estimate of about 34 m.y. for the maximum duration of karst development (Fig. 12.2).

Time-Slice Maps

A time-slice map of Wyoming and adjacent states just before the onset of the Amsden–Big Snowy transgression (Fig. 12.3A) shows the subaerially exposed Madison carbonate plain bordered seaward by a marginal sabkha. The climate was arid, and rivers were poorly developed.

A later time-slice map, at about 341 m.y. BP (time-slice 2 on Figs. 12.2 and 12.3B), shows a marginal eastward-transgressing belt of beach and bar sand (Darwin Sandstone) in Wyoming resulting from extensive development of rivers flowing westward from the Transcontinental arch on the east, where a highland area contributed tremendous quantities of siliceous detritus derived from multicyclic lower Paleozoic sandstones and Precambrian crystalline rocks. In southwesternmost Montana, red shale, siltstone, fine sand, and minor evaporite (Kibbey Formation) was deposited in tidal flats and lagoons at this time. A major change to a humid climate in Wyoming is inferred for this initial stage of karst development, whereas an arid climate prevailed in most of Montana.

As the Amsden–Big Snowy transgression proceeded from late Meramecian into Morrowan time, the red tidal-flat and lagoonal sediments spread eastward across Montana, forming a northern prong of the Cordilleran sea that was separated from the coarser and cleaner beach and bar sand facies that spread eastward across Wyoming (Fig. 12.3B). These two facies form essentially continuous but diachronous sheets that cover most of the karstified surface of the Madison Limestone, except for local topographic highs on the bedrock surface,

where the bedrock protruded above the surface of maximum sedimentary infilling (Fig. 12.2).

A final time-slice map shows the paleogeography of the area near the end of the transgression, which terminated karst development in the middle Morrowan at about 312 m.y. BP. (Fig. 12.3C). At this time, the region was dominated by shallow-water terrigenous and carbonate facies that were seaward and younger than the basal transgressive facies depicted in earlier maps.

Areal Distribution of Karst Fillings

The pattern of previously reported ancient solution cavities in the Madison Limestone is restricted mostly to the outcrop area of the Madison, where such features are readily recognized and studied (Fig. 12.3D). The distribution of red, predominantly fine-grained terrigenous karst filling (Kibbey facies) and coarser grained terrigenous karst fillings (Darwin facies) (Fig. 12.3D) coincides with the distribution of basal transgressive facies overlying the paleokarst surface, which confirms the karst history outlined above. No paleokarst features were previously reported in the subsurface of the Williston and Powder River Basins, but extensive paleosolution effects were described in the Elk Basin field (Bighorn Basin) of north-central Wyoming (McCaleb and Wayhan 1969).

Stratigraphy of Madison Paleokarst

The best known examples of Madison paleokarst features are in southern Montana and central and southeastern Wyoming, where the Madison is well exposed on the flanks of Laramide uplifts (Fig. 12.3D). Solution features are absent in westernmost Wyoming because that area was not emergent during the Madison paleokarst episode (Fig. 12.3A). Madison paleokarst in western Montana has not received as much attention as that in southern Montana and Wyoming, where most of the observations described below were made (see detailed catalog for north-central Wyoming by Sando, 1974, Table 1). A schematic cross section of the Madison paleokarst (Fig. 12.4) illustrates features

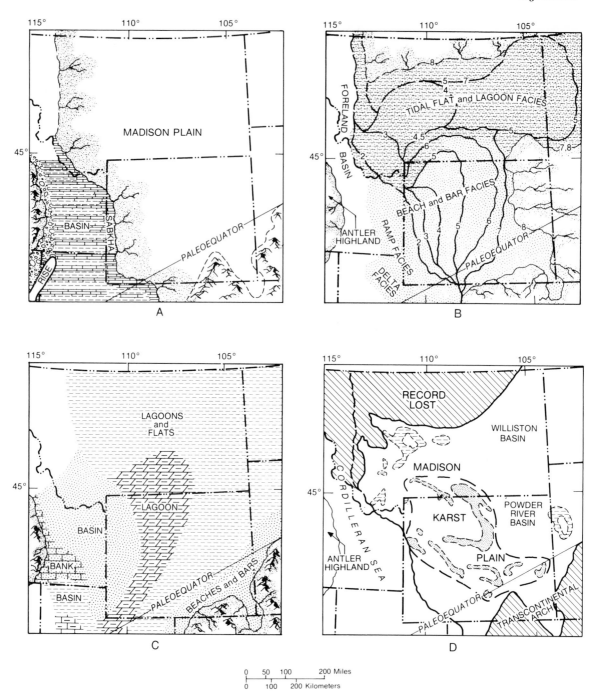

FIGURE 12.3. Maps showing paleogeography of time-slices shown on Figure 12.2 and map showing distribution of terrigenous fillings of solution cavities in Madison paleokarst bedrock. *A*. Paleogeographic map prior to Amsden–Big Snowy transgression (middle composite biozone 14, about 345 m.y. BP). *B*. Map showing successive shoreline positions at time-slices 2 through 8 and progressive distributions of beach and bar facies and tidal flat and lagoon facies directly overlying paleokarst bedrock. *C*. Paleo-geographic map near the end of the Amsden-Big Snowy transgression (middle Morrowan, about 312 m.y. BP). *D*. Map showing distribution of two kinds of material filling solution cavities in paleokarst bed-rock: dashed horizontal lines = red siltstone–fine-grained sandstone, stippled = white to red medium-to coarse- grained sandstone. Area outlined in dashed lines includes localities where karst features have been studied most intensively.

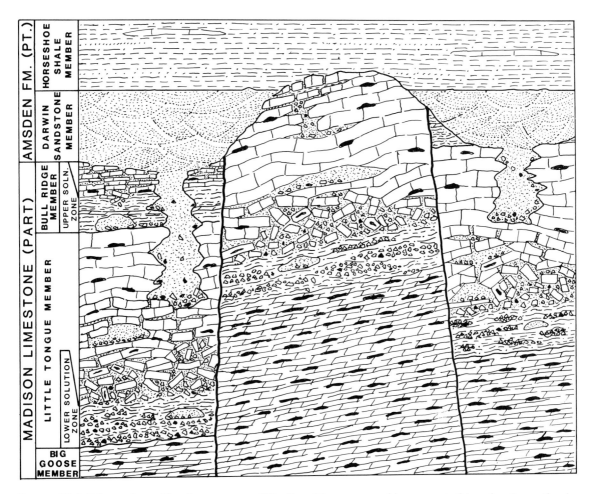

FIGURE 12.4. Schematic profile of upper part of Madison Limestone and lower part of Amsden Formation in north-central Wyoming showing paleokarst features.

observed in outcrop in north-central Wyoming, where the Madison is overlain unconformably by the Darwin Sandstone Member of the Amsden Formation.

In most of Wyoming and southern Montana, paleokarst features are restricted to the upper two members of the Madison Limestone: the Bull Ridge Member (Sando 1968) at the top, underlain by the Little Tongue Member (Sando 1982) (see Sando 1979, for summary of stratigraphy). In the thrust belt of western Wyoming, equivalents of these two members are found in the Mission Canyon Limestone of the Madison Group.

Bull Ridge Member

The Bull Ridge Member, as much as 120 feet (36 m) thick, consists of an upper sequence of

subtidal to supratidal limestone and dolomite and a lower unit of dolomitic siltstone, silty dolomite, and red and green shale. The lower unit has been called the *upper solution zone* (Sando 1967) because collapse breccias in the overlying limestone at some localities betray dissolution of evaporites from the unit. Moreover, evaporite beds occur at this level in some subsurface sections and at Hoback Canyon in western Wyoming.

Little Tongue Member

The Little Tongue Member, which conformably underlies the Bull Ridge Member, ranges from about 80 to 285 feet (24 to 85 m) thick and consists mostly of cherty, medium to thick-bedded, predominantly subtidal limestone and dolomitized limestone. The lower 6 to 45 feet

(2 to 14 m) of the member, which consists of green and red shale and siltstone and minutely brecciated fine grained carbonate and chert has been called the *lower solution zone* (Sando 1967). The lower solution zone has a widespread distribution in southeastern and northern Wyoming and southern Montana, where it is invariably overlain by collapse breccia. Bedded evaporites occur at this level at Hoback Canyon in western Wyoming and in the subsurface of the Williston basin. The Little Tongue Member is underlain conformably by the Big Goose Member of the Madison (Sando 1982), which consists mainly of intensely brecciated, cherty fine grained dolomite.

Paleokarst Relief

The thickness of the Bull Ridge Member is extremely variable due to post-Madison, pre-Amsden erosion, and the member was completely removed from most of central and southern Wyoming, where the Little Tongue Member forms the top of the Madison at most localities. Local erosional relief, exclusive of sinkholes, on the top of the Madison is as much as 16 feet (5 m), and maximum regional erosional relief is about 200 feet (60 m). In most places, the Darwin Sandstone Member of the Amsden, Casper, and Hartville Formations generally disconformably overlies the Madison, except over local pre-Darwin topographic highs where the Horseshoe Shale Member of the Amsden rests directly on the Madison (see Sando et al. 1975 for a summary of contact relations).

Morphology and Origin of Madison Paleokarst Features

Solution features cataloged by Sando (1974, Table 1) for north-central Wyoming form the main part of the discussion to follow, although these features have been seen by the writer throughout the outcrop area of the Madison paleokarst. All paleokarst features described below are distinguished from later solution features by having fillings of terrigenous material (mostly sand) that are related to the over-

lying basal transgressive unit of the Amsden Formation.

Enlarged Joints

Joint passages perpendicular or parallel to bedding in the carbonate bedrock are common in the upper 93 feet (28 m) of the Madison (Fig. 12.5A, B, D). These passages range from less than an inch to as much as a foot (.3 m) wide and are filled with red fine- to medium-grained sandstone, siltstone, and shale. Reticulate networks of joint passages are common. Continuity with the post-Madison, pre-Amsden erosion surface is readily established at most localities.

The joint passages were probably formed during the early stage of karst development in the vadose zone by enlargement of fractures produced by post-Madison, pre-Amsden uplift. Although terrigenous fillings of these openings in the bedrock may have been derived partly from insoluble residues leached from the bedrock and residual soil overlying the bedrock surface, they are composed mainly of sand that entered the joints during the deposition of the fluvial or marine phases of the overlying Darwin Sandstone.

Sinkholes

Large, essentially vertical cavities having demonstrable continuity with the disconformable top of the Madison were observed at eight localities in north-central Wyoming (Sando 1974, Table 1) (see also Fig. 12.6A–C). The sinkholes, found mainly in the Bull Ridge Member, tend to terminate at the base of this unit; but some extend down into the Little Tongue Member, and others are developed in the Little Tongue where the Bull Ridge Member is absent. The cavities are as much as 50 feet (15 m) wide at the top and may extend as far as 90 feet (27 m) below the paleokarst surface. Sinkholes are also present at outcrop localities in southeastern Wyoming and central and western Montana. Anomalous thicknesses of the Darwin Sandstone in subsurface logs are also interpreted as sinkhole fillings.

Although the exact shapes and dimensions of the sinkholes are difficult to determine, they

FIGURE 12.5. A–D. Paleokarst features in upper part of Madison Limestone in Wyoming. *A.* Quartz sandstone, weathered in relief, filling bedding plane joints in upper part of Little Tongue Member at Brown Spring Draw, near center sec 35, T 41 N, R. 92 W, Hot Springs County. *B.* Bedding plane surface in upper part of Bull Ridge Member, showing quartz sandstone, weathered in relief, filling vertical cross joints that form a reticulate pattern at Din- woody Canyon, NE 1/4 sec 1, T 4 N, R 6 W, Fremont County. *C.* Tube (T) filled by quartz sandstone in upper solution zone of Bull Ridge Member at Tensleep Canyon, sec 27, T 48 N, R 87 W., Washakie County. *D.* Bedding plane surface in upper part of Little Tongue Member showing quartz sandstone, weathered in relief, filling vertical joints at Brown Spring Draw, near center sec 35, T 41 N, R 29 W., Hot Springs County.

appear to be characterized by steep, nearly parallel walls, commonly having reentrants and overhangs. Angular blocks of carbonate similar to the adjacent wallrock occur at various levels in the cavities at some localities. Cavity fillings are principally unbedded or poorly bedded fine- to medium-grained red sandstone and lesser amounts of red shale and siltstone.

A combination of both vadose and phreatic processes may account for the origin of the sinkholes. Although outcrops of these features do not present conclusive evidence of their origin by collapse of the roofs of caves, the sink-holes are located mainly in the area of the lower solution zone, which was probably the main paleophreatic conduit and may have provided subterranean open space that promoted roof collapse, resulting in sinkhole formation. The sinkholes were partly filled by residual products of the solution process and partly by breccia derived by collapse from the adjacent bedrock, but the principal fill material is sand deposited during the fluvial or marine phase of the overlying Darwin Sandstone Member. The lack of deformational features like those observed by Bretz (1940) in cavities in the Silurian Joliet

FIGURE 12.6. A–D. Paleokarst features at top of Madison Limestone in Wyoming. A. Madison Paleokarst profile at Shoshone Canyon, NE 1/4 sec 5, T 52 N, R 102 W, Park County, BG = Big Goose Member, BR = Bull Ridge Member, USZ = upper solution zone, D = Darwin Sandstone Member, A = Amsden Formation above Darwin Sandstone. Sinkholes are filled with friable, unbedded quartz sandstone underlying cross-bedded Darwin Sandstone. B. Madison paleokarst profile in gorge below Guernsey Reservoir, NE 1/4 sec 27, T 27 N, R 66 W, Platte County. LT = Little Tongue Member, D = Darwin Sandstone Member, H = Hartville Formation above Darwin

Sandstone. C. Sinkhole about 50 ft (15 m) wide and 50 ft deep in Little Tongue Member (LT) filled with carbonate and chert breccia in matrix of Darwin Sandstone (D) at Tepee Creek, SE 1/4 sec 28, T 54 N, R 85 W, Sheridan County. Beds dip gently away from observer. Approximate position of sinkhole wall indicated by dashed line. D. Cave (C) in upper part of Madison filled with carbonate and chert breccia and quartz sandstone at Little Tongue River (U.S. Highway 14), SE 1/4 SW 1/4 sec 15, T 56 N, R 87 W, Sheridan County. HW = hillwash, BR = Bull Ridge Member, USZ = upper solution zone, LT = Little Tongue Member.

Limestone of Illinois seems to rule out post-Darwin origin of the cavities and filling by sediment injection.

Caves

Irregularly shaped solution cavities lacking obvious connections to the paleokarst surface and ranging from less than a foot to tens of feet in maximum observed dimensions are particularly common in the Bull Ridge and Little Tongue Members to a depth of 200 to 350 feet (60 to 105 m) below the top of the Madison (Figs. 12.5C, 12.6D, 12.7). Constraints imposed by the present outcrop topography make precise determination of the geometry and areal extent

of these caves difficult. The geometry of later caves was undoubtedly overprinted on these ancient features (Elliot 1963, Campbell 1977). Studies by Quinones (1985) have established that many of the caves have the morphology of tubes that are strongly controlled by bedding planes in the host rock. Circular, semicircular, and elliptical tubes have been identified by Quinones. Angular blocks of limestone, dolomite, and chert derived from the wallrock or roofrock by collapse are common in the caves, but the filling is predominantly red sandstone.

A phreatic history for most of the paleocaves is indicated by the absence of flowstone and dripstone deposits. Moreover, the maximum

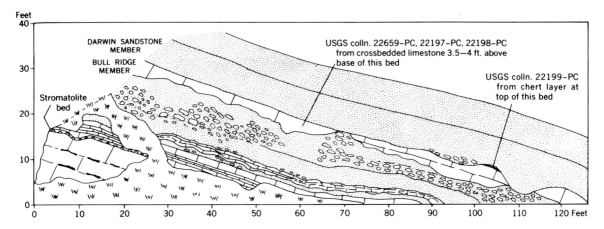

FIGURE 12.7. Sketch of roadcut exposure showing tube devloped along bedding plane in Bull Ridge Member at Little Tongue River (U.S. Highway 14), NW 1/4 NE 1/4 sec 22, T 56 N, R 87 W., Sheridan County, Wyoming. Fossil collections confirm upper limestone bed as Bull Ridge Member of Madison Limestone, not a bed in overlying Darwin Sandstone Member of Amsden Formation. From Sando (1974, Fig. 3).

depth of stream valleys cut into the paleokarst surface was an effective limit to the maximum depth of the vadose zone. Maximum topographic relief was about 200 feet (60 m), estimated from Darwin isopachs, whereas the deepest caves are about 150 feet (45 m) below this level. Most of the caves occur below the deepest local base levels of the post-Madison, pre-Amsden erosion surface. Undoubtedly, some of the highest caves were in the vadose zone for a short period of time, but even these were filled by Darwin Sandstone before characteristic vadose features were developed.

Evaporite-Solution Zones

Widespread leached evaporite zones occur at the base of the Bull Ridge Member (upper solution zone) and at the base of the Little Tongue Member (lower solution zone) in northern Wyoming and southern Montana (Fig. 12.8). The lithology of these zones was described by Sando (1974), who also listed earlier published descriptions of the evaporite-solutions zones.

Collapse breccia in the limestone overlying the upper solution zone is generally thin and areally discontinuous, which suggests limited paleocavity formation in this unit during the paleokarst episode. On the other hand, collapse breccia above the lower solution zone is areally persistent and extends well up into the Little Tongue Member, suggesting that this unit was a significant paleohydrologic conduit during the paleokarst episode. The absence of vadose features in the lower solution zone and its position below stream base levels indicate a phreatic origin. The occurrence of leached evaporite-solution breccias in outcrop at the same stratigraphic levels as preserved evaporite intervals in the subsurface of southwest Montana (Andrichuk 1955) led Roberts (1966) to conclude that leaching of the evaporites occurred during or after Late Cretaceous and early Tertiary uplift, an event superimposed on the Madison paleokarst. By this reasoning, subsurface occurrences of the evaporites were sheltered from the solution process by deep burial. Moreover, Roberts (1966) found a predominance of kaolinite in the sinkholes formed during the paleokarst episode and a predominance of illite in the evaporite-solution breccia zones. He concluded that the kaolinite formed during development of an ancient soil associated with the paleokarst episode, whereas the illite was deposited during marine deposition and was not changed during uplift and brecciation. Severson (1952) had earlier postulated a post-Laramide origin for evaporite-solution breccias on the basis of collapse features in rocks as young as Cretaceous overlying the Madison in southwest Montana. A highly imaginative hypothesis of Laramide-Tertiary

FIGURE 12.8. Evaporite-solution zones in the Madison Limestone in Wyoming and Montana. *A*. Characteristic topographic expression of Little Tongue Member (LT) showing later cave formation in collapse breccia and lower solution zone at base of member at Fremont Canyon, SE1/4 SE 1/4 sec 16, T 29 N, R 83 W, Natrona County, Wyoming. BG = Big Goose Member, C = Casper Formation. *B*. Outcrops of Madison Limestone at Blue Creek in secs 1 and 2, T 42 N, R 85 W, Johnson County, Wyoming, showing characteristic expression of Little Tongue (LT) and Big Goose (BG) Members beneath Amsden Formation (A). *C*. Evaporite-solution breccia (B) in lower part of upper solution zone (USZ) of

Bull Ridge Member ovelying Big Goose Member (BG) at Clarks Fork Canyon, NE 1/4 SE 1/4 sec 6, T 56 N, R 103 W, Park County, Wyoming. *D*. Lower solution zone (LS) overlying Big Goose Member (BG) and overlain by collapse breccia (CB) at South Rock Creek, SE 1/4 NW 1/4 sec. 25, T 52 N. R 84 W, Johnson County , Wyoming. Note undulant beds in Little Tongue Member (LT) directly above collapse breccia (CB). *E*. Evaporite-solution breccia in lower solution zone (LSZ) overlain by collapse breccia (CB) at Little Dryhead Canyon, NW 1/4 SE 1/4 sec 23, T 7 S, R 27 E, Carbon County, Montana. *F*. Evaporite-solution breccia in lower solution zone at Blue Creek, sec 3, T 42 N, R 85 W, Johnson County Wyoming.

subsurface solution (Bridges 1982) to explain the Madison paleokarst features lacks verifiable evidence.

Field observations of the Madison paleokarst over the past 30 years in Wyoming and adjacent states lead me to conclude that most of the leaching of Madison evaporite intervals was one of the effects of the post-Madison, pre-Amsden–Big Snowy paleokarst episode. The matrix of collapse breccias directly overlying the solution breccia zones is invariably the same composition as that of the transgressive terrigenous unit that overlies the paleokarst (Fig. 12.3D). The occurrence of illite in the solution zones in Montana is expectable because illite is present in the overlying Kibbey Formation and was probably also present in the shale that occurs as a primary sedimentary component of the solution zones. Continuity of solution channels from the upper surface of the paleokarst down into the solution breccia zones can be established at many outcrops in both Wyoming and Montana. The Elk Basin occurrence (McCaleb and Wayhan 1969) is an excellent example of brecciation in the subsurface, and I have found evidence of sand matrix in breccia zones in well logs from the Powder River, Bighorn, Wind River, and Denver Basins in Wyoming. Although traces of evaporites are rare in the outcrop area of the Madison, a well-preserved evaporite sequence occurs at the surface at Hoback Canyon, western Wyoming (Sando 1977), which was just west of the shoreline of the Madison karst plain and was protected from solution during the paleokarst episode.

Evidence of Tertiary and Holocene solution overprint an outcrops of Madison paleokarst is indisputable (Elliot 1963, Campbell 1977, McEldowney et al. 1977), but the effects of the earlier episode, including the solution breccias, are clear and pervasive. The preservation of evaporites in the subsurface at various localities in Montana, particularly in the Williston Basin, needs further investigation. Biostratigraphic studies indicate that these areas were exposed to subaerial exposure at about the same time as the Madison karst plain in Wyoming (Sando 1978, Sando and Mamet 1981). A regional paleohydrologic study may explain why certain areas in Montana were sheltered from the pervasive effects of the Madison paleokarst episode.

Topography of Madison Paleokarst

Paleotopographic Model

The petrography, regional geometry, and thickness variation of the Darwin Sandstone in Wyoming support the interpretation that the quartz sand in the unit was originally transported by rivers flowing westward across the Madison karst plain from sources composed of multicyclic lower Paleozoic sandstones and Precambrian crystalline rocks on the Transcontinental Arch highland in southeast Wyoming and Nebraska (Sando et al. 1975, p. A21–23). During eastward transgression of the Cordilleran sea (Amsden–Big Snowy depositional episode), the fluvial sand was reworked and more sand was deposited along a transgressing shoreline, resulting in a sheet of sand that covered all but the highest areas of the erosion surface (Fig. 12.3B). Therefore, variations in the thickness of the sandstone can be used to map the topography of the ancient erosion surface. Mallory (1967, p. G7, pl. 2A) presented an isopach map of the Darwin and plotted on it the course of an ancient river system that he called the Wyoming River, using Darwin isopachs as a measure of the ancient topography.

Data for plotting an isopach map for the Darwin (Fig. 12.9) are now much more abundant and permit the construction of a paleotopographic map (Fig. 12.10) that reveals more details than the map made 20 years ago by Mallory. The new map (Fig. 12.10) shows three essentially parallel river systems that flowed eastward to the shoreline of the Cordilleran sea from an upland area in eastern Wyoming. A series of offshore sand bars and barrier islands compares favorably in size, geographic position, and geometry to those seen today along the Texas Gulf Coast. On the karst plain, isopach highs (thicks) define topographic lows. The mature topography consists of low, rounded hills and broad river and stream valleys. Inferred minimum topographic relief was between 100 and 150 feet (30 to 45 m), and maximum relief was probably about 200 feet (60 m) except for a mountainous area in southeast Wyoming. Maximum heights of hills are unknown because the hills protrude above the datum used for topographic analysis. Sinkholes,

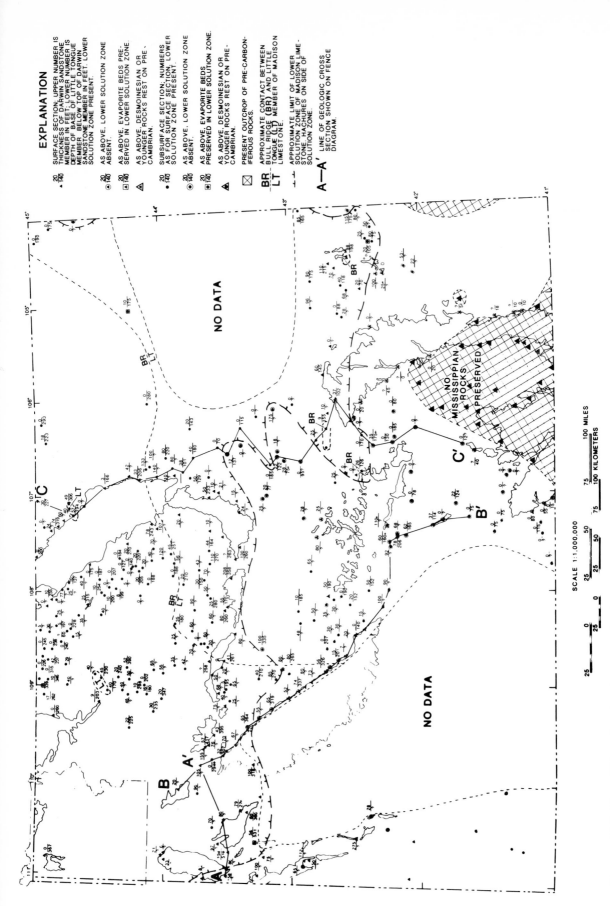

FIGURE 12.9. Map of Wyoming showing control points for plotting paleotopographic map (Fig. 12.10), paleogeologic map (Fig. 12.11), and geologic profiles (Fig. 12.12) of the Madison paleokarst bedrock.

EXPLANATION

20
▲140 SURFACE SECTION: UPPER NUMBER IS THICKNESS OF DARWIN SANDSTONE MEMBER IN FEET; LOWER NUMBER IS DEPTH OF BASE OF LITTLE TONGUE MEMBER BELOW TOP OF DARWIN SANDSTONE MEMBER IN FEET. LOWER SOLUTION ZONE PRESENT.

20
◉140 AS ABOVE, LOWER SOLUTION ZONE ABSENT.

20
◩140 AS ABOVE, EVAPORITE BEDS PRE-SERVED IN LOWER SOLUTION ZONE.

20
▲140 AS ABOVE, DESMOINESIAN OR YOUNGER ROCKS REST ON PRE-CAMBRIAN.

20
•140 SUBSURFACE SECTION: NUMBERS AS FOR SURFACE SECTION, LOWER SOLUTION ZONE PRESENT.

20
◉140 AS ABOVE, LOWER SOLUTION ZONE ABSENT.

20
◩140 AS ABOVE, EVAPORITE BEDS PRESERVED IN LOWER SOLUTION ZONE.

20
▲140 AS ABOVE, DESMOINESIAN OR YOUNGER ROCKS REST ON PRE-CAMBRIAN.

⊠ PRESENT OUTCROP OF PRE-CARBON-IFEROUS ROCKS.

BR
—— ——
LT APPROXIMATE CONTACT BETWEEN BULL RIDGE (BR) AND LITTLE TONGUE (LT) MEMBER OF MADISON LIMESTONE.

APPROXIMATE LIMIT OF LOWER SOLUTION ZONE OF MADISON LIME-STONE. HACHURES ON SIDE OF LOWER SOLUTION ZONE.

A—A' LINE OF GEOLOGIC CROSS SECTION SHOWN ON FENCE DIAGRAM.

NO DATA

NO DATA

NO MISSISSIPPIAN ROCKS PRESERVED

SCALE 1:1,000,000

25   0   25   50   75   100 MILES

25   0   25   50   75   100 KILOMETERS

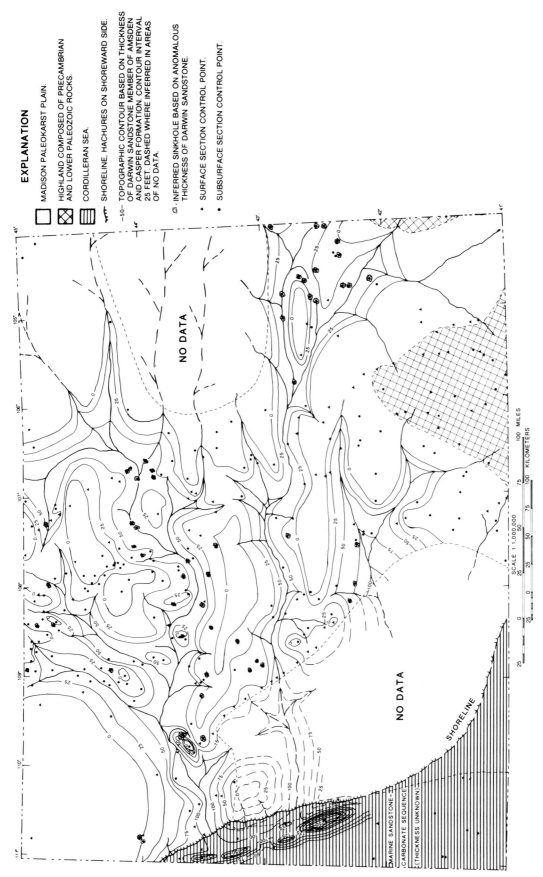

EXPLANATION

MADISON PALEOKARST PLAIN.

HIGHLAND COMPOSED OF PRECAMBRIAN
AND LOWER PALEOZOIC ROCKS.

CORDILLERAN SEA.

SHORELINE, HACHURES ON SHOREWARD SIDE.

—50— TOPOGRAPHIC CONTOUR BASED ON THICKNESS
OF DARWIN SANDSTONE MEMBER OF AMSDEN
AND CASPER FORMATION. CONTOUR INTERVAL
25 FEET. DASHED WHERE INFERRED IN AREAS
OF NO DATA.

⬡ INFERRED SINKHOLE BASED ON ANOMALOUS
THICKNESS OF DARWIN SANDSTONE.

▲ SURFACE SECTION CONTROL POINT.

• SUBSURFACE SECTION CONTROL POINT.

FIGURE 12.10. Map of Wyoming showing topography of Madison paleokarst surface interpreted from isopachs of Darwin Sandstone Member. Control data shown on Figure 9.

269

identified by field observations and by anomalous thicknesses of the sandstone, are concentrated in valleys. Although rivers and streams are shown as continuous, many sinkholes are located at low points in the valleys and could have served as funnels that channeled the surface drainage underground.

Temporal Distortion

The paleotopographic map does not portray an instant in time because the datum used to construct the map was time-transgressive. The shoreline location shown on the map is the location at the beginning of the transgression, whereas the topographic features represent successively younger periods of time from west to east. Hence, the model presents a composite picture of the topography, which is the best picture possible using the available information.

Geology of the Madison Paleokarst

Paleogeologic Model

A map of the paleogeology of the paleokarst bedrock (Fig. 12.11) can be made from data used to construct the paleotopographic map and information on the distribution of rock units directly beneath the paleokarst surface (Fig. 12.9). On the paleogeologic map, the peripheral distribution of successively younger Madison rock units around an area of Precambrian and lower Paleozoic rocks in southeastern Wyoming defines an uplift whose core was a prong of the Transcontinental Arch. The regional dip was essentially radial with respect to the core of this uplift.

Paleostructure

Control data for structure contours consist of measurements of the thickness of the interval from the base of the Little Tongue Member to the top of the Darwin Sandstone (Fig. 12.9). When these structure contours are compared to the Darwin isopachs, it can be seen that the thickness of the Darwin does not consistently vary inversely with the combined thicknesses of the upper two members of the Madison, as would be expected if the paleovalleys had been

cut into horizontal strata. Moreover, the structure contours derived from the data (Fig. 12.11) reveal a discontinuous pattern that does not support the simple folding interpretation presented previously on the basis of a few structural cross sections (Sando et al. 1975, p. A17–18, pl. 2). The new structural interpretation was made by using the paleotopography (Fig. 12.10) as a guide to determine the location of faults that truncate the structure contour lines. Regional dip and regional variation in total isopachs for the Madison (Sando 1979, Fig. 4) were used as guides for extrapolation of isopachs where primary data were lacking or sparse.

The inferred fault pattern is similar to lineament patterns noted by other geologists in the Madison and other rocks in Wyoming and adjacent states. Brown (1978) derived a series of block faulting patterns for Paleozoic rocks in the Powder River Basin of Wyoming and adjacent states from isopach and facies maps and recognized a northeast–southwest primary shear direction and a northwest–southeast complementary shear direction. His structural map for rock units at the top of the Madison (Brown 1978, Fig. 17) is dominated by these two principal shears. Brown et al. (1984, p. B5–6) suggested that these shear directions may have been of Precambrian origin, reactivated during the Paleozoic. They also pointed out that vertical movement was more probable than horizontal movement along these shears. Downey (1984, p. G15, Figs. 16–18) noted the similarity of the patterns of ancient lineaments derived by Brown and his colleagues to lineament patterns inferred from Landsat imagery.

The similarity of the block-faulting pattern herein inferred for the Madison paleokarst bedrock to those derived by Brown and his colleagues strongly suggests a similar origin. Reactivation of Precambrian shear zones during post-Madison, pre-Darwin uplift seems the most probable explanation for the origin of this block fault pattern.

Hydrology of Madison Paleokarst

Paleohydrologic Conduit

Synthesis of the geology and topography of the Madison paleokarst lays the foundation for an analysis of the paleohydrology. The lower so-

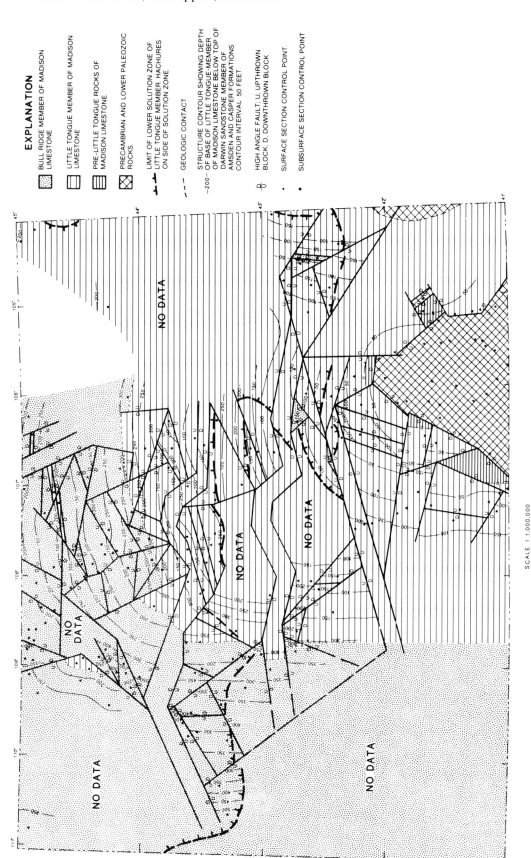

FIGURE 12.11. Map of Wyoming showing paleogeology at the time of paleokarst development. Control data shown on Figure 12.9.

lution zone at the base of the Little Tongue Member may have been the principal conduit for flow in the Madison paleoaquifer system. The widespread distribution (Fig. 12.9) and lateral continuity of this zone and its restriction to the paleophreatic zone support this interpretation. Extensive development of a thick zone of collapse breccia directly overlying the solution zone and sagging of limestone beds above the collapse breccia (Fig. 12.4) indicate removal of a significant amount of rock from within the zone, thus providing considerable open space for the movement of water through the aquifer. Water flow through caves developed in the paleophreatic zone at higher levels of the bedrock was more localized because of less lateral continuity of available open space.

Paleoaquifer System

My own field observations and those of Quinones (1985) suggest that paleohydrologic flow was controlled mainly by bedding planes. Because the regional dip is radial from the uplift in southeast Wyoming (Fig. 12.11), a simple model of the paleoaquifer system would predict radial paleohydrologic flow from a center in southeast Wyoming. However, the paleogeologic map (Fig. 12.11) and geologic profiles (Fig. 12.12) indicate that block faulting and the areal limitation of the principal paleohydrologic conduit (lower solution zone) combined to produce a complex paleoaquifer system. Original distribution of the lower solution zone was controlled by the limits of an evaporite basin of late Osagean age in northern and southeastern Wyoming (Fig. 12.9). The southern limit of this basin was fragmented by prekarst block faulting (Fig. 12.11). Because the evaporite basin did not extend to the core of the uplift in southeast Wyoming, the uplifted edges of the lower solution zone never reached the paleokarst surface. Recharge into the principal hydrologic conduit was probably through fractures scattered throughout the karst plain. Moreover, the lateral continuity of the conduit was broken by the block faulting into local areas bounded by the faults. Bedding plane attitudes in the blocks seem to have been dominated by the regional dip pattern, but faulting caused deviations from this pattern in some blocks.

The conduit was perched in many places (Fig. 12.12).

Inferred constraints on the paleohydrologic model suggest that subterranean paleokarst features should be concentrated in the areas of the principal paleohydrologic conduit, that the influence of regional structural gradient on their distribution should be minimal, and that local controls of distribution should be most important. A synthesis of these relationships (Fig. 12.13) supports the concept of a complex paleoaquifer system. Most sinkholes seen in outcrop or inferred from anomalous Darwin thicknesses are located in the area of the lower solution zone, the postulated principal hydrologic conduit. The sinkholes are concentrated in topographic lows. Most of the deepest observed paleosolution cavities are also located in the region of the lower solution zone; only two deeper than 35 feet (11 m) below the paleokarst surface are outside of that area.

Evidently, the influence of the lower solution zone, local topography, and local structure of the bedrock were the most important factors controlling the distribution of these features. Lack of data on the orientation of tubes and flow structures precludes a final assessment of the importance of the regional gradient. However, abundant evidence of phreatic-zone open space is proof of the existence of paleoflow gradients and the inferred structure of the bedrock suggests that local gradients prevailed. Available data do not provide an obvious correlation between the solution features and regional gradient patterns.

Preserved Evaporites

Preserved evaporite beds at the base of the Little Tongue Member or equivalent strata are recorded at only three localities in Wyoming (Fig. 12.13). Preservation of evaporites in the outcrop section at Hoback Canyon is explained by their position west of the shoreline of the Madison plain, where evaporites were probably sheltered from solution. Two subsurface localities, one in the Powder River Basin in northeast Wyoming and one in the Denver Basin in southeast Wyoming, are not explained by the paleohydrologic model; these may be the result of some local impedance to paleohydrologic flow.

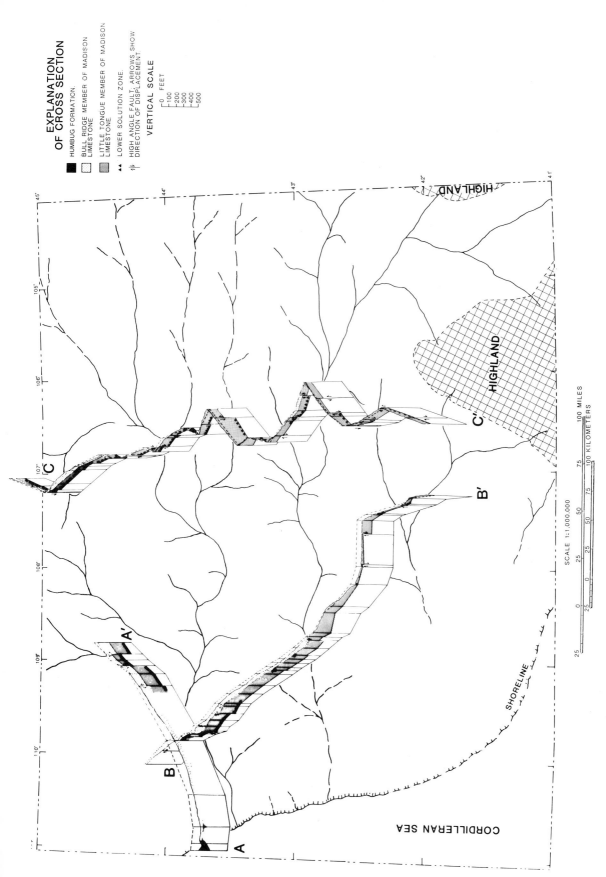

FIGURE 12.12. Map of Wyoming showing geologic profiles through the Madison paleokarst bedrock. Control data shown on Figure 12.9.

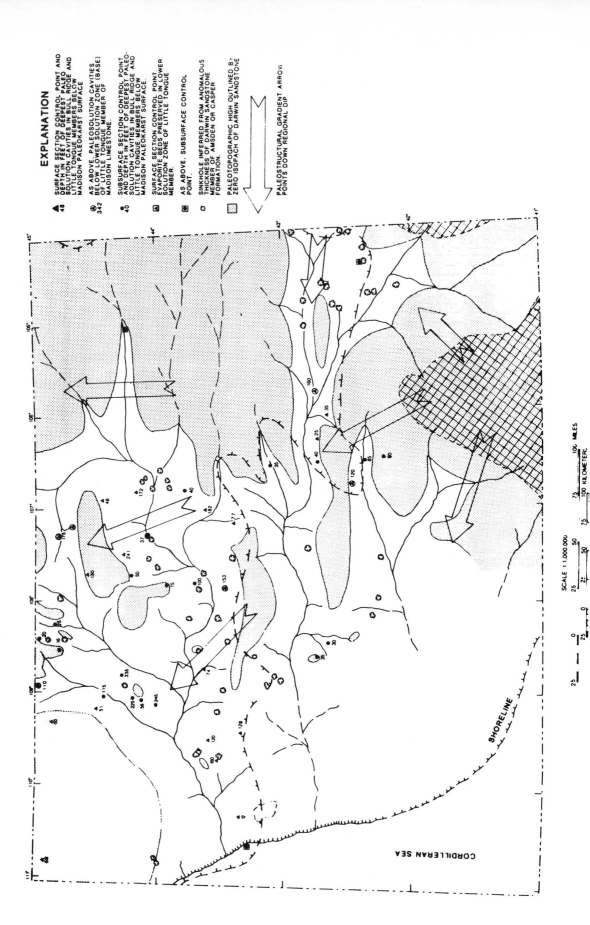

FIGURE 12.13. Map of Wyoming showing distribution of important paleokarst features in relation to factors that probably controlled their distribution.

Future Research and Exploration Perspectives

Madison Limestone paleokarst holds great potential for future scientific research and economic exploration. The synthesis presented here provides only a broad paleokarst model that needs to be scrutinized, tested, and refined by more detailed studies.

Karst Process

Outcrops of the Madison paleokarst afford an excellent outdoor laboratory for testing models of karst formation and details of karst process. The estimated maximum 34 m.y. duration of exposure of Madison carbonate rocks to karst processes may be among the longer of known examples. Modern karst examples are of much shorter duration (e.g., maximum duration of about 5 m.y. for the Yucatan karst.) Little or no work has been done on the petrography, cement stratigraphy, and isotope geology of the Madison paleokarst. Much work remains to be done on the morphology of the paleokarst features. The relationship of Tertiary and Holocene karst to the Madison episode has barely been described.

Sulfide Ore Deposits

The economic potential of the Madison paleokarst has not been completely evaluated and exploited. Although economically important sulfide mineralization of the paleokarst has not been reported in the Madison, there may be some mineralized areas that have been overlooked because of insufficient attention to the possibility of paleokarst control. Recently discovered evidence of Pennsylvanian emplacement of sulfide ores by paleohydrologic flow controlled by the paleokarst, rather than by Laramide-Tertiary intrusive activity, in the nearby Leadville Limestone (De Voto, this volume), may stimulate exploration for ore deposits in the Madison. Although most of the open space in the Madison paleokarst seems to have been filled soon after it was created, there may be some areas where sulfides were emplaced by paleohydrologic flow.

Uranium Ore Deposits

Uranium ore was emplaced in cave systems in the Madison Limestone of the Pryor and Bighorn Mountains during Pliocene-Pleistocene time (McEldowney et al. 1977). Tertiary development of the cave systems was guided by the Madison paleokarst. There may be other mineralized areas where the Madison paleokarst model is useful for exploration of ore bodies localized by later solution episodes.

Petroleum Reservoirs

Economically significant petroleum reservoirs were developed in the Madison paleokarst by selective pore plugging in the Elk basin field of northern Wyoming (McCaleb and Wayhan 1969) where the producing intervals are sealed by the overlying Horseshoe Shale Member of the Amsden Formation and the underlying Big Goose Member of the Madison. The Madison paleokarst needs to be explored in other basins in Wyoming and Montana, where similar conditions may exist. There may be karst reservoir development in some areas of the Williston basin, where exploration geologists have assumed that pre-Big Snowy karstification was not operative. Early filling of the solution cavities and protection of some evaporite areas from the solution process are limiting factors for reservoir development, but as petroleum becomes scarcer in the future, it may be economically feasible to explore for these limiting factors of the paleokarst model.

Groundwater

Recent studies in the Powder River Basin and adjacent areas have shown that porous Madison limestones are confined by an underlying seal of impervious Devonian and Early Mississippian terrigenous rocks and an overlying seal of Late Mississippian terrigenous, carbonate, and evaporite rocks, forming the present Madison aquifer system (Downey 1984). Regional geologic structure controls the regional hydrologic flow of this aquifer system. However, solution cavities in the Madison related directly or secondarily to the Madison paleokarst may influence local hydrologic patterns within the aquifer.

Acknowledgments. I am very grateful to P.W. Choquette and N.P. James for encouraging me to write this paper. I am also indebted to J.M. Berdan, J.T. Dutro, Jr., and W.J. Perry, Jr. for their constructive and stimulating reviews of the manuscript.

References

Andrichuck, J.M., 1955, Mississippian Madison Group stratigraphy and sedimentation in Wyoming and southern Montana: American Association of Petroleum Geologists Bulletin, v. 39, no. 11, p. 2170–2210.

Bretz, J.H., 1940, Solution cavities in the Joliet Limestone of northeastern Illinois: Journal of Geology, v. 48, no. 4, p. 337–384.

Bridges, L.W.D., 1982, Rocky Mountain Laramide–Tertiary subsurface solution vs. Paleozoic karst in Mississippian carbonates: Wyoming Geological Association Guidebook, Thirty-third Annual Field Conference, p. 251–264.

Brown, D.L., 1978, Wrench-style deformational patterns associated with a meridional stress axis recognized in Paleozoic rocks in parts of Montana, South Dakota, and Wyoming: Montana Geological Society Guidebook, Twenty-fourth Field Conference, p. 17–31.

Brown, D.L., Blankenagel, R.K., MacCary, L.M., and Peterson, J.A., 1984, Correlation of paleostructure and sediment deposition in the Madison Limestone and associated rocks in parts of Montana, North Dakota, South Dakota, Wyoming and Nebraska: U.S. Geological Survey Professional Paper 1273-B, 24 p.

Campbell, N.P., 1977, Possible exhumed fossil caverns in the Madison Group (Mississippian) of the northern Rocky Mountains: a discussion: National Speleological Society Quarterly Journal, v. 38, no. 2, p. 43–54.

DeJarnett, J.G., 1984, Stratigraphy, sedimentation, corals, and paleogeographic significance of the Mississippian sequence of Mt. Darby, Overthrust Belt, Wyoming: Unpubl. M.S. thesis, Univ. of Wyoming, Laramie, WY, 231 p.

Downey, J.S., 1984, Geohydrology of the Madison and associated aquifers in parts of Montana, North Dakota, South Dakota, and Wyoming: U.S. Geological Survey Professional Paper 1273-G, 47 p.

Elliott, J.K., 1963, Cave occurrences in the Mississippian Madison Limestone of the Pryor Mountains, Montana: Billings Geological Society Guidebook, p. 1–13.

Gutschick, R.C., and Sandberg, C.A., 1983, Mississippian continental margins of the conterminous United States: Society of Economic Paleontologists and Mineralogists, Special Publication 33, p. 79–96.

Gutschick, R.C., Sandberg, C.A., and Sando, W.J., 1980, Mississippian shelf margin and carbonate platform from Montana to Nevada, *in* Fouch, T.D., and Magathan, E.R., eds., Paleozoic paleogeography of west-central United States, Rocky Mountain Section, Society of Economic Paleontologists and Mineralogists, West-central United States Paleogeography Symposium 1, Denver, Colorado, June, 1980, p. 111–128.

Mallory, W.W., 1967, Pennsylvanian and associated rocks in Wyoming: U.S. Geological Survey Professional Paper 554-G, 31 p.

McCaleb, J.A., and Wayhan, D.A., 1969, Geologic reservoir analysis, Mississippian Madison Formation, Elk Basin field, Wyoming-Montana: American Association of Petroleum Geologists Bulletin, v. 53, no. 10, pt. 1, p. 2094–2113.

McEldowney, R.C., Abshier, J.F., and Lootens, D.J., 1977, Geology of uranium deposits in the Madison Limestone, Little Mountain area, Big Horn County, Wyoming: Rocky Mountain Association of Geologists, Exploration Frontiers of the Central and Southern Rockies, p. 321–336.

Quinones, M., 1985, Study of the ancient karstification and brecciation in the Madison Limestone (Mississippian) in Wyoming: M.S. thesis, State University of New York at Stony Brook, 208 p.

Roberts, A.E., 1966, Stratigraphy of Madison Group near Livingston, Montana, and discussion of karst and solution-breccia features: U.S. Geological Survey Professional Paper 526-B, 23 p.

Rose, P.R., 1976, Mississippian carbonate shelf margins, western United States: U.S. Geological Survey Journal of Research, v. 4, no. 4, p. 449–466.

Sando, W.J., 1967, Madison Limestone (Mississippian), Wind River, Washakie, and Owl Creek Mountains, Wyoming: American Association of Petroleum Geologists Bulletin, v. 51, no. 4, p. 529–557.

Sando, W.J., 1968, A new member of the Madison Limestone (Mississippian) in Wyoming: Geological Society of America Bulletin, v. 79, no. 12, p. 1855–1858.

Sando, W.J., 1974, Ancient solution phenomena in the Madison Limestone (Mississippian) of north-central Wyoming: U.S. Geological Survey Journal of Research, v. 2, no. 2, p. 133–141.

Sando, W.J., 1976, Mississippian history of the northern Rocky Mountains region: U.S. Geological Survey Journal of Research, v. 4, no. 3, p. 317–338.

Sando, W.J., 1977, Stratigraphy of the Madison Group (Mississippian) in the northern part of the

Wyoming-Idaho overthrust belt and adjacent areas: Wyoming Geological Association Guidebook, Twenty-ninth Annual Field Conference, p. 173–177.

Sando, W.J., 1978, Coral zones and problems of Mississippian stratigraphy in the Williston basin: Montana Geological Society, Williston Basin Symposium, Billings, Montana, September 24–27, 1978, p. 231–237.

Sando, W.J., 1979, Lower part of the Carboniferous, *in* Lageson, D.R., Maughan, E.K, and Sando, W.J., The Mississippian and Pennsylvanian (Carboniferous) Systems in the United States—Wyoming, U.S. Geological Survey Professional Paper 1110-U, 38 p.

Sando, W.J., 1982, New members of the Madison Limestone (Devonian and Mississippian), north-central Wyoming and southern Montana, *in* Sohl, N.F., and Wright, W.B., Changes in stratigraphic nomenclature by the U.S. Geological Survey, 1981,

U.S. Geological Survey Bulletin 1529-H, p. H125–130.

Sando, W.J., 1986, Revised Mississippian time scale, western interior region, conterminous USA: U.S. Geological Survey Bulletin 1605A, 18 p.

Sando, W.J., Gordon, M. Jr., and Dutro, J.T., Jr., 1975, Stratigraphy and geologic history of the Amsden Formation (Mississippian and Pennsylvanian) of Wyoming: U.S. Geological Survey Professional Paper 848-A, 78 p.

Sando, W.J., and Mamet, B.L., 1981, Distribution and stratigraphic significance of Foraminifera and algae in well cores from Madison Group (Mississippian), Williston basin, Montana: U.S. Geological Survey Bulletin 1529-F, 12 p.

Severson, J.L., 1952, A comparison of the Madison Group with its subsurface equivalents in central Montana: Unpubl. Ph.D. thesis, University of Wisconsin, Madison, WI (available on microfilm).

13

Late Mississippian Paleokarst and Related Mineral Deposits, Leadville Formation, Central Colorado

RICHARD H. DE VOTO

Abstract

A Late Mississippian paleokarst is developed on the Leadville Formation and underlying carbonate units in central Colorado. Caverns, sinkholes, solution-enlarged vertical joints (cutters), channelways, and breccia-rubble soil zones typify the karst-solution features which occur extensively throughout central Colorado. The subterranean channelways generally occur semiconcordant to bedding. Multiple cavern levels occur in the Leadville (Mississippian), Dyer (Devonian), and Manitou (Ordovician) carbonate rocks adjacent to major (100 to 200 m deep) paleovalleys on the karst landscape, apparently developed progressively with valley incision in the Late Mississippian. The patterns of karst-solution features, paleovalleys, and erosional thinning of the Leadville strata reflect Mississippian movement of local tectonic blocks.

The subterranean karst-solution features are filled dominantly with carbonate breccia and carbonate sand with varying amounts of black chert breccia. In close proximity to the Late Mississippian karst unconformity surface, significant amounts of black shale and gray clay from the overlying Pennsylvania strata (Belden and Molas formations, respectively) also occur within the karst-solution features. Increased thickness of the regolithic residuum of the Molas gray clay with black chert fragments overlies karst-solution-thinned areas of the underlying Leadville strata.

Hundreds of lead-zinc-silver-barite deposits occur over a large area of central Colorado within karst-solution features related to the Late Mississippian landscape. These features are the primary controls on the location of the ore deposits within the Lower and Middle Paleozoic carbonate rocks in central Colorado. The pattern of mineralized karst-solution features was strongly controlled by incision of surface drainages and a strong northeast-trending Mississippian joint system. In the Leadville and Aspen districts, the principal areas of mineralization occur immediately adjacent to deeply incised paleovalleys.

Introduction

The Late Mississippian paleokarst in central Colorado is part of a regionally extensive karst plain that developed on Lower Mississippian shelf carbonates in the Rocky Mountain states from Montana to New Mexico in the Meramecian and Chesterian (Craig 1972, Sando et al. 1975, De Voto 1980a, Gutschick et al. 1980) (Sando, Chap. 12, this volume). The preserved thickness of Mississippian strata beneath the Late Mississippian paleokarst generally ranges from zero to 60 m (200 ft) in central Colorado and thickens toward the northwestern and southeastern portions of the state (Figs. 13.1 and 13.2) (De Voto 1985a).

Mississippian Stratigraphy and Paleogeography

Kinderhookian

The Lower Mississippian strata are comprised of the dominantly Kinderhookian Redcliff and Osagean Castle Butte Members of the Leadville Formation. The Redcliff Member in the Mosquito Range is comprised dominantly of thin-bedded, stromatolitic dolomite boundstone, intraclast wackestone-packstone, and dolomite breccia (De Voto 1980a, Samsela 1980, Dorward 1985). To the west and north in the area

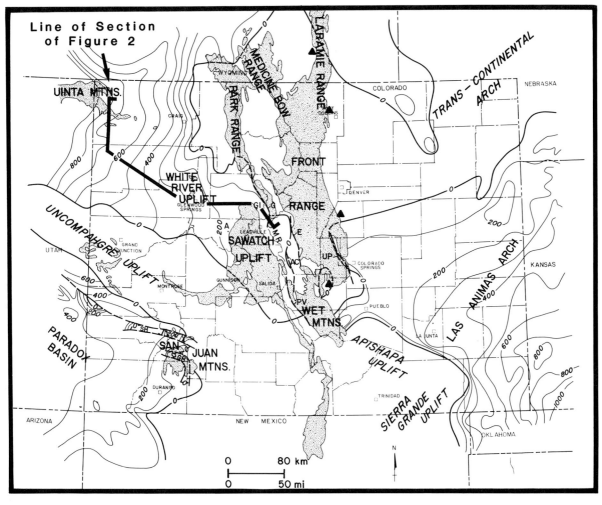

FIGURE 13.1. Map showing thickness of Mississippian rocks and major geologic features in Colorado (De Voto 1980a). Letters indicate: A, Aspen; G1, Gilman Mine; MR, Mosquito Range; G, Gore fault; E, Elkhorn fault; AC, Agate Creek fault; PV, Pleasant Valley fault; UP, Ute Pass fault.

of the northern Sawatch Range and White River uplift, these intertidal lithologies are overlain by lime and dolomite mudstone and pelletal wackestone of probable restricted-circulation subtidal environments (Nadeau 1971 and 1972, Conley 1972). An abrupt facies change from intertidal stromatolitic boundstone in the southeast to mixed-skeletal packstone and wackestone to the northwest occurs within the Redcliff strata in northeastern Gunnison County (Fig. 13.3) (Wittstrom 1979). Locally developed flat-floored bodies of carbonate

breccia, angular masses of coarsely crystalline dolomite in dolomite mudstones, and rare pseudomorphs of halite and gypsum (Conley 1972, Horton 1985b) suggest the possible former existence of evaporites within the Redcliff Member.

The Williams Canyon Limestone, which occurs east of the Front Range, is comprised dominantly of pinkish, thin-bedded, stromatolitic, lime and dolomite boundstone, intraclast packstone and breccia, and well-rounded, fine- to medium-grained, quartz sandstone (Ramirez

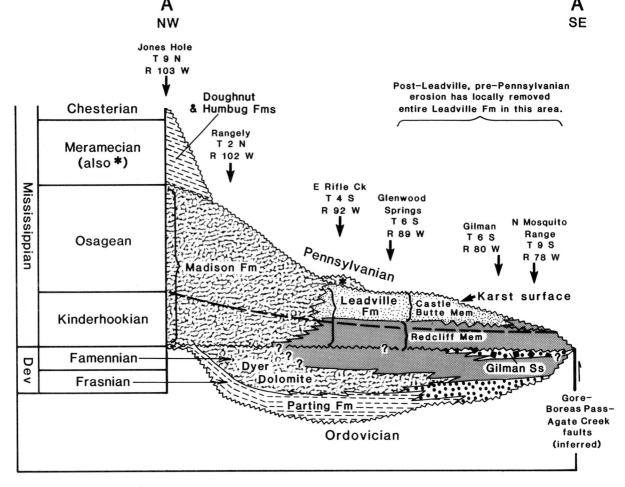

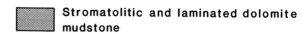

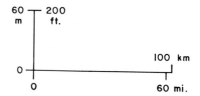

FIGURE 13.2. Restored section of Mississippian and Devonian rocks, northwestern to central Colorado. Data from Bloom (1961), Conley (1964), Campbell (1970), and De Voto (1980a). Line of section is shown on Figure 13.1.

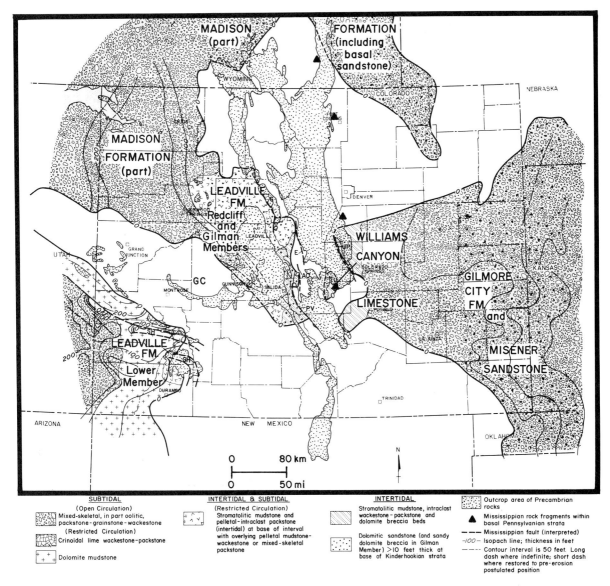

FIGURE 13.3. Thickness and lithofacies of rocks mostly of Kinderhookian age (De Voto 1980a). Note: Gilman Sandstone now considered as separate formation, with unconformity between it and Leadville Formation. GC, Gunnison County.

1973 and 1974, Hill 1983). These rocks change facies to mixed-skeletal and oolitic, lime and dolomite packstone and grainstone, and glauconitic sandstone, indicative of an open-circulation subtidal environment, in eastern Colorado (Fig. 13.3) (De Voto 1980a).

The facies relationships, the intertidal deposits of the Redcliff and Williams Canyon strata on opposite sides of the Front Range and Wet Mountains, and the presence of sandstone in the Williams Canyon and basal part of the

Madison Formation in southern Wyoming (Maughan 1963), all suggest that these strata were deposited on tidal flats adjacent to, but on opposite sides of, an ancestral Front Range emergent area. This emergent area probably extended from the Wet Mountains northward to the Wyoming border and constituted an Early Mississippian continental divide.

Throughout central Colorado a low-relief solution-erosion unconformity occurs at the top of the Redcliff Member (Tweto and Lovering

1977, Dorward 1985) (Fig. 13.2). The unconformity truncates faults that offset the Redcliff
Member (Fig. 13.4). Sandstone-filled channel
scours occur locally at the unconformity surface
(De Voto and Maslyn 1977, Maslyn 1976,
Tschauder and Landis 1985). In general, however, a 1- to 5-m (3- to 16-ft) thick dolomite
breccia, locally with a black shaly and sandy
matrix, overlies the unconformity. The breccia
occupies solution openings, fractures, pockets,
and depressions in the Redcliff strata below the
unconformity. Locally at Spring Creek in the
northern Mosquito Range, the dolomite breccia
is light gray and contains dolomite mudstone
breccia fragments in a matrix of dolomite mud
and sand, which include crinoid fragments and
burrows. This breccia unit, directly overlying
the Redcliff Member, is locally referred to as
the *pink breccia* in the Gilman mining district
(Lovering et al. 1978), the *MC shale* at the Sherman Mine (Tschauder and Landis 1985), and
the *contact fault* in the Aspen area (Spurr 1898).
The breccia is well developed at many places
in central Colorado and is generally not present
or apparent in the area of the White River
uplift. This breccia unit appears to be a depositional breccia deposited on the intraformational karst-erosion surface that developed in
central Colorado on the Redcliff strata.

Sucrosic and coarsely crystalline dolomite
occurs subregionally within the Redcliff strata
in areas around the Sawatch uplift, and limestone of the Castle Butte Member occurs above
the unconformity (De Voto 1985). Mixing-zone
dolomitization and karstification of the Redcliff
Member developed as emergence occurred in
central Colorado (Horton 1985a and 1985b).
The emergence was probably caused subregionally by a eustatic sealevel drop and locally
by Early Mississippian tectonic activity in the
area of the ancestral Sawatch uplift and possibly
the ancestral Front Range uplift (De Voto
1985a).

Osagean

The Osagean Castle Butte Member of the
Leadville Formation rests unconformably on
the Redcliff Member throughout central Colorado. The Castle Butte Member consists
dominantly of pelletal, oolitic, and mixed-skeletal grainstone and packstone with minor
amounts of syringoporid coral bafflestone and
skeletal wackestone (Nadeau 1971 and 1972,
Conley 1972, Wittstrom 1979, Samsela 1980).
The abundance of mud-supported textures
decreases upward within the Castle Butte
Member. The mudstone and wackestone lithologies increase in abundance from the western
to eastern side of the Sawatch uplift as the zero
edge of the Osagean strata is approached (Fig.
13.5). "Zebra structure"—alternating dark gray,
microcrystalline and white, coarsely crystalline
dolomite bands—is commonly developed by
solution enlargement along bedding-plane
laminations and open-space precipitation of the

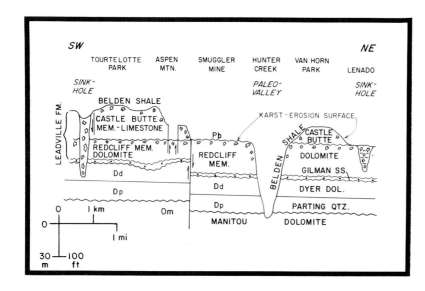

FIGURE 13.4. Schematic restored section of Leadville
Formation, karst-breccia features, paleovalley, and interpreted Mississippian faults,
Aspen area (data from Spurr
1898, Bryant 1971, Maslyn
1976). Line of section is shown
on Figure 13.7.

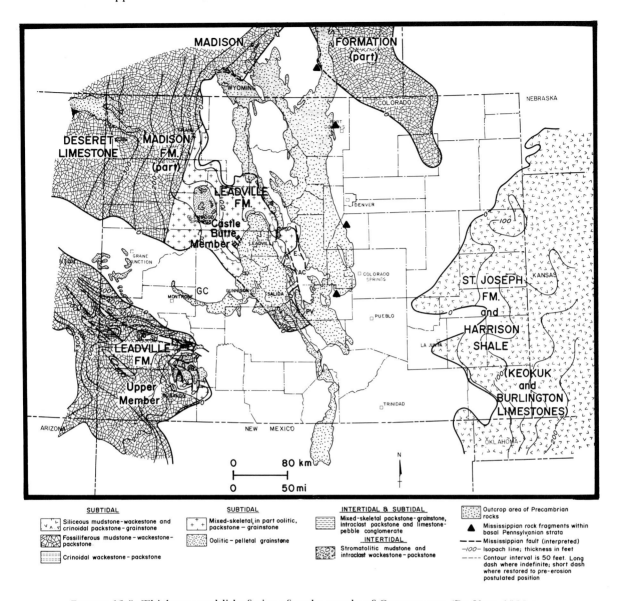

FIGURE 13.5. Thickness and lithofacies of rocks mostly of Osagean age (De Voto 1980a).

coarsely crystalline dolomite within intertidal, stromatolitic, dolomite boundstones of both the Castle Butte and Redcliff members on the east flank of the Sawatch uplift (De Voto 1985a, Horton 1985b). Southeast of Salida, several limestone-pebble conglomerates, intraclast packstones, and sandy zones within the Castle Butte (Samsela 1980) suggest proximity to a contemporaneous erosional area.

A major widespread, oolitic grainstone facies with oolitic-grainstone intraclasts is developed encased within the mixed-skeletal grainstone

strata of the Castle Butte in the area of the White River uplift (Fig. 13.5) (Conley 1972). Detailed facies changes from oolitic grainstones to crinoidal grainstones to mixed-skeletal, pelletal grainstones on the northwestern flank of the Sawatch uplift (Fig. 13.5) appear to be related to subtle seafloor topographic relief induced by local contemporaneous Osagean faulting (De Voto 1980a and 1985b).

The facies patterns within the Castle Butte strata suggest that the north-trending ancestral Front Range–Wet Mountain area continued to

be a low-relief emergent area during the Osagean. The White River uplift and local fault
blocks of the northwestern Sawatch uplift
underwent tectonic activity and created shoal-
water areas on the Osagean sea floor.

Probable eustatic sealevel drop in the Meramecian caused emergence of central Colorado
and development of an extensive karst plain
which extended from Colorado to Canada, including most of western Colorado and the states
of Wyoming and Montana (Sando et al. 1975,
Gutschick et al. 1980, De Voto 1980a, Sando,
this volume). Sucrosic and coarsely crystalline
dolomite has been intensely developed in the
Castle Butte Member below the karst landscape
in the Gilman–northern Mosquito Range area
and in areas of the southwestern flank of the
Sawatch anticline (De Voto 1985a). This dolomitization probably developed in a brackish-
water mixing zone associated with the Late
Mississippian emergence (Horton 1985a and
1985b, Beaty 1985). Zebra dolomite and the
other dolomite types occur as breccia clasts in
cave sediment, and therefore all originated
prior to karst solution-erosion. As the sea retreated from central Colorado, bedded and
nodular chert, sucrosic and coarsely crystalline
dolomite, zebra dolomite, and karst solution-
erosion and brecciation could have occurred
sequentially in the Late Mississippian.

Late Mississippian Paleokarst

Paleokarst Landscape

The topography of the Late Mississippian paleokarst landscape in central Colorado was
generally a broad plain with low hills and towers, abundant dolines (sinkholes) and solution-
enlarged joints (cutters), and occasional low to
moderate relief valleys. The preserved Leadville and correlative Madison strata thin generally from 120 m to 180 m (400-600 ft) in the
northwestern and southeastern corners of the
state to generally less than 60 m (200 ft) in central Colorado (Figs. 13.1 and 13.2). Within
central Colorado, the thickness of the Leadville
generally ranges from 23 m to 60 m (75 to 200
ft), with extreme local variation (Fig. 13.6). Local thickness variations are due to sinkholes,
paleovalleys, and erosional removal of Leadville
strata on faulted horst blocks. Details of these

features and the abrupt thickness variations of
the resulting Leadville strata in the Aspen area
are shown on Figures 13.4 and 13.7 (Maslyn
1976). Table 13.1 lists the major types of surficial and subsurface karst features described
and illustrated in this text.

The sinkholes are generally elliptical to circular in plan view and funnel-shaped or cylindrical in cross-section. They commonly are 100
m (310 ft) but can be up to 450 m (1500 ft) in
diameter and up to 45 to 75 m (150 to 250 ft)
in depth. Several sinkholes have been observed
which penetrate through the entire Leadville
Formation and much or all of the underlying
Dyer Dolomite (Figs. 13.4 and 13.8) (Maslyn
1976, Banks 1967). Solution-enlarged joints
(cutters) are more difficult to observe in two-
dimensional outcrops, but can be interpreted
as abundant from descriptions of vertical-
walled, breccia-filled, tabular bodies perpendicular to bedding within the Leadville Formation in underground mines (Blow 1889,
Spurr 1898, Emmons et al. 1927, Lovering et
al. 1978).

Local towers up to 15 to 30 m (50 to 100 ft)
high occur on the karst landscape (Figs. 13.8
and 13.9). Such tower karst topography has
locally affected the facies of the overlying Lower Pennsylvanian strata (Fig. 13.9) (Dupree
1979).

Several paleovalleys on the karst landscape
have been observed around the edge of the Sawatch uplift (Fig. 13.6). The paleovalley at
Hunter Creek north of Aspen has been eroded
to the stratigraphic level of the Ordovician
Manitou Dolomite, more than 120 m (400 ft)
below the top of the Leadville Formation (Fig.
13.4) (Maslyn 1976, De Voto 1983). An east-
trending paleovalley with more than 75 m (250
ft) of erosional relief occurs at Leadville. Lower
Pennsylvanian rocks rest directly on Devonian
strata where the Leadville strata have been
completely removed within this paleovalley
(Fig. 13.10). A Late Mississippian paleovalley
with 200 m (650 ft) of erosional relief occurs
in the Tincup–Tomichi area (Fig. 13.6). The
Lower Pennsylvanian strata rest directly on the
Ordovician Manitou Dolomite in this paleovalley (Dings and Robinson 1957, De Voto
1983).

The Late Mississippian paleokarst terrain in
central Colorado was affected by Late Missis-

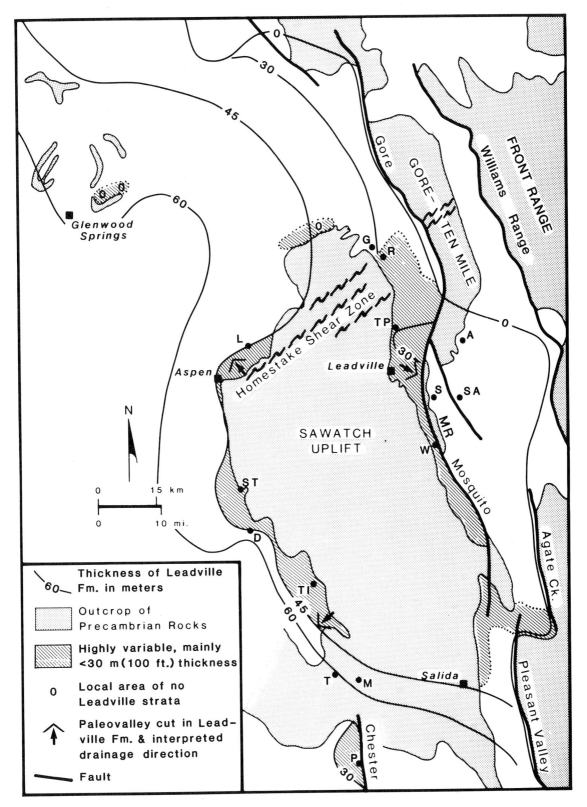

FIGURE 13.6. Thickness of Leadville Formation in central Colorado. MR, Mosquito Range. Mines and mining districts: A, Alma district; D, Doctor mine; G, Gilman (Eagle) mine; M, Monarch district; P, Pitch mine; R, Redcliff district; SA, Sacramento mine; S, Sherman mine; ST, Star mine; TP, Tennessee Pass district; TI, Tincup district; T, Tomichi district; W, Weston Pass district.

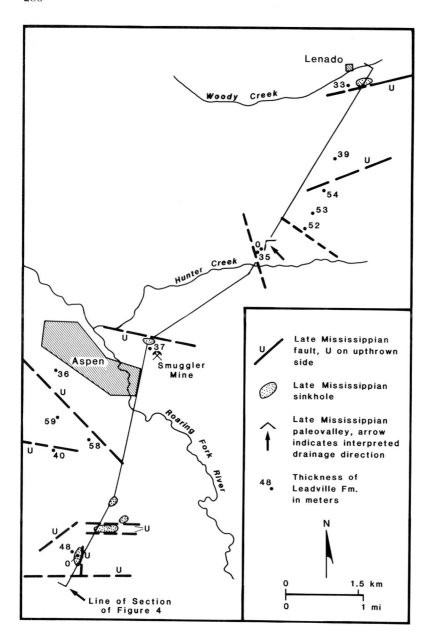

FIGURE 13.7. Mississippian faults, sinkholes, and paleovalley and thickness of Leadville strata, Aspen area (Maslyn 1976).

sippian tectonism and uplift. Local pre-Pennsylvanian fault offsets of Mississippian rocks of up to 20 m (65 ft) have been mapped in the Aspen area (Fig. 13.7) (Maslyn 1976) and in the Sherman Mine (Tschauder and Landis 1985). Moreover, several features in the northern Sawatch Range and Mosquito Range attest to the fact that the northern Sawatch Range was uplifted in the Late Mississippian and Early Pennsylvanian, so that the surface and subsurface drainage on the karst landscape flowed downhill to the east and west off the ancestral

Sawatch uplift. These features include, in the Leadville-northern Mosquito Range area:

1. The east-trending paleovalley at Leadville (Fig. 13.10).
2. Deeper erosion within the paleovalley and generally of Leadville strata to the west.
3. Tributary-like aspect and orientation of major ore-filled subterranean solution features at Leadville (Fig. 13.10).
4. Morphology of individual sinkholes and caverns suggesting recharge on southwest

TABLE 13.1. Karst features described and illustrated in text figures.

Karst features	Figures
Surficial	
Karst plain	13.1, 13.2, 13.3, 13.6
Tower karst	13.8, 13.9
Paleovalley	13.4, 13.7, 13.10, 13.16, 13.20
Paleosol (Molas Fm.)	13.8, 13.9, 13.12, 13.13, 13.14, 13.16
Subsurface	
Dolines (sinkholes)	13.4, 13.7, 13.8, 13.11, 13.15, 13.16, 13.17
Cutters (solution-enlarged joints)	13.13, 13.16, 13.18, 13.28
Caverns (caves)	13.11, 13.15, 13.16, 13.21
Channelways	13.10, 13.11, 13.18, 13.20, 13.27
Chimneys	13.11, 13.15, 13.16, 13.18, 13.19, 13.25
Multiple cavern levels	13.20
Integrated cave system	13.15, 13.18, 13.20
Sinkhole (cutter) breccia	13.4, 13.8, 13.11, 13.13, 13.17
Cave (channelway) breccia	13.11, 13.21, 13.22, 13.24, 13.26
Extraformational sediment	13.9, 13.23
Ore deposits in karst features	13.10, 13.11, 13.13, 13.14, 13.16, 13.18, 13.19, 13.20, 13.21, 13.25, 13.26, 13.27, 13.28

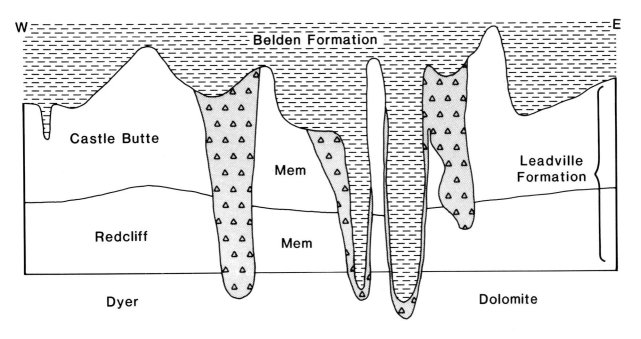

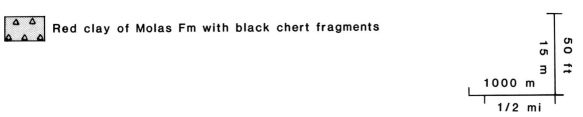

FIGURE 13.8. Sketch cross-section of karst features in outcrops along north side of Glenwood Canyon 15 km (9 mi) east of Glenwood Springs (Banks 1967).

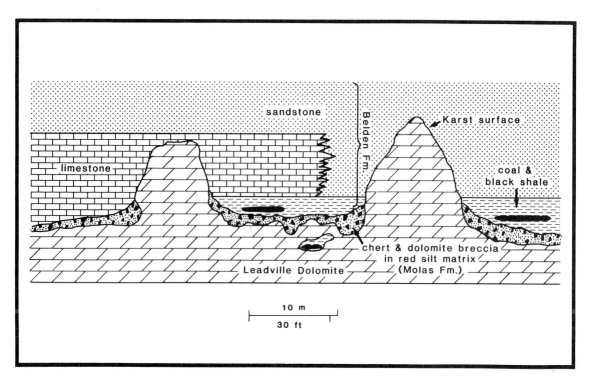

FIGURE 13.9. Tower karst landscape and effects on Lower Pennsylvanian facies, Indian Creek–Pitch mine area, Gunnison County, Colorado (Dupree 1979).

ends and movement of groundwater to the northeast along northeast-trending joints (Fig. 13.11) (De Voto 1983 and 1985b, Tschauder and Landis 1985).

5. West-dipping cave sediment in east-dipping Leadville rocks (Fig. 13.11) (De Voto 1985a, Tschauder and Landis 1985).

6. Onlapping relationship and angular unconformity between Lower Pennsylvanian strata and underlying Leadville Formation (De Voto 1983 and 1985c).

7. Eastward-directed sediment transport of the Lower Pennsylvanian micaceous and arkosic, coarse-grained sediments deposited on the karst-erosion unconformity surface (De Voto 1985a).

Cave morphologies at Gilman and Aspen also suggest groundwater flow during the Late Mississippian away from a high area in the northern Mosquito Range (De Voto 1983 and 1985a, Tschauder and Landis 1985).

Surficial Deposits

A residual soil or insoluble residuum, the Molas Formation, occurs discontinuously and, where present, ranges from centimeters to up to 30 m (100 ft) at the top of the Leadville Formation at some places in central Colorado. Where preserved, the uppermost meter (3 ft) contains red, hematitic quartz silt and kaolinitic clay with rare breccia fragments, a B-zone soil horizon. This zone generally grades downward into a zone of variable thickness and a matrix-supported texture of angular black chert and solution-rounded limestone and/or dolomite fragments in a red mud matrix. This zone grades into unaltered bedrock with red mud filling solution-enlarged fractures and bedding planes. In southwestern Colorado, the red *terra rossa* mudstone has been mapped in the topographically low areas around hills and towers on the karst landscape (Maslyn 1977). In the area of the White River uplift and locally in Gunnison County, the red mudstone of the Molas occurs discontinuously at the top of the Leadville Formation in paleotopographic lows and filling sinkholes that extend into the Dyer Dolomite, as much as 50 m (170 ft) below the top of the Leadville Formation (Figs. 13.8 and 13.9) (Banks 1967, Dupree 1979). The red mud and limestone breccia fragments also occupy caves and solution-enlarged joints and bedding

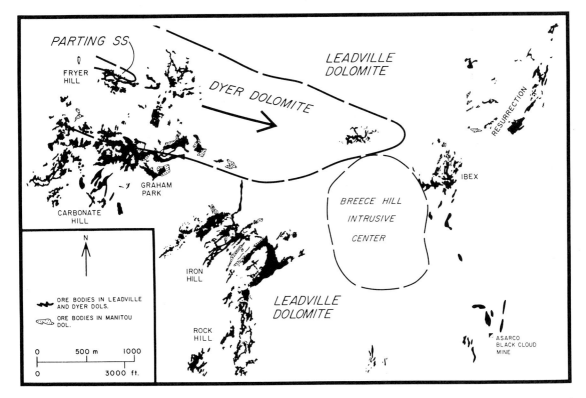

FIGURE 13.10. Ore deposits and subcrop geology below Late Mississippian erosion surface, Leadville mining district (De Voto 1983). Many ore bodies occupy subterranean karst channelways. Pennsylvanian strata rest directly with depositional contact on rock units shown on map. Inferred drainage direction in paleovalley on Late Mississippian erosional landscape shown by arrow.

planes within the limestones of the Leadville Formation (De Voto 1985c).

Throughout most of central Colorado, however, the residual soil and residuum that is called the Molas Formation is dominantly gray claystone with black chert and carbonate breccia fragments (Fig. 13.12). It also occurs discontinuously at the top of the Leadville Formation, most often and with greatest thickness in topographic depressions, and in caves or along fractures and bedding planes shallow beneath the karst surface (Fig. 13.13) (De Voto 1985b).

FIGURE 13.11. Schematic cross-section of Sherman Mine, northern Mosquito Range, with collection sinkhole, channelway, and outlet chimney (De Voto 1983, Tschauder and Landis 1985). Cross section is along length of a mineralized trend.

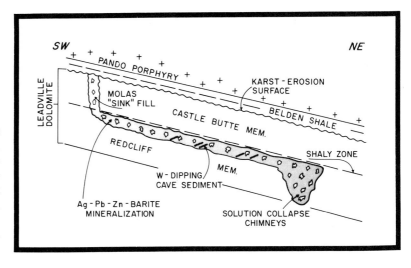

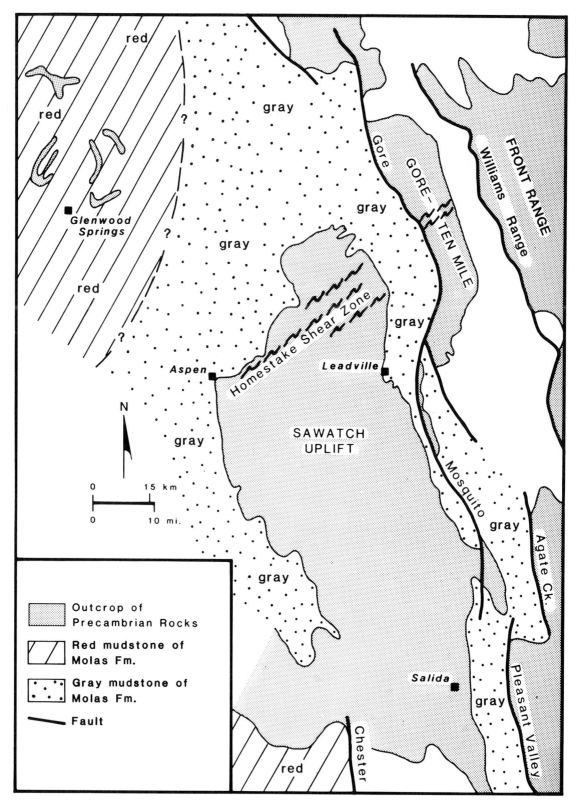

FIGURE 13.12. Distribution of Molas lithologies in residual soil on Late Mississippian karst landscape, central Colorado.

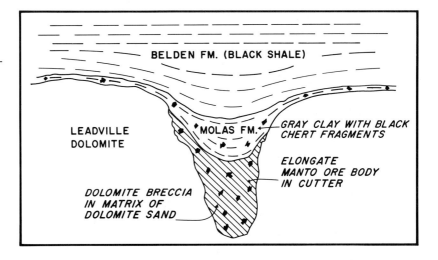

FIGURE 13.13. Schematic cross-section of typical development of Molas residual soil and subjacent manto ore body, Gilman and Leadville areas (De Voto 1983 and 1985b).

In the Aspen area, the matrix of the upper portion of the paleosol is carbonaceous, black clay, similar lithologically to the black shales of the overlying Belden Formation (Fig. 13.14) (Maslyn 1976).

Subsurface Karst-Solution Features

The solution features developed beneath the karst landscape include subterranean channelways, caverns, collapse chimneys, sinkholes, cutters (solution-enlarged, vertical joints), and integrated cave systems with collection sinkholes, subhorizontal channels, and outlet chimneys (Figs. 13.11, 13.15, and 13.16) (De Voto 1983 and 1985b, Tschauder and Landis 1985).

In the Aspen area, sinkholes are circular to elongate in plan view, range from 30 to 500 m

(100 to 1500 ft) in length, and are up to 75 m (250 ft) in depth. They generally are adjacent and elongate parallel to Late Mississippian faults (Maslyn 1976). The Castle Butte sinkhole in Tourtellotte Park on Aspen Mountain (Fig. 13.17) is typical of those in the Aspen area. The sinkhole extends across the entire thickness of the Leadville and Dyer formations and has its base in the Parting Quartzite. The sinkhole is filled dominantly with angular limestone and dolomite fragments derived from the Leadville and Dyer Formations in a matrix of tan, calcareous silt. Individual breccia fragments range from sand size to blocks up to 3 m (10 ft) across. Large blocks of Belden black shale occur within the sinkhole. These were deposited in the then-unfilled sinkhole in the Early Pennsylvanian. The sinkhole has a discontinuous silicified (jasperoid) rim from 0.7 to 4 m (2 to 14 ft) thick

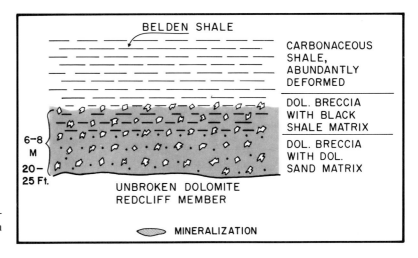

FIGURE 13.14. Late Mississippian paleosol profile, Aspen area (after Maslyn 1976).

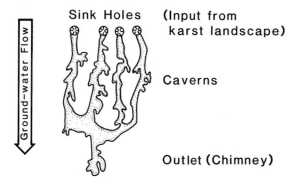

FIGURE 13.15. Schematic plan-view diagram of integrated drainage system (Tschauder and Landis 1985). Note resemblance of Gilman Mine (Figs. 13.19 and 13.20) and Iron Hill area of Leadville district (Fig. 13.10) to this pattern.

(Maslyn 1976). A sinkhole at Lenado, 8 km (5 mi) northeast of Aspen, is filled with carbonate breccia in alternating layers of tan-brown, calcareous silt and black, carbonaceous, kaolinitic clay (Maslyn 1976). This fill sequence suggests deposition under conditions of alternating ponding and draining of the sinkhole.

At the Sherman Mine in the northern Mosquito Range, 29 sinkholes have been found at the up-gradient ends of six cavern systems (Tschauder and Landis 1985). A cavernous channelway subparallel to bedding begins at the bottom of each sinkhole (Fig. 13.11). The channelways follow faults or joints, so that they

commonly are subparallel (Fig. 13.15). The channelways ultimately become tributary to and merge into large caverns, some of which are connected to deeper, less extensive levels by near-vertical chimneys (Tschauder and Landis 1985). The underground cavern and channelway systems occupied by ore bodies at Gilman, Aspen, and Leadville to a great extent display similar subsurface morphologies (Figs. 13.10, 13.16, 13.18, and 13.19). The sinkholes are dominantly vadose zone features, and the subparallel to bedding channelways represent phreatic zone solution features probably developed just below the water table. The pattern of collection sinkholes, tributary channelways, and outlet chimneys permits reconstruction of the paleogroundwater flow beneath the Late Mississippian karst landscape. Such cave patterns indicate that the area of the northern Sawatch uplift was a groundwater divide and presumably a topographic high area in the Late Mississippian.

Multiple stratigraphic levels of breccia-filled underground channelways occur within the Dyer and Leadville strata at outcrops in the northern Mosquito Range and in mines in the Leadville, Gilman, and Aspen mining districts (Figs. 13.16 and 13.20) (De Voto 1983 and 1985b). The morphology, orientation, and proximity of these features to deeply incised paleovalleys at Leadville and Aspen suggest that they were developed progressively as multiple

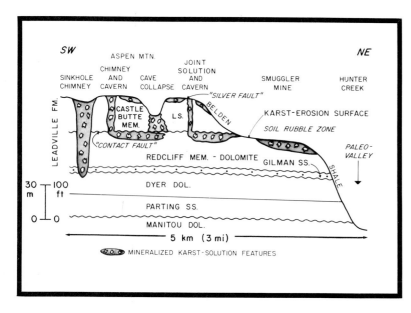

FIGURE 13.16. Schematic restored section, Late Mississippian landscape, subterranean solution features, and ore bodies, Aspen area (De Voto 1983, data from Spurr 1898).

FIGURE 13.17. Photo of Castle Butte sinkhole, Aspen. Bedded limestone at left is at type locality of Castle Butte Member of Leadville Formation. White line marks vertical contact between bedded limestone and sinkhole. Entire area to right of line is breccia of sinkhole fill. Breccia composed of up to 3m (10 ft) blocks of limestone and dolomite.

SINKHOLE BRECCIA

cavern levels as the valleys were progressively incised into the Late Mississippian terrain. Thus the subterranean cavern system in the Leadville strata probably developed while the valley was shallow and poorly developed; those in the Dyer formed as the valley cut downward so that

the groundwater table tributary to the river in the valley existed in the Dyer strata.

The strong northeast trend of many ore-filled channelways and cutters in the Leadville Formation and other units beneath the karst landscape suggests control by a prominent

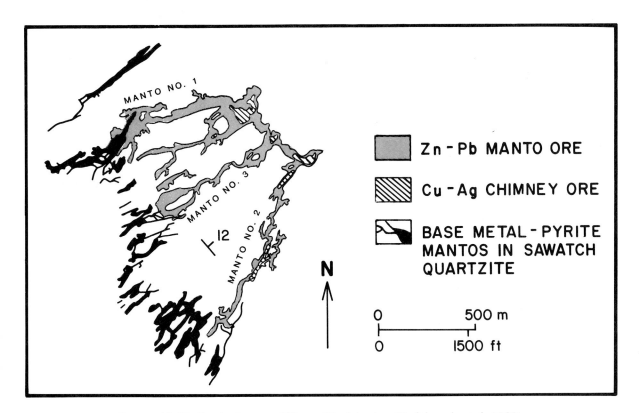

FIGURE 13.18. Ore body map, Gilman (Eagle) mine (Radabaugh et al. 1968).

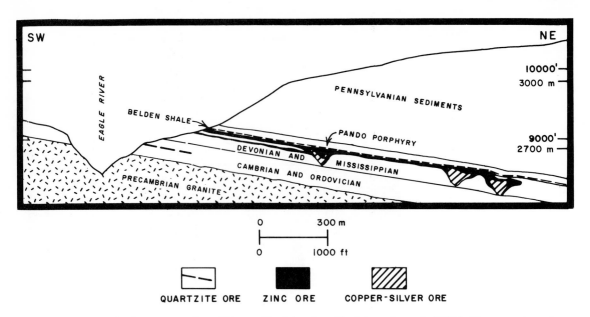

FIGURE 13.19. Geologic cross-section, Gilman (Eagle) mine (Radabaugh et al. 1968). Cross-section along length of Manto No. 2, shown on Figure 13.18.

northeast-trending joint system (Figs. 13.10 and 13.18).

Underground Deposits

The Late Mississippian subsurface karst-solution features (sinkholes, caverns, chimneys, cutters) are most commonly filled with carbonate breccia, carbonate sand and silt, and some black chert breccia (Fig. 13.21). The carbonate sand and silt has resulted from grain-boundary dissolution of the wallrock and sedimentation of the loose carbonate sand and silt grains in solution openings (De Voto 1985b). The sand is occasionally stratified and rarely cross-stratified. Other sedimentary structures include

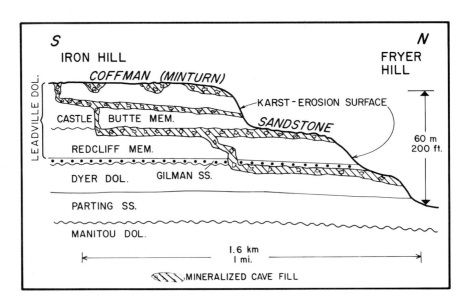

FIGURE 13.20. Schematic restored section, multiple underground channelways, and Late Mississippian paleovalley, Leadville mining district (De Voto 1983 and 1985b). Locations of Iron Hill and Fryer Hill are shown on Figure 13.10.

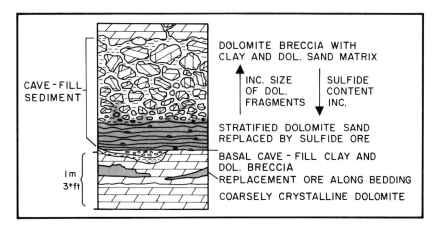

FIGURE 13.21. Sketch of cave sediment and mineralization, Moose mine, Mt. Bross, Alma district (Johansing 1982).

desiccation cracks, mud drapes, and load casts (Tschauder and Landis 1985).

The nature of the carbonate breccia fragments in the breccia bodies ranges from monolithologic breccia fragments of only the lithology of the roof rock of the solution openings (Fig. 13.22), to heterolithic, a diverse mix of colors and lithologic types of the carbonate fragments. Commonly, karst breccias in the Gilman Sandstone contain clasts of sandstone and have matrix of both quartz sand and carbonate sand and silt. Carbonate breccia fragments are chaotically oriented (rubble breccia) at the bases of caves and chimneys and are virtually unrotated (mosaic or crackle breccia) near the tops of these features (Tschauder and Landis 1985).

As proximity to source of extraformational siliciclastic material is approached, that is, immediately beneath the top of the Leadville Formation and immediately beneath the unconformity on top of the Redcliff Member, then some locally available siliciclastic material is common in the underground karst solution features. In the area of the White River uplift where red mud of the Molas Formation occurs locally above the karst surface, red mud occurs within near-surface underground karst openings (De Voto 1985c). Throughout most of central Colorado, however, gray claystone of

FIGURE 13.22. Photo of karst breccia with monolithologic dolomite breccia fragments, some with zebra structure, in matrix of dolomite sand and silt, Redcliff Member, Leadville Formation, Sacramento mine area, northern Mosquito Range.

the Molas Formation or black, carbonaceous shale of the Belden Formation overlies the karst surface (Fig. 13.12) and occurs within near-surface underground karst openings. Locally in the Leadville–northern Mosquito Range area, sandstones of the Coffman Member of the Minturn Formation rest directly on the karst surface and occupy near-surface underground karst openings (Fig. 13.23). Bedding, sometimes cross-bedding in the sandstones, commonly occurs in these extraformational infill sediments. Abundant evidence exists for multiple development of underground karst breccias, with cross-cutting breccia relationships and brecciation of semilithified, dolomite–sand cave sediment (De Voto 1985b).

Most of the underground karst-solution breccia bodies are parallel to bedding. On outcrop, they appear locally tabular although the upper boundaries are commonly undulatory and they enlarge and contract along bedding. The bodies range from centimeters to up to 2 to 3 m (inches to 7 to 10 ft) in thickness. At places, collapse of coherent roof rock has occurred directly over bedded breccia bodies (Fig. 13.24).

Generally throughout central Colorado, vadose and phreatic cave cements, such as flowstone, isopachous cement rims, and colloform crusts, are conspicuously absent within the surficial and underground karst deposits. At

Sherman Mine, however, Tschauder and Landis (1985) have described dolomite isopachous cements (flowstone crusts) around breccia fragments of dolomite and fragments of polymineralogic sulfide ore. They have also described colloform dolomite flowstone intermixed with dolomite sand in cave sediments.

The dominance of bedding-parallel breccia bodies and the rare occurrence of vadose cements indicate that most of the preserved underground karst deposits were phreatic in origin. The monolithologic breccia bodies were derived locally by roof collapse of bedding-parallel channelways and the heterolithic breccia bodies were formed as karst solution features cut vertically across bedding.

Mineral Deposits

Hundreds of combined base and precious metal deposits, primarily lead-zinc-silver-barite deposits, occur within the Late Mississippian karst-solution features in the Leadville and Dyer formations in central Colorado. Although the deposits have a wide variety of paragenetic sequences, dominant minerals, apparent temperatures of formation, and possible fluids of origin (De Voto 1983, Thompson et al. 1985, Beaty et al. 1985, Tschauder and Landis 1985), most deposits occur within the cave sediment

FIGURE 13.23. Photograph showing Pennsylvanian quartz sandstone (light color and shown by arrows) within solution-enlarged bedding features, fractures, and caverns in Castle Butte Member of the Leadville Formation, within 5 m (16 ft) of karst surface, Mosquito Pass, northern Mosquito Range.

FIGURE 13.24. Photo showing roof collapse over cave filled with bedded "pink breccia," Gilman. Arrows denote edge of breccia-filled cave.

and breccia rubble of sinkholes, subterranean channelways, caverns, caves, solution-enlarged vertical-walled joints (cutters), and surface soil-rubble zones related to the Late Mississippian karst landscape. Such features are apparent today in outcrop exposures and in what is acces-

sible of underground workings in the Alma district (Moose mine, Sacramento mine, and others) (Fig. 13.21) (Johansing 1982, De Voto 1985c), at the Sherman mine (Fig. 13.11) (Tschauder and Landis 1985), in the Leadville district (Emmons 1886, Blow 1889, Tweto

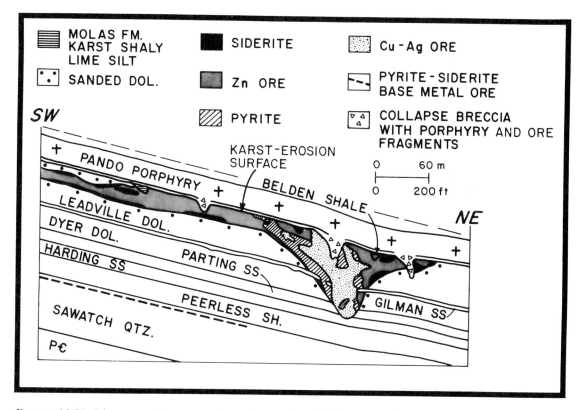

FIGURE 13.25. Diagrammatic cross-section of a portion of Gilman (Eagle) mine (Radabaugh et al. 1968).

1968), at Gilman (Figs. 13.19, 13.25, and 13.26) (Radabaugh et al. 1968, Lovering et al. 1978), in the Aspen district (Figs. 13.14 and 13.16) (Emmons 1887, Spurr 1898, Vanderwilt 1935, Maslyn 1976), at Lenado north of Aspen (Maslyn 1976), at the Star and Doctor mines in Gunnison County (Garrett 1950), to a less apparent extent in the Tincup district (Dings and Robinson 1957), and in hundreds of other prospect pits and dumps throughout central Colorado.

Ore bodies within the upper portion of the Leadville Formation commonly coincide with areas of local or regional solution-erosion thinning of the Leadville strata. Figure 13.13 is a sketch of a cross-section of a typical manto ore body in the upper Leadville strata. Such mineralized features are common at Leadville, Gilman, and Aspen (Emmons 1886, Emmons et al. 1927, Spurr 1898, Radabaugh et al. 1968, Lovering et al. 1978). The manto ore bodies are elongate parallel to bedding (perpendicular to the plane of the sketch of Fig. 13.13) and occur subjacent to a solution-thinned area of the Leadville strata. These manto ore bodies typically occupy karst-solution features that are filled with dolomite breccia, disaggregated dolomite sand, and gray clay, with minor amounts of black chert fragments.

In a more regional and subregional sense, anomalously thin areas of Leadville strata occur to some extent in or adjacent to each of the major ore districts. At Leadville, Tweto (1968),

Emmons (1883), and Emmons et al. (1927) have documented thinning of the Leadville strata to the west within the district, and also the total removal of Leadville strata and even the Dyer Dolomite locally within a major paleovalley developed on the Late Mississippian landscape (Fig. 13.10) (De Voto 1983). The location of this thinning and removal of the Leadville strata is coincident with intense development of ore bodies in the immediately subjacent Dyer Dolomite and is adjacent to intense development of ore bodies in subterranean and surface karst-solution features in the Leadville strata that were tributary to the paleovalley (Figs. 13.10, 13.20, and 13.27) (De Voto 1983 and 1985b). At Gilman, the Leadville strata are regionally thin, as well as locally in the immediate vicinity of some of the ore bodies (Radabaugh et al. 1968, Lovering et al. 1978). The entire 5-km- (3-mi)-long Aspen district occurs immediately adjacent to, and on the south side of, a paleovalley in which the Leadville and underlying Dyer and Parting strata have been removed (Fig. 13.16) (Maslyn 1976). Moreover, in areas of partial removal and thinning of the Leadville strata, extensive ore bodies have been developed in the immediately subjacent remaining Leadville strata, such as at the Smuggler mine (Fig. 13.14) (Spurr 1898, Volin and Hild 1950).

The major paleovalleys on the Late Mississippian landscape in the Leadville, Aspen, and Tincup areas (Figs. 13.6, 13.16, and 13.20) occur adjacent to the few areas in central Colo-

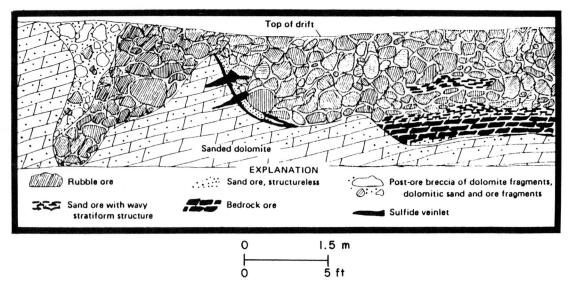

FIGURE 13.26. Mine workings, Manto No. 3, sketch of 15 level, Gilman (Eagle) mine (Lovering et al. 1978).

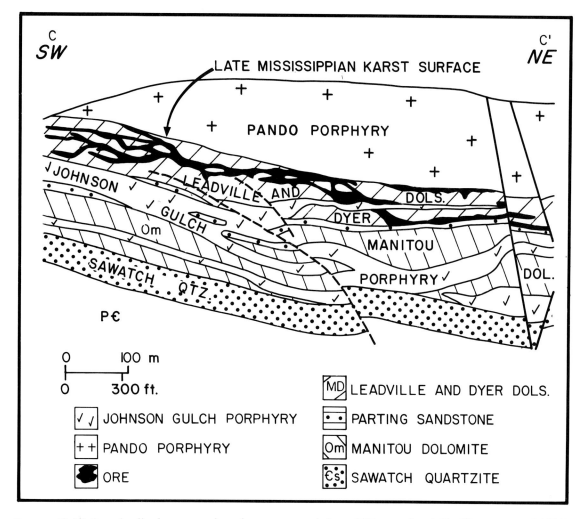

FIGURE 13.27. Longitudinal cross-section along ore trend, Iron Hill area, Leadville (Emmons et al 1927). Location of Iron Hill is shown on Figure 13.10.

rado where stratabound ore bodies have been developed at significantly lower stratagraphic levels than elsewhere (De Voto 1983 and 1985b). At Leadville particularly, ore bodies are extensively developed in the Dyer and Manitou Dolomites, as well as in the Leadville strata, in subterranean karst-solution features (Fig. 13.20) (Emmons 1883 and 1886, Emmons et al. 1927, Tweto 1968). The orientation, location, and development of these karst-solution features which host ore deposits suggest that they were developed progressively as multiple cavern levels as the valley was progressively incised into the Late Mississippian terrain.

Many of the ore bodies in the Lower and Middle Paleozoic carbonate rocks occur commonly semiconcordant to bedding. These features are well displayed in deposits at the Alma district, Sherman Mine area (Fig. 13.11) (Bloom 1965, Tschauder and Landis 1985), Weston Pass, Leadville district, Gilman (Figs. 13.19 and 13.25), Aspen district, in the Tincup, Tomichi, and Monarch districts, and in hundreds of other mines and mineralized occurrences throughout central Colorado. This parallelism to bedding occurs in deposits within the Castle Butte Member, notably at Leadville, Gilman, and, to a lesser extent, Aspen; within the Redcliff Member, notably at the Sherman Mine, Aspen, and Tincup; within the Dyer Dolomite, notably at Leadville and, to a lesser extent, Tincup; within the Fremont Dolomite to a minor extent at Tincup; and within the Manitou Dolomite at Leadville and Monarch (De Voto 1983 and 1985b). The preferred orientation of these deposits parallel to bedding is a function of the control of bedding planes, lithologic differences, and the groundwater table on the development of subterranean karst-solution features.

The ore bodies occur dominantly in textures that are relict of the karst-solution features that they occupy. The best-developed mineralization generally occurs in well-sorted dolomite sand (disaggregated and partly corroded grains of wallrock dolomite) or in rather open-textured, dolomite-breccia cave-fill material (Figs. 13.21 and 13.26). As these textures grade into cave fill with more shale constituents, the extent of mineralization usually decreases. In most deposits, very little ore occurs as replacement of dolomite or limestone that has not been attacked by the karst-solution activity, the so-called bedrock ore of Lovering et al. (1978). As a result of the strong control of the dolomite breccia and sand-filled solution features on mineralization, the ore bodies most commonly have sharp wallrock contacts (Fig. 13.26). However, in the few areas subject to the highest-temperature mineralizing fluids (300°C), such as that proximal to the Breece Hill intrusive stock at Leadville (Thompson et al. 1983) and in the gold-rich, high-temperature mineralization in the Tennessee Pass area (Beaty

et al. 1986), the ore deposits replace the carbonate rocks and are apparently unrelated to karst-solution features.

Many of the individual ore deposits and clusters of deposits have a strong northeast trend. This trend is well displayed by the ore bodies at Mt. Sherman, the Iron Hill area of the Leadville district (Figs. 13.10 and 13.28), and to a certain extent at Aspen (Spurr 1898). The strong northeast trend is a joint-system control of solution channelways within the carbonate strata which have been subsequently mineralized. At the Iron Hill area at Leadville, northeast-trending ore bodies are developed in the Manitou Dolomite directly beneath northeast-trending ore bodies within the stratigraphically higher Leadville Dolomite (Fig. 13.10). At Gilman, northeast-trending manto ore bodies are developed within the upper part of the Castle Butte Member, the upper part of the Redcliff Member, and within the dolomitic Rocky Point zone of the Sawatch Quartzite along the same northeast-trending joint system (Fig. 13.18).

A strong northeast-trending joint system developed in the Late Mississippian, providing high-permeability conduits for groundwater flow. During the solution erosion and devel-

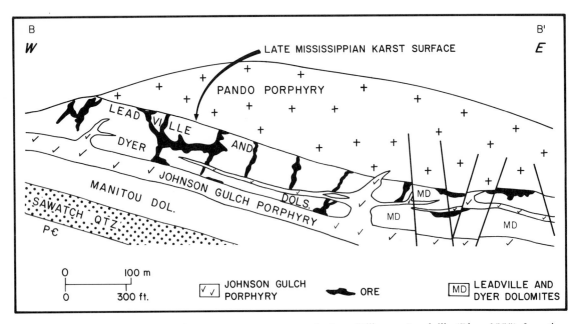

FIGURE 13.28. East–west cross-section across manto ore trends, Iron Hill area, Leadville (Blow 1889). Location of Iron Hill is shown on Figure 13.10.

opment of the karst topography in the Late Mississippian, the joints were solution-enlarged and became steep-walled openings into which dolomite breccia and sand collapsed and were deposited. The northeast-trending manto ore bodies occupy these solution-enlarged joints. Those at the top of the Leadville were open to the ground surface, whereas those in stratigraphically lower units were developed in the subsurface.

The base and precious metal deposits within the Lower and Middle Paleozoic carbonate rocks in central Colorado have been attributed to a Tertiary magmatic hydrothermal origin by most workers (Emmons 1883, 1886, 1889, Spurr 1898, Emmons et al. 1927, Radabaugh et al. 1968, Tweto 1968, Lovering et al. 1978, Thompson et al. 1983 and 1985, Beaty et al. 1986). Recent workers and detailed descriptions of ore deposits and mines somewhat distant from major Tertiary intrusive centers have described lead-zinc-silver-barite deposits in karst-solution features in the Leadville Formation which have an apparent preintrusive origin (De Voto 1983 and 1985b, Tschauder and Landis 1985, Beaty et al. 1985). Tschauder and Landis (1985) have described sphalerite-tetrahedrite-barite mineralization of apparent Late Mississippian age, interbedded with cave sediment at the Sherman Mine. Beaty et al. (1985) have documented an early-stage, presumably Late Paleozoic galena-sphalerite-acanthite-chalcocite-barite (lead-zinc-silver-copper) deposit within a paleokarst cave in the Leadville Formation at Redcliff. Ralph Stegen (pers. comm. 1986) has described two preintrusive episodes of base metal-silver mineralization at the Smuggler Mine at Aspen. The first mineralizing episode was apparently derived from sedimentary basinal fluids. Johansing (1982) has attributed the lead-silver-barite ores at the Sherman Mine to a sedimentary basinal source. De Voto (1983 and 1985b) has suggested that many of the widespread karst-hosted, lead-zinc-silver-barite ore deposits throughout central Colorado were introduced into the karst features during the Pennsylvanian when rifting and basinal subsidence occurred along with an anomalous heat flow (De Voto 1980b, De Voto et al. 1986). Abundant oil staining and overmature petroleum source rocks occur in the Pennsylvanian sediments that overlie the paleokarst erosional surface. Oil staining also occurs coating galena and barite crystals in several mines in the northern Mosquito Range. Thus, the early suite of ore deposits could have been emplaced by metalliferous brines that were expelled from overpressured petroleum source rocks during Pennsylvanian rifting, basin subsidence, and sedimentation (De Voto 1983 and 1985b).

Tertiary magmatic-hydrothermal activity emplaced additional mineral deposits in karst features and high-temperature replacement bodies in the Lower and Middle Paleozoic carbonate rocks adjacent to major intrusive centers (Emmons et al. 1927, Tweto 1968, Thompson et al. 1983). These deposits have generally been of higher-temperature origin (250 to 300°C), high iron content, and higher gold content, and, in these intrusive centers, have obscured evidence of the earlier suite of deposits (Beaty et al. 1986a, Beaty et al. 1986b). In any case, the existence of porous and permeable channelways and openings in highly reactive carbonate rock with a large surface area of cave-fill sediment immediately below relatively impermeable shale and siltstone has provided favorable conditions for hydrothermal fluid flow and ore deposition.

Conclusions

During the Late Devonian and Early Mississippian, central Colorado was an area of shallow subtidal and tidal flat carbonate deposition. The broad, shallow sea covered most of Colorado except for extensive tidal flats that existed on the west and east sides of a low relief emergent area roughly coincident with the north-trending Front Range and Wet Mountains. Eustatic sealevel changes in the Late Devonian and Early Mississippian caused sandstones to be deposited occasionally on the tidal flats (Gilman Sandstone), and caused subaerial erosion to occur within central Colorado on top of the Gilman Sandstone and on top of the Redcliff Member of the Leadville Formation (Fig. 13.2). Subaerial exposure caused the development of a low-relief solution-erosion unconformity and locally extensive solution breccia on top of the Redcliff Member and local subsurface karst-solution breccia bodies within the Redcliff Member. The distribution of sucrosic and coarsely crystalline dolomite within the Redcliff strata and lime-stones of the overlying Castle Butte Member indicates that mixing-zone dolomitization and karstification of the Redcliff Member developed as emergence occurred after Redcliff deposition in the vicinity of the ancestral Sawatch uplift and the ancestral Front Range uplift in central Colorado.

The Castle Butte strata indicate deposition in more open-circulation, high-energy shelf environments across most of central Colorado. Facies patterns indicate the continued presence of the low-relief emergent Front Range–Wet Mountains area and the existence of tectonically induced shoal water areas on the Osagean sea floor in the area of the White River uplift and locally in the northwestern Sawatch uplift.

Substantial sealevel drop in the Meramecian caused emergence of central Colorado and development of an extensive karst plain extending from central Colorado to Canada. As the sea retreated from central Colorado, bedded and nodular chert, sucrosic and coarsely crystalline dolomite, zebra dolomite, and karst-solution erosion occurred sequentially in the Late Mississippian.

The topography of the Late Mississippian landscape was generally a broad plain with low hills and towers, abundant dolines and cutters, and occasional low to moderate relief valleys. Significant uplift of the northern Sawatch Range area occurred, so that the surface and groundwater drainage on the karst landscape flowed downhill from it to the east and west. As paleovalleys were eroded into the karst plain, integrated cave systems with collection sinkholes (dolines), subhorizontal channels, and outlet chimneys formed in the subsurface. As the valleys were incised more deeply, multiple cavern levels developed and the vadose sinkholes deepened and expanded. Paleovalleys up to 200 m (700 ft) deep and sinkholes up to 75 m (250 ft) deep occur in the Sawatch Range area.

A terra rossa regolith developed on the karst plain in southwestern Colorado and in the White River uplift area. A residuum of gray claystone with chert and carbonate breccia occurs discontinuously on top of the Leadville Formation in central Colorado. The underground solution openings (sinkholes, cutters, caverns, channelways, and chimneys) became filled with carbonate breccia and a matrix of disaggregated carbonate sand and silt with a minor amount of black chert breccia and insoluble residuum. The dominance of bedding-parallel breccia bodies and the rare occurrence of vadose cements indicate that most of the preserved underground karst deposits were phreatic in origin.

Karst solution erosion continued in central Colorado from the Meramecian through the end of the Mississippian and into the Early Pennsylvanian. Block-fault tectonism, basinal rifting, deposition of thousands of meters of sediments, and a possible anomalous geothermal gradient occurred in central Colorado in the Pennsylvanian (De Voto et al. 1986). As the northwest-trending Pennsylvanian basin subsided and 3000 to 4000 m (10,000 to 13,000 ft) of sediment accumulated in it, the organic-rich shales in the Belden Formation at the base of the Pennsylvanian section became heated sufficiently to generate hydrocarbons (Nuccio and Schenk 1986). The thermal maturation of the source rocks and generation of hydrocarbons overpressured the fluid pressures so that hydrocarbons and hot waters were expelled from the source rocks. The preintrusive (pre-Laramide) and basinal fluid origin of a widespread, early suite of lead-zinc-silver-barite deposits hosted in the Late Mississippian karst-solution features suggests that the deposits may have been emplaced by metalliferous brines derived from the overpressured Pennsylvanian basinal sediments that migrated along basin-margin and intrabasinal faults during the Pennsylvanian (De Voto 1983). These metalliferous, basinal fluids also enlarged previously formed karst-solution cavities, and created additional disaggregation and "sanding" of carbonate grains (John Hall pers. comm. 1986).

Tertiary magmatic-hydrothermal activity emplaced additional mineral deposits in the earlier karst features adjacent to major intrusive centers. These hydrothermal fluids have also enlarged some of the previously formed karst-solution cavities and caused additional disaggregation of carbonate grains (Beaty et al. 1986a, Thompson et al. 1983).

The porous and permeable channelways and openings of the Late Mississippian karst-solution features in the Leadville Formation beneath relatively impermeable shale and siltstone of the Belden Formation provided suitable local conduits for hydrothermal fluid flow. The highly reactive and large surface area of carbonate cave-fill sediment provided favorable conditions for ore deposition by these different later hydrothermal fluids.

References

Banks, N.G., 1967, Geology and geochemistry of the Leadville Limestone (Mississippian, Colorado) and its diagenetic, supergene, hydrothermal and metamorphic derivatives: Unpubl. Ph.D. thesis, Univ. of California, San Diego, 298 p.

Beaty, D.W., 1985, The oxygen and carbon isotope geochemistry of the Leadville Formation, in De Voto, R.H., ed., Sedimentology, dolomitization, karstification, and mineralization of the Leadville Limestone (Mississippian), central Colorado: Guidebook for Field Trip No. 6, Midyear Meeting, Soc. Economic Paleontologists and Mineralogists, p. 6–71 to 6–78.

Beaty, D.W., Naeser, C.W., and Lynch, W.C., 1986b, Geology and significance of the auriferous manto deposits at Tennessee Pass Colorado (abstr.): Geol. Soc. America Annual Meeting, in press.

Beaty, D.W., Lynch, W.C., and Solomon, G.C., 1986a, Origin of the ore deposits at Gilman, Colorado; oxygen and hydrogen isotopic constraints (abstr.): Geol. Soc. America Annual Meeting, in press.

Beaty, D.W., 1985, Two episodes of sulfide deposition in paleocaves in the Leadville Dolomite at Red Cliff, Colorado, in De Voto, R.H., ed., Sedimentology, dolomitization, karstification, and mineralization of the Leadville Limestone (Mississippian), central Colorado: Guidebook for Field Trip No. 6, Midyear Meeting, Soc. Economic Paleontologists and Mineralogists, p. 6–127 to 6–136.

Bloom, D.N., 1965, Geology of the Horseshoe district and ore deposits of the Hilltop Mine, Park County, Colorado: Unpubl. D.Sc. thesis, Colo. School of Mines, Golden, CO 210 p.

Bloom, D.N., 1961, Devonian and Mississippian stratigraphy of central and northwestern Colorado, in Berg, R.R. and Rold, J.W., eds., Symposium on lower and middle Paleozoic rocks of Colorado: Rocky Mountain Association of Geologists, p. 25–35.

Blow, A.A., 1889, The geology and ore deposits of Iron Hill, Leadville, Colorado: Amer. Inst. Mining and Metal. Engineers, Trans. v. 18, p. 145–181.

Bryant, Bruce, 1971, Geologic map of the Aspen Quadrangle, Pitkin County, Colorado: U.S. Geol. Survey GQ-933.

Campbell, J.A., 1970, Stratigraphy of the Chaffee Group (Upper Devonian), west-central Colorado: Am. Assoc. Petroleum Geologists Bull., v. 54, no. 2, p. 313–325.

Conley, C.D., 1964, Petrology of the Leadville Limestone (Mississippian), White River Plateau, Colorado: Unpubl. Ph.D. thesis, Univ. of Wyoming, Laramie, WY 122 p.

Conley, C.D., 1972, Depositional and diagenetic history of Mississippian Leadville Formation, White River Plateau, Colorado: Colo. School of Mines Quarterly, v. 67, no. 4, p. 102–135.

Craig, L.C., 1972, Mississippian system, in Mallory, W.W., ed., Geologic atlas of the Rocky Mountain region: Rocky Mountain Assoc. Geologists, p. 100–110.

De Voto, R.H., 1980a, Mississippian stratigraphy and history of Colorado, in Kent, H.C., and Porter, K.W., eds., Colorado geology: Rocky Mountain Assoc. Geologists, p. 57–70.

De Voto, R.H., 1980b, Pennsylvanian stratigraphy and history of Colorado, in Kent, H.C., and Porter, K.W., eds., Colorado geology: Rocky Mountain Assoc. Geologists, p. 71–101.

De Voto, R.H., 1983, Central Colorado karst-controlled lead-zinc-silver deposits (Leadville, Gilman, Aspen, and others), a Late Paleozoic Mississippi Valley-type district, in The genesis of Rocky Mountain ore deposits: changes with time and tectonics: Denver Region, Soc. Exploration Geologists, p. 51–70.

De Voto, R.H., 1985a, Mississippian stratigraphy and history of central Colorado, in De Voto, R.H., ed., Sedimentology, dolomitization, karstification, and mineralization of the Leadville Limestone (Mississippian), central Colorado: Guidebook for Field Trip No. 6, Midyear Meeting, Soc. Economic Paleontologists and Mineralogists, p. 6-3 to 6-37.

De Voto, R.H., 1985b, Stratigraphic controls and Late Paleozoic origin of the karst-hosted lead-zinc-silver deposits in central Colorado, in De Voto, R.H., ed., Sedimentology, dolomitization, karstification, and mineralization of the Leadville Limestone (Mississippian), central Colorado: Guidebook for Field Trip No. 6, Midyear Meeting, Soc. Economic Paleontologists and Mineralogists, p. 6-93 to 6-126.

De Voto, R.H., 1985c, Field trip guide: Sedimentology, dolomitization, karstification, and mineralization of the Leadville Limestone (Mississippian), central Colorado, in De Voto, R.H., ed., Guidebook for Field Trip No. 6, Midyear Meeting, Soc. Economic Paleontologists and Mineralogists, p. 6-143 to 6-180.

De Voto, R.H., Bartleson, B.L., Schenk, C.J., and Waechter, N.B., 1986, Late Paleozoic stratigraphy and syndepositional tectonism, northwestern Colorado, in Stone, D.S., ed., New interpretations of northwest Colorado geology: Rocky Mountain Assoc. of Geologists, p. 37–49.

De Voto, R.H., and Maslyn, R.M., 1977, Sedimentology and diagenesis of the Leadville Formation and controls of lead-zinc-silver deposits, central Colorado: The Mountain Geologist, v. 15, no. 1, p. 27–28.

Dings, M.G., and Robinson, C.S., 1957, Geology and ore deposits of the Garfield Quadrangle, Colorado: US Geol. Survey Prof. Paper 289, 110 p.

Dorward, R.A., 1985, Sedimentation and diagenesis of the Devonian Dyer and Mississippian Leadville Formation, central Colorado: Unpubl. M.S. thesis, Colo. School of Mines, Golden, CO 201 p.

Dupree, J.A., 1979, Stratigraphic control of uranium mineralization at the Pitch Mine, Saguache County, Colorado: Unpubl. M.S. thesis, Colo. School of Mines, Golden, CO 111 p.

Emmons, S.F., 1883, Atlas on the geology and mining industry of Leadville, Colorado: U.S. Geol. Survey.

Emmons, S.F., 1886, Geology and mining industry of Leadville, Colorado: U.S. Geol. Survey Mon. 12, 770 p.

Emmons, S.F., 1887, Preliminary notes on Aspen, Colorado: Colo. Scientific Soc. Proc., v. 2, pt. 3, p. 251–277.

Emmons, S.F., Irving, J.D., and Loughlin, G.F., 1927, Geology and ore deposits of the Leadville mining district, Colorado: U.S. Geol. Survey Prof. Paper 148, 368 p.

Garrett, H.L., 1950, The geology of Star Basin and Star Mine, Gunnison County, Colorado: Unpubl. M.S. thesis, Colo. School of Mines, Golden, CO 45 p.

Gutschick, R.C., Sandberg, C.A., and Sando, W.J., 1980, Mississippian shelf margin and carbonate platform from Montana to Nevada, in Fouch, T.D., and Magathan, E.R., eds., Paleozoic paleogeography of the west-central United States, Rocky Mountain section: Soc. Economic Paleontologists and Mineralogists, p. 111–128.

Hill, V.S., 1983, Mississippian Williams Canyon Limestone Member of the Leadville Limestone, south-central Colorado: Unpubl. M.S. thesis, Colo. School of Mines, Golden, CO 112 p.

Horton, R.A., Jr., 1985a, Dolomitization of the Leadville Limestone, in De Voto, R.H., ed., Sedimentology, dolomitization, karstification, and mineralization of the Leadville Limestone (Mississippian), central Colorado: Guidebook for Field Trip No. 6, Midyear Meeting, Soc. Economic Paleontologists and Mineralogists, p. 6-57 to 6-70.

Horton, R.A., 1985b, Dolomitization and diagenesis of the Leadville Limestone (Mississippian), central Colorado: Unpubl. Ph.D. thesis, Colo. School of Mines, Golden, CO 153 p.

Johansing, R.J., 1982, Physical-chemical controls of dolomite-hosted Sherman-type mineralization, Lake and Park Counties, Colorado: Unpubl. M.S. thesis, Colo. State Univ., Ft. Collins, CO, 158 p.

Lovering, T.S., Tweto, O., and Lovering, T.G., 1978, Ore deposits of the Gilman district, Eagle County, Colorado: U.S. Geol. Survey Prof. Paper 1017, 90 p.

Maslyn, R.M., 1976, Late Mississippian paleokarst in the Aspen, Colorado area: Unpubl. M.S. thesis, Colo. School of Mines, Golden, CO, 96 p.

Maslyn, R.M., 1977, Fossil tower karst near Molas Lake, Colorado: The Mountain Geologist, v. 14, no. 1, p. 17–25.

Maughan, E.K., 1963, Mississippian rocks in the Laramie Range, Wyoming and adjacent areas: U.S. Geol. Survey Prof. Paper 457-C, p. C23–C27.

Nadeau, J.E., 1971, The stratigraphy of the Leadville Limestone, central Colorado: Unpubl. Ph.D. thesis, Washington State University, Pullman, WA, 144 p.

Nadeau, J.E., 1972, Mississippian stratigraphy of central Colorado: Colo. School of Mines Quarterly, v. 67, no. 4, p. 77–101.

Nuccio, V.F., and Schenk, C.J., 1986, Thermal maturity and hydrocarbon source-rock potential of the Eagle Basin, northwestern Colorado, in Stone, D.S., ed., New interpretations of northwest Colorado geology: Rocky Mountain Assoc. of Geologists, p. 259–264.

Radabaugh, R.E., Merchant, J.S., and Brown, J.M., 1968, Geology and ore deposits of the Gilman (Red Cliff, Battle Mountain) district, Eagle County, Colorado, in Ridge, J.D., ed., Ore deposits of the United States, 1933–1967: Amer. Inst. Mining, Metal and Petroleum Engineers, Inc., v. 1, p. 641–664.

Ramirez, L.M., 1973, Stratigraphy of the Mississippian System, Las Animas arch, Colorado: Unpubl. M.S. thesis, Colo. School of Mines, Golden, CO, 86 p.

Ramirez, L.M., 1974, Stratigraphy of the Mississippian System, Las Animas arch, Colorado: The Mountain Geologist, v. 11, no. 1, p. 1–32.

Samsela, J.J., 1980, Sedimentology of the Leadville Limestone (Mississippian) and the Chaffee Group (upper Devonian), Chaffee, Fremont' and Saguache Counties, Colorado: Unpubl. M.S. thesis, Colo. School of Mines, Golden, CO 168 p.

Sando, W.J., Gordon, M., Jr., and Dutro, J.T., Jr., 1975, Stratigraphy and geologic history of the Amsden Formation (Mississippian and Pennsylvanian) of Wyoming: U.S. Geol. Survey Prof. Paper 848-I, 83 p.

Spurr, J.E., 1898, Geology of the Aspen mining district, Colorado, with Atlas: U.S. Geol. Survey Mon. 31, 260 p.

Thompson, T.B., Archart, G.B., Johansing, R.J., Osborne, L.W., Jr., and Landis, G.P., 1983, Geology and geochemistry of the Leadville district, Colorado, *in* The genesis of Rocky Mountain ore deposits: changes with time and tectonics: Denver Region Exploration Geologists Soc., p. 101–115.

Thompson, T.B., Beaty, D.W., Naesen, C.W., Cunningham, C.G., 1985, Origin of the ore deposits at Gilman and Leadville, Colorado, *in* De Voto, R.H., ed., Sedimentology, dolomitization, karstification and mineralization of the Leadville Limestone (Mississippian) central Colorado: Guidebook for Field Trip No. 6, Midyear Meeting, Soc. Economic Paleontologists and Mineralogists, p. 6-137 to 6-142.

Tschauder, R.J., and Landis, G.P., 1985, Late Paleozoic karst development and mineralization in central Colorado, *in* De Voto, R.H., ed., Sedimentology, dolomitization, karstification, and mineralization of the Leadville Limestone (Mississippian), central Colorado: Guidebook for Field Trip No. 6, Soc. Economic Paleontologists and Mineralogists, p. 6-79 to 6-91.

Tweto, O., 1968, Leadville district, Colorado, *in* Ridge, J.D., ed., Ore deposits of the United States, 1933–1967: Amer. Inst. Mining, Metal., and Petroleum Engineers, Inc., v. 1, p. 681–705.

Tweto, O., and Lovering, T.S., 1977, Geology of the Minturn 15-minute quadrangle, Eagle and Summit Counties, Colorado: U.S. Geol. Survey Prof. Paper 956, 96 p.

Vanderwilt, J.W., 1935, Revision of structure and stratigraphy of the Aspen district, Colorado, and its bearing on the ore deposits: Econ. Geol, v. 30, p. 233–241.

Volin, M.E., and Hild, J.H., 1950, Investigation of Smuggler Lead-Zinc Mine, Aspen, Pitkin County, Colorado: U.S. Bur. Mines. Rept. of Invest. 4696, 47 p.

Wittstrom, M.D., 1979, Sedimentology of the Leadville Limestone (Mississippian), northeastern Gunnison County, Colorado: Unpubl. M.S. thesis, Colo. School of Mines, Golden, CO, 159 p.

14
Paleokarstic Features in Mississippian Limestones, New Mexico

WILLIAM J. MEYERS

Abstract

Paleokarst features in Mississippian limestones of southern New Mexico occur beneath the sub-Pennsylvanian unconformity over an area greater than 20,000 km². These paleokarst features pervade the upper few meters to few tens of meters of the crinoidal calcarenites of the Lake Valley and Kelley formations, and occur in a vertical profile that shows progressively more intense dissolution upward. This profile consists of, in ascending order: (1) etching of pre-Pennsylvanian freshwater phreatic syntaxial cements; (2) micrite and microspar plugging of intergranular pores, filling around crinoids, bryozoans, and pre-Pennsylvanian freshwater syntaxial cements; (3) clay and detrital quartz silt plugging intergranular pores and filling around etched crinoids and etched pre-Pennsylvanian freshwater syntaxial cements; (4) fragmented host limestone forming rubble-and-fissure fabrics composed of nodules of host rock that are surrounded by anastomosing veinlets and fissures filled with "weathering calcarenite," clay, and detrital quartz silt; and (5) chert breccia composed of chert fragments derived from the underlying Lake Valley/Kelley formations within a matrix of clay, quartz silt, and sand-sized chert fragments.

The paleokarst profile is interpreted as having developed mainly in the vadose zone on crinoidal limestones that were partly cemented but had high intergranular porosity and permeability. The chert breccias are interpreted as insoluble residues from intense weathering of Lake Valley/Kelley formations limestones plus extraformational clay and detrital quartz. The rubble-and-fissure fabrics resulted from dissolution fragmentation, and the intergranular clay and quartz comprise eluviated sediment carried downward by percolating vadose waters. The intergranular micrite and microspar are interpreted as a combination of neomorphic microspar and eluviated carbonate sediment. This eluviated sediment was possibly derived from chemical disintegration of host rock carbonate in the rubble-and-fissure zone and from remobilized primary depositional mud.

Karstification was most intense in the western and northwestern parts of the study area, a regional trend that correlates with intensity of other diagenetic features. These regional trends attest to a chemically more aggressive groundwater system in the west and northwest than in the east during pre-Pennsylvanian freshwater diagenesis and karstification.

The restriction of much of the alteration in the Lake Valley/Kelley formations to intergranular pores and relatively small dissolution cavities within a relatively thin interval (meters to tens of meters), contrasts with the large bedding- and joint-controlled caves and passages extending over thick intervals (hundreds of meters or more) characteristic of conventional extant karst. This difference is due in part to the high intergranular permeabilities and porosities of the Lake Valley/Kelley formations crinoidal calcarenites during karstification. As such, the Lake Valley/Kelley formations may provide a model applicable to karstification of other carbonates during early stages of their diagenetic histories while they still have high intergranular permeabilities.

Introduction

The study and description of karst features have historically been dominated by research on extant karst largely carried out by groundwater hydrologists and geomorphologists. In more recent years, there has been an increasing number of studies of paleokarst by carbonate petrologists (e.g., Walkden 1974, Sando 1974, Kobluk et al. 1977, Chafetz 1982, Wright 1982, Esteban and Klappa 1983, and studies in this volume) in an attempt to characterize the important features of paleokarst and compare them with those of extant karst. These studies are important in understanding paleokarst as a distinctive diagenetic facies (Esteban and Klappa 1983) and in evaluating the importance of paleokarst terranes both for supplying dissolved constituents required for regional cementation and dolomitization and for supplying chemically aggressive groundwaters that can promote chemical compaction and generation of secondary porosity.

The study of the Lake Valley and Kelley formations adds to the portfolio of detailed descriptions of paleokarst terranes, and is perhaps unusual in that many of the karst processes occurred on the intergranular scale, due largely to high intergranular porosities and permeabilities during karstification. The specific purposes of this paper are to (1) describe paleokarst features in the context of a karst profile and within the framework of the cementation and chertification histories; (2) describe the regional and stratigraphic distribution of these features; (3) refine the timing of karstification; (4) present a model of karstification related to late Mississippian/early Pennsylvanian sealevel changes; and (5) briefly compare the key characteristics of paleokarst in the Lake Valley and Kelley formations with those in classic karst.

The subaerial-related features described herein from the Lake Valley/Kelley formations differ from most of those described from modern karst terranes, and therefore in many workers' views may not fit the strict definition of karst. In this study I have adopted the broader definition used by most carbonate petrologists and expressed by Esteban and Klappa (1983): "Karst is a diagenetic facies, an overprint in subaerially exposed carbonate bodies, produced and controlled by dissolution and migration of calcium carbonate in meteoric waters, occurring in a wide variety of climatic and tectonic settings, and generating a recognizable landscape."

Stratigraphy of Mississippian Rocks

The Mississippian limestone within the study area of southern New Mexico (Fig. 14.1) consists of the Caballero Formation (Kinder-

hookian), the Lake Valley and Kelley formations (Osagean), and the Rancheria Formation (Meramecian–Chesterian) (Laudon and Bowsher 1941 and 1949, Pray 1961, Armstrong 1962, Lane 1974 and 1982, Yurewicz 1977). The stratigraphy is interpreted to represent a lower transgressive interval and an upper progradational interval (Meyers 1978). The lowermost interval consists mainly of echinoderm-bryozoan limestone of the Andrecito Member of the Lake Valley Formation overlain by deeper-water lime mudstones of the Alamogordo Member. This in turn is overlain by argillaceous echinoderm-bryozoan calcarenites of the Nunn Member, which represent a transition into progressively shallower-water echinoderm-bryozoan packstones and grainstones of the overlying Tierra Blanca Member. The Tierra Blanca represents progradation and depositional shallowing of totally subtidal shallow-water skeletal sand shelf facies. The Alamagordo, Nunn, and Tierra Blanca Members grade to the echinoderm-bryozoan calcarenites of the Kelley Formation in the northwest part of the study area.

In the San Andres and Sacramento Mountains a younger deep-water lime mudstone-echinoderm packstone couplet, the Arcente and Dona Ana Members, was deposited on top of the Tierra Blanca. The Rancheria Formation occurs only in the San Andres and Sacramento Mountains. It consists of lime mudstones to fine-grained grainstones (Yurewicz 1977) and is separated from the Lake Valley Formation by an erosional unconformity. In the central and northern Sacramento Mountains there is evidence that this unconformity represents a surface of subaerial weathering and erosion (Meyers 1977), and is therefore relevant to timing of karstification, as discussed later.

Mississippian limestones throughout most of the region are overlain by a major sub-Pennsylvanian unconformity that is marked by karstified limestones and karst breccias that are the focus of this study (Fig. 14.1 and 14.2). These karst features occur dominantly in the Lake Valley and Kelley formations and pervade the upper few tens of meters of Mississippian strata throughout the several tens of thousands km^2 of study area (Fig. 14.1). This karstified surface extends well into central western and central Arizona and into northern New Mexico; how-

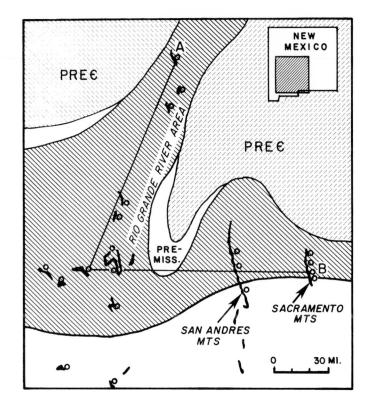

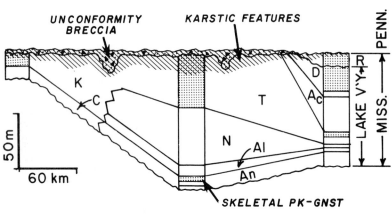

FIGURE 14.1. Sub-Pennsylvanian paleogeologic map showing distribution of paleokarst on top of the Lake Valley and Kelley formations, Mississippian outcrops (black), localities of measured sections used in study (open dots), line of cross-section. Cross-section shows stratigraphic units and shows schemati- cally the distribution of paleokarst. Members of the Lake Valley Formation: An = Andrecito, Al = Ala- mogordo, N = Nunn, T = Tierra Blanca, Ac = Arcente, D = Dona Ana. In the northwest: K = Kelley Formation, C = Caloso Formation. In the east: R = Rancheria Formation.

FIGURE 14.2. Outcrop of sub-Pennsylvanian paleokarst on top of the Lake Valley Formation crinoidal limestones (CL), Sacramento Mountains. F = claystone-filled fissure, R = clasts of host limestone in rubble-and-fissure fabric, CB = chert breccia with silty claystone matrix. Scale marked in feet.

ever, this study reports only on that area, south central New Mexico, that has been studied in detail.

Diagenetic Framework

Relevant to understanding the karst features is the cementation history of the Mississippian limestones, particularly the prekarst cementation. An extensive and detailed cement stratigraphy has been established for the calcite cements within the echinoderm-bryozoan packstones and grainstones of the Lake Valley and Kelley formations (Meyers 1974 and 1978). Throughout the study area, Mississippian limestones have been cemented mainly by syntaxial, inclusion-free calcites on echinoderms which make up about 85% of the total intergranular cements (Fig. 14.3). These inclusion-free, syntaxial cements, which comprise about 26% by volume of the skeletal packstones and grainstones, contain as many as seven compositional zones distinguishable with cathodoluminescence, of which four zones are volumetrically important (Fig. 14.3A) and are mappable over all or most of the study area. The oldest three, zones 1, 2, 3, are nonferroan, and the youngest, zone 5, is generally ferroan. Petrography of the limestones immediately adjacent to sub-Pennsylvanian and sub-Permian unconformities demonstrates that cement zones

1, 2, and 3 are pre-Pennsylvanian, and zone 5 is post-Mississippian and probably pre-Permian (Meyers 1974 and 1978, and this paper.) Based on their petrography and chemistry, inclusion-free zones 1, 2, and 3 are interpreted as freshwater phreatic cements, and zone 5 is interpreted as a burial cement (Meyers 1974 and 1978, Meyers and Lohmann 1985).

Similarly, the petrography of chert-cement relationships and of cherts immediately adjacent to the sub-Pennsylvanian unconformity indicates that most of the chertification preceded the sub-Pennsylvanian erosion surface (Meyers 1977, and this paper).

In summary, approximately 57% of the intergranular cements had been emplaced (Fig. 14.3B), and nearly all chert nodules and masses had formed by the time pre-Pennsylvanian karstification occurred. This indicates that most of the Mississippian skeletal packstones and grainstones had relatively high intergranular porosities during karstification. Considering that these calcarenites had initial intergranular porosities of about 40% (Meyers and Hill 1983), the average intergranular porosity during karstification may have been in the range of 10% to 20%. Uncertainties in timing of intergranular compaction (Meyers and Hill 1983) prevent more definitive determination of prekarst porosities. Considering the coarse grain size of these rocks, this further implies high prekarst permeabilities.

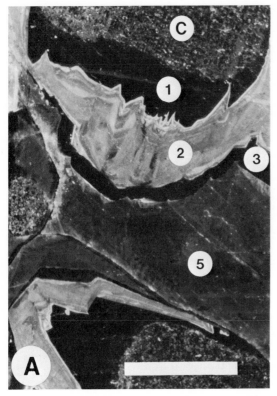

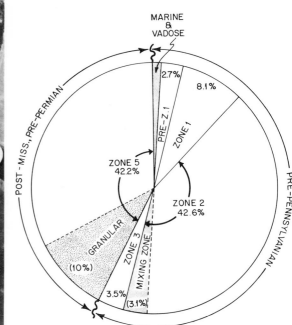

B PROPORTIONS OF MAJOR CEMENTS MAKING UP
   TOTAL INTERGRANULAR CEMENT

FIGURE 14.3.*A*. Cathodoluminescent photomicrograph showing the four major regionally extensive cement zones. C = crinoid. 1, 2, 3, 5 = zones 1, 2, 3, 5. Scale = 0.5 mm. *B*. Pie diagram showing mean percentages of major cements based on point counts of 88 thin sections. Cements comprise inclusion-free syntaxial cements except those marked *marine, mixing zone* (microdolomite-bearing syntaxial zone 2), and *granular* (zone 5 inclusion-free cements). Note that about 55% of the cements had been emplaced before the major pre-Pennsylvanian episode of karstification.

Paleokarst Profile

The paleokarst features in the Lake Valley and Kelley formations tend to occur in a vertical sequence depicted in Fig. 14.4. The ideal complete sequence comprises, from bottom to top, (1) chemical etching of prezone 5 calcite cements; (2) an interval in which micrite and microspar fill partly cemented intergranular pores; (3) a zone of clay, iron oxides, and quartz silt filling partly cemented intergranular pores, often with small solution pockets; (4) a fissure-and-rubble zone; (5) a chert breccia zone. This profile becomes less well developed toward the south, and in general it is poorly developed and incomplete where the Pennsylvanian rests on the Rancheria Formation. In addition, the profile is poorly developed or absent in about one

fourth of the measured sections where Pennsylvanian overlies the Lake Valley and Kelley formations (Figs. 14.5 and 14.6). The following sections describe the key features of the karst profile in the above order.

Etching of Prezone 5 Calcite Cements

Sharp and irregular contacts between zone 5 and pre-5 cements are common throughout the region, and have been interpreted as chemical etching before precipitation of zone 5 cements (Fig. 14.7A). Presumed etching at interzonal contacts has been recognized between most of the Lake Valley/Kelley cements; however, it is most common between zone 5 and older cements. Specifically, greater than 50% of the contacts between zone 5 and prezone 5 cements

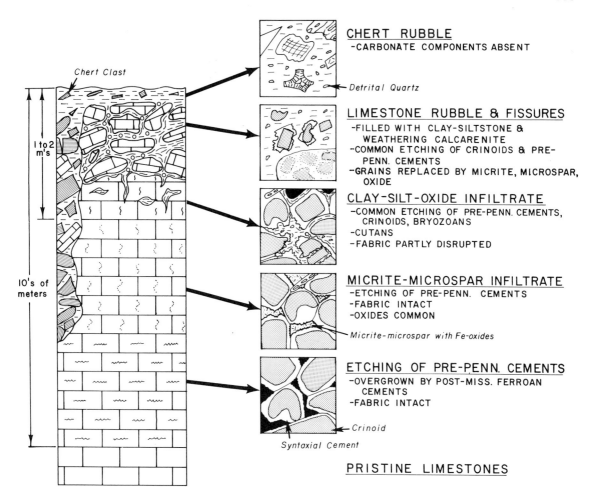

FIGURE 14.4. Summary diagram of paleokarst profile developed on the crinoidal calcarenites of the Lake Valley and Kelley formations.

show evidence of etching, whereas about 30% to 35% of the contacts between pre-5 zones show evidence of etching (Meyers 1978). In addition, etching between zone 5 and older cements is confined mainly to the upper part of the formations (Fig. 14.5), and the most intense etching of cements occurs in the upper parts of the karst profile associated with the clay–oxide interstitial fills and with the rubble-and-fissure zones.

Intergranular Micrite and Microspar

Intergranular pores of crinoidal grain-supported rocks in the paleokarst profile are often filled with micrite or microspar, which fills around calcite cement zones 1, 2, and 3 (Figs. 14.7B, 7C, 7D, 14.8A), but not around zone 5.

Two types of interstitial micrite/microspar can be distinguished. The *first* type is a microspar, often containing iron oxide. This microspar contains poorly preserved bryozoan fragments (Fig. 14.7D), and semicircular reentrants in syntaxial cements (Fig. 14.8A). The *second* type fills around bryozoans and prezone 5 cements with little or no etching or replacement of either bryozoans or cements (Fig. 14.7B, C). This second type of fill rarely is in geopetal arrangement within interstices (Fig. 14.7C), in which case it is overlain by zone 5 cement. The contrasting petrographic characteristics of these two types indicates that they formed by two different processes.

The first type, comprising microspar (often impregnated with oxides) is interpreted as a replacement of bryozoans and possibly of dep-

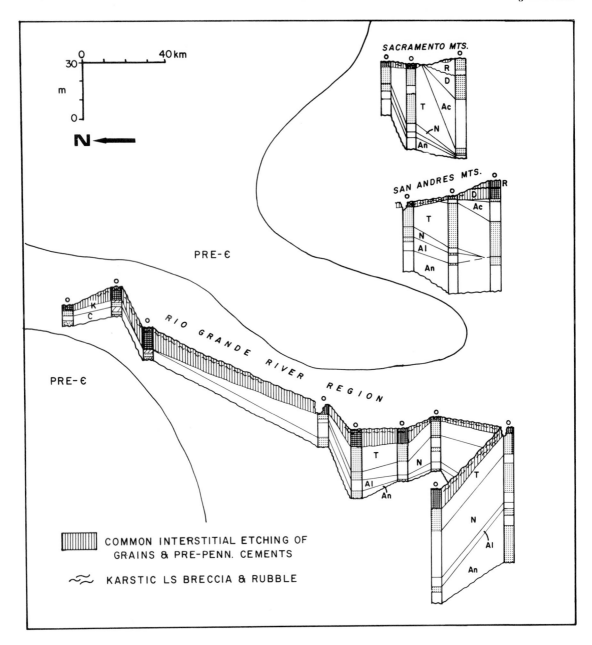

FIGURE 14.5. Fence diagram showing regional and stratigraphic distribution of etching of pre-Pennsylvanian cements and of rubble-and-fissure fabrics. Note that north is to the left. Letters denote members of the Lake Valley Formation (see Fig. 14.1).

ositional lime mud. Evidence for this is the fact that the encased bryozoans show a range of degrees of preservation (Fig. 14.7D), in contrast to nonkarst facies in which bryozoans are virtually always well preserved (e.g., Meyers 1978, Fig. 4A). A second type of evidence is the circular reentrants in associated syntaxial cements (Fig. 14.8A). These reentrants are the size and shape of bryozoan fragments in adjacent rocks, and are thus interpreted as replaced bryozoans. These microspars are also accompanied by minor etching of encased calcite cements and crinoids (Fig. 14.7A).

The second type of micrite and microspar fill (Fig. 14.7B,C) contains well-preserved bryozoans, and this condition plus the rare geo-

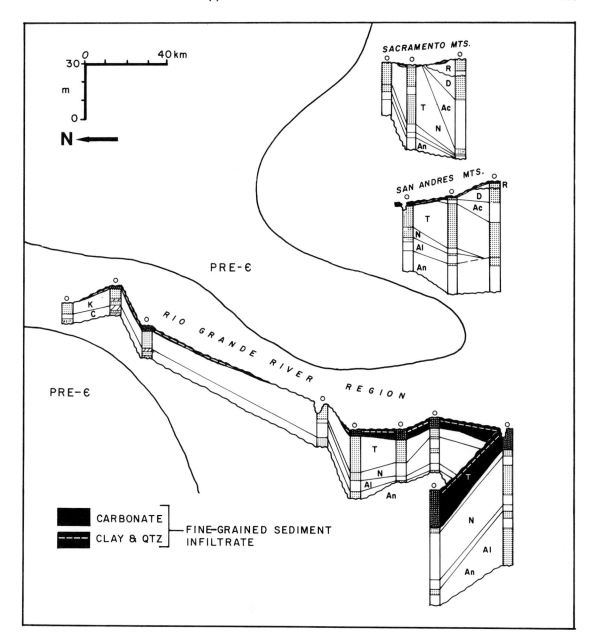

FIGURE 14.6. Fence diagram showing regional and stratigraphic distribution of intergranular karst-related micrite/microspar and of eluviated clay. North is to the left. Letters denote members of the Lake Valley Formation (see Fig. 14.1).

petals indicate that this is a mechanically infiltered sediment. This sediment is unlikely to be original marine sediment deposited with the skeletal grains because it overlies prezone 5 syntaxial cements as young as zone 3. These cements have been previously interpreted as freshwater phreatic calcites, on the basis of a variety of petrographic and geochemical ar-

guments (Meyers 1978, Meyers and Lohmann 1985). Furthermore, there is no evidence of subaerial exposure within the Lake Valley/Kelley formations interval, which makes it highly unlikely that there was a period of freshwater cementation followed by marine mud infiltration that preceded the pre-Pennsylvanian unconformity. In addition, if these infiltrates were

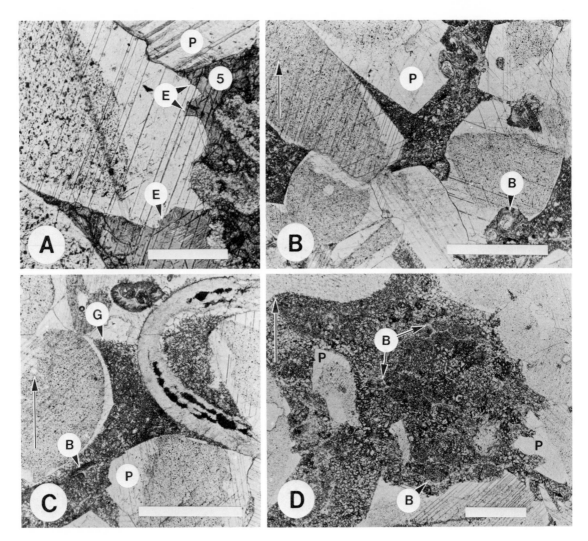

FIGURE 14.7A–D. A. Sharp and irregular contact between zone 5 (5) and prezone 5 cements (P) interpreted as microetched contact (E). Zone 5 is stained with potassium ferricyanide. Up direction is to the left. Scale = 0.25 mm. B. Micrite/microspar filling intergranular pores and filling around pre-Pennsylvanian cements (P) composed of zones 1, 2, 3. Note well-preserved bryozoans (B). Carbonate mud is interpreted as eluviated sediment. Scale = 1 mm. C. Micrite/microspar filling around zones 1, 2, 3, (P). Note geopetal structure (G) and fragment of well-preserved bryozoan (B). Carbonate mud is interpreted as eluviated sediment. Scale = 1 mm. D. Intergranular microspar filling around pre-Pennsylvanian cements (P) and containing poorly preserved bryozoans (B). Microspar is interpreted as a neomorphic fabric that replaced bryozoans and probably primary lime mud. Scale = 0.5 mm.

marine, they should be preferentially associated with known marine cements (microdolomite-rich radiaxial and columnar cements) that occur sporadically in the upper Lake Valley/Kelley formations (Meyers 1978), but they do not occur in these rocks.

Ruling out a primary marine origin, these infiltrates are interpreted to be eluviated carbonate sediments moved downward within the karst profile by percolating vadose water, an interpretation supported by their restriction to the upper part of the Lake Valley/Kelley formations (Fig. 14.5). The origin and precise processes of transport are unknown, but the infiltrates may be analogous to the vadose silt of Dunham (1969), or to the "moon milk" of

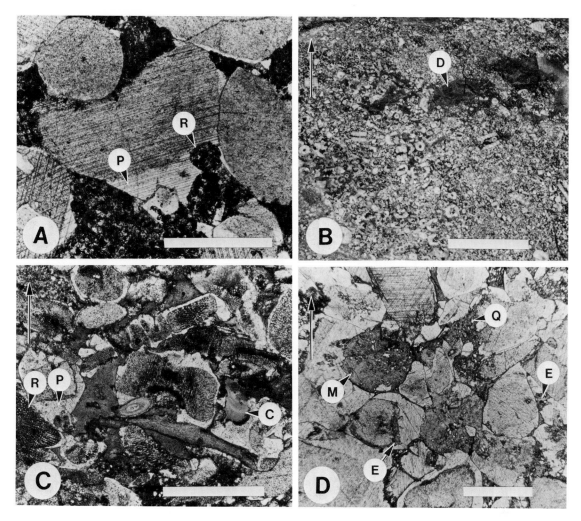

FIGURE 14.8*A–D*. *A*. Intergranular microspar with common Fe oxide (opaque) filling intergranular pores and around pre-Pennsylvanian cements (P). Note reentrant in syntaxial cement (R), interpreted as replaced or leached bryozoan fragment. Microspar is interpreted as a neomorphic fabric. Scale = 1 mm. *B*. Outcrop surface of crinoidal packstone from the Lake Valley Formation showing plugging of intergranular pores and small dissolution pockets (D) filled by dark silty claystone. Scale = 5 cm. *C*. Clay filling intergranular pores and filling around pre-Pennsylvanian cements (P). Note the meniscus-shaped clay cutan structure (C), and partial replacement of crinoid stereom by clay (R). Clay and silt are interpreted as eluviated sediment. Scale = 1 mm. *D*. Clay filling intergranular pores and filling around pre-Pennsylvanian cements. Note detrital quartz silt (Q) in clay, wholesale leaching of some crinoids and filling of molds with clay (M), etching of crinoids intercrystalline boundaries and crinoid cement boundary (E). Scale = 1 mm.

Esteban and Klappa (1983), having formed as microcrystalline precipitates of calcite or by disintegration of host-rock carbonate. The disintegration of host-rock carbonate to micrite and microspar may have occurred at the weathering surface or within the upper part of the karst profile. There is direct evidence of chemical disintegration of crinoids within the karst profile, which resulted in their conversion to delicate chalkified grains (Meyers 1980), and to microspar (Fig. 14.11C). It is also quite possible that the dissolution resulting in the veins and fissures in the rubble-and-fissure fabrics produced micrite and microspar sediment. In addition to this "diagenetic sediment," poorly lithified (or chalkified) primary marine lime muds may have been mechanically remobilized during vadose eluviation.

In any case, the eluviation must have been effected by vigorous vadose flow as indicated by the occurrence of infiltrate carbonate sediment to depths of 10 m below the pre-Pennsylvanian unconformity (Fig. 14.5). Furthermore, there are no known paleosinkholes or solution-enlarged fissures and cavities associated with these areas of deep eluviation, suggesting that the eluviation took place mainly within intergranular pores.

Interstitial Clay, Iron Oxide, and Quartz Silt

One of the most distinctive petrographic features of the paleokarst interval is the plugging of interstices by clay, often containing detrital quartz silt and iron oxides (Fig. 14.8B,C). The clay commonly fills the intergranular pores, surrounding grains and pre-Pennsylvanian syntaxial cements (zones 1, 2, 3) (Fig. 14.8C). In addition, the clay often fills etch pits within syntaxial cements and crinoids and fills solution-enlarged intercrystalline and intergranular contacts (Fig. 14.8D). Commonly, the clay fills solution-modified stereoms of crinoids (Fig. 14.8C) and solution-enlarged boundaries between crinoid and its syntaxial overgrowth (Fig. 14.8D). Bryozoans within these zones are typically poorly preserved, being partly replaced by iron oxides and microspar. In some cases, the interstices are partly filled by a combination of clay, iron oxide, quartz silt, and calcisilt, in which the clay and silt may comprise geopetal fabrics (Fig. 14.9A). Other features seen in the clays are meniscus morphologies (Figs. 14.8C, 14.9A), which are interpreted as clay cutans.

Within this interval of clay plugging are common dissolution cavities. These cavities range in size from slightly larger than intergranular pores to tens of centimeters, but typically are a few centimeters across (Fig. 14.8B). The margins of some of the cavities typically are irregular, wrapping around grains, but in some cases cavity margins truncate bryozoans, crinoids, and, more rarely, primary lime mud. Most cavities are filled with clay, oxide, and quartz silt, often with irregular laminations;

however, some of the smaller cavities are filled with iron oxides and zone 5 calcite.

The depositional fabrics of most of the host rocks are intact, i.e., the grains have not been physically moved apart (Fig. 14.8C,D). However, adjacent to the margins of the larger solution cavities, and in the transition areas to the rubble-and-fissure zone, the host rock contains disrupted fabrics. Specifically, in these rocks, the crinoids, with their syntaxial overgrowths, show a range of disruption from minor rotation and separation (Fig. 14.9B) to complete separation of crinoid from adjacent syntaxial cement (Fig. 14.9C). Within these rocks, the disrupted fabrics may be patchily distributed among areas of intact fabrics.

The clay–quartz silt fills described above are interpreted as eluviated sediment transported from the sub-Pennsylvanian unconformity by percolating vadose waters. The absence of detrital quartz silt in nonkarstified host-rock calcarenites, and the restriction of these features to the upper meter or two of calcarenite below the pre-Pennsylvanian unconformity indicate that the clays and quartz silts derived from the unconformity. This interpretation is also supported by the abundant clay and quartz silt in the rubble-and-fissure zone and in the overlying chert breccia. Furthermore, the clay cutans and iron oxides argue for vadose zone environments.

In summary, the intergranular micrite and microspar described in the previous section, the etching and solution pockets, and the clay, detrital quartz silt, and iron oxides are all interpreted as vadose zone features. Accepting this interpretation, it is not clear why the clay did not penetrate as deeply into the Mississippian calcarenites as did the eluviated micrite and microspar. One possibility is that at any locality, the eluviated carbonate sediment plugged much of the intergranular pore space before the clay-quartz sediment was transported to the locality. This assumes that the clay and detrital quartz were transported rather than being in-place insoluble residues. In this scenario, the eluviated carbonate sediment plugged interstices and thus prevented deep eluviation of clay and quartz silt. It is also possible that the water table was nearer the unconformity (thinner vadose zone) during clay–quartz eluviation than during carbonate eluviation.

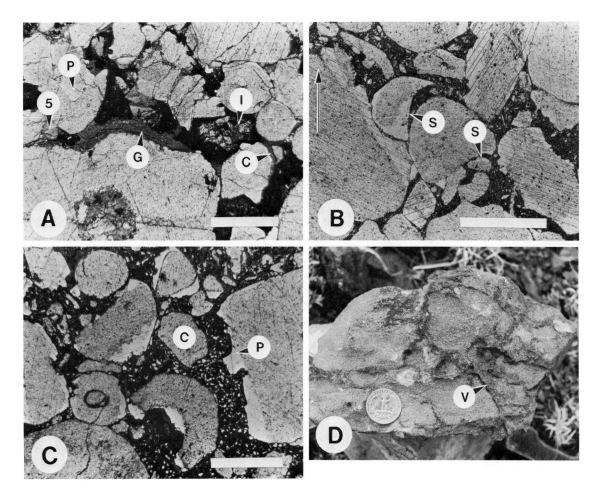

FIGURE 14.9A–D. A. Intergranular pores partly filled with clay that forms geopetal structure (G) and cutan structure (C). Opaque = iron oxides, P = Pre-Pennsylvanian cements. 5 = zone 5 calcite stained with potassium ferricyanide. Zone 5 calcite replaced oxides and clays as indicated by inclusions of both (I) dispersed within calcite 5 crystals. Scale = 1 mm. B. Grain fabric showing minor physical rotation and separation from neighboring grain (S). Cements are pre-Pennsylvanian. Scale = 1 mm. C. Dispersed grain fabric showing complete separation of crinoids (C) from its adjacent pre-Pennsylvanian cement (P). Matrix consists of silty oxide-rich claystone. Scale = 1 mm. D. Outcrop photograph of rubble-and-fissure fabric composed of nodules of host limestone surrounded by anastomosing veinlets of clay (V). Quarter for scale.

Rubble-and-Fissure Zone

The rubble-and-fissure interval comprises the top and most intensely karstified portion of the remaining skeletal limestone bedrock, the overlying chert breccia being composed of noncarbonate constituents. The rubble consists of nodules of host-rock skeletal calcarenites, separated by anastomosing veins and fissures (Figs. 14.2, 14.9D, 14.10A). These nodules comprise skeletal wackestones to grainstones and consist mainly of rounded clasts of limestone and minor, more angular clasts of chert (Figs. 14.9D, 14.10B). These nodules range in size from less than a centimeter to several tens of centimeters and generally have irregular margins that wrap around grains (Fig. 14.10B, C). In some cases the limestone nodules are more or less in place as implied by their intact bedding orientations (Fig. 14.9D). In others, bedding in nodules is at various orientations, and nodules are of mixed lithologies, including

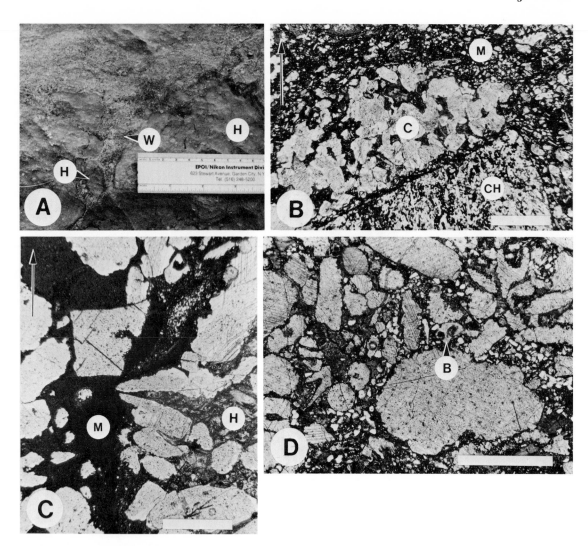

FIGURE 14.10A–D. A. Outcrop photograph of rubble-and-fissure zone showing fissures filled with "weathering calcarenite" (W). H = host limestone as clast in fissure and as in-place wallrock. Part of 6-inch rule is scale. B. Clast of calcarenite (C) and chert (CH) surrounded by matrix of silty claystone containing crinoids (M). Boundaries of limestone clast wrap around skeletal grains. Scale = 1 mm. C. Edge of small vertical fissure filled with oxide-rich silty claystone (M) containing crinoid grains. Note that edge of fissure wraps around grains in clast of host limestone (H). Scale = 1 mm. D. Crinoidal packstone that fills fissures and areas around clasts of host limestone. Crinoidal calcarenite is interpreted to have been derived by weathering disintegration of host rock. Note general absence of bryozoan fragments except those that have been silicified (B). Serrated and irregular margins of crinoids are interpreted as etching of grains. Matrix is silty claystone. Scale = 1 mm.

chert clasts. This latter situation implies that nodules were probably transported at least short distances.

The veins separating the nodules are composed of clay, detrital quartz silt, iron oxide, and crinoidal calcarenite (Figs. 14.9D, 14.10A). Although the most common occurrences are small anastomosing veinlets (Fig. 14.9D), some larger fissures extend below the rubbly zone as much as several meters into the host calcarenite and are filled with clay (Fig. 14.2), limestone clasts, and calcarenite (Fig. 14.10A). The material filling veinlets and separating nodules and clasts contains variable amounts of clay, quartz silt, and oxide and crinoid grains. One of the most common textures of this fill is crinoidal packstone, with clay, detrital quartz silt, and iron oxide matrix (Fig. 14.10D). The texture of this clayey fill, however, ranges to crinoid-dominated wackestone with silty claystone matrix (Fig. 14.11A), to silty claystone with rare crinoids (Fig. 11B). In all these types, bryozoans are largely absent except where they have been silicified (Fig. 14.10D), and crinoids are often severely etched (Fig. 14.10D).

The rubble-and-fissure fabrics just described are interpreted to have resulted from weathering and dissolution of bedrock calcarenites in the upper vadose zone, probably in a subsoil setting. The precise processes and timing of vein and fissure formation and their filling are not clear, but the dissolution and filling processes were probably closely related and may have occurred at the same time. Had the veinlets, fissures, and dissolution cavities (discussed previously) been passively filled well after their formation, there would likely have been more horizontal layering of the claystone and calcarenite fills, and more geopetal fabrics. These are rare in these former cavities.

The rubble may be comparable to that described by Wright (1982) from the Carboniferous Oolite Group of southern Wales, where the nodules are interpreted as soil zone features

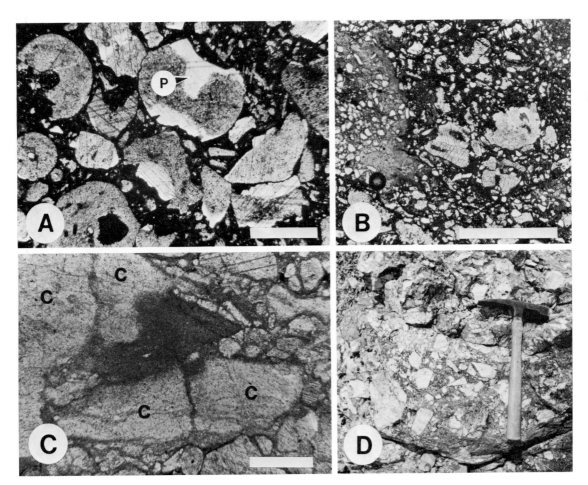

FIGURE 14.11A–D. A. Weathering calcarenite composed of crinoids, many with pre-Pennsylvanian cement overgrowths (P) floating in matrix of oxide-rich silty claystone. Scale = 1 mm. B. Quartz-silty claystone fill of fissures containing relatively few floating crinoids. Scale = 1 mm. C. Crinoids partially converted to microspar around edges and within patches and veinlets within grains. Portions marked C all belong to the same large crinoid. Scale = 1 mm. D. Outcrop photograph of chert breccia composed of angular fragments of Lake Valley/Kelley cherts in matrix of clayey siltstone.

formed on a partly lithified ooid grainstone. These types of features appear to be rare in modern karst terranes, judging by the literature. One possible exception is the nodular structures, reported by Ireland (1979), formed on poorly lithified limestones from Puerto Rico. In any case, the processes are poorly understood, and the nodules are not comparable to those described in calcrete profiles, in that the Lake Valley nodules comprise host-rock lithologies rather than microcrystalline carbonate typical of calcrete glaebules (Esteban and Klappa 1983).

The crinoidal calcarenite filling fissures and veinlets is interpreted to have derived from the weathering and disintegration of the Lake Valley/Kelley formations host rock, rather than from marine sedimentation. This is evidenced by the common presence of freshwater calcite cements of zones 1, 2, and 3 on crinoids surrounded by clay and detrital quartz. These are interpreted as grains that have been cemented, then reworked during weathering. In addition, the rubble-and-fissure zone is directly overlain by chert breccia (in some areas with paleosinkholes) in most areas, with no intervening marine limestone. The few rare exceptions to this are in the central and northern Sacramento Mountains, where the Lake Valley Formation is overlain by small patches, a few feet thick, of fine-grained skeletal calcarenite with chert fragments at its base (Meyers 1977). These calcarenite patches are interpreted as remnants of the Rancheria Formation. Rubble-and-fissure features are absent beneath these outliers of Rancheria, indicating that these features were probably formed after the Rancheria and during pre-Pennsylvanian weathering.

The weathering origin for the "weathering calcarenites" is further supported by the etching of crinoids and cements (zones 1, 2, 3), and by the absence of bryozoans. In addition, this weathering calcarenite has modern analogs in the loose crinoids typically found on crinoidal limestones beneath modern soils.

Other Microscopic Features of Karstified Limestones

A number of other replacement and leaching features that are restricted to the paleokarst profile occur in the Lake Valley/Kelley formations. Specifically, crinoids in nodules and host rock are often altered to microspar (Fig. 14.11C) (Meyers 1980, Fig. 3B) or, more rarely, are pseudomorphically replaced by iron oxide. The alteration to microspar generally preferentially affects margins of crinoids but also occurs as veinlets and patches within crinoids (Fig. 14.11C). The conversion to microspar probably involves internal etching or "chalkification" of the grains, thus forming a fragile "chalky" grain. The evidence of this is that the microspar crinoids often show intense mechanical compaction (Fig. 14.11C) (Meyers 1980, Fig. 3B). Some brachiopods were probably also "chalkified" near the unconformity, as indicated by their susceptibility to compactive fragmentation (Meyers 1980, Fig. 3C).

The chalkification of crinoids has not been reported from other karst facies, but similar alteration fabrics have been reported in crinoids from Cretaceous chalks, where they have been attributed to intragranular etching during early burial diagenesis (Neugebauer 1978). In addition, intercrystal etching within modern carbonate skeletal grains occurs in modern high latitude seawater to form "chalky" grains (Alexandersson 1972). It is also possible that the chalkification of the Lake Valley/Kelley crinoids may be analogous to some of the intragranular dissolution that presumably comprises part of the micritization of grains in calcretes (e.g., James 1972, Coniglio and Harrison 1983, Kahle 1977). However, in the latter case dissolution is accompanied by precipitation so that the grain remains internally well lithified.

In addition to the alteration of crinoids, primary lime mud in the Lake Valley/Kelley paleokarst is commonly replaced by iron oxides and more rarely occurs on tops of pores that are filled with zone 5 cement. This latter situation is interpreted as intergranular dissolution of some lime mud before emplacement of zone 5 cement.

Another common feature in the paleokarst profile is the occurrence of zone 5 calcite crystals surrounded by iron oxide and clay, and often containing inclusions of iron oxide, clay, and more rarely quartz silt (Fig. 14.9A) (Meyers 1978, Fig. 6A). This is interpreted as zone 5 calcites having replaced iron oxides and clay after burial of the paleokarst surface by Pennsylvanian strata. This interpretation is supported by the presence of zone 5 calcites filling fractures that cut grains and karst-profile clays, iron oxides, and host-rock clasts, and by calcite 5 occurring as a cement in basal Pennsylvanian sandstones (Meyers 1978).

Chert Breccias

Chert breccia is widespread in the study area, occurring as a distinctive bed up to several meters thick separating the Lake Valley/Kelley formations from the overlying Pennsylvanian strata (Figs. 14.2, 14.11D). More rarely, the chert breccia occurs as a massive fill of paleo-sinkholes up to 10 m deep. The chert breccia is present in about three-fourths of the measured section where Pennsylvanian rocks directly overlie the Lake Valley/Kelley limestones. It is absent as a distinct stratum between the Lake Valley and Rancheria formations, but scattered chert clasts derived from the Lake Valley Formation occur in the basal Rancheria Formation calcarenites in about one-third of the measured section examined (Meyers 1974 and 1977). The chert breccia crops out as a massive or poorly bedded unit (Fig. 14.11D) (Meyers 1977, Fig. 12B), or occurs as a zone of loose varicolored chert rubble composed of angular fragments of white, red, and green chert, within the modern weathering zone. The chert clasts

range from sand-sized to about one-half meter across and are angular to rounded.

The chert clasts often contain former molds of crinoids now filled with clay, detrital quartz silt, small chert fragments, and ferroan calcite (Fig. 14.12A). In some cases the clay occurs as multiple laminar coatings on the walls of the molds or as apparent geopetal fabrics. The laminar clay coatings are interpreted as clay cutans (Brewer 1964). In other chert clasts the crinoid molds are brecciated and filled with clay containing angular fragments of replacement chert and chert cement (chalcedony and megaquartz) (Meyers 1977, Fig. 14C). In addition, in some outcrops chert clasts are cut by fractures filled with basal Pennsylvanian sandstone (Meyers 1977, Fig. 14A). Typically, there are no limestone clasts associated with the chert fragments, except in some cases in the transition to the rubble-and-fissure zone.

The matrix of the chert clasts consists of green to red claystone containing detrital quartz silt and sand (Fig. 14.2), sand- and granule-sized chert fragments, and iron oxides. This matrix contains no carbonate grains; instead, the sand- and granule-sized material is dominated by silicified portions of crinoids, shells, fragments of chalcedony and megaquartz chert cements (Fig. 14.12B), and chertified skeletal packstones and wackestones.

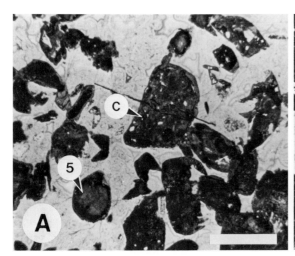
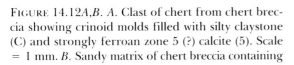
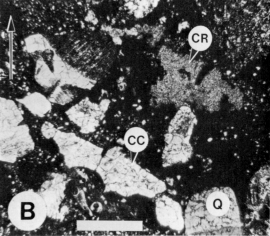

FIGURE 14.12A,B. A. Clast of chert from chert breccia showing crinoid molds filled with silty claystone (C) and strongly ferroan zone 5 (?) calcite (5). Scale = 1 mm. B. Sandy matrix of chert breccia containing sand-sized grains of chalcedony cement (CC), silicified portions of crinoids (CR), and detrital quartz (Q) within a matrix of oxide-rich silty claystone. Scale = 1 mm.

In addition to the chert breccias, chert clasts occur at the base of the Rancheria in a few localities in the Sacramento Mountains. Evidence for these being clasts rather than in situ chert nodules is the fact that they are rotated from horizontal (Meyers 1973), and some are fractured, with fractures filled with Rancheria sediment (Meyers 1977, Fig. 13A). In addition, margins of some clasts are scalloped, the scallops representing former crinoid grain boundaries and now filled with Rancheria siltstone (Meyers 1977, Fig. 13C). Most clasts are chertified crinoidal packstones distinctly coarser than the host-rock Rancheria, which comprises fine-grained quartz sandstones and siltstones and fine-grained calcarenites. Another feature seen in some of these clasts is chalcedony pseudomorphed by clay containing detrital quartz silt (Meyers 1977, Fig. 13B). The layering of the clay resembles clay cutans (Brewer 1964), implying a soil zone feature. These data indicate that the clasts were derived from the underlying Lake Valley Formation and that they were chert when incorporated into the Rancheria Formation.

To summarize, the features described above from the breccia cherts and from the Rancheria chert clasts are never seen in the in situ chert nodules and masses in the main body of the Lake Valley and Kelley formations. The chert breccias are interpreted as the insoluble residue of intense subaerial weathering of the Lake Valley limestones, which was mixed with allochthonous quartz silt and clay and perhaps transported modest distances. Furthermore, the described features from basal Rancheria chert clasts indicate that the subaerial weathering of the Lake Valley Formation began before the Rancheria sediments were deposited.

Distribution of Paleokarst and Correlation with Other Diagenetic Trends

As indicated previously, the paleokarst features are widespread, and their distributions are summarized on Figs. 14.5 and 14.6. The etching of pre-Pennsylvanian cements occurs nearly everywhere and extends to depths of about 20 m below the sub-Pennsylvanian unconformity

(Fig. 14.5). The microspar and micrite plugging of interstices is also widespread but not as ubiquitous as the etching (Fig. 14.6). The micrite and microspar in most of the area extends to shallower depths than the etching, except in the southern Rio Grande valley region where they penetrate the Lake Valley to 40 m below the sub-Pennsylvanian unconformity (Fig. 14.6). In these areas where the interstitial carbonate extends well below the unconformity, the micrite and microspar occur sporadically, i.e., they do not occur in all samples throughout the interval of penetration indicated on Fig. 14.6. The clay–quartz silt infiltrate (Fig. 14.6) is as widespread geographically as the other features but is restricted to the upper 0.5 to 3.0 m beneath the sub-Pennsylvanian unconformity. Similarly, the rubble-and-fissure zone approximately coincides with the clay-infiltrate zone (Fig. 14.5).

The chert breccia occurs throughout most of the area, being present in about four-fifths of the sections where Pennsylvanian overlies the Lake Valley and Kelley formations. Its precise thickness is often difficult to determine due to poor exposures of its upper contact, but it probably ranges from a fraction of a meter to 3 m, except in paleosinkholes where it gets up to 10 m thick.

Regionally, the paleokarst features are most common and most extensively developed in the north and the west, and are generally absent between the Pennsylvanian strata and the Rancheria Formation and between the Rancheria and Lake Valley formations. Rarely, rubble-and-fissure and clay-infiltrate fabrics are developed on the top of the Rancheria Formation in the southern Sacramento Mountains, and rarely (as discussed above), weathered chert clasts occur in basal Rancheria strata. Furthermore, the chert breccias die out southward in the Sacramento Mountains (Meyers 1973), where no paleosinkholes have been recognized. In contrast, the paleokarst features, including sinkholes, occur more extensively in the Rio Grande Valley region, where they are more common, and extend deeper into the upper Lake Valley/Kelley limestones.

This regional variation in intensity of pre-Pennsylvanian karstification is paralleled by regional variations in other diagenetic features. Specifically, the Lake Valley/Kelley skeletal grainstones and packstones from the Rio Grande Valley region show distinctly greater intergranular chemical compaction, and crinoids contain significantly less Mg (now microdolomite inclusions) compared to the San Andres Mountains and Sacramento Mountains rocks (Fig. 14.13A) (Meyers and Hill 1983, Leutloff and Meyers 1984). Furthermore, crinoids from the Rio Grande Valley region have distinctly "lighter" $\delta^{18}O$ values (Fig. 14.13B).

These regional patterns attest to a chemically more aggressive groundwater system in the Rio Grande Valley region (Meyers and Hill 1983, Leutloff and Meyers 1984). In this scenario, the Rio Grande Valley region was flushed more frequently and with chemically more aggressive groundwaters than areas to the east; this resulted in greater alteration of marine chemistries (Mg and ^{18}O loss), and greater chemical etching and/or less freshwater cementation of calcarenites in the west. The greater karstification in the Rio Grande region thus may have been due to this preconditioning of the Lake Valley/Kelley sediments, or it may be due to the inheritance of a topography which favored more intense diagenesis in the west before karstification.

Timing of Karstification

The foregoing discussions indicate that weathering of the Lake Valley and Kelley formations began during a pre-Rancheria relative sealevel drop and resultant subaerial exposure (Fig. 14.14A). Evidence for this is the presence of Lake Valley Formation chert nodules in the base of the Rancheria that contain leached chalcedony cements, which were pseudomorphed by clay cutans. In addition, some of cement zone 1 may have predated the Rancheria (Meyers 1974). It is probable that most of the study area north of the present wedge-edge of the Rancheria was subaerially exposed at this time, resulting in the initiation of freshwater cementation (Fig. 14.14A). The relatively minor karstification at the Rancheria/Lake Valley contact is probably due in part to the later overprinting by pre-Pennsylvanian karsting, and may be due in part to short duration of exposure and low paleoslope, or relatively dry climate.

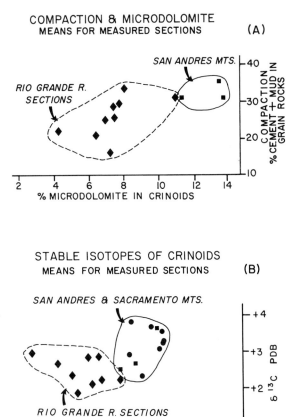

FIGURE 14.13A,B. A. Cross-plot of microdolomite content of crinoids and compaction intensity form Lake Valley/Kelley formation packstones and grainstones from the San Andres Mts. and Rio Grande River regions. Each dot or diamond represents the mean for a single measured section. Each mean represents 11 to 63 samples per measured section for microdolomite measurements and 5 to 14 samples per measured section for compaction measurements. B. Cross-plot of stable isotopes of crinoids from the Rio Grande River area and the San Andres and Sacramento Mts. Each dot or diamond represents the mean for a single measured section, which in turn is based on 4 to 21 samples per measured section.

The next stage in the history of the unconformity was the inundation by the Rancheria sea, incorporating clasts of chert from the Lake Valley Formation (Fig. 14.14B). This, in turn, was followed by a major relative sealevel drop after the Rancheria deposition and before the

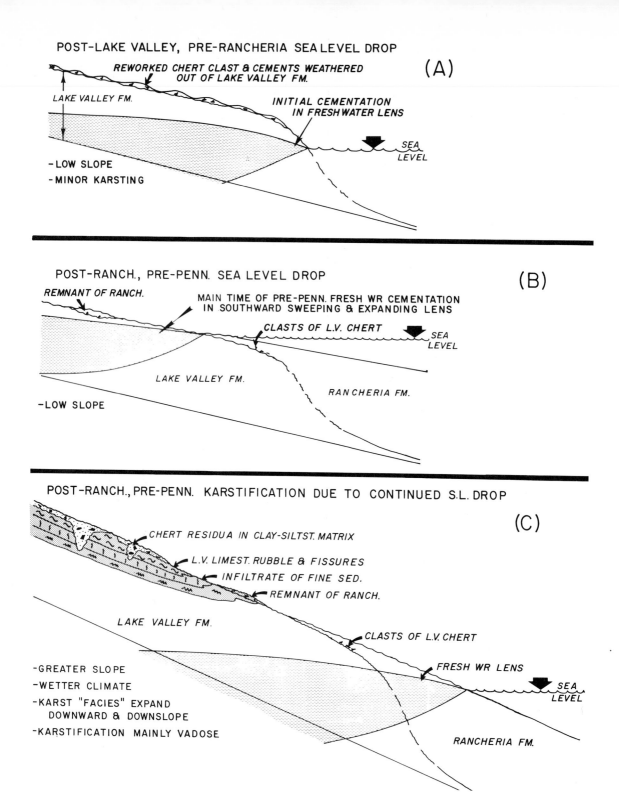

POST-LAKE VALLEY, PRE-RANCHERIA SEA LEVEL DROP (A)

REWORKED CHERT CLAST & CEMENTS WEATHERED
OUT OF LAKE VALLEY FM.

LAKE VALLEY FM.

INITIAL CEMENTATION
IN FRESHWATER LENS

SEA
LEVEL

-LOW SLOPE
-MINOR KARSTING

POST-RANCH., PRE-PENN. SEA LEVEL DROP (B)

REMNANT OF RANCH.

MAIN TIME OF PRE-PENN. FRESH WR CEMENTATION
IN SOUTHWARD SWEEPING & EXPANDING LENS

CLASTS OF L.V. CHERT

SEA
LEVEL

LAKE VALLEY FM.

RANCHERIA FM.

-LOW SLOPE

POST-RANCH., PRE-PENN. KARSTIFICATION DUE TO CONTINUED S.L. DROP

(C)

CHERT RESIDUA IN CLAY-SILTST. MATRIX

L.V. LIMEST. RUBBLE & FISSURES

INFILTRATE OF FINE SED.

REMNANT OF RANCH.

LAKE VALLEY FM.

CLASTS OF L.V. CHERT

FRESH WR LENS

SEA
LEVEL

-GREATER SLOPE
-WETTER CLIMATE
-KARST "FACIES" EXPAND
 DOWNWARD & DOWNSLOPE
-KARSTIFICATION MAINLY VADOSE

RANCHERIA FM.

FIGURE 14.14A–C. A. Model for beginning of karstification during post-Lake Valley pre-Rancheria subaerial exposure. B. Model for diagenesis during early phases of post-Rancheria pre-Pennsylvanian subaerial exposure. This period resulted in establishment of major freshwater lenses that resulted in most of the pre-Pennsylvanian cementation. C. Model for main phase of karstification during pre-Pennsylvanian exposure. It is suggested that the karst facies expanded southward and possibly stratigraphically downward with time and followed the main phase(s) of freshwater cementation.

basal (Morrowan) Pennsylvanian. This resulted in the establishment of an extensive fresh-groundwater system throughout the region that precipitated the bulk of the pre-Pennsylvanian calcite cements (zones 1, 2, 3).

The major phase of karstification followed the extensive freshwater cementation (Fig. 14.14C). The reasons for the change from dominantly cementing mode to dominantly karst modes are problematic, but they probably involved a number of factors. It is possible that this change reflects a change to a wetter climate and an increase in paleoslope. Taken together these may have led to more vigorous weathering dominated by vadose dissolution, eluviation, and neomorphism. The model presented in Figure 14.14C suggests that the karst microfacies displayed in vertical succession at any one locality may be diachronous on a regional scale. In this scenario, the eluviated microfacies preceded the fissure-and-rubble facies, which preceded the chert breccia during the southward expansion of the karst plain. Unfortunately, the petrographic evidence for the relative timing of micrite/microspar eluviation, clay-silt eluviation, and development of fissure-and-rubble fabrics is lacking or equivocal.

Finally, in regard to karsting environment, the karstification processes took place in a subaerial or subsoil setting, rather than an intrastratal setting. The main evidence against in-trastratal karsting is the absence of collapse features in Pennsylvanian strata overlying paleosinkholes and the absence of fragments of Pennsylvanian strata in the chert rubble. In addition, there are no subaerial karren features on top of the Lake Valley or Kelley formations, where the formation tops are now exposed beneath the chert breccia; this suggests subsoil dissolution. This absence could, however, also be a function of poor preservation due to recent weathering.

Comparison with Classic Karst

The Lake Valley/Kelley formations paleokarst differs from classic extant karst described in the literature (Jennings 1971, Sweeting 1973, Jakucs 1977), and their comparison offers a convenient summary of the salient Lake Valley/Kelley karst features (Fig. 14.15). First, most classic karst is dominated by solution-enlarged bedding planes and joints commonly within well-lithified limestones having little or negligible intergranular or intercrystalline porosity and permeability. This in turn generally resulted in large (meters to hundreds of meters) caves and passages. In contrast, in the Lake Valley/Kelley formations, dissolution processes were concentrated in intergranular pores and small dissolution cavities, in irregular veins, and

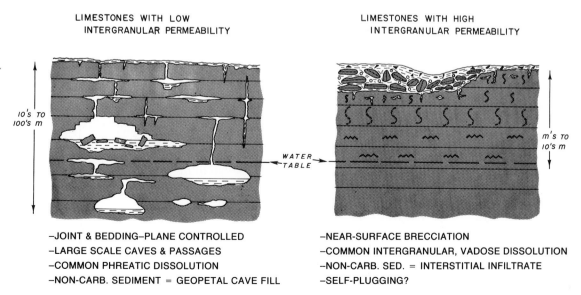

LIMESTONES WITH LOW INTERGRANULAR PERMEABILITY

LIMESTONES WITH HIGH INTERGRANULAR PERMEABILITY

10's TO 100's m

m's TO 10's m

WATER TABLE

–JOINT & BEDDING–PLANE CONTROLLED
–LARGE SCALE CAVES & PASSAGES
–COMMON PHREATIC DISSOLUTION
–NON-CARB. SEDIMENT = GEOPETAL CAVE FILL

–NEAR-SURFACE BRECCIATION
–COMMON INTERGRANULAR, VADOSE DISSOLUTION
–NON-CARB. SED. = INTERSTITIAL INFILTRATE
–SELF-PLUGGING?

FIGURE 14.15. Comparison of key features of the Lake Valley/Kelley formations paleokarst profile with those of extant classic karst.

in fissures with no apparent bedding or joint control. Dissolution ranged from subcentimeter (intergranular) to meters (sinkholes and larger vertical fissures).

Second, in extant karst, the depth of penetration into the host carbonate strata may be on the scale of tens to hundreds of meters, or in some cases thousands of meters. In the Lake Valley/Kelley case, karst features typically extend a few meters to a few tens of meters, with the intense karstification (fissure-and-rubble, chert breccia) restricted to the upper meter or two.

Third. several of the features common in the Lake Valley/Kelley formations paleokarst are largely absent or undescribed in extant karst. Specifically, the nodular fabric of the rubble-and-fissure zone has not been described in any detail in existing karst. Similarly, eluviated carbonate mud has apparently not been described in modern karst settings in terms of its petrographic characteristics, origin, and infiltration mechanisms. This is particularly puzzling considering the volumetric importance of carbonate mud as matrix in many paleokarst breccias (e.g., the Leadville Formation, De Voto, Chap. 13, this volume). Intergranular eluviation products and processes within the host limestones in general have not been systematically described in modern settings.

Fourth, dissolution in classic karst is interpreted to occur mainly in the phreatic, especially the upper phreatic zone. The Lake Valley/Kelley karst, on the other hand, is interpreted to have taken place mainly in the vadose zone.

Conclusions

1. Paleokarst features at the pre-Pennsylvanian unconformity occur over an area of at least 20,000 km^2 in southern New Mexico.

2. The paleokarst features developed on skeletal limestones, mainly crinoidal calcarenites, of the Mississippian Lake Valley and Kelley formations.

3. These features developed on partly cemented crinoidal limestones with high intergranular porosities and permeabilities.

4. The karst processes generally affected the upper meter or two of the Lake Valley and Kel-

ley formations, but in some cases penetrated to several tens of meters below the unconformity.

5. Paleokarst features occur in a vertical profile of progressively more intense alteration upward. This profile consists of, in ascending order:

a. Etching of pre-Pennsylvanian calcite cements.

b. Plugging of partly cemented intergranular pores by micrite and microspar, which are interpreted to be a combination of eluviated carbonate sediment and neomorphic microspar.

c. Plugging of etched, partly cemented intergranular pores and small dissolution cavities by eluviated clay and detrital quartz silt.

d. Fragmented host limestones forming rubble-and-fissure fabrics in which nodules of host rock are surrounded by anastomosing veinlets and fissures filled with "weathering calcarenite," clay, and detrital quartz silt.

e. Chert breccia composed of chert fragments weathered from the Lake Valley/Kelley formations within a matrix of clay and detrital quartz silt.

6. Karstification was most common and best developed in the western and northwestern parts of the study area. This correlates with intensity of other diagenetic features and indicates a chemically more aggressive groundwater system in the west and northwest than in the east during late Mississippian/early Pennsylvanian freshwater diagenesis and karstification.

7. Karstification began during pre-Rancheria subaerial exposure, but the main phase of karstification was after the Rancheria Formation and before the Pennsylvanian.

8. The restriction of much of the alteration processes in the Lake Valley/Kelley formations to intergranular pores and relatively small dissolution cavities within a relatively thin interval (meters to tens of meters), contrasts with the large bedding- and joint-controlled caves and passages extending over thick intervals (hundreds of meters) characteristic of conventional extant karst.

9. This difference is in part due to the high intergranular permeabilities and porosities of the Lake Valley/Kelley formations' crinoidal calcarenites during karstification. As such, the Lake Valley/Kelley formations may provide a broadly applicable model for karstification of carbonates with high intergranular permeability.

Acknowledgments. This research was part of a broader study of regional diagenesis in Mississippian limestones in New Mexico and Arizona, the major parts of which were supported by NSF grants EAR7412253 and EAR7713133. I wish to thank the White Sands Missile Range for permission to enter the range, and to thank Phil Choquette and Noel James for organizing the paleokarst symposium and editing of manuscripts.

References

Alexandersson, T., 1972, Micritization of carbonate particles: processes of precipitation and dissolution in modern shallow-marine sediments: Bull. Geological Institutions Univ. Uppsala, New Series, v. 3, p. 201–236.

Armstrong, A.K., 1962, Stratigraphy and paleontology of the Mississippian System in southwestern New Mexico and adjacent southeastern Arizona: New Mexico Bureau of Mines and Mineral Resources Memoir 8, 79 p.

Brewer, R., 1964, Fabric and mineral analysis of soils: New York, John Wiley and Sons, 470 p.

Chafetz, H.S., 1982, The Upper Cretaceous Beartooth Sandstone of southwestern New Mexico: a transgressive deltaic complex on silicified paleokarst: Jour. Sed. Petrology, v. 52, p. 157–169.

Coniglio, M., and Harrison, R.S., 1983, Holocene and Pleistocene caliche from Big Pine Key, Florida: Bull. Canadian Petroleum Geology, v. 31, p. 3–13.

Dunham, R.J., 1969, Early vadose silt in Townsend mound (reef), New Mexico, *in* Friedman, G.M., ed., Depositional environments in carbonate rocks: Soc. Econ. Paleontologists Mineralogists Spec. Pub. No. 14, p. 139–181.

Esteban, M., and Klappa, C.F., 1983, Subaerial exposure environments, *in* Scholle, P.A., Bebout, D.G., Moore, C.H., eds., Carbonate depositional environments: Am. Assoc. Petroleum Geologists Memoir No. 33, p. 1–54.

Ireland, P., 1979, Geomorphological variations of "case hardening" in Puerto Rico: Z. Geomorph. N.R., Suppl.—Bd., v. 32, p. 9–20.

Jakucs, L., 1977, Morphogenetics of karst regions: New York, John Wiley and Sons, 284 p.

James, N.P., 1972, Holocene and Pleistocene calcareous crust (caliche) profiles: criteria for subaerial exposure: Jour. Sed. Petrology, v. 42, p. 817–836.

Jennings, J.N., 1971, Karst: Cambridge, MA, MIT Press, 252 p.

Kahle, C.F., 1977, Origin of subaerial Holocene calcareous crusts: role of algae, fungi and spar micritisation: Sedimentology, v. 24, p. 413–435.

Kobluk, D.R., Pemberton, S.G., Karolyi, M., Risk, M.J., 1977, The Silurian–Devonian disconformity in southern Ontario: Bull. Canadian Petroleum Geology, v. 25, p. 1157–1186.

Lane, H.R., 1974, The Mississippian of southeastern New Mexico and West Texas—a wedge-on-wedge relation: Am. Assoc. Petroleum Geologists Bull., v. 58, p. 269–282.

Lane, H.R., 1982, The distribution of Waulsortian facies in North America as exemplified in the Sacramento Mountains of New Mexico, *in* Bolton, K., Lane, H.R., LeMone, D.V., eds., Symposium on the paleoenvironmental setting and distribution of the Waulsortian facies: El Paso, TX, El Paso Geological Society, p. 96–114.

Laudon, L.R., and Bowsher, A.L., 1941, Mississippian formations of the Sacramento Mountains, New Mexico: Am. Assoc. Petroleum Geologists Bull., v. 25, p. 2107–2160.

Laudon, L.R., and Bowsher, A.L., 1949, Mississippian formations of southwestern New Mexico: Geol. Soc. America Bull., v. 60, p. 1–88.

Leutloff, A.H., and Meyers, W.J., 1984, Regional distribution of microdolomite inclusions in Mississippian echinoderms from southwestern New Mexico: Jour. Sed. Petrology, v. 54, p. 432–446.

Meyers, W.J., 1973, Chertification and carbonate cementation in the Mississippian Lake Valley Fm., Sacramento Mts., New Mexico: Unpubl. Ph.D. thesis, Rice University, Houston, 353 p.

Meyers, W.J., 1974, Carbonate cement stratigraphy of the Lake Valley Formation (Mississippian), Sacramento Mts., New Mexico: Jour. Sed. Petrology, v. 44, p. 837–861.

Meyers, W.J., 1977, Chertification in the Mississippian Lake Valley Formation, Sacramento Mts., New Mexico: Sedimentology, v. 24, p. 75–105.

Meyers, W.J., 1978, Carbonate cements: their regional distribution and interpretation in Mississippian limestones of southwestern New Mexico: Sedimentology, v. 25, p. 371–400.

Meyers, W.J., 1980, Compaction in Mississippian skeletal limestones southwestern New Mexico: Jour. Sed. Petrology, v. 50, p. 457–474.

Meyers, W.J., and Hill, B.E., 1983, Quantitative studies of compaction in Mississippian skeletal limestones, New Mexico: Jour. Sed. Petrology, v. 53, p. 231–242.

Meyers, W.J., and Lohmann, K.C., 1985, Isotope geochemistry of regionally extensive calcite cement zones and marine components in Mississippian limestones, New Mexico, *in* Schneidermann, N., and Harris, P.M., Carbonate cements: Soc. Econ.

Paleontologists and Mineralogists Spec. Pub. No. 36, p. 223–240.

Neugebauer, J., 1978, Micritization of crinoids by diagenetic dissolution: Sedimentology, v. 25, 267–283.

Pray, L.C., 1961, Geology of the Sacramento Mts. Escarpment: Bull. New Mexico Bureau of Mines and Mineral Resources 35, 144 p.

Sando, W.J., 1974, Ancient solution phenomena in the Madison limestone (Mississippian) of north-central Wyoming: Jour. Research US Geol. Survey, v. 2, no. 2, p. 133–141.

Sweeting, M.M., 1973, Karst landforms: London, MacMillan Publishing Co., 362 p.

Walkden, G.M., 1974, Paleokarstic surfaces in the upper Visean (Carboniferous) limestones of the Derbyshire Block, England: Jour. Sed. Petrology, v. 44, p. 1232–1247.

Wright, V.P., 1982, The recognition and interpretation of paleokarsts: two examples from the Lower Carboniferous of south Wales: Jour. Sed. Petrology, v. 52, p. 83–94.

Yurewicz, D.A., 1977, Basin-margin sedimentation, Rancheria Formation, Sacramento Mts., New Mexico, in Pray, L.C., ed., Guidebook to the Mississippian shelf-edge and basin facies carbonates, Sacramento Mts. and Southern New Mexico Region: Dallas Geological Society, p. 67–86.

15
Paleokarsts and Paleosols as Indicators of Paleoclimate and Porosity Evolution: A Case Study from the Carboniferous of South Wales

V. PAUL WRIGHT

Abstract

A study of subaerial surfaces in the Carboniferous limestones of South Wales has revealed two paleokarst associations. The first consists of densely piped and rubbly solution horizons lacking features such as calcrete crusts, rhizocretions, and needle-fiber calcite but with meniscus cements and extensive phreatic blocky calcite cements. The paleokarst morphology is comparable to some present-day humid karsts. The other association shows horizons of less well-developed karst which are associated with calcrete crusts, rhizocretions, and needle-fiber calcite. Thick calcrete horizons also occur indicating formation under a semi-arid climate. These subaerial surfaces are associated with only minor early meteoric cementation.

These sets of features can be directly compared to the Quaternary eolianites of Yucatan. Here the Upper Pleistocene deposits, subaerially exposed under a more arid climate, while exhibiting little early cementation, do contain calcrete crusts, rhizocretions, and needle-fiber calcite. The Holocene equivalents, now exposed under the present more humid climate, exhibit extensive blocky sparry calcite cementation but lack calcrete crusts and their associated features.

The recognition of different suites of subaerial features not only allows the assessment of paleoclimates but also provides a means of predicting the degree of cementation concurrent with each subaerial exposure phase. This may prove a useful tool in reservoir evaluation in carbonates.

Introduction

Paleokarsts are now widely recognized in ancient carbonate sequences but they have been commonly described for their novelty or as simple indicators of subaerial exposure. Karst features are sensitive indicators of hydrological regimes, and it should be possible to apply the data-base on modern karstic processes and their products to interpreting ancient forms. A study of paleokarsts and related subaerial features in the Lower Carboniferous (Mississippian) limestones of South Wales provides evidence of climatic oscillations between pluvial and semi-arid phases comparable to changes which have been recognized in the Pleistocene and Holocene eolianites of Yucatan by Ward (1978). The climate not only influenced the types of subaerial weathering and soil formations but, like the Quaternary deposits of Yucatan, also profoundly influenced the early cementation of the carbonate sediments.

The types of paleokarst and paleosol developed in the Lower Carboniferous are indicators or the degree of early cementation associated with each subaerial phase and may provide a useful guide for cement prediction in other carbonate sequences. However, the history of subaerial exposure for some ancient paleokarstic surfaces is extremely complex and requires a detailed understanding of the processes, products, and preservation potential of various weathering features. As in the case of detailed studies of Quaternary geomorphology, all observations must be set within a local microstratigraphy which takes into account complex phases of landscape stability and instability. In addition the geological dimension involves using diagenesis and cement stratigraphy as an added aid to interpreting the phases of meteoric alteration of the host sediments.

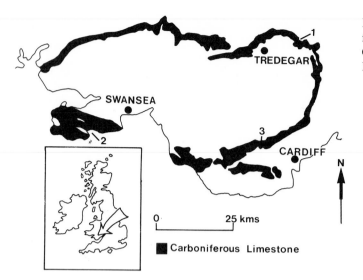

FIGURE 15.1. Locality map of Carboniferous paleokarst horizons. 1: Clydach Gorge; 2: Three Cliffs ay, Gower; 3: Miskin.

Geological Setting

The Lower Carboniferous (Mississippian) limestones of South Wales (Fig. 15.1) were mainly deposited on a carbonate ramp bounded to the south (in what is now Devon and Cornwall) by a complex basin (Issac et al. 1982) and to the north (mid-Wales) by a land mass (St. George's Land) (Wright 1986a). The present northern outcrops form the rim of the South Wales Coalfield synclinorium and represent a thin, upper ramp sequence consisting predominantly of oolitic and peritidal carbonates with numerous subaerial exposure surfaces and some flu-vial intercalations (Fig. 15.2). The southern outcrops are predominantly midramp deposits (Wright 1986a) containing a number of cycles from bioclastic wackestones (with event beds and hummocky cross-stratification) to shallowing-upwards oolitic grainstones capped by subaerial exposure surfaces. These grainstones are commonly stacked with up to three or possibly four individual sand bodies in each oolite formation (Waters 1984).Peritidal units were deposited during transgressive phases (Riding and Wright 1981) but are not a significant component of the midramp sequence. The southernmost outcrops, in Dyfed, consist of

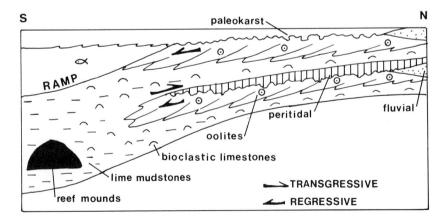

FIGURE 15.2. Simplified cross-section of early Carboniferous ramp in South Wales. Paleokarsts cap regressive (prograding?) oolitic shoal deposits and are capped by transgressive peritidal deposits. Further details are in Wright (1986a).

thick bioclastic wackestones and mudstones with Waulsortian reef mounds (Fig. 15.2) but lack any subaerial exposure surfaces (Wright 1986a).

In the southern outcrops, detailed work by Waters (1984) and Ramsay (1987) has shown that each individual shallowing oolitic sand body passes from offshore or lower shoreface deposits up into beach facies with prominent subaerial features such as rhizocretions and calcrete crusts (Fig. 15.3). The oolites on the northern outcrop have not as yet revealed such fine detail but show a similar trend ranging from bioclastic-rich at the base to oolitic at the top, and are capped by subaerial surfaces. These individual oolitic cycles range from 6 to 15 m in thickness, and are thicker in the mid-ramp zone.

Biostratigraphic control is poor, and the complex upper ramp zone sequences make detailed correlation with the midramp sequences only tentative. Individual units, where dated, show marked diachronism, and it seems likely that subaerial surfaces in the northern outcrops represent longer periods of exposure than their southern counterparts. This geomorphic longevity is clearly shown by the more complex

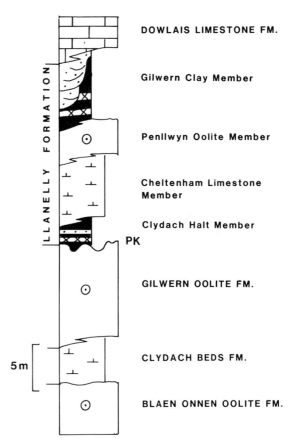

FIGURE 15.4. Stratigraphic log of the Llanelly Formation and associated units. PK = major paleokarst. For symbols see Figure 15.6.

histories of the northern paleokarstic surfaces (see below).

Two paleokarstic surfaces will be described. The first is developed at the top of the Oolite Group in the northern outcrop, specifically between the Gilwern Oolite and the Llanelly Formation (Fig. 15.4). The second surface occurs in the southern outcrops and separates the Gully Oolite (synonymous with the Caswell Bay Oolite) from the peritidal Caswell Bay Mudstone (Fig. 15.9). The northern succession provides evidence of a change in the climate through a single sequence, while the southern outcrop example provides additional evidence that specific subaerial features can provide a guide to the degree of early meteoric cementation.

The following descriptions are based on a review of the existing although scattered and often locally oriented literature, and on unpublished data.

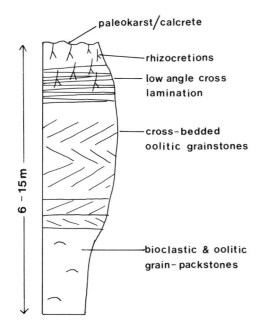

FIGURE 15.3. Simplified log of a shallowing-up oolitic unit based on data in Waters (1984). Such units have been interpreted as shallowing from lower shoreface to beach or eolian.

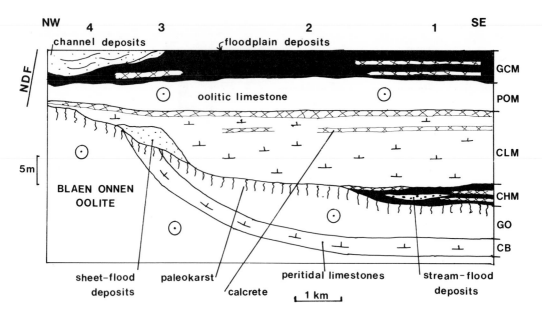

FIGURE 15.5. Schematic diagram showing facies associations in the Llanelly Formation and its overstep of the Oolite Group (Gilwern Oolite–Blaen Onnen Oolite). NDF = Neath Disturbance Fault zone. CB = Clydach Beds, GO = Gilwern Oolite,

CHM = Clydach Halt Member, CLM = Cheltenham Limestone Member, POM = Penllwyn Oolite Member, GCM = Gilwern Clay Member. 1: Clydach Gorge. 2: Craig-y-Gaer. 3: Daren Cilau. 4: Blaen Onneu. (Details of localities in Wright 1981.)

Subaerial Features Associated with the Gilwern Oolite and Llanelly Formation

A paleokarst occurs at the top of the Oolite Group in the northern outcrop zone and is exposed in a large number of exposures stretching from near Abergavenny to Penderyn (see Wright 1982a). The paleokarst caps progressively older formations in the Oolite Group as it is traced northwest along the outcrop, reflecting a northward directed overstep (Fig. 15.5). This is due to intra-Carboniferous uplift related to an active fault zone (the Neath Disturbance) north of the present outcrop (George 1954, Wright 1981).

The paleokarstic surface is similar throughout the outcrop and consists of two types of solutional features (Wright 1982a). The most common type consists of a horizon up to 5 m thick with the appearance of a boulder rubble (Fig. 15.6A). It consists of fluted blocks of cemented oolitic grainstone of pebble to boulder size surrounded by green clay-filled fissures and pipes. Locally large rafts of unfissured oolitic limestone several meters in all dimen-

sions are set within this rubble. The dense fissure system has so riddled the limestone that the outcrop face is very unstable, and this highly jumbled mass has presumably resulted, in part, from solution collapse.

A second but less common set of karstic features also occurs (Fig. 15.6B), which consist of clay-filled sinuous solution pipes up to 40 cm wide and 1 m to 2 m in length. Locally these karstic zones are underlain by a narrow fissure running perpendicular to the pipes, i.e., approximately horizontally, as the local dip is only a few degrees. The horizontal pipe pinches and swells, does not follow any bedding planes or stylolites, and connects with the pipes being filled with the same green clay.

The upper surface of oolite shows no laminar calcrete crusts, and the piped zone lacks such features as rhizocretions (see below), although meniscus cements have been found by Raven (1983) with the use of cathodoluminescence.

There are several subaerial surfaces within the Oolite Group, with the result that the cement stratigraphy is complex (Raven 1983 and 1984). In order to assess the diagenetic effects of the phase of meteoric diagenesis relating to the paleokarst it is necessary to look at the last

FIGURE 15.6. Paleokarstic features developed at the top of the Oolite Group. *A.* Rubbly horizon with numerous irregular clay-filled fissures and pipes. *B.* Solution pipes (lens cap for scale, diameter 6 cm).

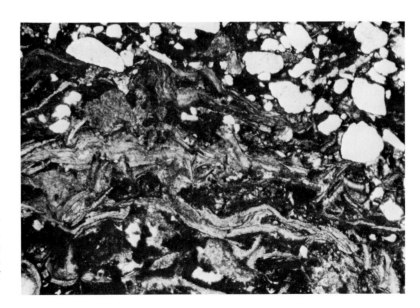

FIGURE 15.7*A,B.* Compacted vermiform gastropod shells in a compacted bioclastic–quartz grainstone. The limestone lacks any cement. Field of view is 3.5 mm wide.

oolitic unit only, the Gilwern Oolite exposed in the Clydach Gorge (Fig. 15.1 and 15.5). There has been extensive cementation of the Gilwern Oolite in this area, of which at least some pre-dated karst formation. In a detailed study, including the use of cathodoluminescence and stable isotopes, Raven (1983 and 1984) has shown that the cementation at the top of the Gilwern Oolite relates to the passage from marine phreatic to meteoric phreatic to meteoric vadose environments, as a result of the lowering of sealevel. The most common cement is a clear blocky calcite.

Interpretation

The paleokarstic features were interpreted by Wright (1982a) as being true surface paleokarst, and not interstratal karst, for a variety of reasons including the lack of karstic features in overlying units. The intense level of piping has been compared to *kavornossen karren* (Jennings 1971, Gams 1973), and such karst is developed on the Aymamon Limestone in Puerto Rico and has been described, although only briefly, by Ireland (1979). Here the limestone is riddled with solution pipes to such an extent that the outcrops appear like a mass of small boulders and cobbles, honeycombed by solution holes. Such a phenomenon may have occurred at the top of the Oolite Group, associated with local small-scale collapse.

The sinuous horizontal fissure was compared by Wright (1982a) to the water table "caves" described by Land (1973) from exposed Pleistocene limestones on Jamaica. The absence of pipes in the Oolite Group below this fissure seemed to confirm this explanation, and Raven (1983) in detailed work in the Clydach Gorge area confirmed this interpretation by finding meniscus cements limited to above the fissure.

The Llanelly Formation

The overlying Llanelly Formation contains a different suite of subaerial features and consists of four members (Fig. 15.4 and 15.5). The Clydach Halt Member is a complex unit showing marked lateral variability, with sandstones and conglomerates which have been interpreted as sheet- and stream-flood deposits possibly of a

distal alluvial fan association (Wright 1981). Within these deposits there are a number of blocky, fine-grained, laterally extensive carbonate beds, individually up to 0.75 m thick. These horizons have bases of coalesced nodules and overlie green clays with similar but less densely packed nodules. The profile characteristics and presence of a variety of calcrete microfabrics led to their interpretation as pedogenic horizons comparable to the petrocalcic horizons found in present-day Aridisols (Wright 1982b). At some localities as many as six petrocalcic horizons occur, one above the other, in a thickness of 6 m between the paleokarst and the overlying peritidal limestones. None of these stacked horizons shows any

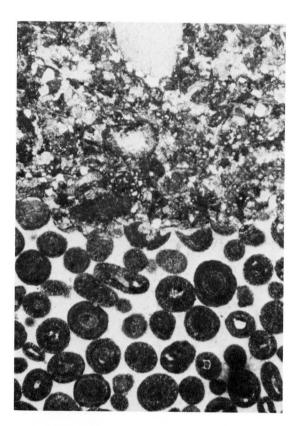

FIGURE 15.8. Contact between the base of the Cheltenham Limestone Member (Llanelly Formation) and the Gilwern Oolite. The contact is a low-amplitude stylolite. The Oolite was extensively cemented, while the overlying bioclastic-intraclastic limestone lacks cement and has a fitted fabric with every grain defined by a stylolite. From Craig-y-Gaer, locality 2, Figure 15.6. Field of view is 4.8 mm wide.

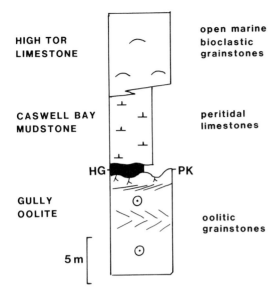

FIGURE 15.9. Stratigraphic log of the Caswell Bay Mudstone–Gully Oolite. HG = Heatherslade Geosol (Heatherslade Bed, see Wright 1984), a prominent calcrete developed on the Gully Oolite. PK = paleokarst.

karstic effects of the types seen at the top of the underlying oolite.

The overlying Cheltenham Limestone Member contains a variety of peritidal limestones including replaced evaporites, petrocalcic horizons, and calcrete crusts with rhizocretions, as well as peloidal, bioclastic, and intraclastic grainstones (Wright 1982b, 1983, 1986a). Most of these grainstones show evidence of compaction and grain suturing (Fig. 15.7). Precompaction nonferroan calcites are irregularly distributed stratigraphically and laterally within the member. Most cementation consists of ferroan calcite precipitated after compaction and pressure solution had occurred (Wright 1981). The contrast between the lack of cement in this member and the high degree of cementation at the top of the Gilwern Oolite is shown clearly in Fig. 15.8, which represents the junction where the Clydach Halt Member is missing due probably to erosion.

Paleokarst on the Gully Oolite

A paleokarstic surface separates the Gully Oolite from the peritidal Caswell Bay Mudstone (Fig. 15.9). Outcrops have been described from

Miskin near Cardiff (Fig. 15.1) by Riding and Wright (1981) and from the Gower by Wright (1982a and 1984). The exact age of the Gilwern Oolite is not known, but the Gully Oolite is Chadian in age, and the overlying Caswell Bay Mudstone Arundian (Wright 1986a). The Gully Oolite is therefore older than the Gilwern Oolite and Llanelly Formation, which are probably both Arundian in age (Wright 1981), and therefore corresponds in age to the Blaen Onnen Oolite (Fig. 15.5).

The top of the Gully Oolite may be only slightly irregular as at Miskin (Riding and Wright 1981, Fig. 10) or may exhibit a mammillated surface as in the Gower (Wright 1982a, Fig. 12a). Solution rubbling or piping is absent in both areas. The mammillated surface, which is black in contrast to the white or light grey of the underlying oolitic grainstone, consists of numerous small pits averaging 20 cm to 40 cm in diameter and 50 cm to 100 cm deep. The pits are separated by irregular domes veneered in a thin, discontinuous crust (Fig. 15.10), less than 2 cm thick containing abundant alveolar textures (Fig. 15.11A). These consist of well-preserved needle-fiber calcite (Wright 1984) (Fig. 15.11B). Other fabrics present include irregular micritic grain coats around ooids. The crust is overlain by a thin green-red mottled clay up to 20 cm thick which contains black rounded pebbles and cobbles of oolite. This horizon is truncated by the peritidal limestones of the Caswell Bay Mudstone. This crust, and the overlying clay, have been named the Heatherslade Bed by George (1978), and the same horizon at Miskin has been described in detail by Riding and Wright (1981). Here a thick paleosol occurs with abundant calcrete fabrics within an irregular oolitic breccia-conglomerate interpreted as a regolith. Alveolar textures are particularly abundant, and the regolith is overlain by mottled clays with a dense limestone interpreted as a petrocalcic calcrete horizon of a paleosol, comparable to those in the Clydach Halt Member of the Llanelly Formation. Replaced evaporites also occur in the Caswell Bay Mudstone.

The underlying oolitic grainstones show little early cementation (Hird et al. 1987). The earliest phase of cementation consists of isopachous fibrous fringes around lithoclasts in the oolite representing early marine cements. The

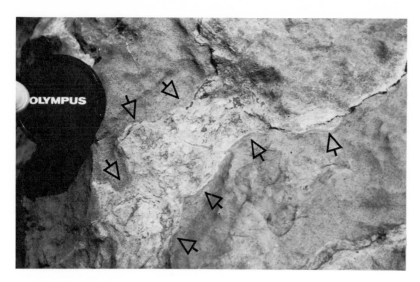

FIGURE 15.10. Irregular calcrete-crust veneer (plan view, arrowed) on top of the Gully Oolite at Three Cliffs Bay. The crust consists of alveolar textures set in an oolitic matrix with micritic grain coats and intergranular pores filled by degraded needle-fiber calcite. Lens cap is 6 cm. in diameter.

bulk of the oolite has a blocky nonferroan calcite as the first cement which postdates significant compaction and pressure solution. This cement sequence has been confirmed by more detailed studies currently in progress by Alison Searl (pers. comm. 1986).

Climatic Considerations

If the interpretation of the paleokarst at the top of the Oolite Group as karvornossen karren is correct, karst formation probably occurred under a high-rainfall, tropical regime (Jennings 1971). The Puerto Rican karst forms in an area with an annual rainfall of 1500 mm (Ireland 1979). While calcrete and karst features are not mutually exclusive (Esteban and Klappa 1983), the nature of the overlying calcrete paleosols seems incompatible with a tropical, humid environment. The paleosols have been interpreted as typical semiarid aridisols (Wright 1982b) and could not have formed contemporaneously with the paleokarsts. Such horizons usually form at the maximum depth of seasonal setting in a soil profile (Weider and Yaalon 1974) and effectively would not have survived if soil water had been able to percolate through them into the underlying oolitic limestones. Furthermore, in the examples where up to six profiles are stacked none of the lower profiles show any evidence of solution of the type seen in the limestone beneath. It was on these grounds that

the author earlier inferred a climatic change from humid to semi-arid between the phase of karsting and calcrete formation (Wright 1980). I here stress that while calcrete and karst are not mutually exclusive, kavornossen karren and the type of petrocalcic horizon seen in the Clydach Halt Member are indeed incompatible under a single climate. The presence of petrocalcic horizons and replaced evaporites in the overlying peritidal unit confirm a continuation of semi-arid conditions.

A direct comparison can be made here with the Quaternary eolianites of Yucatan described by Ward (1978) and by McKee and Ward (1983). The Upper Pleistocene eolianites in Yucatan underwent little cementation during formation and exhibit features such as finely crystalline cements, rhizocretions, needle-fiber calcite, and calcrete crusts. In the younger, Holocene eolianites, coarse, sparry calcites are common including even the youngest deposits locally being completely cemented while the above listed features are absent. The Holocene (present-day) climate is humid, whereas Ward suggested that the late Pleistocene climate had been more arid. As a result of a drier climate less dissolution occurred, less through-flow of water, and subsequently less cementation. Similar climatic controls on diagenesis and porosity evolution have been described by Calvet (1979) and Calvet et al. (1980) for the Pleistocene of Mallorca and by Harrison (1975) for Barbados.

The similarities with the Yucatan deposits are

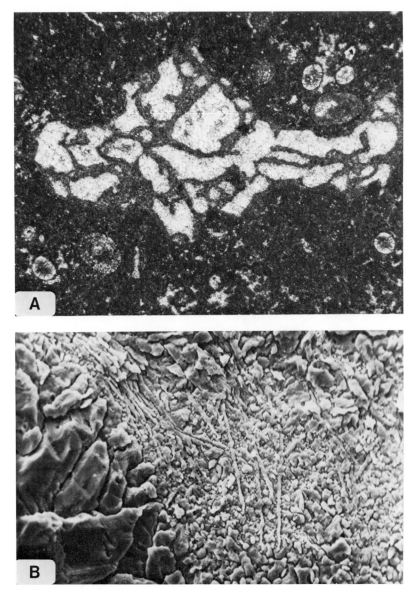

FIGURE 15.11*A,B*. *A*. Alveolar texture from the crust at the top of the Gully Oolite. Field of view is 4.5 mm wide. These rhizocretions are abundant at this level but are absent from the top of the Gilwern Oolite despite meniscus cements occurring there indicating vadose conditions. The septae are composed of needle-fiber calcite which also occurs in the dense micritic matrix. *B*. S.E.M. photomicrograph showing needle-fiber calcite from one of the septae in the rhizocretions. Sparry calcite occurs to the left of the needles. For further details see Wright (1984 and 1986b). Field of view is 210 μm wide.

striking. The Oolite Group, with its humid karst features, lacks any of the features found in the Pleistocene eolianites of Yucatan. Erosion at the top of the Oolite Group probably occurred, yet the removal of the vadose zone was not great for the occurrence of meniscus cements above the water table cave indicates that at least 1 to 2 m of the original vadose zone is preserved. Rhizocretions are noticeably absent. The high degree of cementation (100% locally, Fig. 15.9) is also similar to the humid diagenesis of the Holocene of Yucatan.

However, the overlying Llanelly Formation with its aridisols, calcrete crusts, rhizocretions, and evaporites is similar to the more arid upper Pleistocene deposits. The lack of early cementation is another feature in common. No karst features occur in the Llanelly Formation despite evidence of several subaerial phases during its deposition (Wright 1982b, 1983, 1986).

Other explanations may be offered for these differences. For example the absence of paleokarstic features in the Llanelly Formation might be a function of lack of sufficient time. This seems unlikely for mature Aridisols occur in the peritidal unit and in the overlying Gilwern Clay Member (Fig. 15.4) (Wright 1982b), yet no paleokarstic features underlie them. The well-developed solution features may be more marked in the Gilwern Oolite because it had already undergone some marine phreatic and meteoric phreatic cementation, while the Llanelly Formation deposits were more porous and did not undergo localized solution in pipes. This seems an unlikely explanation because when the paleokarst is developed, because of overstep, on other units in the Oolite Group, the karstic effects are identical to those in the Gilwern Oolite despite the lower oolites having had a different cementation history (Raven 1983).

The explanation favored here is that the dif-ferences represent subaerial exposure under different climates, as summarized in Fig. 15.12. The paleokarst and paleosols are diagnostic of humid and arid or semi-arid climates, respectively, and the differences in diagenesis of the associated sediments, by analogy with the Quaternary of Yucatan, substantiate this view.

Some features of the Gully Oolite—its calcrete crust, alveolar textures, which represent rhizocretions (e.g., Esteban and Klappa 1983), and needle-fiber calcite—are found in the Upper Pleistocene eolianites of Yucatan, and by analogy suggest a more arid climate. Furthermore, the presence at Miskin of a well developed petrocalcic soil horizon is further evidence of at least a semi-arid climate. By analogy with the upper Pleistocene of Yucatan and also with the limestone of the Llanelly Formation, the underlying oolites might be expected to show little early cementation, as appears to be the case.

Discussion

The analogies between the subaerial diagenesis of these early Carboniferous sequences and the Quaternary of Yucatan are striking. The dis-

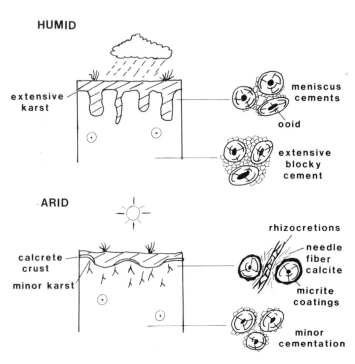

FIGURE 15.12. Interpretation of the diagenetic effects of humid and arid phases (see text).

tribution of pedogenic features (rhizocretions, needle-fiber calcite, and calcrete crusts) and the limited degree of early cementation parallel features found in Yucatan by Ward (1978) and agree with additional observations made by workers on other Pleistocene sequences (Harrison 1975, Calvet 1979, Calvet et al. 1980). In addition, the interpretations are qualified in part by additional information from the specific karst type (Tropical) capping the Gilwern Oolite and by the presence of petrocalcic horizons in units interpreted as being more arid. During the "arid" phases, karsting was less well developed within the Llanelly Formation and at the top of the Gully Oolite.

The cause of these climatic shifts is unknown, but the striking cyclicity which occurs throughout the Lower Carboniferous limestones in Britain has been claimed to be due to eustatic sealevel changes (Ramsbottom 1981). Whether such sealevel changes were the effect of climatic changes and glaciations is unknown.

The examples from Quaternary deposits and from the early Carboniferous suggest that it should be possible to use suites of subaerial features (paleokarsts and paleosols) to predict the degree of early cementation of their associated limestones. This may prove a useful tool in hydrocarbon exploration. One striking feature of the Gilwern Oolite subaerial horizon is that while an excellent secondary pore system is developed within the oolite under a humid climate, extensive meteoric phreatic cementation has also occurred, resulting in extensive blocky calcite cement. This confirms the view expressed by Longman (1981, p. 146) that most major oolite shoal reservoirs in the geological record have developed under arid climates, whereas comparable reservoirs developed under humid climates are virtually unknown.

A rider should be added to this point, for in both the Gully Oolite and the Llanelly Formation, even though early cementation is volumetrically minor, compaction and pressure solution have locally caused considerable loss of porosity. Minor early cementation can prevent or retard later porosity loss by compaction-pressure solution and may preserve porosity (Becher and Moore 1976, Purser 1978). Phases of subaerial exposure with at least some cementation may be more favorable to the for-mation of diagenetic traps than prolonged periods of humid karstification.

It must be stated that in the examples of different degrees of cementation and karstification a key factor is time, and it is not possible at present to categorically exclude this factor as a control. The explanations offered here must all be viewed with this proviso, but the added information from specific karst and soil types does indicate that climatic changes did occur.

Conclusions

The types of subaerial exposure features developed within the Carboniferous limestones in South Wales can be compared directly with Quaternary examples. During humid phases prominent karst horizons developed analogous to those formed in tropical regions such as Puerto Rico. Rhizocretions, needle-fiber calcite, and calcrete crusts are absent, although meniscus cements occur. Early cementation was extensive, locally apparently 100%. These features represent high dissolution rates and a flux of carbonate-rich waters.

In contrast, during more arid phases karstic features are poorly developed or absent. Subaerial surfaces may exhibit calcrete crusts, rhizocretions, needle-fiber calcite, and even mature petrocalcic horizons, associated with sheet and stream flood deposits. Evaporites also formed and early cementation was minor or absent. Later burial resulted in a significant loss of porosity by compaction and pressure solution.

Paleokarsts, used in conjunction with other subaerial exposure features, not only can provide useful indicators of paleoclimates but also may prove useful as guides to the early cementation histories of carbonate reservoirs.

Acknowledgments. I should especially like to thank Madeleine Raven (Robertson Research Int., Llandudno) for many discussions on the diagenesis of the Oolite Group. I also thank Maurice Tucker (Durham) and Alison Searl (Cambridge) for information on the Gully Oolite. Mavis Hardiman typed the manuscript and

Simon Powell prepared the photographs. My thanks go to both. The Nuffield Foundation is thanked for their financial support.

References

Becher, J.W., and Moore, C.H., 1976, The Walker Creek Field: a Smackover diagenetic trap: Transactions of Gulf Coast Association of Geological Societies, v. 26, p. 34–56.

Calvet, F., 1979, Evolucio diagenetica en els sediment carbonates del Pleistoceno Mallorqui: Unpubl. Ph.D. thesis, University of Barcelona, Spain, 273 p.

Calvet, F., Plana, F., and Traveria, A., 1980, La tendecia mineralogica de las eolianites de Pleistocene de Mallorca, mediante la aplicacion del metodo de Chung: Acta Geologica Hispanica, v. 15, p. 39–44.

Esteban, M., and Klappa, C.F., 1983, Subaerial exposure environment, in Scholle, P.A., Bebout, D.G., and Moore, C.H., eds., Carbonate depositional environments: American Association of Petroleum Geologists Memoir 33, p. 1–54.

Gams, I., 1973, Forms of subsoil karst: International Speleological Congress, 1973, II subsection Ba., p. 169–179.

George, T.N., 1954, Pre-Seminulan Main Limestone of the Avonian series in Breconshire: Geological Society of London, Quarterly Journal, v. 110, p. 282–322.

George, T.N., 1978, Mid-Dinantian (Chadian) limestones in Gower: Philosophical Transactions of the Royal Society, Series B., v. 282, p. 411–462.

Harrison, R.H., 1975, Porosity in Pleistocene grainstones from Barbados: some preliminary observations: Canadian Petroleum Geologists Bulletin, v. 23, p. 383–392.

Hird, K., Tucker, M.E., and Waters, R., 1987, Petrology, geochemistry and origin of Dinantian dolomites from southeast Wales, in Adams, A.E., Miller, J., and Wright, V.P., eds., European Dinantian environments: Special Publication of the Geological Journal, New York, Wiley. no. 12, p. 359–377.

Ireland, P., 1979, Geomorphological variations of "case-hardening" in Puerto Rico: Z Geomorph. N.F., Suppl.—Bd., v. 32, p. 9–20.

Issac, K.P., Turner, P.J., and Stewart, I.J., 1982, The evolution of the Hercynides of central S.W. England: Journal of the Geological Society of London, v. 139, p. 521–531.

Jennings, J.N., 1971, Karst: Cambridge, MA, MIT Press.

Land, L.S., 1973, Holocene meteoric dolomitization of Pleistocene limestones, North Jamaica: Sedimentology, v. 20, p. 411–424.

Longman, M.W., 1981, Carbonate diagenesis as a control on stratigraphic traps: Am. Assoc. Petroleum Geologists, Education Course Note Ser: v. 21, 159 p.

McKee, E.D., and Ward, W.C., 1983, Eolian environment, in Scholle, P.A., Bebout, D.G., and Moore, C.H., eds., Carbonate depositional environments: American Association of Petroleum Geologists Memoir 33, p. 131–170.

Purser, B.H., 1978, Early diagenesis and the preservation of porosity in Jurassic limestones: Journal of Petroleum Geology, v. 1, p. 83–94.

Ramsay, A.T.S., 1987, Depositional environments in the Dinantian limestones of Gower, in Adams, A.E., Miller, J., and Wright, V.P., eds., European Dinantian environments: Special Publication of the Geological Journal, New York, Wiley. no. 12, p. 265–308.

Ramsbottom, W.H.C., 1981, Eustacy, sea level and local tectonism, with examples from the British Carboniferous: Proceedings of the Yorkshire Geological Society, v. 43, p. 473–482.

Raven, M., 1983, The diagenesis of the Oolite Group between Blaen Onneu and Pwll Du, Lower Carboniferous, South Wales: Unpubl. Ph.D. thesis, University of Nottingham, England.

Raven, M., 1984, The cement sequence occurring at the top of the Oolite Group (Lower Carboniferous, South Wales): the passage from marine phreatic to meteoric phreatic and then meteoric vadose diagenetic environments (abstr.): European Dinantian Environments, First Meeting, Open University, Milton Keynes, UK, 1984, Abstracts, p. 53–55.

Riding, R., and Wright, V.P., 1981, Paleosols and tidal flat/lagoon sequences on a Carboniferous carbonate shelf: Journal of Sedimentary Petrology, v. 51, p. 1323–1339.

Ward, W.C., 1975, Petrology and diagenesis of carbonate eolianites of north eastern Yucatan Peninsula, Mexico, in Wantland, K.F., and Pusey, W.C., eds., Belize shelf: carbonate sediments, clastic sediments and ecology: American Association of Petroleum Geologists, Studies in Geology 2, p. 500–571.

Ward, W.C., 1978, Indicators of climate in carbonate dune rocks, in Ward, W.C., and Weidie, A.E., eds., Geology and hydrogeology of north eastern Yucatan: New Orleans Geological Society, p. 191–208.

Waters, R.A., 1984, Some aspects of the Black Rock limestone and gully oolite in the eastern Vale of Glamorgan: Proceedings of the Geologist's Association, U.K., v. 95, p. 391–392.

Wieder, M., and Yaalon, D.H., 1974, Effect of matrix composition on carbonate nodule crystallization: Geoferma, v. 11, p. 95–121.

Wright, V.P., 1980. Climate fluctuation in the Lower Carboniferous: Naturwissenschaften, v. 67, p. 252–253.

Wright, V.P., 1981, Stratigraphy and sedimentology of the Llanelly Formation between Blorenge and Penderyn, South Wales: Unpubl. Ph.D. thesis, University of Wales, 409 p.

Wright, V.P., 1982a, The recognition and interpretation of paleokarsts: two examples from the Lower Carboniferous of South Wales: Journal of Sedimentary Petrology, v. 52, p. 83–94.

Wright, V.P., 1982b, Calcrete paleosols from the Lower Carboniferous Llanelly Formation, South Wales: Sedimentary Geology, v. 33, p. 1–33.

Wright, V.P., 1983, A rendzina from the Lower Carboniferous of South Wales: Sedimentology, v. 30, p. 159–179.

Wright, V.P., 1986a, Facies sequences on a carbonate ramp: the Carboniferous limestone of South Wales. Sedimentology, v. 33, p. 221–241.

Wright, V.P., 1986b, The role of fungal biomineralization in the formation of Early Carboniferous soil fabrics: Sedimentology, v. 33, p. 831–838.

16
Caves and Other Features of Permian Karst in San Andres Dolomite, Yates Field Reservoir, West Texas

DEXTER H. CRAIG

Abstract

Cave cements, breccias, and several kinds of internal sediment are lithologic features of a Late Permian (Guadalupian, Kazanian) karst recognized in cores from the San Andres Formation in the reservoir of the Yates oil field in the Permian basin of west Texas. Other indications of a paleokarst system at Yates have been bit drops, sudden rushes of oil during drilling, extremely high flow rates recorded for some early field wells, and fragments of dissolution rubble produced with the oil from some wells. Karst is also indicated by 285 unfilled caves in 142 of the 898 wells drilled within the boundaries of the field to the end of year 1983. The caves range in height from 1 foot (0.3 m) to 21 feet (6.4 m), average 2.9 feet (0.9 m), and have a modal height of 2 feet (0.6 m). They are most numerous in the eastern part of the field, where permeable shelf-edge skeletal sands in the upper San Andres made the rock more susceptible to meteoric infiltration and solution. The agents of karstification were probably dynamic freshwater lenses which developed beneath a cluster of low-relief limestone islands produced by tectonic uplift of the field or by one or more low stands of the Late Permian seas. Support for this island hydrologic model comes from well data on the heights of the San Andres caves and their spatial distribution in the Late Permian. The patterns of cave abundance and height closely fit patterns of physical and chemical dynamics believed to be characteristic of freshwater lenses beneath limestone islands.

Introduction

Since its discovery in October 1926, the Yates oil field, located in Pecos County, west Texas, has shown some remarkable production char- acteristics (Craig and Schoonmaker 1968). Among these, observed from 1926 to 1976 (the year of field unitization) have been (1) bit drops during drilling of some wells, (2) sudden rushes of oil while drilling in the "Big Lime," the principal carbonate producing sequence, (3) quantities of dolomite sand and gravel brought to the surface by the oil through well casing and flow lines, and (4) sustained high flow rates from many of the wells. At least one early observer, Frank Clark, discoverer of the field and chief geologist for Marathon Oil Company, pointed out in testimony before the Railroad Commission of Texas that these unusual features might be explained by assuming that oil in the principal reservoir flowed from a system of caves.

This interpretation of caves or cavelike conduits in parts of the reservoir remained largely unsupported by lithologic evidence of karst until a number of rotary cores were taken during well deepenings in the 1950s. Some of these cores contained collapse breccias, cave cements, and gravity- and water-deposited cave sediments of various kinds. Taken together, these features confirmed that some of the carbonates in the Yates field reservoir had undergone extensive karstification during the Late Permian (middle Guadalupian).

During the period 1960–1970, five wells, which cored the entire reservoir section, were drilled by various operators in the field. Two of the five yielded karst-related lithologies in the San Andres Dolomite and encountered open caves while coring. However, it was only after the field had been unitized and extensive infill drilling had taken place that enough

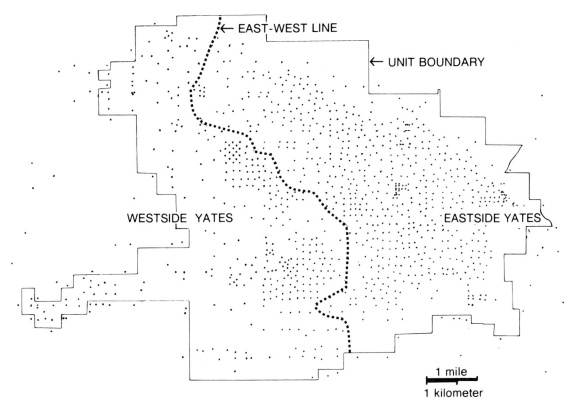

FIGURE 16.1. Yates field unit boundary and wells in early 1983. The east–west line is an engineering line separating eastside from westside Yates. Eastside production is dominated by gravity drainage, and westside by pressure depletion mechanisms. Eastside

Yates has produced approximately 985 million and westside Yates 86 million barrels of oil to October 1, 1986. The field produced its billionth barrel of oil in January 1985, during its 59th year of production.

wireline log and rock data became available to permit a systematic analysis of the caves and karst lithologies. This paper describes the lithologic evidence for karst and summarizes other geologic elements of a statistical study aimed at identifying and describing the hydrogeologic model which best explains the varying size and spatial arrangement of caves in the San Andres Formation in the Yates field reservoir.[1]

The Yates field and its unit boundary, the subdivision of the field for reservoir-engineering purposes into "eastside" and "westside," and the approximately 900 wells that had been drilled by 1983 are shown in Figure 16.1. The productive area within the field unit covers ap-

proximately 35 mi^2 (91 km^2); and in 60 years the principal reservoir, which includes the San Andres, has produced more than 1.07 billion barrels of oil.

Regional Geologic Setting of Yates Field

The Yates field lies at the southern tip of the Central Basin platform, one of the major structural and stratigraphic elements of the Permian basin of west Texas and southeast New Mexico (Fig. 16.2). The field straddles the platform margin. As a result, a platform-interior facies of intertidal and lagoonal mudstones and wackestones dominates in westside Yates, whereas high-energy subtidal grainstones and packstones of the shelf edge facies dominate in eastside Yates. The field area is the highest

---

[1]The results of the statistical study are contained in an unpublished report of research conducted for Marathon Oil Company, dated August 1984, and written by J.L. Carlson, D.H. Craig, and E.A. Nosal.

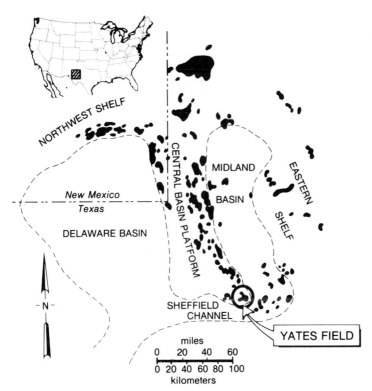

Figure 16.2. Regional paleogeography of the Permian basin of west Texas and southeast New Mexico during Late Permian (Guadalupian–Kazanian) time. Black areas are fields producing oil and gas from San Andres, Grayburg, and Queen formations, the principal reservoir units in the Yates field. (Modified from Shirley, 1987. Published by permission, American Association of Petroleum Geologists.)

structural point on the Central Basin platform at the stratigraphic levels of the field's main reservoir in the San Andres, Grayburg, and Queen formations, and it continues so upward through beds of youngest Permian (Ochoan) age. The persistently positive tendency of the field area and adjoining parts of the southern Central Basin platform may have played a role in development of the karst unconformity at the top of the San Andres Formation, although this role may have been secondary to that of eustatic changes in Permian sealevel.

Reservoir Stratigraphy

The stratigraphic units that comprise the Yates field reservoir are shown in Figure 16.3. They consist of the San Andres Dolomite, which in the field area is up to 750 feet (229 m) thick. The San Andres is unconformably overlain by the Grayburg Formation, a unit mostly of silty dolomite which is 53 feet (16.2 m) thick on the average in the field area. In eastside Yates, the top of the Grayburg displays some evidence of local erosion and a superficial paleokarst. In the western part of the field the Grayburg consists of dolomite and siltstone in about equal amounts, and shows no signs of erosion at its top. The third and youngest of the major formations productive in the field is the Queen Formation, which is 47 feet (14.3 m) thick on the average and consists largely of siltstone and dolomite with minor amounts of sandstone. The Queen is overlain disconformably by the Seven Rivers Formation, a unit averaging 427 feet (130 m) in thickness and composed of anhydrite with minor interbeds of siltstone, silty dolomite, and gypsum. Evaporites in the Seven Rivers are the seal on the Yates reservoir, but some sandstones in the basal part of the formation are locally productive of oil and gas. The San Andres, Grayburg, Queen, and basal Seven Rivers comprise a single reservoir with relatively uniform fluid pressures throughout and no major horizontal or vertical barriers to flow.

Typical wells representing westside and east-side Yates are illustrated in Figure 16.3. The geographic locations of the two wells are shown in Figure 16.4. Of prime importance in the study of San Andres karst is recognition of the major unconformity at the top of the formation,

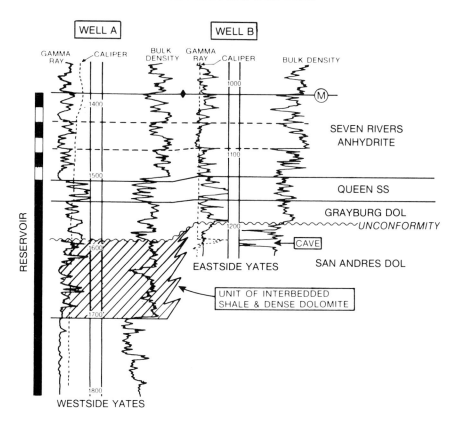

FIGURE 16.3. Principal stratigraphic units of the Yates field trap shown by wells representative of the geology of eastside Yates (well B) and westside Yates (well A). Locations of the wells are shown in Figure 16.4. Over most of the field area, the San Andres, Grayburg, and Queen formations comprise the reservoir, and the Seven Rivers Anhydrite is the seal. The unconformity that separates the San Andres from the Grayburg was an exposure surface below which most of the caves and other karst features in the reservoir were produced. The unit of compact dolomite and bedded shale shown in well A tops the San Andres over the western two-thirds of the field (see Fig. 16.4). A cave in well B, shown in this figure, is indicated by both caliper and bulk density curves.

shown in Figure 16.3. Also important is the facies contrast of the uppermost San Andres Formation in the west as compared to the east. Figure 16.3 shows with hachures on the log of Well A the intertidal and lagoonal facies of the San Andres which consists of compact dolomite mudstones and wackestones with a number of 1- to 2-feet (0.3- to 0.6-m) interbeds of clay shale. This unit of relatively impervious rock is slightly more than 100 feet (30.5 m) thick in this well (Fig. 16.3) but is entirely absent from the section in Well B. The areal distribution of the westside dense facies in relation to the two wells in Figure 16.3 and also in relation to the areal distribution of wells with one or more San Andres caves is shown in Figure 16.4. Most of the wells with caves occur east of the dense facies or close to its thin eastern edge. Wells with caves located within the limits of the dense facies probably coincide with zones of fracturing, which may be related to differential compaction of the shale interbeds. It is clear from the distribution of wells with caves (Fig. 16.4) that, whatever the details of karstification may have been, and regardless of what the patterns of fresh-water movement in the upper San Andres were, the dense westside interior-platform facies tended to resist the development of caves.

Lines of stratigraphic correlation between Wells A and B (Fig. 16.3) show the essentially "layer-cake" stratigraphy that marks the field section from the top of the Grayburg into the tabular evaporites of the Seven Rivers Formation. The high degree of areal correlation

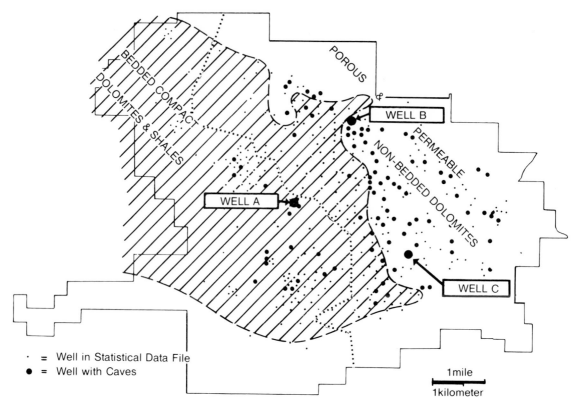

FIGURE 16.4. Wells with caves shown in relation to the paleogeographic distribution of major lithofacies exposed at the unconformity surface of the San Andres Dolomite. Except locally, where thin or excessively fractured, the westside facies (hachured) resisted the karstification which generally affected the eastside upper San Andres. Note the strong tendency for wells with caves to occur east of the limits of the westside facies or close to its edge.

of units in the Seven Rivers Anhydrite and the identification of most of these units as sabkha deposits justify the assumption that they were essentially horizontal and parallel to sealevel at the time of deposition. And thus these units can be used as paleodatums for mapping the paleostructure or paleotopography of appropriate strata beneath them. Horizon M in the Seven Rivers (Fig. 16.3) has been used as a reference datum to restore the San Andres unconformity to its Late Permian configuration. More will be said about reconstruction of the unconformity and its importance to cave interpretation in the section dealing with the statistics of the San Andres caves.

Recognition of the top of San Andres unconformity in Yates is based on observations from slabbed core and from interpretation of wireline logs. Megascopic features seen in cores include erosion, solution, collapse brecciation, cave cementation, and sediment infiltration of cavities at or close below the contact of the San Andres with the Grayburg Formation. At least one of these features occurs in about one-half of the 60 field wells that have been cored through the formation contact. Petrographic evidence of an unconformity from core materials is a subarkosic silt characteristic of the basal Grayburg seen in fractures and internal sediment of the upper San Andres. Detailed bed-by-bed correlation by means of wireline logs (chiefly total gamma ray logs) gives evidence of onlap on the unconformity surface by the basal Grayburg at various points in westside Yates, and especially in areas of transition between paleotopographic lows of westside Yates and paleotopographic highs of eastside Yates. The San Andres unconformity recognized at

Yates may cover much of the southern end of the Central Basin platform, although it probably occurs as a surface of significant karsting only across major paleotopographic highs like those at Yates and at Taylor-Link, a field which is 17 mi (27 km) west of Yates (Kerans and Parsley 1986). It has also been identified in widely separated fields farther north on the Central Basin platform, notably in the Hobbs field in Lea County, New Mexico (DeFord and Wahlstrom 1932) and the Goldsmith field in Ector County, Texas (Young et al. 1939).

Lithologic Features of San Andres Paleokarst

Lithologic features in San Andres cores from Well C, Figure 16.4, which are interpreted as evidence of karst processes, are shown in Figure 16.5. All of the carbonate in these slabs is now dolomite but is believed to have been mostly limestone at the time of karstification. Although thin beds of syngenetic dolomite have been tentatively identified in the upper Andres of westside Yates, most of the dolomitization of San Andres carbonates probably occurred by seepage refluxion during deposition of evaporitic interior platform facies of the Grayburg and Queen and the sabkha and salina evaporites of the basal Seven Rivers.

Clearly apparent in Figure 16.5 are several kinds of cavity- or cave-filling sediments: A, B, C, and D show cave sediment in various relations to the light gray dolomite country rock and to other features such as breccias and flowstone-dripstone (C); E and F demonstrate two different types of fracture enlargement, one by solution (F) and the other (E) possibly by tectonic extension.

Although no attempt has been made to measure the importance of sediment infilling in comparison to solution in the process of karstification in this particular setting, examination of selected cores leaves the impression that development and destruction of porosity were of about equal effect. Clearly, however, there must have been a net loss of rock from the area of karstification over the long term as CaCO$_3$ went into solution and was carried into the nearby Permian sea.

Included in the captions for Figure 16.5 are the depths below the San Andres unconformity of the samples shown in the photographs. It is likely that these samples and the features they show are not from a single karst system, but in fact represent two of the three or four unconformities which can be established in the San Andres Dolomite on the basis of lithology and cave abundance. More will be said about these lesser unconformities below.

Statistical Analysis of San Andres Caves

Greatly increased drilling, coring, and logging in the Yates field following unitization in 1976 added support to the concept of a karst system in the San Andres Dolomite. Among the data gathered systematically were those relating to the presence of caves, that is, bit drops, significant departures from gauge on caliper logs, and bulk density readings significantly below those that would be expected in an environment of matrix porosity, i.e., porosity consisting of small vugs, molds, or voids between sand-size carbonate grains or crystals. The criteria that were used to count the number of caves and the number of cave feet (that is, feet of cave height) in the San Andres Formation are summarized in Table 16.1.

Statistical study of the caves was carried out with a population of 400 wells, all of which had been properly logged with caliper and compensated bulk density tools and had been neither acidized nor shot with nitroglycerin nor otherwise artificially fractured before being logged. The wells in this statistical population, which were drilled before 1982, are shown in Figures 16.4 and 16.6. Beginning in 1982, knowledge of the spatial distribution of caves in the San Andres influenced the choice of drilling locations for infill wells; and although caves continued to be detected, we chose to leave them out of the population rather than deal with the bias they would introduce into statistical analysis. An exception was made of cave heights, since these data were not then known to be related systematically to spatial position and therefore were not considered in the choice of well locations in the San Andres.

FIGURE 16.5. *A–F* Lithologic evidence of karst in slabbed cores from Well C, Figure 16.4. All carbonate shown in these photographs is dolomite, but presumably was calcite at the time of karstification. The scale bars represent 1 cm in length. *A.* Internal or cave sediment, consisting of ooliths and subarkosic silt, filling a depression near the top of the San Andres Formation. The San Andres country rock (lower half) is skeletal packstone. *B.* Another type of internal sediment (a calcilutite, dark gray), 10 feet (3 m) below the San Andres unconformity, filling in around a jumble of skeletal packstone breccia fragments. Later solution of the lithified internal sediment produced vugs (black) rimmed by thin rinds of isopachous cement, probably of phreatic origin. *C.* Banded dolomite flowstone/dripstone (and isopachous phreatic cement?) more than 2 cm thick, resting on a floor of cave sediment. This sample comes from a depth of 11 feet (3.4 m) below the San Andres unconformity. Most of the banded material is flowstone/dripstone (travertine) and consists of cement precipitated in an air-filled vug or cave. *D.* Two gen-

TABLE 16.1. Log/drilling indicators of San Andres caves.

	Cave property		
	Width	Height	Porosity (ϕ)
Property derived from	Differential[a] caliber (in)	Bit drop or caliper or density (ft)	Bulk density (ρ_b) (gm/cm^3)
Case 1	(No deflection required)	≥ 1	≤ 1.75 [$\phi \geq 59\%$][b]
Case 2	≥ 4	≥ 1	$1.75 \leq \rho_b \leq 2.00$ [$\phi = 59\%$ to 45%][b]

[a]Difference between maximum deflection and deflection of in-gauge hole.
[b]Mineral density of dolomite (2.83 gm/cm^3) assumed for calculation of porosity.

Cave Heights

Data on the heights of caves in the San Andres Formation are summarized in Figure 16.7. Included are those caves detected by wireline logs and those reported as bit drops. Here, as in Figure 16.6, the contrast between eastside and westside Yates is immediately apparent. The modal population for both sides of the field centers on 2 feet (0.6 m) and the arithmetic average is 2.9 feet (0.9 m). The range is from 1 foot (0.3 m) (our lower limit of "cave," fixed by logging tools) to 21 feet (6.4 m). The marked difference between bit drop caves and log caves probably reflects (1) the irregular geometry and small size of most of the caves relative to the borehole diameters, and (2) a tendency on the part of drillers not to report small bit drops. It can be seen in the eastside population that as cave height increases, the numbers of bit drop caves and logged caves tend to converge.

The spectrum of cave heights points to a system of rather small caves.[2] This may mean that the time of cave formation was short or that the vigor of the hydrologic system or systems responsible for the caves was on average at a low level.

---

[2]Some readers will note that we have chosen not to observe the definition of cave proposed by Choquette and Pray (1970), which limits application of the word *cave* or *cavern* to a void large enough to permit passage by a man. Because of the significant contribution of smaller voids to the high flows of oil from many wells drilled into the San Andres at Yates, we defined the minimum size (height) at 1 ft (0.3 m), which is approximately the lower limit of detection by both the density log caliper and the bulk density porosity log.

---

erations of cave-filling calcisiltite and calcilutite sediment separated by a sinuous contact (arrow) from 34 feet (10.4 m) below top of San Andres. The apparent sequence of diagenetic events included:

1. solution of country rock (CR)
2. deposition of first generation of internal sediment in cavity (IS1)
3. lithification of IS1
4. solution of IS1 to give another cavity
5. deposition of second generation of internal sediment (IS2)
6. lithification of IS2
7. renewed solution of IS1 and IS2

E. Fractured and brecciated San Andres Dolomite from 47 feet (14.3 m) below the unconformity. The longer vertical fractures have been widened by solution or possibly by extension due to bending of the field structure during Ochoan time. F. A vertical channel dissolved along a fracture about 100 feet (30.5 m) below the San Andres unconformity. Solution at this depth in the formation must be related to one of the earlier intraformation unconformities which affected the middle San Andres rather than to the later San Andres hydrologic events which are the main subject of this paper.

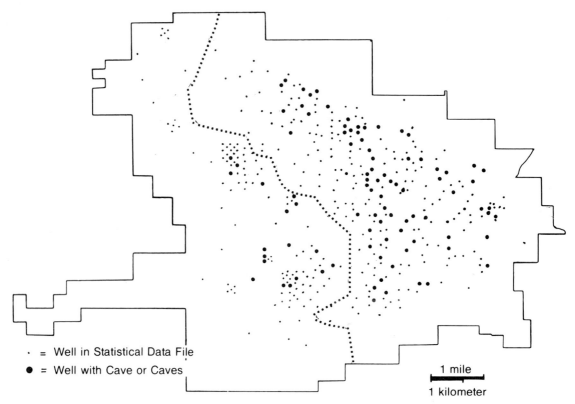

· = Well in Statistical Data File

● = Well with Cave or Caves

1 mile

1 kilometer

Figure 16.6. Four hundred Yates field wells drilled before 1982 which were used in the statistical study of San Andres caves. Ninety-two (23%) of the wells have caves at least 1 foot (0.3 m) high. A total of 179 caves were detected before 1982, 152 in eastside Yates and 27 in westside Yates. It was assumed that wells drilled before 1982 were drilled without certain knowledge of the caves and therefore were drilled randomly with respect to the cave population.

Spatial Distribution of Caves, Present and Past

The San Andres caves in the statistical population have been examined in three ways. The first is in terms of elevation above present sealevel; the second is in terms of elevation relative to a paleodatum (that is, in terms of sealevel and a paleotopographic configuration at some time in the past); and the third is in terms of depth below the San Andres unconformity itself.

The concepts and data that relate the caves to present sealevel are contained in Figures 16.8 and 16.9. Figure 16.8 shows in plan view (16.8A) and in cross-section (16.8B) the structure of the San Andres Dolomite in the field area. The structure is a gentle asymmetrical anticline or dome with dips nowhere in excess of 2° or 3°. (Considerable vertical exaggeration

was necessary in Figure 16.8B to bring out the structure at the lateral scale necessary for the figure to fit on a page.) Figure 16.9 shows the cave population for eastside and westside Yates expressed in cave feet (vertical feet of cave space) and related to present structural elevation. The westside population has a seemingly uniform distribution with depth, but the eastside caves show a pronounced mode over an interval of about 100 feet (30.5 m). The mode centers on +1200 feet (+366 m) above sealevel, and the entire range of cave feet extends roughly 200 feet (61 m) below that point and 200 feet (61 m) above it in a manner which suggests a normal distribution, as shown by the dashed line. The virtual absence of data below +1150 feet (+351 m) is related to the fact that before 1982 almost no wells were drilled deeper than that elevation.

The data that result from relating the San

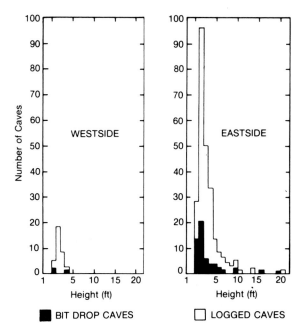

FIGURE 16.7. Heights of San Andres caves determined from logs and drilling records. Data from wire-line logs were derived from caliper and bulk density curves, and bit drop data from drillers' reports. Eighteen wells and 35 caves were represented by the westside Yates population, and 77 wells and 255 caves by the eastside population. Only 3 (9%) of the caves in the west and 52 (20%) of the caves in the east have been reported as bit drops. It is presumed that many more bit drops occurred than were reported, but with a marked tendency for the higher caves in the eastside population to be reported as bit drops.

Andres population to a paleodatum—in this case the M horizon in the Seven Rivers Formation—are summarized in Figures 16.10A and 16.11; M corresponds to the top of a thin unit of silty dolomite which is traceable throughout the field and well beyond the field along the Central Basin platform. It is part of a sequence of sabkha and shallow salina evaporites (the Seven Rivers) (Spencer and Warren 1986) and probably was essentially horizontal and isochronous when deposited. We have assumed that M is a satisfactory substitute for the horizontality of Late Permian sealevel generally and have used it as such in restoring the San Andres carbonates and their caves to their approximate spatial relationships of 250 million years ago.

Histograms of cave incidence expressed as cave feet as a function of depth below the M datum are shown in Figure 16.10A. As in Figure 16.9, the westside population appears to be evenly distributed over roughly 200 feet (61 m) of section. The eastside population, on the other hand, has a distinct mode in the vicinity of 200 feet (61 m) below M and a sharp cutoff 30 feet (9.1 m) below that point. The outline of the histogram looks much like the hypothetical profiles in Figure 16.10B in which LeGrand and LaMoreaux (1975) related circulation and solution to the free water table in exposed limestone. In light of the principles expressed in Figure 16.10B, it seems reasonable to interpret the histogram of Figure 16.10A for the eastside cave population as a reasonable picture of dissolution concentrated below an unconfined water table in a limestone terrane.

While the site of cave development associated with the principal San Andres unconformity is at 200 to 230 feet (61 to 70 m) below M, other zones—with fewer caves—can be identified at 270 to 320 feet (82 to 98 m), 350 to 360 feet (107 to 110 m), 370 to 380 feet (113 to 116 m), and 410 to 420 feet (125 to 128 m) below the M horizon (Fig. 16.10A). The deeper two of these four zones of weaker cave development have been identified with lithologic evidence of unconformity (specifically brecciation and erosion) from cores cut at appropriate depths-below-M in the eastside San Andres.

Use of M as a datum to restore San Andres caves to their original vertical positions can also be used to restore the unconformity surface of the San Andres Formation to its Late Permian (Guadalupian, Kazanian) configuration, so that the paleotopography of that surface can be related to the caves which underlie it. Figure 16.11 represents such a map on which the small local highs may be taken to represent karst towers and the closed depressions may be interpreted as sinkholes; it also represents a flooding of the unconformity to the level of 200 feet (61 m) below M. The result is the cluster of islands shown by the hachure pattern. These San Andres islands are also shown in Figure 16.12 in relation to the areal distribution and contrasting compactibilities of the lithofacies that comprise the upper San Andres Formation in eastside and westside Yates. The eastside is-

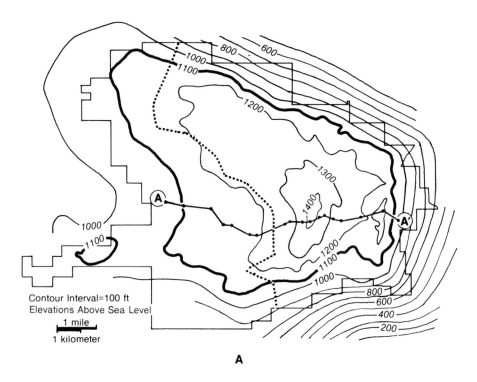

A

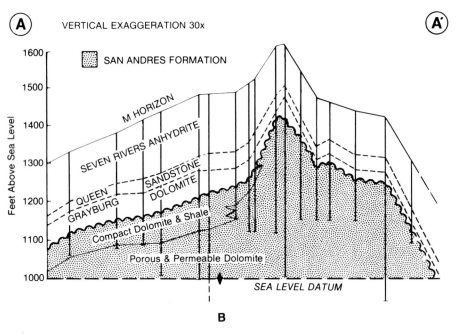

B

Figure 16.8. *A, B. A.* Present structure on top of the San Andres Formation in Yates field. Irregularity of the +1100 ft (+335 m) contour and higher contours is due to differential solution at the San Andres unconformity. Comparison of this map with the map of Figure 16.11 shows the significant differences between present structure and Late Permian structure/ topography on the San Andres. *B.* West to east cross-section along the line A-A' on the map of 16.8A. Approximate parallelism of the Grayburg, Queen, and M horizon surfaces indicates that the field area was warped structurally after deposition of the M horizon. This probably took place during the Ochoan Epoch.

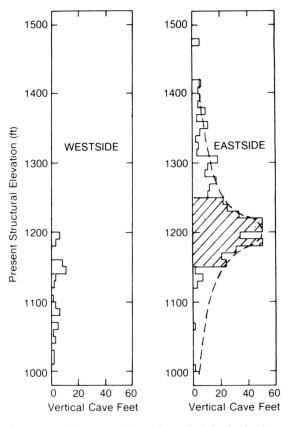

FIGURE 16.9. Cave feet (feet of cave height) in the San Andres Dolomite as a function of present elevation above sealevel. The sharp reduction in cave feet at +1150 feet (+351 m) is a result of the fact that from 1977 to 1981 most infill wells were bottomed at that elevation. The mode of the eastside population appears to center on +1200 feet (+366 m), and were it not for the relative scarcity of wells drilled below elevation +1150 feet (+351 m), the distribution of cave feet about the mode might be statistically normal as shown by the dashed curve. Cave feet versus present structural elevation in westside Yates shows an essentially uniform (random?) distribution between elevations +1000 feet (+305 m) and +1200 feet (+366 m), (as compared to eastside Yates).

lands occur in an area of grain-supported (former) limestones which, being relatively incompactible, stood higher on the surface of the San Andres unconformity. Having the advantages of higher paleotopographic elevation and high porosity and permeability as well, they became preferentially the sites of cave development during exposure to Late Permian rainfall (see Fig. 16.12).

Depending on the length of the low stand and the amount of rainfall in the area, we can presume (the caves and karst lithologies being our evidence) that freshwater lenses developed beneath the larger of these islands, as shown in the cross-section of Figure 16.13. The lenses must not have exceeded 30 feet (9.1 m) in thickness, judging from the histogram for eastside caves in Figure 16.10A; and sealevel and the tops of the freshwater lenses must have been at about 200 feet (61 m) below the M datum. As will be demonstrated later in this chapter, the paleoelevation of the water table relative to the paleotopography of the San Andres is confirmed by data on average cave heights.

Budd (1984, p. 100) has shown that for two islands in the Bahamas the ratio of lens thickness to minimum width of island ranges from 0.5% to 1.5%, "regardless of the island size or whether a steady-state or transient freshwater lens exists." The apparent 30-foot (9.1-m) lens thickness demonstrated in Figure 16.10A is probably an average for the San Andres islands, as indicated above. Therefore, depending on whether the ratio of lens thickness to island width is 0.5% or 1.5%, island width may be from 2000 to 6000 feet (610 to 1829 m). This range of island widths appears to be reasonable when compared to the map of the San Andres islands in Figure 16.11.

Cave abundance expressed as a function of depth below the top of the San Andres Formation is shown in Figure 16.14. From the graph of the eastside population it appears that most caves occur within 50 feet (15.2 m) of the top of the San Andres. This makes sense, because the San Andres islands probably averaged about 20 feet (6.1 m) above paleosealevel (see Fig. 16.11); and the 20 feet (6.1 m) added to the 30 feet (9.1 m) of lens thickness established by Figure 16.10A gives a total of 50 feet (15.2 m), as Figure 16.14 shows.

Joints, Caves, and Sinkholes

Closed depressions (sinkholes) and conical highs (karst towers) are illustrated on the map of Figure 16.11. Figure 16.15 shows again the

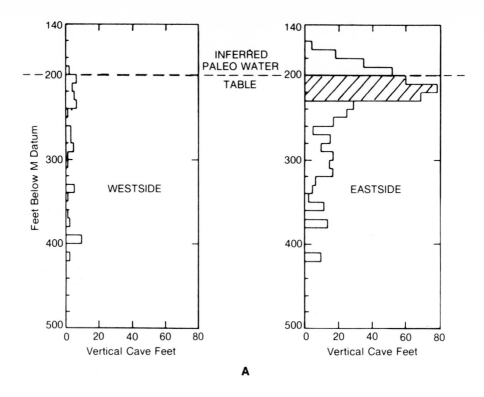

A

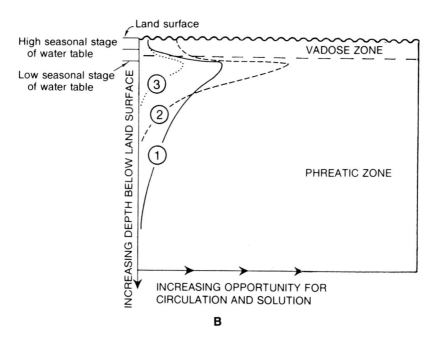

B

FIGURE 16.10 A, B. A. Vertical cave feet (feet of cave) as a function of depth below the M datum, a horizon in the Seven Rivers Anhydrite (see Fig. 16.3). Because of its association with sabkha deposition, M is thought to have been essentially flat and parallel to Late Permian sealevel. Thus the M datum has been used as a reference to restore the San Andres carbonates and caves to their Late Permian spatial relationships. Transformed in this way, cave abundance in eastside Yates is seen to be focused in a relatively thin zone of 30 feet (9 m) (shown by hachures) as compared to the 100 feet (30.5 m) of modal interval which results when present sealevel is used as the datum (Fig. 16.9). B. Relation of water circulation and solution to the water table in a limestone terrane. Profile (1) represents a usual condition and profiles (2) and (3) less common but not unusual conditions related to greater (2) or lesser (3) flow in the upper part of the phreatic zone. Master caves tend to develop and progressively concentrate flow

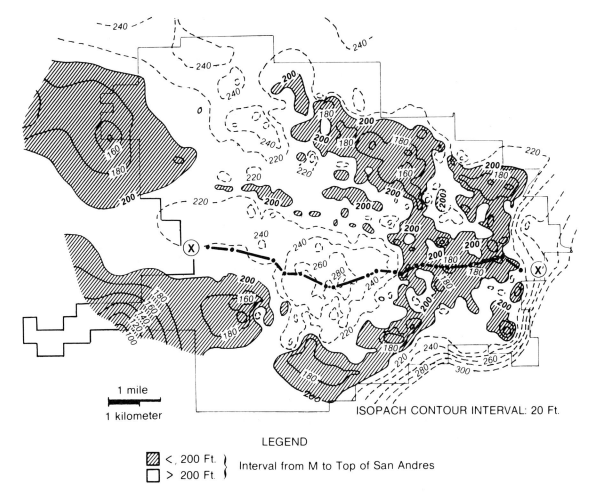

1 mile

1 kilometer

ISOPACH CONTOUR INTERVAL: 20 Ft.

LEGEND

< , 200 Ft. }
> 200 Ft.   } Interval from M to Top of San Andres

FIGURE 16.11. Paleotopography on the San Andres unconformity expressed in terms of the isopach interval from Seven Rivers M to top of San Andres. Contours generated by computer. Thins represent topographic highs and thicks are topographic lows on the unconformity surface. Closed lows (thicks) are interpreted as karst sinkholes and closed highs (thins) as karst towers or knobs. Hachured areas are San Andres islands which were created when the unconformity surface was drained or flooded to the level of 200 feet (61 m) below M, that is, to the elevation of the Late Permian (Guadalupian) sea and also the elevation of the island water tables indicated by histograms of cave feet and cave height in Figures 16.10A and 16.17. M to San Andres intervals greater than 220 ft (67 m) in the interior of the field area probably are due to the physical and chemical compaction of shales and mud-supported limestones present there in the upper San Andres (see Figs. 16.12 and 16.13). Presumably, final compaction of this section took place after island time and in response to the addition of hundreds of feet of sedimentary overburden.

in the upper part of the saturated zone at the expense of other solution passageways farther below the water table. Diagram modified from LeGrand and LaMoreaux (1975). Published by permission, International Association of Hydrogeologists.

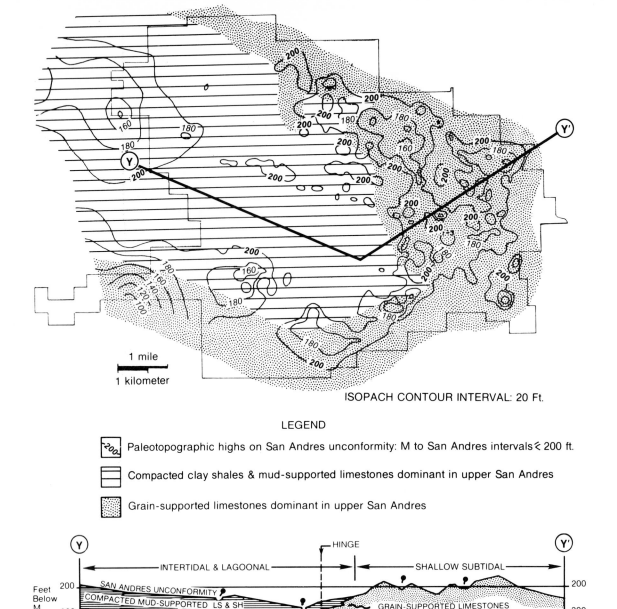

ISOPACH CONTOUR INTERVAL: 20 Ft.

LEGEND

⌐‑200⌐ Paleotopographic highs on San Andres unconformity: M to San Andres intervals ≤ 200 ft.

▤ Compacted clay shales & mud-supported limestones dominant in upper San Andres

▧ Grain-supported limestones dominant in upper San Andres

FIGURE 16.12. Distribution of San Andres islands in relation to major depositional facies of the upper San Andres Formation. Compare with Figure 16.11. Note the tendency for most of the paleotopographic highs (islands) to occupy the band of relatively un-compactible, grain-supported limestones (stipple pattern), the shelf-margin facies. As shown on the map and in cross-section Y-Y' (and in Fig. 16.11), mud-supported limestones and clay shales (lined pattern) of the intertidal–lagoonal facies tend to co-incide with topographic lows on the San Andres unconformity, a characteristic related to physical and chemical compaction of the sequence. The pre-sumption is that compaction of the clay shales alone (not including the effects of stylolites) probably reduced the thickness of the sequence by at least one-third to one-half (Fig. 16.13) and in general accentuated the paleotopographic relief between the mud-rich westside of the field area and the grain-rich eastside. Thus, the eastside stood higher topographically, and the eastside islands became sites of karstification during the lowstand(s) of the Permian sea at 200 feet (61 m) below M. At the same time, the islands of the westside, composed largely of low permeability limestones and shales, developed less extensive systems of caves (Fig. 16.10A) and other

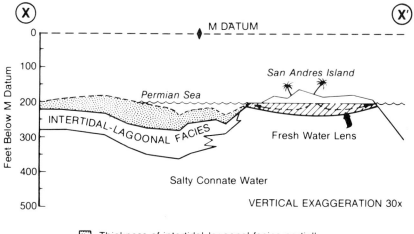

Thickness of intertidal-lagoonal facies partially
restored by decompaction of shale interbeds.

FIGURE 16.13. West to east cross section X-X' in Figure 16.11, showing island and seafloor paleotopographic profile, westside stratigraphy, and sealevel and island hydrology during San Andres island time. The key paleoelevation, 200 feet (61 m) below M, is related to the principal cave population of Figure 16.10A, and it defines the shoreline and freshwater table of the San Andres island. Estimated vertical expansion of the intertidal–lagoonal facies (stipple pattern) by decompaction of shale interbeds suggests the possibility that the sediments of the westside area were awash or only slightly submerged during island time rather than in water depths of up to 80 feet (24.1 m) as might be interpreted from Figure 16.11.

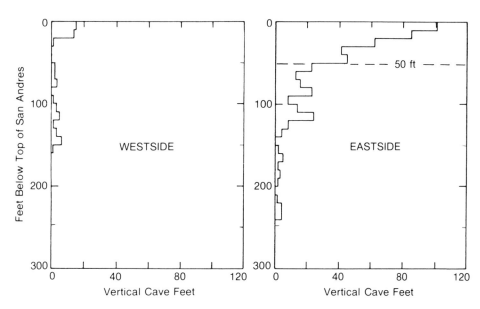

FIGURE 16.14. Cave abundance as a function of depth below the top of the San Andres. Tendency for most caves to occur in the upper 50 feet (15.2 m) of the formation is in harmony with paleotopographic evidence for the low relief of the San Andres islands, as demonstrated in Figure 16.11.

karst features. Interpreted sinkholes in the deepest paleotopographic lows (cross-section Y-Y') probably are genetically related to intraformation unconformities in the middle San Andres, but they may have been drowned and largely inactive during much of upper San Andres island time.

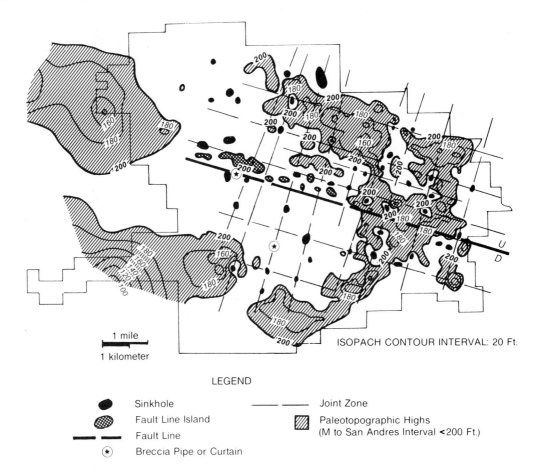

ISOPACH CONTOUR INTERVAL: 20 Ft.

1 mile

1 kilometer

LEGEND

⬤   Sinkhole   ———————   Joint Zone

▨   Fault Line Island

—— —   Fault Line   ▨   Paleotopographic Highs
                                      (M to San Andres Interval <200 Ft.)

(★)   Breccia Pipe or Curtain

FIGURE 16.15. Hypothetical pattern of orthogonal joints with joint intersections related to closed depressions (sinkholes) on the San Andres unconformity. Location of sinkholes was determined from the paleotopography of Figure 16.11. Azimuth of the joints is fixed by trend of a supposed fault line which in turn is related to a rectilinear string of small islands bearing west–northwest across the Yates area. There is a relatively uniform distribution of sinkholes across the field, in spite of the strong contrast between upper San Andres lithofacies in eastside and westside Yates (Figs. 16.4 and 16.12). This suggests

that the joint-sinkhole system may have been established during earlier times of San Andres unconformity (as in the cross-section of Fig. 16.12) and, once in place, tended to propagate upward through added prisms of carbonate rock, irrespective of their lithologic characteristics. Wells cored at the two locations shown by circled stars recovered from the San Andres intensely brecciated dolomite in a matrix of green clay, a lithologic assocation that might be expected at points of solution and collapse at sinkholes or along the planes of joints.

San Andres islands and highlights in black the closed depressions which probably were sinkholes on the surface of the San Andres unconformity. The relationship between sinkholes in limestone terranes and the intersection of joints or fractures is well known (e.g., Bretz 1942, Lattman and Parizek 1964, Davis and DeWiest 1966).

A possible fit of orthogonal joints to the depressions interpreted as sinkholes in the San Andres surface is shown in Figure 16.15. An

azimuth for one of the joint sets was established by accepting a line of small islands trending west-northwestward through the center of the field area as evidence of faulting (the islands being interpreted as prominences on the upthrown [north] side of the fault) and by assuming that the fault began as a joint in a set of joints oriented west-northwestward.

Two wells cored at the locations of the circled stars in Figure 16.15 recovered dolomite breccias in a matrix of clay shale from more than

100 feet (30.5 m) of San Andres section. Both wells appear to have encountered vertically extensive pipes or curtains of collapse debris probably associated with solution at sinkholes or along joint planes.

It seems probable that the sinkholes located off the islands in Figure 16.15 were established on San Andres unconformities other than the island unconformity and that they were drowned and inactive during island time when sinkholes on the islands were part of a developing karst system. More discussion of this problem of interpretation is included in the captions for Figures 16.11, 16.12, and 16.13.

The Island Hydrologic Model

The freshwater lens or island hydrologic model or shoreline model is probably the best understood of hydrologic models. Its behavior has been observed in many places worldwide, and the hydrodynamics and geochemistry of its waters have been described in considerable detail (Hubbert 1940, Thrailkill 1968, Vacher 1974 and 1978, Vacher and Ayers 1980, Fetter 1980, James and Choquette 1984). Some of the consequences of the model in terms of cave development have been observed along shorelines (Back et al. 1984 and 1986) and in vertical shafts penetrating the vadose zone (Bretz 1960). Figure 16.16 shows the basic patterns of flow in and above a hypothetical freshwater lens, the zones of aggressive or reactive waters which accompany the activity of the lens, and the patterns of cave development that result from flow and enhanced solution.

Two important generalizations can be made about the lens system. The first is that the most aggressive waters are produced at sites in the phreatic parts of the lens system where waters of unlike chemistry or temperature are mingled and the mixture is undersaturated with respect to calcite. Those sites (Fig. 16.16B) are near the top of the lens, the upper part of the freshwater phreatic environment (1); in the outflow area along the island shoreline where lens water and marine water mingle by turbulence and tidal flow (2); and in the zone of transition that underlies the lens and marks the mixing of fresh and connate salt water, largely by ionic diffusion (3). In the near surface rock, in the upper part of the vadose zone (site 4), fluctuations in rainfall, surface temperature, and plant metabolism and decay, rather than mixing of waters, control the evolution and aggressiveness of the vadose water. The second general principle is that caves that have developed in response to the flow of aggressive waters tend to be elongated in the direction of flow (Fig. 16.16B). Caves in the vadose zone, where flow or percolation is vertical, tend to be shaft-like, whereas caves developed in the upper part of the lens tend to be horizontally elongated and with tabular or anastomosing geometry.

Cave Statistics and Island Model Dynamics

Random sampling of the San Andres Dolomite by large numbers of wells gives a statistically satisfactory picture of the San Andres island hydrologic environment at Yates. The caves—their numbers, their individual and aggregate heights, and their vertical and horizontal arrangements in the rock mass—make sense when compared to patterns of water flow predicted for lenses of fresh water beneath a cluster of small islands.

In the previous section we observed the theoretical relationships between the physical and chemical dynamics of the island model and what appears to be the likely response of the rock in terms of cave development. It remains only to compare the cave data from Yates wells with the hydrologic theory as it applies to two aspects of the island model. The first is to examine an expected contrast between the average heights of vadose and phreatic caves. The second is to look at the details of cave numbers and heights in the vicinity of the San Andres island shorelines to see if the data there show the intense solution known to be associated with that zone of turbulent mixed waters.

The spectra of average cave feet for various classes of cave height in wells drilled in eastside Yates are shown by histograms in Figure 16.17. Caves of different heights above the interpreted paleowater table are compared with caves below the paleowater table. Clearly evident in all but the class of smallest caves (1 to 2 ft) (0.3 to 0.6 m) is the tendency for average cave height in wells to be greater in the paleovadose zone of

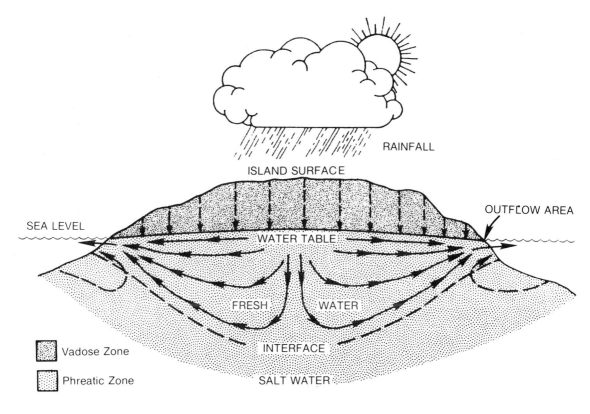

A

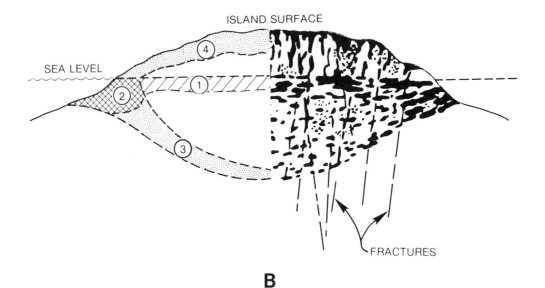

B

Figure 16.16. A, B. The island hydrologic model expressed in terms of water flow, zones of heightened solution, and the caves that might be expected to develop beneath a limestone island. A. Patterns of water flow (streamlines) in the island hydrologic model B. Sites of aggressive waters (left) and resulting caves (right) in a hypothetical limestone island. Caves tend to be elongated in directions of flow and are largest and most numerous at sites of mixed waters as at (1), (2), and (3), and in association with soil processes as at (4). Strong contrasts in water salinity, temperature, and CO_2 content, coupled with high flow rates, make sites (1) and (2) the most significant in terms of cave development.

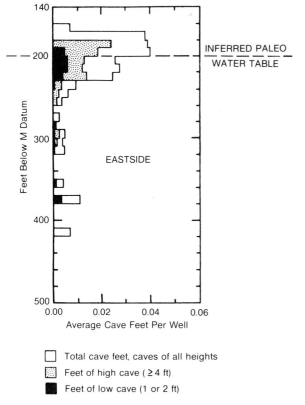

FIGURE 16.17. Average cave height in wells with caves in eastside Yates in relation to the paleoelevation of the principal water tables under the San Andres islands. The marked shift in average cave feet (or average cave height) at 200 feet (61 m) below M reflects the contrast between "high" shaft-like caves in the vadose zone of the islands and "low" tabular or tubular caves in the phreatic zone. These data confirm the choice of 200 feet (61 m) below M established in Figure 16.10A and used to define the San Andres islands in Figure 16.11.

the island than in the paleophreatic zone. Since the total number of vertical cave feet reaches a maximum below the paleowater table (Fig. 16.10A), and average cave feet per well reaches a maximum above the water table (Fig. 16.17), it follows that higher caves tend to occur above the water table, that is, in the vadose zone. This conclusion is in harmony with many observations (Davis 1930, Bretz 1942, Bruckner et al. 1972, Bögli 1980, and others) that vertical shafts or pipes are usually found in the vadose zones of unconfined carbonate aquifers where water movement is dominated by gravity drainage.

The results of analyzing cave abundance and size in relation to the San Andres island shorelines are shown diagrammatically in Figure 16.18. Histograms A, B, and C of the figure show distinct modes corresponding to the 500-foot (152-m) strip of island bounded by the interpreted paleoshoreline. More than one-third of the caves, one-third of the wells with caves, and one-third of the cave feet lie within this

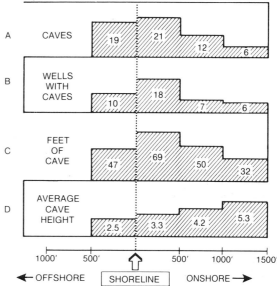

FIGURE 16.18. Histograms demonstrating a shoreline effect on the development of San Andres caves in eastside Yates. The caves are those from 200 to 230 feet (61 to 70 m) below M, which was the interval occupied by lenses of freshwater in Late Permian time. The interval is that shown by hachures in Figure 16.10A. Cave data are from 41 wells drilled before 1982 which are thus known to have been located without reference to what was later known about the distribution of the cave population. Histograms A, B, and C show clearly the relationship of cavernous conditions to the onshore zone within 500 feet (152 m) of the shoreline. This zone, with a freshwater lens present, was the site of mixing of fresh and ocean waters by turbulence and by tidal ebb and flow. Large and numerous caves were the result. The gradual increase in average cave height with increasing onshore distance from the shoreline (shown in histogram D) is a reflection of increasing tendency to vertical flow toward the center of the freshwater lens (Fig. 16.16A). Normalizing the cave data for lens volume or cross-section area might be expected to shift the modes in the upper three histograms to the offshore part of the lens system.

strip. The histogram of average cave height (histogram D), however, does not show a response to the shoreline as do the other measures of cave abundance and size. The reason for this may lie in the fact that toward the center of the freshwater lens in the island environment (see Fig. 16.16A) flow streamlines take on more of a vertical component; this presumably produces more tendencies for elongation of caves in the vertical direction. It seems likely that the modes in A, B, and C would be shifted to the offshore if the data represented were normalized for the lessening of lens volume in the offshore direction. Nevertheless, whether left as they are or adjusted for lens volume, the data support a clear picture of enhanced cave development in the vicinity of the island shorelines, which is in keeping with observations by Back et al. (1984 and 1986) and the author along the eastern coastline of the Yucatan peninsula in Mexico.

Conclusions

Paleotopography; sediment types, sedimentary structures, and cements associated with porosity development and destruction beneath the surface of the formation; and the presence of numerous unfilled caves—all of these testify that the San Andres Formation underwent significant karstification in the Yates field area during the Late Permian. Although cores and the distribution of caves indicate that there were several episodes of subaerial exposure during building of the San Andres carbonate pile, the principal unconformity marks the top of the San Andres and its contact with the Grayburg Formation.

The karst system probably developed beneath a cluster of low-relief limestone islands which were emergent by tectonic uplift or by one or more low-stands of the Permian seas. The principal agents of karstification are likely to have been dynamic freshwater lenses and associated zones of mixed fresh and marine waters. Well data on the numbers, heights, and spatial distributions of caves with respect to the topography of the San Andres islands yield patterns of a sort that would be predicted, given the hydrodynamics and geochemistry of the island model.

Acknowledgments. It has been almost 30 years since the first efforts were made by Marathon Oil Company to bring together rock samples, wireline logs, and production data from the Yates field to develop a synthesis of the reservoir geology. L.C. Pray, R.J. Lickus, J.T. Morgan, P.W. Choquette, W.K. Stenzel, W.J. McMichael, R.M. Williams, and the author were involved in this work during the 1960's and early 1970's. Much of our present understanding of the Yates field, on which the present paper depends, is the fruit of their individual and collaborative studies.

Closer to the present time and to the subject of this paper was a statistical study of San Andres caves during the period 1980 to 1984 by E.A. Nosal, J.L. Carlson, and C.A. Meeder. The author expresses his thanks to these coworkers for their many contributions; to R.A. Pfile, former manager of Yates Unit Engineering, for encouraging the cave study; and to the management of Marathon Oil Company for permission to publish this paper. T.J. Melton prepared the illustrations, and L.M. Lockrem typed the several drafts of the text.

References

Back, W., Hanshaw, B.B., Herman, J.S., and Van Driel, J.N., 1986, Differential dissolution of a Pleistocene reef in the groundwater mixing zone of coastal Yucatan, Mexico: Geology, v. 14, p. 137–140.

Back, W., Hanshaw, B.B., and Van Driel, J.N., 1984, Role of groundwater in shaping the eastern coastline of the Yucatan Peninsula, Mexico, *in* LaFleur, R.G., ed., Groundwater as a geomorphic agent: Winchester, MA, Allen and Unwin, p. 157–172.

Bögli, J., 1980, Karst hydrology and physical speleology: Berlin and Heidelberg, Springer-Verlag, 285 p.

Bretz, J.H., 1942, Vadose and phreatic features of limestone caverns: Journal of Geology, v. 50, p. 675–811.

Bretz, J.H., 1960, Bermuda, a partially drowned late mature, Pleistocene karst: Geological Society of America Bulletin, v. 71, p. 1729–1754.

Bruckner, R.W., Hess, J.W., and White, W.B., 1972, Role of vertical shafts in the movement of groundwater in carbonate aquifers: Ground Water, v. 10, p. 5–13.

Budd, D.A., 1984, Freshwater diagenesis of Holocene ooid sands, Schooner Cays, Bahamas: Ph.D.

dissertation, University of Texas, Austin, Texas, 410 p.

Choquette, P.W., and Pray, L.C., 1970, Geologic nomenclature and classification of porosity in sedimentary carbonates: American Association of Petroleum Geologists Bulletin, v. 54, p. 207–250.

Craig, D.H., and Schoonmaker, G.R., 1968, Yates oil field, Pecos County, Texas: Abstr., American Association of Petroleum Geologists Bulletin, v. 52, p. 523.

Davis, W.M., 1930, Origin of limestone caverns: Geological Society of America Bulletin, v. 41, p. 475–628.

Davis, S.N., and DeWiest, R.J.M., 1966, Hydrogeology: New York, London, Sydney, John Wiley & Sons, 463 p.

DeFord, R.K., and Wahlstrom, E.A., 1932, Hobbs field, Lea County, New Mexico: American Association of Petroleum Geologists Bulletin, v. 16, p. 51–90.

Fetter, C.W., Jr., 1980, Applied hydrogeology: Columbus, OH, Charles E. Merrill Publishing Co., 488 p.

Hubbert, M.K., 1940, The theory of groundwater motion: Journal of Geology, v. 48, p. 785–944.

James, N.P., and Choquette, P.W., 1984, Diagenesis 9. Limestones—the meteoric diagenetic environment: Geoscience Canada, v. 11, p. 161–194.

Kerans, C., and Parsley, M.J., 1986, Depositional facies and porosity evolution in a karst-modified San Andres reservoir—Taylor Link West San Andres, Pecos County, Texas, in Bebout, D.C., and Harris, P.M., eds., Hydrocarbon reservoir studies, San Andres/Grayburg Formations, Permian Basin: Proceedings, Research Conference, October 7–9, 1986, Midland, Texas; The Bureau of Economic Geology (The University of Texas at Austin) and The Permian Basin Section, Society of Economic Paleontologists and Mineralogists; PBS–SEPM Publication No. 86-26, p. 133–134.

Lattman, L.H., and Parizek, R.R., 1964, Relationship between fracture traces and occurrence of groundwater in carbonate rocks: Journal of Hydrology, v. 2, p. 73–91.

Legrand, H.E., and Lamoreaux, P.E., 1975, Hydrogeology and hydrology of karst: in A. Burger and L. Dubertret, eds., Hydrogeology of Karstic Terrains, International Union of Geological Sciences, Series B-Number 3, International Association of Hydrogeologists, 8 Rue Buffon, 75005 Paris, France, p. 9–19.

Shirley, K., 1987, Colorful history, odd geology—Yates field celebrates 60 years: AAPG Explorer, v. 8, n. 1, p. 4–5.

Spencer, A.W., and Warren, J.K., 1986, Depositional styles in the Queen and Seven Rivers Formations—Yates field, Pecos Co., Texas, in Bebout, D.G., and Harris, P.M., eds., Hydrocarbon reservoir studies, San Andres/Grayburg Formations, Permian Basin: Proceedings, Research Conference, October 7–9, 1986, Midland, Texas; The Bureau of Economic Geology (The University of Texas at Austin) and The Permian Basin Section, Society of Economic Paleontologists and Mineralogists; PBS–SEPM Publication No. 86-26, p. 135–137.

Thrailkill, J., 1968, Chemical and hydrologic factors in the excavation of limestone caves: Geological Society of America Bulletin, v. 79, p. 19–45.

Vacher, H.L., 1974, Groundwater hydrology of Bermuda: Public Works Department, Government of Bermuda, Hamilton, 87 p.

Vacher, H.L., 1978, Hydrogeology of Bermuda—significance of an across-the-island variation in permeability: Journal of Hydrology, v. 39, p. 207–226.

Vacher, H.L., and Ayers, J.F., 1980, Hydrology of small oceanic islands—utility of an estimate of recharge inferred from the chloride concentration of the freshwater lenses: Journal of Hydrology, v. 45, p. 21–37.

Young, A., David, M., and Wahlstrom, E.A., 1939, Goldsmith field, Ector County, Texas: American Association of Petroleum Geologists Bulletin, v. 23, p. 1525–52.

17
Paleokarst and Related Pelagic Sediments in the Jurassic of the Subbetic Zone, Southern Spain

JUAN-ANTONIO VERA, P.A. RUIZ-ORTIZ, M. GARCIA-HERNANDEZ, and J.M. MOLINA

Abstract

Several paleokarst terranes occur in the Jurassic of the Subbetic Zone (Betic Cordillera, southern Spain). They are related to stratigraphic breaks which affected the Subbetic basin during the Carixian (Middle Liassic), earliest Middle Jurassic, and the latest Bathonian. These paleokarst surfaces have been buried by pelagic sediments. They are interpreted to be related to the temporal and local emergence of pelagic swells resulting from both listric faulting and eustatic sealevel movement.

The following criteria were used to recognize paleokarst: (1) cavity morphology, (2) speleothems partially covering cavity walls, (3) collapse breccias with speleothem-enveloped clasts, (4) laminated continental cavern sediments, (5) freshwater phreatic and vadose cements in cavity host rocks, (6) age of pelagic sediments in neptunian dikes, and (7) karstic bauxites.

A model is proposed for the evolution of the pelagic swells which includes temporal emergence. Subaerial exposure surfaces are correlated with hardgrounds and condensed sequences on some swells; all are related to stratigraphic breaks which resulted from sudden global tectonic events and/or eustatic sealevel falls. The subsequent sealevel rise caused submergence and burial, cavity filling, and finally paleokarst fossilization. The proposed model of paleogeographic evolution is applicable to other, similar Mediterranean alpine realms.

Introduction

The term *paleokarst* is used, following Wright (1982 and 1986), to describe karst features formed by dissolution associated with a landscape of the past. Two types of paleokarst, relict and buried, have been recognized in the geological record. *Relict paleokarst,* following Jennings (1971) and Sweeting (1972), corresponds to present landscapes formed in the past and not covered with younger sediments. *Buried paleokarst* corresponds to karst landscapes formed in the past and later covered by sediment. Paleokarst surfaces are locally exposed due to erosion, in which case the term *exhumed paleokarst* is used (Buchbinder et al. 1983, Walkden 1974, Wright 1986). Buried paleokarst described in the geologic literature (e.g., Pirlet 1970, Read and Grover 1977, Van der Lingen et al. 1978, Farinacci et al. 1983, Cherns 1983, Esteban and Klappa 1983, Arnaud and Monleau 1984, Wright 1985, Garcia-Hernandez et al. 1986a and 1986b, Vera et al. 1986) is generally distinguished by such features as sinkholes, irregular surfaces, speleothems, cavern sediments, and various textural or fabric characteristics.

The examples of paleokarst reported in this paper formed at different times in the Jurassic and were buried by Jurassic sediments. The features are related to three different stratigraphic breaks affecting Jurassic sediments in a geosynclinal province known as the Subbetic Zone. Special attention is paid to the development of paleokarst on features interpreted to be pelagic swells, an association which to the authors' knowledge has not been documented before. The subaerial paleokarst surfaces pass laterally into hardgrounds and condensed marine sequences. All of these aspects are integrated in a model explaining the paleogeographic evolution of the pelagic swells.

Geological and Stratigraphic Setting

The Subbetic Zone is one of the main paleogeographic realms of the External Zones of the Spanish Betic Cordillera. Jurassic sedimentation in the Subbetic began with shallow marine carbonate deposition. During the Middle Liassic (Carixian), the shelf began to break up and founder, resulting in pelagic sedimentation (Garcia-Hernandez et al. 1976 and 1980, Azema et al. 1979) similar to that which occurred in other Alpine Mediterranean ranges (Bernoulli and Jenkyns 1974). This Carixian event has been interpreted to be the initiation of a continental margin (intracontinental rifting stage, Garcia-Hernandez et al. 1986a). During the rest of the Jurassic the Subbetic was a pelagic basin that underwent marked differential subsidence, resulting in troughs and swells. Three realms were formed as a result of this paleogeographic differentiation. The realms to the north and south, external and internal Subbetic, respectively, were mainly pelagic swells during the Middle and Upper Jurassic and covered by deposits of the "Ammonitico Rosso" facies and condensed sequences (Vera 1981). The third realm received deposits of marl, radiolarian marl, limestone, limestone turbidites intercalated with submarine volcanic and subvolcanic rocks (Garcia-Hernandez et al. 1980).

Information about Jurassic paleokarst in the Subbetic Zone can be found in papers dealing with various aspects of the regional stratigraphy (Busnardo 1979, Seyfried 1978 and 1979, Garcia-Hernandez et al. 1979, Dabrio and Polo 1985, Molina et al. 1985, Ruiz-Ortiz et al. 1985). Some studies have focused, however, on paleokarst events and related features such as neptunian dikes and bauxite deposits (Seyfried 1979 and 1981, Martin-Algarra et al. 1983, Garcia-Hernandez et al. 1986a and 1986b, Vera et al. 1984, Vera et al. 1986). In a monographic paper, Vera (1984) relates these features to sealevel fluctuations on sedimentary swells and makes comparisons with other Alpine Mediterranean cordilleras.

The paleokarst examples presented here are associated with three stratigraphic breaks that have been recognized in different parts of the Subbetic Zone. The first (Inter-Carixian) corresponds to the breakup of the extensive Liassic carbonate shelf, and was accompanied by local episodes of emergence and karstification of the underlying shallow marine limestones. The second break is a discontinuity of Early Middle Jurassic age which developed at the top of a shallowing-upward sequence on some of the sedimentary swells (Molina et al. 1985). The third event occurred in the Late Bathonian–Early Callovian and likewise gave rise to local emergence of the sedimentary swells, exposing both pelagic and shallow-marine limestones to subaerial conditions. Other episodes of emergence and karstification occurred in the Upper Jurassic and Lower Cretaceous but are not considered in this study.

Ten localities in southern Spain make up the data base for this study. Locations of the 10 sections, together with the paleogeographic setting in which they lie and the karstification stages observed in each of them, are shown in Figure 17.1. The Jurassic sequences in the sections in which stratigraphic breaks and related paleokarst surfaces occur are shown and correlated in Figure 17.2.

Paleokarst

The Jurassic paleokarst terranes are recognized on the basis of a variety of features: the morphology of their surfaces; the nature of the materials which fill cavities, specifically vadose sediments and speleothems; collapse breccias; soils and other edaphic features; and some diagenetic features of the wallrocks. Several of these criteria were used to confirm a paleokarst episode, and where criteria were not clear, interpretations were conservative. Thus, many erosion surfaces and cavities filled with pelagic sediments (neptunian dikes) without continental fillings or speleothems were not considered to be related to subaerial karst events, because cavities with very similar morphologies to karstic voids can be generated by submarine erosive processes, as some authors (Wendt 1971, Hallam 1975) have suggested in other Mediterranean Alpine regions. Where fossiliferous marine sediments fill karst cavities, it is possible to roughly date the time of paleokarst burial or fossilization.

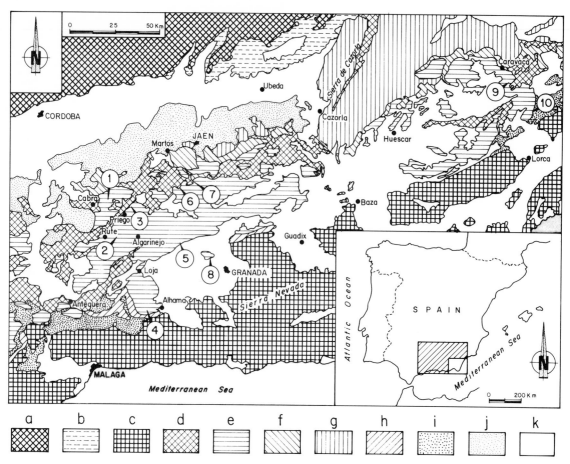

FIGURE 17.1. Geological and geographical position of the studied localities. KEY: (a) Iberian Hercynian fold-belts (Spanish Meseta). (b) Tabular cover of the Spanish Meseta (Triassic and Jurassic). (c) Internal Zones. (d) Triassic (mainly Keuper) of the External Zones. (e) Subbetic Zone. (f) Intermediate Units between the Prebetic and the Subbetic. (g) Subbetic Zone. (h) Ultra-internal Subbetic, Dorsal and related units. (i) Campo de Gibraltar units (Numidian flysch and other turbidite formations making up several nappes). (j) Guadalquivir basin area, with Subbetic nappes and olistostromes (Guadalquivir units and Carmona nappe). (k) Neogene-Quaternary of mainly postorogenic basins.

Studied localities (in parentheses the paleokarst terranes at each locality, C—Carixian, M—Base of the Middle Jurassic, B—Latest Bathonian): (1) Sierra de Cabra, External Subbetic (B). (2) Loma de las Ventanas, south of Rute, Medium Subbetic (C). (3) Las Angosturas, north of Priego, External Subbetic (C,M?,B). (4) Sierra Gorda, south of Loja, Internal Subbetic (C,M,B). (5) Illora, Medium Subbetic (C). (6) Gracia, east of Castillo de Locubin, External Subbetic (C,M). (7) Noguerones, south of Valdepenas de Jaen. (8) Sierra Elvira, west of Granada, Medium Subbetic (C). (9) Sierra de Quipar, south of Cehegin, External Subbetic (M,B). (10) Zarzadilla de Totana, Internal Subbetic (C,M,B).

Karst Host Rocks

Rocks of different ages were karstified at several times during the Jurassic. These rocks are mainly shallow marine limestones that were deposited on wide shelves (Gavilan Formation, Lower Liassic) or on carbonate banks (Camerena Formation, Middle Jurassic and locally Ze-gri Formation, Upper Liassic). The more common facies types are: (1) fine-grained, light-colored limestones (tidalites), locally dolomitized, with pisolitic beds in the Gavilan Formation; (2) oolitic and oncolitic cross-bedded limestones in the Camarena Formation; (3) crinoidal limestones interpreted as sand-waves in the upper member of the Gavilan Formation

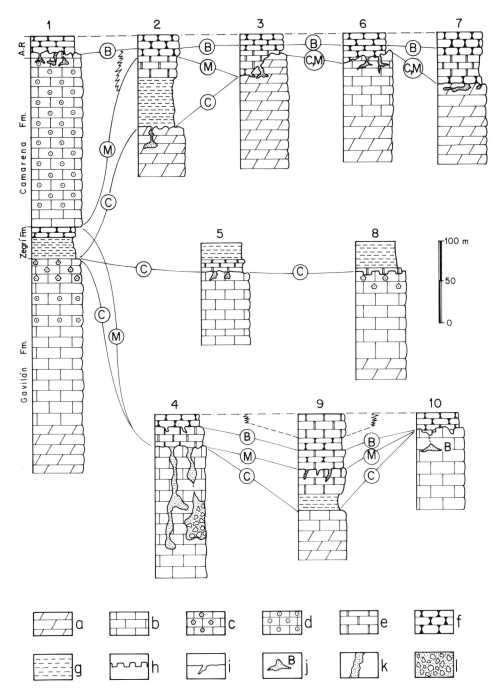

FIGURE 17.2. Jurassic stratigraphic sections of the localities studied. KEY: (a) Dolostones. (b) Neritic and tidal limestones. (c) Crinoidal limestones. (d) Oolite limestones. (e) Pelagic (locally cherty) limestones. (f) Nodular limestones (Ammonitico Rosso). (g) Pelagic marls and limestone-marl rhythmites. (h) Kamenitzas. (i) Karst sinkholes. (j) Karst bauxites. (k) Fillings of neptunian dikes (pelagic sediments and locally spelean sediments). (1) Collapse breccias. Uppermost horizontal line: Jurassic/Cretaceous boundary (datum) stratigraphic breaks: B—Latest Bathonian, M—Base of the Middle Jurassic, C—Carixian.

and locally in the top of the Camarena Formation. These formations are organized in shallowing-upward sequences (see Fig. 17.2) with paleokarst at the top; thus, the stratigraphic and sedimentological criteria are consistent with the development of an emergent stage.

A fourth group of host rocks consists of pelagic limestone. These include condensed sequences of red pelagic limestone, mainly of the Middle Jurassic, Ammonitico Rosso facies, and grey pelagic limestone capping a limestone-marl rhythmite (Zegri Formation). Both types of pelagic limestone were deposited on pelagic sedimentary swells that later became emergent and were karstified.

Morphology

It is possible to observe the paleokarst morphology both in horizontal section and, more commonly, in cross-section. Fracture-controlled, funnel-shaped features of various sizes with locally flat bottoms (localities 1, 3, 10) are interpreted as sinkholes (Fig. 17.3C, D). At lo-

cality 8 there are good examples of medium-sized karren with solution basins ranging in depth from 10 cm to 25 cm depth and with diameters on the order of a few meters. The karren have flat tops and bottoms (Fig. 17.4). Solution basins or kamenitzas like these are considered characteristic of coastal karst (Esteban and Klappa 1983). Similar features have been described in the Penibetic Cretaceous paleokarst (Company et al. 1982, Gonzalez-Donoso et al. 1983).

Cave systems in the Jurassic rocks show varied development (Fig. 17.2). At a level 100 m below the stratigraphic break at the top of the Gavilan Formation (Lower Liassic), there are speleothem-covered caves with geopetal internal sediments and collapse breccias (Garcia-Hernandez et al. 1986b). The caves form a complex network of openings controlled by fractures in the upper part. The same karst event is present at locality 1 (Vera et al. 1984). Bauxite-filled caves vertically related to sinkholes (Vera et al. 1986) occur in the Gavilan Formation, (locality 10). At locality 2, cavities penetrate 36 m to 42 m down from the

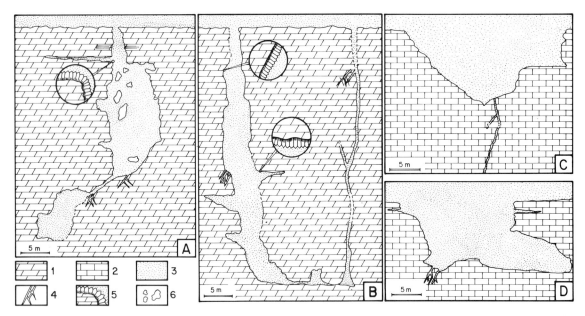

FIGURE 17.3. *A, B* Morphology of some paleokarst cavity walls, drawn directly from field observation. *A, B.* Neptunian dikes in the uppermost part of the Gavilan Formation (Lower Liassic) at locality 2; the fillings are pelagic limestones of Toarcian age and locally speleal sediments and blocks of wallrock. *C, D.* Two examples of paleodolines at locality 10; the wall-rocks are limestones of the Gavilan Formation (Lower Liassic), and the fillings consist of pelagic limestones and Ammonitico Rosso facies of Oxfordian age. KEY: (1) Dolostones and dolomitic limestones. (2) Neritic and tidal limestones. (3) Pelagic sediments. (4) Small neptunian dikes with fillings of pelagic materials. (5) Speleothems. (6) Blocks of wallrock.

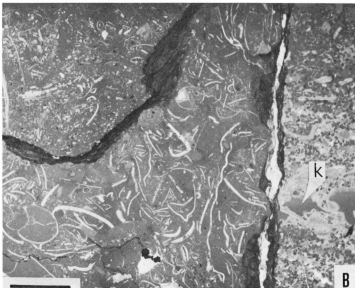

FIGURE 17.4. *A, B* Kamenitza morphologies. *A.* Field photograph of locality 8. *B.* Thin section; bar is 0.5 cm long. The wallrock (on the right) is a crinoidal limestone of Carixian age, and the cavity filling (center) is a biomicrite with abundant sections of Lower Domerian ammonoids; minor breaks can be observed in the pelagic sediment. Note the small karstic cavities (k) with internal sediments in the wallrock.

top of the Gavilan Formation (Fig. 17.3A and B, respectively), are filled by pelagic sediments (neptunian dikes), and have partial coverings of speleothems on their walls (Garcia-Hernandez et al. 1986a). At locality 9, fracture-controlled cavities with speleothems occur in the upper member of the Zegri Fm. (Upper Liassic).

Penetration depth of the karst cavities is considered to be a crude measure of the amplitude of emergence of the pelagic swells. The presence of cavities is to be expected in the vadose zones of gravity percolation (James and Choquette 1984), where flat-bottomed cavities are commonly elongate parallel to the water table or, in the case of these paleokarsts, with sealevel. Therefore, it can be assumed that the emergence of the pelagic swells could have been as much as 100 m or more, with different values at different positions on a given swell. The kamenitza-shaped cavities probably correspond to flat emergent areas close to sealevel, as presently seen in modern coastal karsts in tropical regions such as Florida and the Yucatan.

Speleothems

Speleothems are among the more useful features in recognizing emergence and karstification. The term *speleothem* is used here in the sense of Thrailkill (1976), Folk and Assereto (1978), and Esteban and Klappa (1983), and is equivalent to spelean sediments of White (1978). It refers to every type of precipitated calcium carbonate deposit covering the walls, roofs, and floors of dissolved cavities, including stalactites and stalagmites.

In many cases, the Jurassic speleothems are flowstones on cavity walls. They have variable thicknesses ranging from a few millimeters to a few centimeters (Fig. 17.5). They show the greatest development at localities 1, 4, 6, and 7, and are of different ages according to the age of the exposure. Where speleothems fill an entire cavity, without marine fossiliferous sediments, it is difficult to differentiate Jurassic speleothems from modern ones.

Speleothems are quite commonly the first materials coating the karstic cavities. Under the microscope (Fig. 17.5), several layers of spe-

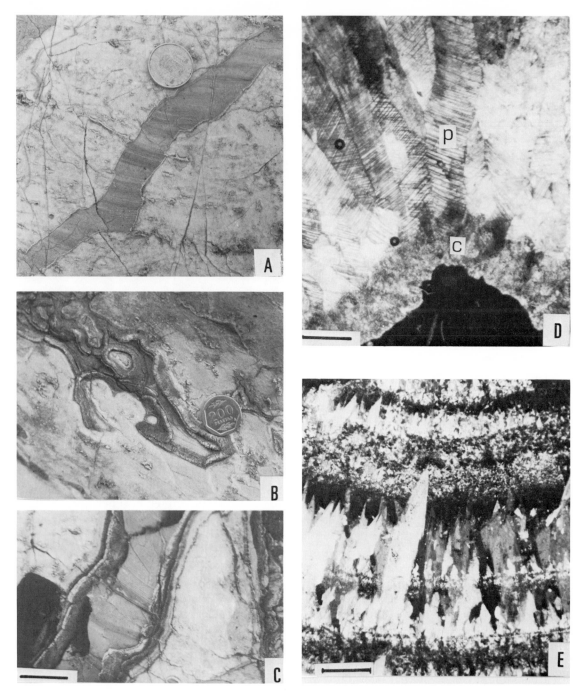

Figure 17.5. *A–G* Speleothems on the karst cavity walls. *A.* Cavity at locality 4 with speleothem cover and filled with geopetal internal sediments. *B.* Similar example with a thick speleothem (locality 4). *C.* Spelothem-covered cavity filled with geopetal internal sediment; bar is 2 cm long. *D.* Thin section of a speleothem showing length-slow "coconut-meat" calcite (c), and columnar crystals of palisade calcite (p); bar is 1 mm long. *E.* Closeup in thin section of a speleothem outer margin; bar is 0.5 mm long. *F.* Thin section of a speleothem with an inner cavity filled with geopetal internal sediment; bar is 0.2 cm long. *G.* Thin section of a cavity partly filled with speleothems showing their internal structure (c and p as in Fig. 17.5*D*, V = geopetal vadose sediment); bar is 0.2 cm long.

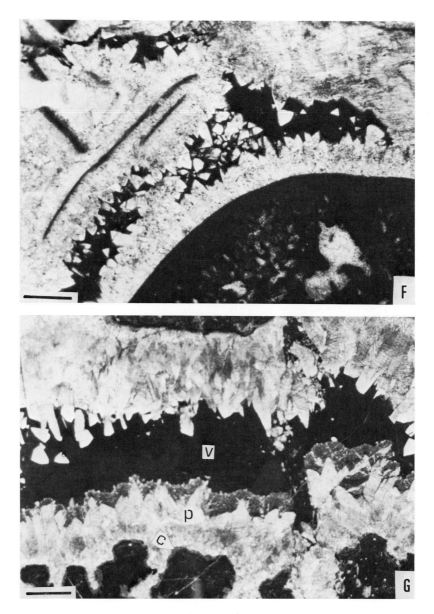

FIGURE 17.5

leothem growth, similar to those described by Folk and Assereto (1976) and Kendall and Broughton (1978), can be distinguished. The best examples comprise laminar flowstone growing directly on host rock and associated vadose geopetal sediment. The cement consists of several layers and is from numerous studied samples. An ideal zonation of speleothem fabrics within a cavity is proposed as follows.

1. A layer, 0.5 to 1.0 mm thick, of small (0.5 mm) fibrous calcite crystals, length slow, with a turbid yellowish appearance, which have nucleated directly on the host rock and in optical continuity with overlying coarser crystals. This layer, which is common in our samples, is similar to the "coconut-meat" calcite described by Folk and Assereto (1976) from speleothems. When present, this type of layer may recur between layers of longer columnar crystals described below.

2. A layer of palisade calcite consisting of clear and transparent radiaxial crystals without inclusions or equant calcite crystal aggre-

gates. The crystal length averages 0.3 mm but may range from 0.1 to 0.8 mm.

3. A second type of palisade calcite composed of large columnar crystals (2–3 mm long and 0.2 to 0.5 mm wide) that form centimeter layers and are separated by thin laminations of microgranular calcite. Crystals grow on the type 2 layers towards the center of the cavity, are radiaxial, and exhibit good undulose extinction. Crystal terminations are pyramidal and best seen directly beneath geopetal vadose sediment.

4. A layer of laminated calcite.

Internal vadose and pelagic sediments are deposited successively over these precipitates (Fig. 17.6). Where sediment is absent, coarse blocky calcite cement occludes the void. According to Folk and Assereto (1976), Kendall and Broughton (1978), and Assereto and Folk (1980), these attributes are typical of meteoric speleothem fabrics developed in response to variations in fluid composition and oscillations of the water table.

Breccia clasts are coated by speleothems (Fig. 17.7) and speleothems occur on top of, or intercalated with, continental internal sediments. The first features described are interpreted as flowstones deposited on breccia clasts, which were eventually covered completely; the second features are interpreted to be speleothems formed on temporary sediment "floors" of cavities.

Vadose Geopetal Internal Sediments

These occur either as partial fillings on floors of karst cavities or as matrix sediment in collapse breccias, most commonly in the lower

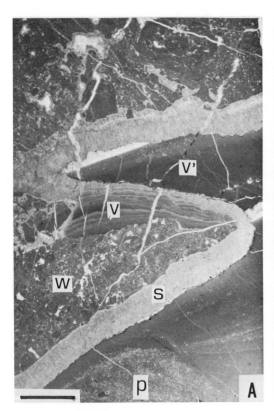
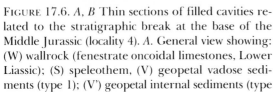
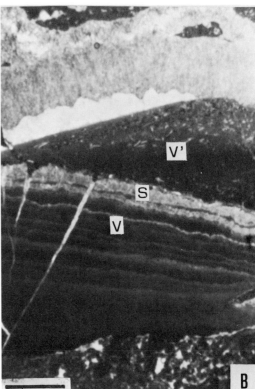

FIGURE 17.6. *A, B* Thin sections of filled cavities related to the stratigraphic break at the base of the Middle Jurassic (locality 4). *A*. General view showing: (W) wallrock (fenestrate oncoidal limestones, Lower Liassic); (S) speleothem, (V) geopetal vadose sediments (type 1); (V') geopetal internal sediments (type 2); (p) pelagic biomicrite with remains of thin pelecypods ("filaments") of Middle Jurassic age; bar is 0.5 cm long. *B*. Closeup of a similar sample in which the two types of geopetal internal sediments and the speleothem are clearly distinguished; bar is 0.2 cm long.

FIGURE 17.7. Field closeup photo of a collapse breccia with a speleothem-covered clast and matrix of geopetal vadose sediments.

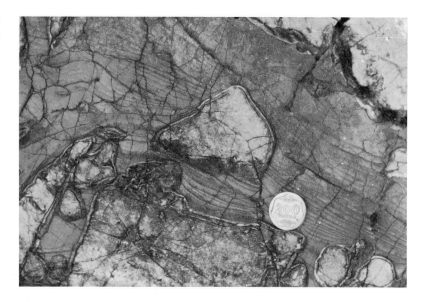

levels of caves. The sediments are present at localities 1, 2, 6 and most notable at locality 4. They consist of laminated, greyish-yellow, silty calcareous sediments that are commonly interlayered with calcite beds. Laminations are locally deformed as a result of postdepositional movement in the cavity interior.

Two types of geopetal internal sediments are distinguished at localities 4 and 6 (Fig. 17.6A, points V and V'). The first corresponds to vadose mechanical sediment, a finely laminated silt with layers of calcite which are more abundant toward the interior of the cavity. The layers are isopachous calcite, formed on top of beds of mechanical vadose sediment during pauses in sedimentation; they are similar to examples described by Esteban (1976) and Assereto and Folk (1980). This type of laminated sediment commonly overlies a layer of isopachous rhombic calcite (speleothem) (Fig. 17.6A, location s) believed to have been generated in the shallow freshwater phreatic environment. The second type of geopetal internal sediment (Fig. 17.6A, location V') is developed at the top of the speleothem, over a ferruginous surface of probable corrosion origin. This type has coarser laminations without interlayered cements, and consists of peloidal micrite with small bioclasts (pelecypods, ostracodes, and echinoderms) arranged in fining-upward, millimeter-thick sequences. Pelagic sediment (Fig. 17.6A, location p), or, in its ab-

sence, blocky calcite filling a void, overlies this type of geopetal internal sediment and is thought to represent former influx of overlying marine mud.

Host Rock Diagenetic Modifications

The cavity host rock, notably lime grainstone, was altered in a meteoric environment which formed during the early stages of karst development. Corroded grains in the cavity walls and rock dissolution, chiefly affecting ooids, are common features. Fabrics typical of freshwater phreatic origin are developed after this stage of corrosion and dissolution. Among these are (Fig. 17.8B,s): isopachous rhombic calcite coating ooids and bioclasts, and equant calcite coarsening towards pore centers (blocky calcite) and syntaxial cement overgrowths on echinoderm fragments. Because these fabrics are covered and so fossilized by speleothems and internal sediments in the paleokarst void (Fig. 17.8), they are additional evidence for dating the meteoric events. The timing of such events is discussed in a later section of this paper.

Neptunian Dikes

We use this term for cavities of any origin that are filled with younger marine sediment. Smart et al. (this volume) define *neptunian dikes* as bodies of younger marine sediment filling cav-

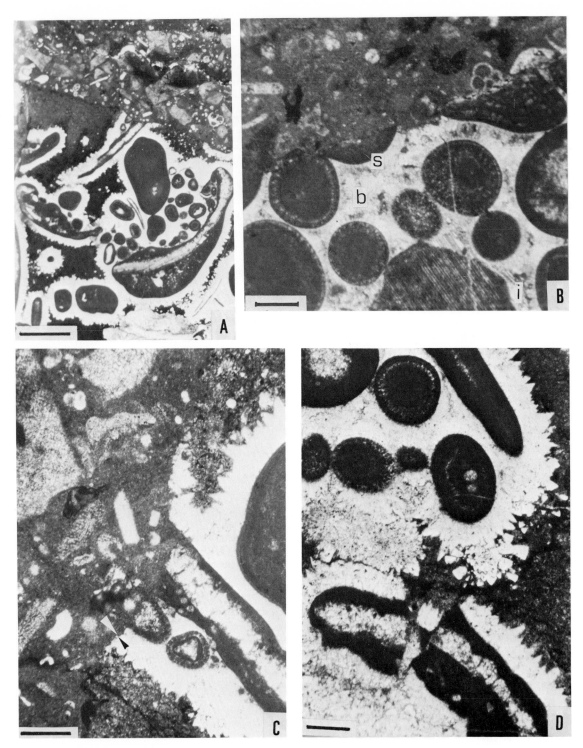

Figure 17.8. *A–D* Diagenetic features on the neptunian dike walls as the result of emergence, locality 4. *A.* Contact between wallrock (Camarena Formation Bathonian grainstones) and dike filling (Callovian); the dike filling is composed of condensed pelagic limestones with abundant crinoid bioclasts; bar is 2 mm long. *B.* Thin section showing isopachous rhombic calcite (i) around ooids and bioclasts (s), blocky calcite (b), coarsening towards pore centers and syntaxial overgrowths (i) on echinoderm fragments; bar is 0.2 mm long (C). Closeup of the contact (arrows); note the isopachous rhombic calcite covered (fossilized) by the pelagic sediment; bar is 0.4 mm long. *D.* Closeup of 17.8A, showing isopachous rhombic calcite around ooids and bioclasts; bar is 0.3 mm long.

ities in rocks exposed on the sea floor and *fissure fills* as continental sediments filling subaerial cavities. Repetition of both types has been observed in Subbetic outcrops. The many cavities that have mixed fillings of continental and marine sediments are considered neptunian dikes. Following Wendt (1971), we distinguish two types of neptunian dikes, on the basis of the geometric relationship between dikes and bedding: *Q fissures*, which are oblique to bedding and may extend as much as 100 m below the surface; and *S Fissures* (Fig. 17.9), which are parallel to bedding.

The neptunian dikes allow us to date more precisely the age of karstification as they are obviously younger than the host rocks but older than the cavity fillings. The ages of the fillings differ according to the related stratigraphic break. At locality 1, for example, dikes in the Middle Jurassic oolitic limestones (Camarena Formation) extend 50 m below the top of the formation and are filled with Callovian pelagic limestones (Fig. 17.9). At locality 2, the pelagic limestones filling dikes in Lower Liassic limestones and dolostones contain local brachiopod accumulations of Toarcian age. At localities 3, 6, and 7 limestones with remains of thin-shelled

pelecypods *(filaments)* and small ammonoids, fill cavities developed in Lower-Middle Liassic limestones. At locality 4, fillings of different ages, Carixian, Upper Liassic, and Middle Jurassic, occur in the Lower Liassic Gavilan Formation (Garcia-Hernandez et al. 1986a). At localities 5 and 8, the neptunian dikes in the Gavilan Formation are filled with fossiliferous pelagic limestones of early Domerian (Middle Liassic) age (Garcia-Hernandez et al. 1986a). At locality 9 Middle Jurassic pelagic sediments fill karst cavities in the pelagic limestones of the Upper Liassic. Finally, at locality 10, red nodular limestones (Ammonitico Rosso) of Oxfordian age occur in the cavities in Lower Liassic limestones (Vera et al. 1986). Details of some of these occurrences are described in Vera et al. (1984 and 1986), Ruiz-Ortiz et al. (1985), Molina et al. (1985), Garcia-Hernandez et al. (1986a and 1986b).

Marine Sediments in Cavities

These make up the bulk of the cavity fillings and generally represent the basal sediments of the deepening-upward sequences which overlie each stratigraphic break and paleokarst. All the

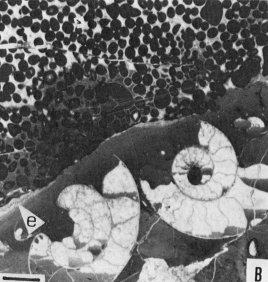

FIGURE 17.9. *A, B* Neptunian dikes at locality 1 related to the latest Bathonian stratigraphic break. The wallrock is Bathonian ooid limestone and the dike filling is pelagic limestone of Callovian age. *A.* Field photograph of an S-type neptunian dike (arrows) 2

m thick. *B.* Thin section showing a dike wall; accumulations of small ammonoids make up the filling and thin speleothems (e) and blocky calcite line in contact with wallrock; bar is 1.0 mm long.

marine sediments filling such cavities are limestones of the following facies types:

1. *Crinoidal grainstones* occur locally as initial fillings in cavities (e.g., locality 4). These internal sediments seem to be the same composition as the shoals which occur as stratified bodies on top of the sedimentary cycle on which the karst is developed. In the rare cases where this first filling does occur it is generally fossilized by geopetal vadose sediments and speleothems (Fig. 17.10). The crinoidal sediments also make up the matrix and some clasts in collapse breccias. In these cases they appear to have been affected by early freshwater diagenesis which resulted in syntaxial overgrowths on the crinoid clasts that predate subsequent internal sediments. This lithology occurs in cavities associated with all three Jurassic stratigraphic breaks but mainly in the Lower Jurassic.

2. *Poorly fossiliferous lime mudstones* occur within large cavities (Fig. 17.3A, B) as thick, massive, unstratified limestone showing no evidence of stratigraphic condensation. This type of filling is best developed at locality 2. Fossils are generally absent and when present are only brachipods and ammonoids. According to Seyfried (1979) and Garcia-Hernandez et al. (1986a), these deposits are thought to be pelagic sediments of insular platforms.

3. *Fossiliferous pelagic limestones* in cavities consist of cephalopod accumulations, in which the shells have undergone neomorphism. Other organisms, such as echinoderms (crinoids), brachiopods, pelecypods, some planktonic foraminifera, ostracodes, and rare calcitized radiolarians, among others, are also present. The sediments range in color from beige to red. They are a thin, condensed facies characterized by omission surfaces, local hardgrounds, and stylolites. They commonly have limonite crusts and associated "stromatolites." They fill small dikes and generally make up the first sediment deposited over hardgrounds, into which they pass laterally. These pelagic sediments are the most typical deposits of submarine swells.

4. *Ammonitico Rosso* facies occur, in the form of brecciated red nodular limestones filling neptunian dikes and paleodolines related to the Late Bathonian stratigraphic break. These limestones differ from the aforementioned types because they are not so condensed, have fewer fossils, and they have only molds of ammonoids. They are deposits of pelagic swells and adjacent slopes.

5. *Limestones with nodular chert* occur in only one section (locality 4), associated with the Early-Middle Jurassic paleokarst. They are fine-grained limestones which contain remains of pelecypods ("filaments"), radiolarians, and locally older limestone clasts, and are interpreted as deposits of more deeply subsident areas which were tectonically segregated by submarine swells (Garcia-Hernandez et al. 1986b).

Figure 17.10. Field photograph of a small neptunian dike hosted in the Gavilan Formation (shallow-shelf limestones, Lower Liassic) at locality 4. Speleothems cover the walls. From bottom to top, the filling is composed of crinoidal grainstone (c), geopetal vadose sediment (v), speleothem (s), and Middle Jurassic pelagic sediments (p).

Breccias

Breccias related to karst cavities occur in several studied outcrops. The best examples are at locality 4 (Fig. 17.11), where they are believed to be collapse breccias generated by rock-fall. The clasts, with sharp boundaries and in varying sizes, come mainly from the wallrocks, although clasts from previous dike-fillings and from vadose sediments are also present. Some clasts have diameters of more than a meter and locally may be covered by speleothems (Fig. 17.7). The breccia matrix has a varied composition. In the lowest parts of cavities, laminated vadose sediments prevail, while in the uppermost parts pelagic sediments make up the final cavity filling. The local occurrence of vadose sediment clasts may be evidence for several stages of dike filling interrupted by periods of partial erosion.

FIGURE 17.11. Field photograph of collapse breccia at locality 4. The clasts are of white Liassic limestone (Gavilan Fm.) and the matrix is composed mainly of laminated geopetal vadose sediment.

Bauxites

Bauxite is perhaps the most persuasive evidence of emergence and karstification of these rocks. Although only one body of bauxite has been mined recently (locality 10, Vera et al. 1986), many smaller masses are found in the region (see Fig. 17.2). The bauxites fill cavities (caves in this case) in shallow marine limestones of the Lower Liassic (Gavilan Formation). The caves are vertically connected along fissures and neptunian dikes to dolines developed on the top of the Gavilian Formation and covered by Oxfordian red pelagic limestones of the Ammonitico Rosso facies. Thus, karstification took place between the Carixian and the Oxfordian, probably during Middle Jurassic time. Bardossy (1982) presents evidence of numerous karstic bauxite bodies in Jurassic and Cretaceous sequences from other Mediterranean Alpine areas, related to extensive periods (5 m.y. minimum) of emergence and karstification.

Relationships Between Paleokarst and Stratigraphic Breaks

The episodes of emergence and karstification are related to three stratigraphic breaks which are present throughout the Subbetic basin. The first of these events took place in the Carixian, the second at the beginning of the Middle Jurassic, and the third towards the end of the Bathonian. According to the time scale of Westermann (1984) they correspond to 195, 180, and 170 m.y., respectively, or to 183, 173, and 157 m.y., following Vail et al. (1984). The first and third are major breaks that can be easily recognized anywhere in the basin, while the second is a minor break which outcrops only locally. The three breaks coincide with episodes of sudden sealevel fall (Vera 1984) in the External Zones of the Spanish Betic Cordillera, which were caused by local tectonic events and eustatic changes. All the described events are best understood by postulating shallow submarine swells, far off the continent. The presence of such swells is based on the type of pelagic sediments which were deposited. Similar interpretations have been made by D'Argenio (1974) and Farinacci et al. (1981) for the Apennines, and Wendt et al. (1984) for the Devonian of the Anti-Atlas (Morocco).

The Carixian break probably corresponds to a stage in the evolution of intracontinental rifting in the Subbetic zone (Garcia-Hernandez et al. 1986a), and signals the end of sedimentation on the large Liassic shallow-marine carbonate shelf which occupied the alpine realms of the Mediterranean region (Bernoulli and Jenkyns 1974) prior to the start of pelagic sedimentation. Specifically, the upper part of the underlying formation (Gavilan Fm.) is a shallowing-upward sequence ending either locally with emergence and karstification or, more generally, with a hardground. The beginning of pelagic sedimentation subsequently drowned and eventually fossilized the paleokarst. Emergence and karstification events of this age have been described from the Alps (Baud and Masson 1975, Baud et al. 1979) and the Apennines (Farinacci et al. 1981, Fazzuoli et al. 1981). The sealevel fluctuation curves proposed by different authors (Hallam 1978 and 1981, Vail and Todd 1981, Vera 1984) show a sudden sealevel fall which coincides approximately with this Carixian break.

The paleokarst of Early Middle Jurassic age cannot be timed precisely; it could have developed in either the Early or Late Aalenian. Regional stratigraphic evidence favors the Early Aalenian, which coincides with one of the major eustatic sealevel falls postulated by Hallam (1981) and Vail et al. (1984). This is, however, one of the less apparent stratigraphic breaks in the Subbetic Zone. In only 2 of the 10 localities discussed (4 and 9) has this paleokarst stage been recognized. Nevertheless, it is an interesting event because the emergence and karstification affect pelagic limestones. Similar stages of emergence related to a sudden sealevel fall in the Late Toarcian of North Africa have been described by Elmi (1981).

The stratigraphic break and resulting paleokarst in the Latest Bathonian, are among the most spectacular in the Subbetic region. This break is present at numerous localities, affecting either Middle Jurassic shallow limestones (locality 1) or Bajocian–Bathonian pelagic limestones containing pelecypod remains (localities 4, 6, and 9). Analogs have been described in Provence and the Maritime Alps of

France (Arnaud and Monleau 1984, Monleau 1986), in the Pre-Alps (Brianconnais) (Septfontaine 1983), and in the Apennines (Farinacci et al. 1981). This event is believed to be the result of a sudden sealevel fall (Vail and Todd 1981, Vera 1984) which caused emergence of earlier formed pelagic swells. On the basis of geochemical anomalies detected in the Carpathians (Poland) and in the Iberian ranges (Spain) some authors (Brochwica-Lewinski et al. 1984 and 1986) have suggested a catastrophic event, specifically an asteroid impact, to explain this event.

Thus, the described phenomena are not unique to the Subbetic Zone but are also found, with the same apparent ages, in other alpine ranges of the Mediterranean region (Alps, Apennines, Carpathians, North Africa) and in other related realms (Iberian ranges, Provence). Analogous events, but without evidence of emergence, have been described by Bernoulli and Kalin (1984) along the northwest margin of Africa (e.g., DSDP site 546). These phenomena are therefore interpreted to be the result of events which affected the whole Mediterranean Alpine realm during the Jurassic (Vera 1981).

The role of tectonics, however, must also be considered. Lemoine (1985) has suggested a faulting phase of the western Alpine margin during which some tilted blocks were emergent. Livnat et al. (1986), using information from eustatic fluctuations similar to those documented here, have also identified the influence of local faulting. Several phases of faulting, generating faults parallel and transverse to the margin, have been recognized in the Subbetic Basin (Garcia-Hernandez et al. 1976 and 1980, Ruiz-Ortiz 1983, Vera et al. 1984), and so movements of varying amplitude and taking place at various times could have affected these rocks. Specifically, Vera et al. (1984) have concluded, on the basis of fractures filled with Jurassic sediments at locality 1, that wrench faults formed in the latest Middle Jurassic. This correlates with sinestral movement along transform faults which determined the relative position of the Iberian and African plates during the Jurassic.

Thus, there were both tectonic events, commonly involving listric faults of different scope in each paleogeographic-paleotectonic range, and eustatic changes related to global sealevel fluctuations.

Evolution of the Pelagic Swells

From the evidence presented, once interpreted in its paleogeographic context of the continental margin of the Subbetic Zone, the following model is proposed to explain the development, emergence, karstification, and drowning of the pelagic swells (Fig. 17.12). This model may also apply to the outer margins of other geosynclinal ranges of other ages during their rifting and spreading phases.

The proposed model began with an initial stage (Fig. 17.12A) of marine sedimentation, either on a shallow shelf or on shallow pelagic swells, as proposed by Wendt et al. (1984). Subsequent faulting (Fig. 17.12B) along with a drop in sealevel caused the isolation of a sedimentary swell and its partial emergence. This was followed by karstification of the emergent pelagic swell, in an area that might have been a large carbonate island (Fig. 17.12C). It seems likely that karstification took place mainly above the water table (*sensu* Esteban and Klappa 1983, James and Choquette 1984) under warm and humid conditions. In areas close to sealevel, kamenitza and other surface karst features of very low relief developed, while at higher elevations karstification extended down into the underlying sequence to depths similar to the magnitude of emergence. In submerged parts of the swell, adjacent to the weakly karstified zones, hardgrounds were generated as a result of nondeposition. It is also possible that a bypass zone may have developed along the opposite margin of the swell, as suggested by Bice and Stewart (1985) in the Apennines. No exhumed bypass margin has been recognized to date in the Subbetic Zone. Finally, pelagic sedimentation, mainly in the form of limestone-marl rhythmites, took place on the sunken fault block.

Sedimentary filling and fossilization of the uplifted block began with a later sealevel rise (Fig. 17.12D). At the same time, on the higher parts of the block, karst processes continued, in the form of cavity filling. The hardgrounds

(Fig. 17.12D, closeup 1) were covered by condensed pelagic deposits showing some sedimentary breaks in the form of omission surfaces and minor hardgrounds. The kamenitza relief (Fig. 17.12D, closeup 2) is fossilized by the same sediments (see Fig. 17.6). In areas with minor karst development (Fig. 17.12D, closeup 3) cavities were filled by marine sediments, either pelagic limestones with ammonoids, pelecypod remains ("filaments"), radiolarians, or shallow-water crinoidal limestones, giving rise to numerous neptunian dikes. Speleothems were precipitated on cavity walls towards the interior of the carbonate island, in areas slightly more uplifted and therefore exposed for longer periods of time (Fig. 17.12D, closeup 4). The remaining open spaces in the cavities were subsequently filled with pelagic marine sediments.

Fossilization of the swell continued (Fig. 17.12E) when it was completely submerged as a consequence of sealevel rise. Sedimentation was mostly pelagic, except on the highest areas of the swell where erosion prevailed due to high-energy conditions and there was minor sediment deposition in cavities, resulting in local neptunian dikes. In still higher areas (Fig. 17.12E, closeup 5) speleothems were precipitated and continental vadose sediment filled the cavities. These were then covered locally by second generation speleothems. Finally, in areas of greatest karst development (Fig. 17.12E, closeup 6) collapse megabreccias accumulated in the cavities. Speleothems commonly covered both the clasts and cavity walls. The matrix of these megabreccias is of continental origin in the lowest parts of the cavities and of marine origin in the remainder; the presence of vadose sediment clasts points to alternating phases of erosion and filling. The "blue holes" of the Bahamas are a good model for these neptunian dikes being filled by marine sediments. After cavity formation during times of emergence in the mixing zone at the base of freshwater lens, the waves and currents as well as tidal action controlled the style of cavity filling (Smart et al. this volume).

The swells were eventually completely buried (Fig. 17.12F) beneath a flat sea floor similar to that in the initial stage (Fig. 17.12A). At this point a new cycle of sedimentation and development of a swell could begin, either by reju-

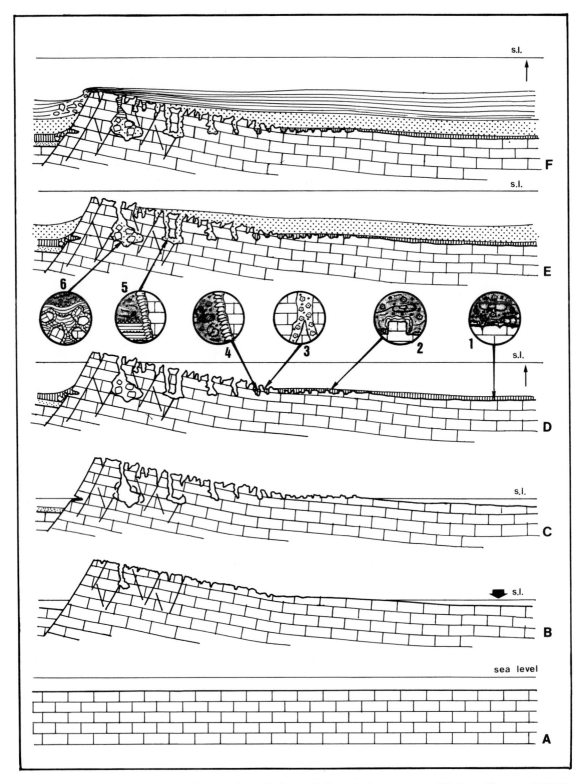

FIGURE 17.12. A model for the evolution of a pelagic swell. Stages of emergence (B), karstification (C,D,E), and burial (F) are distinguished. General orientation of diagram: NW to the left. For explanation see text.

venation along the same fault or movement along a new faults.

This theoretical model shows specific variations because of local factors such as: (1) soil and bauxite formation in caves during long periods of emergence (Vera et al. 1986); (2) karstification phases superimposed in such a way that the results of the older phases are masked (Vera et al. 1984, Garcia-Hernandez et al. 1986b); (3) development of only part of the model on some swells (e.g., only the areas very near sealevel; and (4) recurring minor karstification and sedimentation events, producing complex overprinting.

The lateral and temporal relationship between karstification and hardground formation in the proposed model is of particular importance. Both events are connected to stratigraphic breaks, caused by sealevel falls which affected areas that were uplifted relative to the adjacent sea floor. Karstification took place on the emerged areas while hardgrounds formed in the submerged areas. Fursich (1979, Fig. 21) proposed a hardground model for seamounts temporarily situated in vadose zones that may apply to the swells discussed here. Lateral variations in the magnitude of fault slip resulted in irregular distribution of the paleokarst and hardground areas.

Conclusions

1. We propose three stages of emergence and karstification (Carixian, Early Middle Jurassic, and latest Bathonian) on pelagic swells which developed in the Subbetic basin coincident with times of sealevel fall.

2. The emergence and karstification events have been deduced from a variety of features including cavities with karst morphologies, internal sediments of continental as well as marine origin and speleothems within the cavities, bauxites, and textural modifications of meteoric origin in the cavity host rocks.

3. The timing and age of emergence and karstification are established from study of fossiliferous marine sediments that fill the cavities.

4. A model for the evolution of pelagic swells is proposed in which the stages of emergence and hardground formation are contemporaneous and spatially related on a given swell.

Tectonic and eustatic factors are thought to have controlled all of these processes.

Acknowledgments. The authors gratefully acknowledge Drs. P.W. Choquette, N.P. James, and V.P. Wright for comments and critical review of this study. We also wish to thank Dr. M. Esteban and L. Pomar for their suggestions and comments. Financial support was provided from the CAICYT (1224/84).

References

Arnaud, M., and Monleau, C., 1984, Emersion et karstification dans le Bathonien de Provence. Implications climatiques et paléogeographiques: Geologie de la France, v. 1–2, p. 179–180.

Assereto, R., and Folk, R.L., 1980, Diagenetic fabrics of aragonite, calcite, and dolomite in an ancient peritidal-spelean environment: Triassic calcare rosso, Lombardia, Italy: J. Sediment. Petrol., v. 50, p. 371–394.

Azema, J., Foucault, A., Fourcade, E., Garcia-Hernandez, M., Gonzalez-Donoso, J.M., Linares, A., Linares, D., Lopez-Garrido, A.C., Rivas, P., and Vera, J.A., 1979, Las microfacies del Jurásico y Cretácico de las Zonas Externas de las Cordilleras Béticas: Secr. Public. Univ. Granada, 83 p.

Bardossy, G., 1982, Karst bauxites: bauxite deposits in carbonate rocks: Developments in Economic Geology 14: Amsterdam, Elsevier, 441 p.

Baud, A., and Masson, H., 1975: Preuves d'une tectonique liasique de distension dans le domaine briançonnais: failles conjuguées et paléokarst á Saint-Triphon (Préalpes médianes, Suisse): Eclogae geol. Helv., v. 68, p. 131–145.

Baud, A., Masson, H., and Septfontaine, M., 1979, Karsts et paléotectonique jurassique du domaine briançonnais des Préalpes: Symp. sediment. jurassique, W-Europ. Paris, 1977, A.S.F. Publ. spec. 1, p. 441–442.

Bernoulli, D., and Jenkyns, H.C., 1974, Alpine, Mediterranean and Central Atlantic Mesozoic facies relation to early evolution of the Tethys, *in* Dott, R.H., Jr., and Shaver, R.H., eds., Modern and ancient geosynclinal sedimentation: Soc. Econ. Paleont. Miner., Spec. publ. 19, p. 129–160.

Bernoulli, D., and Kalin, O., 1984, Jurassic Sediments, site 547, northwest African margin: remarks on stratigraphy, facies, and diagenesis, and comparison with some tethyan equivalents, *in* Init. Rep. Deep Sea drill. Proj. v. 74, p. 437–448.

Bice, D.M., and Stewart, K.G., 1985, Ancient erosional grooves on exhumed bypass margins of carbonate platforms: examples from the Apennines: Geol., v. 13, p. 565–568.

Brochwicz-Lewinski, W., Gasiewicz, A., Krumbein, W.E., Melendez, G., Sequeiros, L., Suffczynski, S., Szatkowski, K., Tarkowski, R., and Zbik, M., 1986, Anomalia irydowa na granigy jury srodkowej i gornej: Przeglad Geol. v. 14, p. 83–88.

Brochwicz-Lewinski, W., Gasiewicz, A., Suffczynski, S., Szatkowski, K., and Zbik, M., 1984, Lacunes et condensations à la limite Jurassique moyen-supèrieur dans le Sud de la Pologne: manifestation d'un phénomène mondial?: C.R. Ac. Sc. Paris, v. 299, p. 1359–1360.

Buchbinder, B., Magaritz, M., and Buchbinder, L.G., 1983, Turonian to Neogene paleokarst in Israel: Palaeogeogr. Palaeoclimatol. Palaeoecol. v. 43, p. 329–350.

Busnardo, R., 1979, Prébetique et Subbetique de Jaen à Lucena (Andalousie): Le Lias: Docum. Lab. Geol. Fac. Sci. Lyon, v. 74, p. 1–140.

Cherns, L., 1982, Palaeokarst, tidal erosion surfaces and stromatolites in the Silurian Eke Formation of Gotland, Sweden: Sedimentol., v. 29, p. 819–833.

Company, M., Gonzalez-Donoso, J.M., Linares, D., Martin-Algarra, A., Rebollo, M., Serrano, F., Tavera, J.M., and Vera, J.A., 1982, Diques neptúnicos en el Cretácico del Penibético: aspectos genéticos y etapas de relleno: Cuader. Geol. Ibérica, v. 8, p. 347–367.

Dabrio, C.J., and Polo, M.D., 1985, Interpretación sedimentaria de las calizas de crinoides del Carixiense Subbético: Mediterranea, v. 4, p. 55–77.

D'Argenio, B., 1974, Le Piattaforme Carbonatiche periadriatiche: una rassegna di problemi nel quadro geodinamico mesozoico dell'area mediterranea. Mem. Soc. Geol. Italia, v. 13, p. 1–28.

Elmi, S., 1981, Sédimentation rythmique et organisation sequentielle dans les Ammonitico-Rosso et les facies associés du Jurassique de la Méditerranée occidentale. Interprétation des grumeaux et des nodules, in Farinacci, A., and Elmi, S., eds., Ammonitico Rosso Symposium, Tecnoscienza, Roma, p. 251–289.

Esteban, M., 1976, Vadose pisolite and caliche: Bull. Am. Ass. Petrol. Geol., v. 60, p. 2048–2057.

Esteban, M., and Klappa, C.F., 1983, Subaerial exposure environment, in Scholle, P.A., Bebout, D.G., and Moore, C.H., eds., Carbonate depositional environments: Am. Ass. Petrol. Geol., Mem. 33, p. 1–54.

Farinacci, A., Mariotti, M., Nicosia, U., Pallini, G., and Schiavinotto, F., 1981, Jurassic sediments in the Umbro-Marchean Apennines: an alternative model: in Farinacci, A. and Elmi, S., eds. Proc. Rosso Ammonitico Symposium, Technoscienza Roma, p 335–398.

Fazzuoli, M., Marcucci-Passerini, M., and Sguazzoni, G., 1981, Occurrence of Ammonitico Rosso and paleokarst sinkholes on the top of the Marni Fm. (Lower Liassic) Apuane Alps, Northern Apennines, in Farinacci, A., and Elmi, S., eds., Ammonitico Rosso Symposium, Tecnoscienza, Roma, p. 399–417.

Folk, R.L., and Assereto, R., 1976, Comparative fabrics of length-slow and length-fast calcite and calcitized aragonite in a Holocene speleothem, Carlsbad caverns, New Mexico: J. Sediment. Petrol. v. 46, p. 486–496.

Fursich, F., 1979, Genesis, environments and ecology of Jurassic hardgrounds: N. Jb. Geol. Palaeont. Ab., v. 158, p. 1–63.

Garcia-Hernandez, M., Gonzalez-Donoso, J.M., Linares, A., Rivas, P., and Vera, J.A., 1976, Characteristics ambientales del Lias inferior y medio en la Zona Subbética y su significado en la interpretación general de la Cordillera, in Reunión sobre la Geodinámica de la Cordillera Bética: Secr. Publi. Univ. Granada, p. 125–157.

Garcia-Hernandez, M., Lopez-Garrido, A.C., Rivas, P., Sanz de Galdeano, C., and Vera, J.A., 1980, Mesozoic paleogeographic evolution of the External Zones of the Betic Cordillera: Geologie in Mijnbouw, v. 59, p. 155–168.

Garcia-Hernandez, M., Lupiani, E., and Vera, J.A., 1986a, La sedimentación liásica en el sector central del Subbético Medio: de la evolución de un rift intracontinental: Acta Geológica Hispánica, (in press).

Garcia-Hernandez, M., Lupiani, E., and Vera, J.A., 1986b, Discontinuidades estratigráficas del Jurásico de Sierra Gorda (Subbético interno, provincia de Granada): Acta Geológica Hispánica, (in press).

Garcia-Hernandez, M., Rivas, P., and Vera, J.A., 1979, Distribución de las calizas de llanuras de mareas en el Jurásico del Subbético y Prebético: Cuad. Geol. Univ. Granada, v. 10, p. 557–569.

Gonzalez-Donoso, J.M., Linares, D., Martin-Algarra, A., Rebollo, M., Serrano, F., and Vera, J.A., 1983, Discontinuidades estratigraficas durante el Cretácico en el Penibético (Cordillera Bética): Estudios geológicos, v. 8, p. 739–758.

Hallam, A., 1975, Jurassic environments: Cambridge, U.K., Cambridge Univ. Press., 369 p.

Hallam, A., 1978, Eustatic cycles in the Jurassic: Palaeogeogr. Palaeoclimatol. Palaeoecol., v. 23, p. 1–32.

Hallam, A., 1981, A revised sea-level curve for the early Jurassic: J. Geol. Soc. London, v. 138, p. 735–743.

James, N.P., and Choquette, P.W., 1984, Diagenesis 9, limestones, the meteoric diagenetic environment: Geoscience Canada, v. 11, p. 161–194.

Jennings, J.N., 1971, Karst: Cambridge, MA, MIT Press, 252 p.

Kendall, A.C., and Broughton, P.L., 1978, Origin of fabrics in speleothems composed of columnar calcite crystals: J. Sediment. Petrol., v. 48, p. 519–538.

Lemoine, M., 1985, Structuration jurassique des Alpes occidentales et palinspastique de la Tethys Ligure: Bull. Soc. Geol. France (8), v. 1, p. 126–137.

Livnat, A., Flexer, A., and Shafran, N., 1986, Mesozoic unconformities in Israel: characteristics, mode of origin and implications for the development of the Tethys: Palaeogeogr. Pàlaeoclimatol. Palaeoecol., v. 55, p. 189–212.

Martin-Algarra, A., Checa, A., Oloriz, F., and Vera, J.A., 1983, Un modelo de sedimentación pelágica en cavidades kársticas: la Almola (Cordillera Bética): X Congreso Nacional de Sediment, Mahón, Univ. Aut. Barcelona, v. 3, p. 21–25.

Molina, J.M., Ruiz-Ortiz, P.A., and Vera, J.A., 1985, Sedimentación marina somera entre sedimentos pelágicos en el Dogger del Subbético externo (sierras de Cabra y Puente Genil, Provincia de Córdoba): Trabajos de Geologia, Univ. Oviedo, v. 15, p. 127–146.

Monleau, C., 1986, Le Jurassique inférieur et moyen de Provence, Sardaigne et Alpes maritimes: corrélations, essai de synthèse paléogéographique: Bull. Geol. Dynam. Geogr. Phys., v. 27, p. 3–11.

Pirlet, H., 1970, L'influence d'un karst sous-jacent sur la sédimentation calcaire et l'interêt de l'étude des paléokarst: Ann. Soc. Geol. Belgique, v. 93, p. 247–254.

Read, J.F., and Grover Jr., G.A., 1977, Scalloped and planar erosion surfaces, Middle Ordovician Limestones, Virginia: analogues of Holocene exposed karst or tidal rock platforms: J. Sediment. Petrol., v. 47, p. 956–972.

Ruiz-Ortiz, P.A., 1983, A carbonate submarine fan in a fault-controlled basin of the Upper Jurassic, Betic Cordillera, Southern Spain: Sedimentol., v. 30, p. 33–48.

Ruiz-Ortiz, P.A., Molina, J.M., and Vera, J.A., 1985, Coral-ooid-oncoid facies in a shallowing-upward sequence of the Middle Jurassic (External Subbetic, Southern Spain): 6th. Europ. Meet. Sedim. I.A.S., Lleida, p. 403–406.

Septfontaine, M., 1983, Le Dogger des Prealpes medianes suisses et françaises: Mem. Soc. Helv. Scienc., v. 97, p. 1–121.

Seyfried, H., 1978, Der Subbetische Jura von Murcia (Sudost-Spanien): Geol. Jabrb., v. 29, p. 3–201.

Seyfried, H., 1979, Ensayo sobre el significado paleogeográfico de los sedimentos del Jurásico de las Cordilleras Béticas orientales: Cuad. Geol. Univ. Granada, v. 10, p. 317–348.

Seyfried, H., 1981, Genesis of "regressive" and "transgressive" pelagic sequences in the Tethyan Jurassic, in Farinacci, A., and Elmi, S., eds., Proc. Rosso Ammonitico Symposium, Tecnoscienza, Roma, p. 547–579.

Sweeting, M.M., 1972, Karst landforms: London, MacMillan, 362 p.

Thrailkill, J., 1976, Speleothems, in Walther, M.R., ed., Stromatolites, developments in sedimentology, 20: Amsterdam, Elsevier, v. 73–86.

Vail, P.R., Hardenbol, J., and Todd, R.G., 1984, Jurassic unconformities, chronostratigraphy and sealevel changes from seismic stratigraphy and biostratigraphy, in Schlee, J.C., ed., Interregional unconformities and hydrocarbon accumulation: Amer. Assoc. Petr. Geol. Mem. 36, Tulsa, Oklahoma, p. 347–369.

Vail, P.R., and Todd, R.G., 1981, Northern North Sea Jurassic unconformities, chronostratigraphy and sealevel changes from seismic stratigraphy, in Illing, L.V., and Hobson, C.D., eds., Petroleum geology of the continental shelf of northwest Europe: London, Heyden, p. 216–236.

Van der Lingen, G.J., Smale, D., and Lewis, D.W., 1978, Alterations of a pelagic chalk below a paleokarst surface, Oxford, South Island, New Zealand: Sedim. Geol., v. 21, p. 45–66.

Vera, J.A., 1981, Correlación entre las cordilleras Béticas y otras cordilleras alpinas durante el Mesozoico, in Programa Internacional de Correlación Geológica, P.I.C.G., Real Acad. Cienc. Exact. Fis. Nat. Madrid, v. 2, p. 125–160.

Vera, J.A., 1984, Aspectos sedimentológicos de los dominios alpinos mediterráneos durante el Mesozoico, in Obrador, A., ed., Libro Homenaje a L. Sanchez de la Torre: Publicaciones de Geologia, v. 20, Univ. Aut. Barcelona, p. 25–54.

Vera, J.A., Molina, J.M., Molina-Diaz, A., and Ruiz-Ortiz, P.A., 1986, Bauxitas kársticas jurásicas en la Zona Subbética (Zarzadilla de Totana, provincia de Murcia, Sureste de España): Acta Geológica Hispánica (in press).

Vera, J.A., Molina, J.M., and Ruiz-Ortiz, P.A., 1984, Discontinuidades estratigráficas, diques neptúnicos y brechas sinsedimentarias en la sierra de Cabra (Mesozoico, Subbético externo), in Obrador, A., ed., Libro Homenaje a L. Sanchez de la Torre: Publicaciones de Geología, v. 20, Univ. Aut. Barcelona, p. 141–162.

Walkden, G.M., 1974, Palaeokarst surfaces in Upper Visean (carboniferous) limestones of the derbyshire block, England: J. Sediment. Petrol., v. 44, p. 1232–1247.

Wendt, J., 1971, Genese und fauna submariner sedimentarer Spaltenfullungen im Mediterranean Jura: Paleontog. A., v. 136, p. 122–192.

Wendt, J., Aigner, T., and Neugebauer, J., 1984, Cephalopod limestone deposition on a shallow pelagic ridge: the Taflait Platform (Upper Devonian), eastern Anti-Atlas, Morocco: Sedimentol., v. 31, p. 601–625.

Westermann, G., 1984, Gauging the duration of stages: a new approach for the Jurassic: Episodes, v. 7, p. 26–27.

White, W.B., 1978, Speleal sediments, *in* Fairbridge R.W., and Bourgois, J., eds., Encyclopedia of sedimentology: Stroudsburg, PA, Dowden, Hutchinson & Ross, p. 754–759.

Wood, A.W., 1981, Extensional tectonics and the bird of Lagonegro Basin (southern Italian Apennines): N. Jb. Geol. Palaeont. Abh. v. 161, p. 93–131.

Wright, S.C., 1985, The origin and nature of karstic fissures in the Sailmhor Formation, Durness Group, N.W. Scotland: 6th Europ. Meet. Sedim. I.A.S., Lleida, 483–486.

Wright, V.P., 1982, The recognition and interpretation of paleokarst: two examples from the Lower Carboniferous of South Wales: J. Sediment. Petrol., v. 52, p. 83–94.

Wright, V.P., 1986, The polyphase karstification of the Carboniferous limestones in South Wales, *in* Paterson, K., and Sweeting, M.M., eds., New directions in karst: Proceed. Anglo-French Karst Symposium, Geoabstracts, Norwich, 569–590.

18
Sedimentation and Diagenesis Along an Island-Sheltered Platform Margin, El Abra Formation, Cretaceous of Mexico

CHARLES J. MINERO

Abstract

The mid-Cretaceous El Abra Formation was studied along the eastern margin of the Valles–San Luis Potosi (SLP) platform to determine (1) environments and processes of platform-margin sedimentation, and (2) the diagenetic sequence and mechanisms of porosity evolution in carbonates subjected to extensive early diagenesis, including subaerial exposure and development of microkarst. Syndepositional, late eogenetic, and telogenetic episodes of karst formation are present.

Eight lithofacies are recognized from textural and faunal attributes and sedimentary structures. Platform-lagoon lithofacies include peloid-miliolid, requienid, and bioclastic limestones. Lime mudstone, laminated lime grainstone, cryptalgal laminites, and fenestral limestone record tidal-flat environments. Perireefal islands consisted of storm-deposited rudistid-skeletal limestones cemented by calcrete.

Numerous subaerial discontinuity surfaces marked by microkarst, calcrete, and penecontemporaneous dolomite horizons provide correlation surfaces for identification of laterally equivalent depositional environments. Islands graded bankward through a mosaic of coalescing tidal flats into the platform lagoon. Whereas islands persisted through time, tidal flats were episodically submerged during rapid relative rises in sealevel of less than 10 m net magnitude. Subsequently, tidal flats prograded into the platform lagoon, resulting in asymmetric, shoaling-upward sequences separated by subaerial discontinuity surfaces.

Early diagenetic environments on tidal flats were ephemeral; but comparable environments were recreated at comparable stages of each depositional cycle. Extensive penecontemporaneous meteoric diagenesis occurred as a consequence of enduring exposure of islands and repeated exposure of tidal flats. Dissolution below microkarst surfaces, calcrete formation, and mineralogic stabilization rapidly transformed the sediment to limestone. Upon submergence, marine cementation by radiaxial fibrous Mg-calcite and internal sedimentation were important porosity-reducing processes, further lithifying the limestone and precluding compaction.

The Valles–SLP platform was locally exposed and subjected to a second episode of karst development during latest Turonian to Santonian time. Major porosity occlusion by meteoric calcite cement occurred at this time.

Limited burial diagenesis included precipitation of saddle dolomite, stylolitization, and fracturing during the Laramide Orogeny. Uplift and erosion during the Cenozoic resulted in a third episode of karst development, evident in the present landscape as subsurface drainage.

Introduction

The El Abra Formation was deposited during mid-Cretaceous time on several steep-sided, shallow-water carbonate platforms in east-central Mexico. It is 1.5 km to 2.0 km thick. Windward platform margins were fringed by a narrow zone of rudist reefs and bioclast grainstones. The platform interior is comprised of muddy lagoonal carbonates. Pelagic lime mudstone of the Upper Tamaulipas Formation accumulated in interplatform basins. Interbedded, fine pelagic limestone and coarse, platform-derived debris were deposited along the basin margins as the Tamabra Formation. Giant oil fields occur in basinal (Ebano-Panuco fields), basin-margin (Poza Rica trend), and platform-margin (Golden Lane trend) carbonates of the Tampico Embayment. Minor production occurs from platform-interior carbonates. Karst-enhanced reservoirs of the Golden Lane trend have yielded 1.6 billion barrels of oil. Regional paleogeography and petroleum geology is reviewed by Enos (1974 and 1983).

The windward margin of the Valles–San Luis Potosi (SLP) platform was studied in the type quarry exposures in the Sierra de El Abra (SEA) (Fig. 18.1). The SEA is an open, N–S trending anticline within the Gulf Coastal plain approximately 25 km east of the Sierra Madre Oriental structural front. Folding occurred during Maastrichtian to Paleocene time (Laramide orogeny). The abrupt eastern escarpment of the SEA largely coincides with the exhumed eastern margin of the Valles-SLP platform (Enos 1983 and 1986). This report discusses depositional environments and diagenesis in the narrow transition zone between the fringing rudist reefs and the platform lagoon.

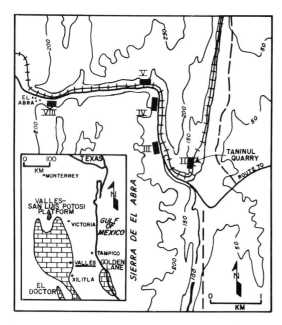

FIGURE 18.1. Location map. Measured sections are quarries III, IV, V, and VIII (numeration after Aguayo 1978), located 8 km to 12 km east of Valles, San Luis Potosi, Mexico (see insert map). The eastern escarpment of the Sierra de El Abra (100 m contour) coincides with the eastern margin of the Cretaceous Valles–San Luis Potosi platform (insert map; modified from Enos 1974). Rudist reefs are well exposed in Taninul Quarry (II).

The El Abra Formation in the SEA has undergone three episodes of karst formation. Syndepositional exposure of island and tidal-flat sediments resulted in extensive meteoric diagenesis and formation of several types of subaerial discontinuity surfaces, including microkarst (Stage 1). This study emphasizes the early episode of karst formation and the close relationship between sedimentary environments and early diagenetic processes. Correlation of discontinuity surfaces facilitated study of lateral variations of depositional environments (Fig. 18.2). A second karst event resulted from Late Cretaceous exposure but is poorly expressed in outcrop (Stage 2). Equant calcite cement precipitated from the associated meteoric aquifer, however, and filled most porosity surviving early diagenesis. Uplift in the Late Tertiary has produced a third karst event evident in the present landscape as large-scale subsurface drainage features (Stage 3). Early and late karst episodes generally enhanced porosity, whereas the intervening Late Cretaceous episode occluded porosity.

Lithofacies and Depositional Environments

Eight lithofacies are recognized on the basis of texture, constituent grains, and sedimentary structures. Major features, including depositional environments, are summarized in Tables 18.1 and 18.2. More detailed descriptions, photographs, and interpretations are presented in Minero (1983a and 1983b). Three broad lithofacies associations are present. Peloid-miliolid, requienid, and bioclastic limestones are gradationally interbedded and most common in sections V and VIII. They constitute the subtidal association. Mudstone, fenestral, and cryptalgal-laminated limestones comprise the tidal-flat association. Rudistid-skeletal limestone and laminated grainstone comprise the island/ beach association which is prominent in sections III and IV.

Two styles of depositional packages are present (Fig. 18.2). Cyclic peritidal deposition occurs in sections V and VIII. Shoaling-upward cycles of peloid-miliolid, requienid, fenestral, and cryptalgal laminite lithofacies are separated by subaerial exposure surfaces. Nearer the platform-margin reefs (III, IV), upper intertidal and supratidal lithofacies are interbedded to form thick sequences. Discontinuity surfaces are more numerous and early subaerial diagenetic features are better developed here than further bankward.

Figures 18.3 and 18.4 illustrate a sedimentary model of the eastern margin of the Valles-SLP platform. Low islands or "sand cays" formed near the platform margin largely from bioclastic debris of rudist reefs. Early vadose cementation, primarily by calcrete, location leeward of an abundant source of reef sediment, and the inability of the daily current and wave regime to rework sediments deposited by storms above sealevel (Enos and Perkins 1979) caused islands to be persistent paleotopographic features. Sedimentation on islands was primarily from storm deposits (rudistid-skeletal limestone). Storm sediments occasionally traversed the islands and were deposited in the leeward lagoon, as recorded by bioclastic limestones in section V. Islands sheltered nearby platform areas from oceanic waves and currents, enabling muddy sediments to accumulate near the platform margin. Shoaling-upward sedimentation ensued, forming tidal flats at least as far bankward as section VIII. Prolonged subaerial exposure of islands and supratidal flats resulted in microkarst, partial calcrete profiles, and penecontemporaneously dolomitized and lithified rock (see section on Syndepositional Meteoric Diagenesis).

Exposure surfaces preferentially occur in intertidal and supratidal lithofacies. This suggests that exposure was a consequence of sedimentary upbuilding and/or that falls in sealevel postdated upbuilding. Enos (1986) infers minor drops in sealevel to explain widespread evidence of an initial meteoric diagenetic stage in the platform-margin rudist reefs. Subsequent net rises in sealevel are estimated at 3 m to 9 m from the thickness of individual shoaling-up sequences.

Islands remained emergent during rising sealevel; flooding was constrained to former tidal passes and flats (cf. Perkins 1977). Renewed shoaling-up lagoonal to tidal-flat sedimentation ensued in the "energy-shadow" of the islands. Thus, hydrodynamic and ecologic conditions along the platform margin were similar through successive incremental rises in relative sealevel. This resulted in an enduring paleogeographic motif of platform-margin rudist reefs with leeward islands and intermittent tidal flats.

Diagenesis

The El Abra Formation has been subjected to five stages of diagenesis (Fig. 18.5): syndepositional meteoric, early marine, late eogenetic meteoric phreatic, mesogenetic, and telogenetic. Excluding telogenesis, the effect of each successive stage was less pervasive than the preceding stage.

Early postdepositional diagenesis extensively modified El Abra sediments. Asymmetric, shoaling-upward depositional packages and island sequences are punctuated by numerous weathering horizons, which include microkarst surfaces, partial calcrete profiles, and dolomitic horizons (Fig. 18.2). These discontinuity surfaces occur individually or in combination in intertidal and supratidal carbonates. Contemporaneous meteoric diagenetic features such as secondary porosity, micritic and bladed calcite cements, and internal sediments coincide with and decrease rapidly in abundance below discontinuity surfaces. Numerous discontinuities within each measured section facilitate correlation (Fig. 18.2) and paleogeographic reconstruction of near backreef and lagoonal limestones.

Early marine diagenesis reduced porosity through introduction of internal sediment and precipitation of radiaxial fibrous calcite. It postdated the initial episode of meteoric diagenesis. Local subaerial exposure during Late Cretaceous time resulted in a second period of karst development. Equant calcite cement precipitated from the Late Cretaceous meteoric aquifer and occluded most remaining porosity.

Minor mesogenetic features include saddle dolomite, replacement by anhydrite, fracturing, stylolitization, and hydrocarbon migration. The El Abra Formation is presently subject to a third period of karst development on outcrop.

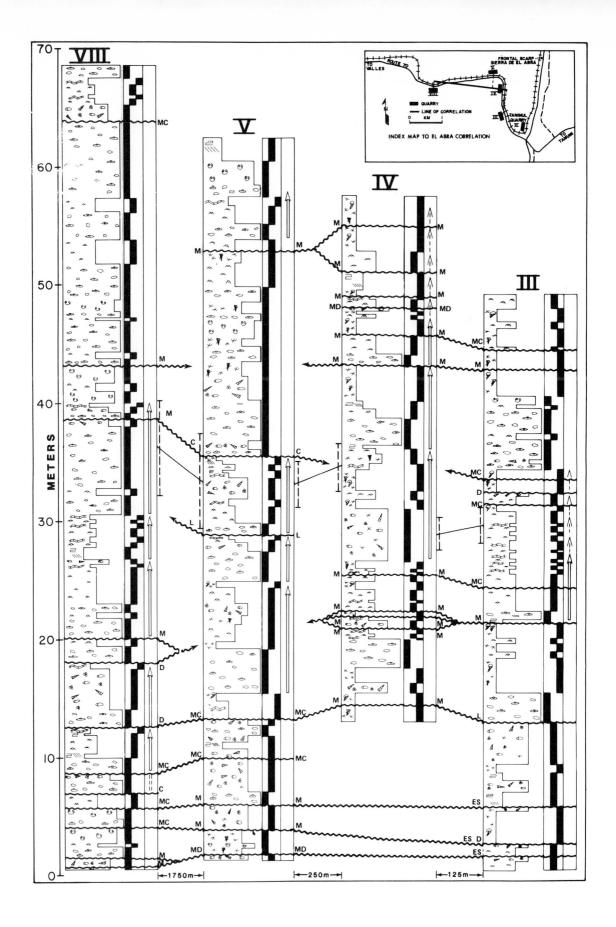

Syndepositional Meteoric Diagenesis and Discontinuity Surfaces

Microkarst Surfaces (Stage 1 Karst)

Pitted truncation surfaces are the most common discontinuity features. They are exposed only in profile and consist of a series of bowl- or funnel-shaped pits and intervening domal to peaked projections (Fig. 18.6A, B). Surfaces may be concordant with bedding for short distances. Pitted surfaces sharply truncate intertidal and supratidal lithofacies. Surfaces are irregular even at the scale of individual grains. Burrowing and encrustation are absent; borings are rare. Blackened lithoclasts of the truncated unit and coarse bioclastic debris fill depressions and are mixed into the overlying rock. Surfaces near the platform margin are overlain by rudistid-skeletal limestone and lime mudstone, whereas further bankward they are overlain by subtidal lithofacies.

Pitted truncation surfaces are continuous with subjacent channels, small caverns, and vugs. Filled vuggy porosity generally comprises less than 10% of rock volume, but locally attains 50% (Fig. 18.6C). Vugs decrease notably within a meter below discontinuity surfaces. Moldic porosity is prevalent in fossiliferous lithologies and locally within internal sediments. Mineralogic selectivity is illustrated by requienid rudists wherein the aragonitic shell wall is invariably preserved as a mold and the calcitic layer retains the original skeletal microstructure. Secondary pores are less common above pitted truncation surfaces except where several surfaces are closely spaced.

Laminated micrite cement is developed discontinuously along pitted truncation surfaces and in associated secondary pores. Commonly, the discontinuity surface is micritized. Vadose cements and internal sediment (see below) partially filled vugs and lithified rubble in the pits. Whereas the initial meteoric diagenetic events are commonly absent in subtidal limestones above discontinuities, later marine diagenesis was prominent on both sides of truncation surfaces.

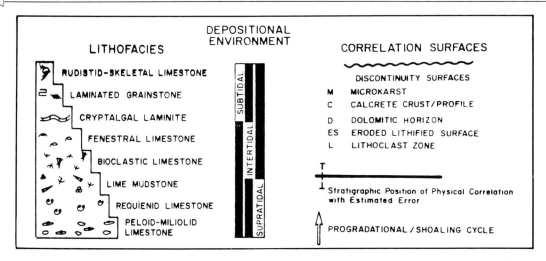

FIGURE 18.2. Lithofacies and correlation of the El Abra Formation. The section is constructed normal to the eastern margin of the Valles–SLP platform (Fig. 18.1) and illustrates the rapid gradation of sediment character across the platform margin. Note especially the change from predominantly coarse subaerially deposited limestone of the near backreef (III and IV) to muddy cyclic lagoon and tidal-flat sedimentation further bankward (V and VIII). Width of lithology column is keyed to lithofacies types.

Physical correlation by walking out beds was confirmed by correlation of numerous discontinuity surfaces in each measured section. Correlative discontinuities change character laterally. Whereas well-developed calcrete profiles and high-relief microkarst form continuous surfaces, minor discontinuities are local features or converge with major discontinuities.

TABLE 18.1. Near backreef and platform lagoon lithofacies.

Lithofacies	Dominant textures	Grains/fossils/bioclasts	Sedimentary structures	Depositional environment
Peloid-miliolid	Wkst/pkst	Miliolid forams, peloids (a) Fecal pellets (c) Benthonic forams, gastropods, requienid rudists, bivalves, green algae, ostracodes (r)	20–200+ cm beds Bioturbation	Subtidal zone Low-energy platform lagoon
Requienid	Wkst/pkst	Requienid rudists (c-a) Peloids, miliolid forams (c) Monopleurid rudists, molluscs, encrusting organisms (r)	<100 cm thick lenses Bioturbation	Subtidal to lower intertidal zone (r) Low-energy platform lagoon
Fenestral	Pkst/wkst	Peloids, miliolid forams (c) Bivalves, gastropods, ostracodes, intraclasts (r)	20–100 cm beds Wavy and flat lamination 10-25% filled fenestral porosity Desiccation cracks, burrows (r)	Upper intertidal zone
Cryptalgal laminite	Grst/pkst with thin mdst/wkst partings	Peloids, miliolid forams (c) Intraclasts (r-c) Molluscs, other bioclasts (r)	5–80 cm beds Wavy to lenticular lamination Erosion scours Desiccation cracks Laminoid fenestrae	Upper intertidal to supratidal zone
Lime mudstone	Mdst	Benthonic forams, ostracodes (r)	5–100 cm beds Bioturbated	Restricted subtidal to intertidal zone "Ponds," platform lagoon
Laminated grainstone	Grst	Peloids, ooids, miliolid forams, bioclasts Individual beds dominated by single grain type	10–50 cm beds Horizontal lamination Well-sorted and rounded Laminoid fenestrae and keystone vugs	Intertidal zone Beach Subtidal zone (r) Channel
Rudistid-skeletal	Grst/pkst (Floatstone)	Caprinid and radiolitid rudists, calcified algae, mollusc bioclasts (c-a) Corals, echinoderms, bryozoa, stromatoporoids, ostracodes, benthonic forams, peloids (r) Lithoclasts with radiaxial fibrous cement (r)	10–75 cm beds Flat lamination (r) Foresets (r) indicate lagoonward paleo-currents Coarse-tail grading and basal erosion surfaces Decreasing abundance and coarseness bankward	Supratidal-island zone Storm deposit Major meteoric vadose diagenesis

Bioclastic	Pkst/wkst	Similar to rudistid-skeletal lithofacies with
		  Miliolid forams, peloids (c)
		  Pellets (r-c)
		  Calcrete lithoclasts (r)
			50–100 cm beds
			Bioturbation
			Poorly sorted
				Subtidal zone
				Storm deposit in platform lagoon

r = rare
c = common
a = abundant

TABLE 18.2. Percent abundance of lithofacies.

Lithofacies	Tidal setting	Section III	IV	V	VIII
Subtidal association					
Peloid-miliolid	Subtidal	16	11	31	49
Requienid	Subtidal/lower intertidal (?)	1	0	7	9
Bioclastic	Subtidal	0	2	10	0
Total		17	13	48	58
Tidal-flat association					
Lime mudstone	Subtidal/intertidal	13	14	22	12
Fenestral	Upper intertidal	25	21	16	22
Cryptalgal	Upper intertidal/supratidal	5	6	6	6
Total		43	41	44	40
Island/beach association					
Laminated grainstone	Intertidal/subtidal (R)	4	5	5	2
Rudistid-skeletal	Supratidal	36	41	3	0
Total		40	46	8	2

Contrasts in depositional environment and diagenetic history across pitted surfaces indicate that the truncated bed was subaerially exposed prior to or during formation of the discontinuity. The steep profile, lithoclasts, and lack of bioturbation across truncation surfaces suggest lithification prior to erosion and subsequent deposition.

Pitted truncation surfaces are interpreted as microkarst formed during subaerial exposure of peritidal carbonates and meteoric lithification. Bowl- and funnel-shaped profiles suggest *rundkarren* or *hohlkarren*—small-scale solution features developed under soil cover. More finely sculpted surfaces are similar to rain pits and solution rills formed on bare limestone (see Jennings 1971, Sweeting 1973).

Presence or absence of soil cover influences karst morphology in that it affects the manner and duration of wetting of a limestone surface, as well as the chemistry of the solution. Soil cover may vary spatially or temporally during formation of microkarst. Once formed, depressions act as basins collecting storm-borne detritus and organic material (Dodd and Siemers 1971). Bare limestone karren thus may have evolved into bowl and funnel forms as soil cover accumulated. Storms probably eroded soils from the El Abra islands (McKee 1959, Stoddart and Steers 1977) as indicated by calcrete lithoclasts in the rudistid-skeletal and bioclastic lithofacies. This may account for

the scarcity of paleosols above rounded (i.e., covered) karst and of the unlithified upper members of calcrete profiles (see Calcrete, below).

Microkarst must have lowered island and tidal flat surfaces only slightly as it forms low-relief horizons which are largely restricted to individual, thin, subaerially deposited lithofacies. Average solution rates of limestone in warm, humid climates (Sweeting 1973) may be cautiously applied to El Abra microkarsts to obtain rough minimum estimates of the subaerial hiatus represented by these surfaces. A typical microkarst with solution relief of approximately 40 cm could have formed in 25,000 to 40,000 years.

Microkarst has important implications for subaerial exposure and meteoric diagenesis. Thus, it must be distinguished from marine hardgrounds or planation surfaces and interstratal karst. Microkarst surfaces are distinguished from hardgrounds by their common occurrence above or between supratidal lithofacies and by associated subaerial diagenetic features. Finely sculpted microkarst is similar to intertidal erosion surfaces but is nearly devoid of bores and encrustation typical of such surfaces (Kennedy 1975). Microkarst formed subaerially, as it is associated with calcrete. There is no evidence of collapse or solution piping that typifies interstratal karst (Wright 1982).

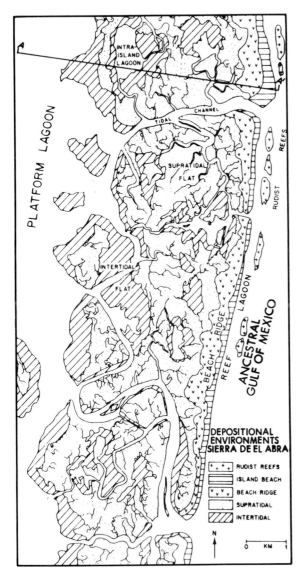

FIGURE 18.3. Sedimentary model of the SEA platform margin. Sand cays and linear rubble islands formed leeward of discontinuous rudist reefs. Tidal flats, sheltered by islands, prograded into the shallow lagoon. Episodic rises in sealevel largely submerged the tidal flats.

Calcrete

Calcrete indicates subaerial exposure and meteoric vadose diagenesis (Esteban and Klappa 1983) and is useful in location correlation (Walkden 1974). Calcrete fabrics (Fig. 18.7A, B, C) include widespread micrite cement and neomorphic micrite, laminated crusts, pisoids, breccia, and alveolar texture. SEA micrite ce-

ment occurs as laminated crusts and reduces intergranular porosity in subaerially deposited carbonates. Micrite cement is distinguished from detrital carbonate mud by contrary-to-gravity disposition and prominence below discontinuity surfaces. The distribution of micrite cement mimics that of vadose water. Features indicating an origin in the vadose zone include meniscus (Dunham 1971) and microstalactitic habit (Müller 1971), preferential cementation of grain contacts and finer pores (capillarity effect), and thickening at the base of larger pores (microponding, Read 1974).

Laminated crusts are broadly conformable with the underlying surface, lining microkarst surfaces (Fig. 18.7A) and associated channels. Crusts are up to 2 cm thick, thickening in depressions and thinning over substrate highs. Laminated crusts are predominantly micrite but also contain thin bands of bladed calcite spar (Fig. 18.7B) and rare lenses of fine bioclastic limestone. Internal laminae are wavy to lenticular with minor solution truncations.

Calcrete is a complex diagenetic fabric resulting from precipitation of micrite and pendant bladed cements, vadose internal sedimentation, replacement, and dissolution. SEA calcrete profiles are up to 2 m thick. A discontinuous laminated crust, typically lining a microkarst surface, delimits the syndepositionally lithified surface. Micrite-coated lithoclasts and calcrete pisoids are locally present above the crust. Subjacent rocks were repeatedly brecciated and cemented by micrite. Fractures and secondary pores were partially occluded by micrite cement, vadose silt, and lithoclasts, including fragments of laminated crust. Cementation and replacement by micrite decrease irregularly downward. Local, highly cemented and replaced mottles up to several centimeters in width mark the base of calcrete profiles.

Micrite-coated grains, the initial stage of profile development in loose sediment (Read 1974), are the most widespread calcrete feature. Micrite-coated aragonite bioclasts, such as codiacian algae (Fig. 18.7C), are preserved in detail; elsewhere they are reduced to molds. Incipient calcrete profiles through thick sequences of intertidal and supratidal sediments in sections III and IV suggest that storm deposition of rudistid-skeletal debris repeatedly interrupted calcrete formation by shifting the

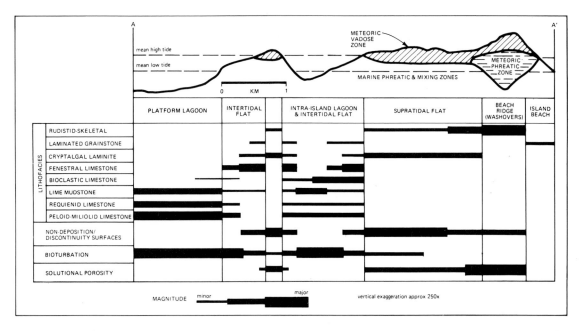

FIGURE 18.4. Schematic cross-section illustrating the relative abundance and lateral distribution of lithofacies among the depositional environments. Syndepositional and early postdepositional modification of the sediment was related to depositional environment. Bioturbation was extensive in the subtidal and lower intertidal zones, whereas dissolution and formation of discontinuity surfaces occurred in the upper intertidal and supratidal zones.

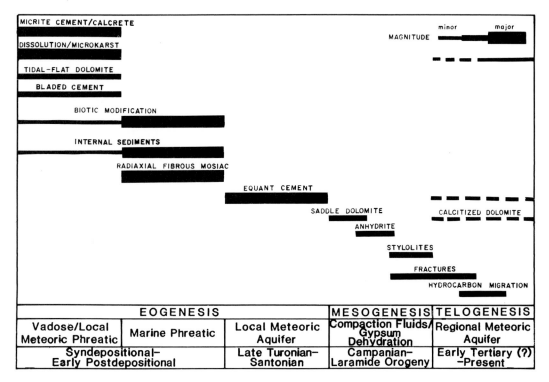

FIGURE 18.5. Chart summarizing the sequence, relative magnitude, and duration of diagenetic events and the inferred diagenetic environment or pore fluid at each stage. Syndepositional vadose diagenesis includes meteoric and marine (tidal-flat dolomite) environments.

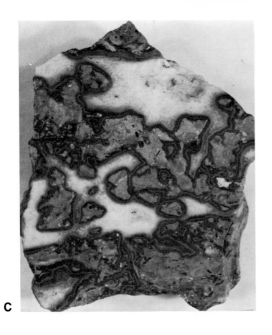

FIGURE 18.6. *A–C. A.* Microkarst surface between two depositional units of rudistid-skeletal limestone. Blackened lithoclasts occur above the truncation surface. Horizontally elongate vugs or recessive weathering zones locally occur several decimeters below microkarst surface. Ruler segment 15 cm. III 33 m. *B.* A slab 16 cm across illustrating microkarst expressed as funnel-shaped solution channels filled by rudistid-skeletal limestone. IV 8.9 m. *C.* A slab, 10 cm across, of vuggy limestone. Note truncation of coarse grains. Porosity was reduced by isopachous radiaxial fibrous cement and filled by equant calcite cement. Sample is located 30 cm below a microkarst surface. III 44.1 m.

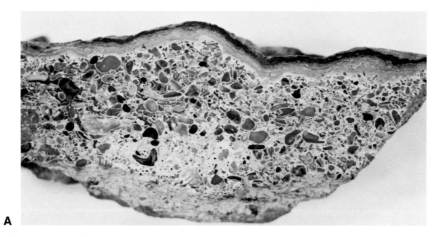

A

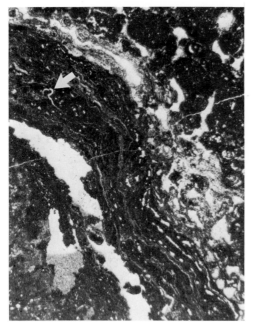

B

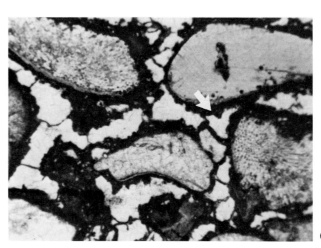

C

FIGURE 18.7. *A–C A.* A slab, 10 cm across in which a laminated calcrete crust lines microkarst-truncated rudistid-skeletal grainstone. Light-colored rinds and matrix are predominantly micrite cement. III 30.6 m. *B*. Photomicrograph of calcrete crust, 6 mm across, illustrating irregular internal lamination of micrite and bladed calcite (arrow) cement. Geopetal vadose silt reduces vugs. III 33.1 m. *C*. Photomicrograph from Figure 18.7A, 7 mm across, demonstrating irregular distribution of micrite cement, alveolar texture, and pendant bladed calcite cement (arrow). Note well-preserved codiacian algae bioclasts (orginally aragonite) in the lower right and left center. III 30.6 m.

locus of calcrete formation upward. In places, the unlithified upper part of calcrete profiles was eroded as indicated by calcrete-coated clasts in storm-derived bioclastic limestone, and rare calcrete pisoids in soil-filled vugs.

Laminated crusts locally line microkarst surfaces in the absence of other calcrete fabrics. Such crusts contain aligned bioclasts and are similar to modern crusts forming on imperme-able or lithified substrates (Multer and Hoffmeister 1968, Harrison 1977). Crusts associated with calcrete profiles contain micritized grains in a fabric similar to the substrate. Thus, in the SEA, crusts containing transported sediment probably formed at the surface, whereas laminated crusts that incorporate and modify the sediment substrate were probably subsoil formations.

Dolomitic Horizons

White-weathering dolomitic bands up to 30 cm thick occur in some muddy supratidal and upper intertidal carbonates (Fig. 18.8). Dolomite is microcrystalline, contains 50 to 55 mole % $CaCO_3$, and has faint ordering peaks (Graf and Goldsmith 1956). The upper surface of dolomitic beds is sharp and is locally truncated by microkarst. Dolomite content decreases gradationally downward. The original sediment and internal clasts (see below) are partially dolomitized, but some fossils and marine internal sediment remain calcite.

Dolomite occurs only in tidal-flat carbonates, suggesting that processes operative in the depositional setting have controlled its distribution. It is mineralogically similar to that present on Holocene tidal flats (Shinn et al. 1965). Selective replacement of depositional elements suggests that dolomitization preceded marine internal sedimentation. Lithoclasts of microcrystalline dolomite occur above dolomitic horizons and in storm deposits which are otherwise entirely calcitic. Microkarst-truncated dolomitic horizons indicate early lithification, although not necessarily by dolomite. These observations suggest that dolomitic horizons formed syndepositionally as lithified supratidal crusts somewhat analogous to those in recent sediments of south Florida (Shinn 1968) and the Bahamas (Shinn et al. 1965). Dolomitic horizons are discontinuity surfaces, as they record breaks in deposition. They formed in a marine-influenced vadose zone and subsequently were modified by meteoric vadose processes.

Bladed Calcite Cement

Bladed cement is minor. Crystals occur either as a single isopachous layer up to 75 μm thick or are longer and preferentially distributed on upper surfaces of pores (pendant). Isopachous cement locally reduces interparticle pores of laminated grainstones and rudistid-skeletal limestones. Pendant bladed cement occurs within one meter below microkarst surfaces and locally in rudistid-skeletal limestones. Where both are present in the same bed, pendant crusts occur several decimeters above isopachous crusts in an identical position in the diagenetic sequence.

Pendant bladed cement precipitated in the meteoric vadose zone below subaerial discontinuity surfaces (Fig. 18.7C). Pendant cement mimics a gravitational distribution of vadose water below grains (Müller 1971). It was contemporaneous with micrite cement and locally occurs in calcrete profiles.

The isopachous distribution of bladed cement indicates uniform growth in water-filled pores of the phreatic zone. This cement is locally truncated by dissolution but also has reduced moldic pores. It precipitated from meteoric pore fluids contemporaneous with formation of secondary pores. Pendant and micritic cements with vadose habits formed synchronously with

Figure 18.8. Slab of light-colored, dolomite cryptalgal laminate, 10 cm across. Dolomite content decreases gradationally downward from the sharp upper contact with peloid-miliolid limestone. Disrupted and curled lamination is suggestive of desiccated, algal-bound sediment. VIII 17.5 m.

bladed isopachous cement, but in stratigraphically higher positions.

Crystalline Internal Sediment and Clasts

Minor geopetal accumulations of crystalline internal sediment (Fig. 18.9) are composed of an equant mosaic of 10 to 25 μm calcite crystals. Crystals generally coarsen upward in a pore, but lamination is rare. The upper contact may be steeply inclined, locally forming terraces of perched sediment. Crumbly aggregates of the adjacent rock, including partially dissolved fossils with attached matrix (Fig. 18.10), occur as rare "internal clasts" (Dunham 1969, p. 164). Crystalline internal sediment reduced secondary pores. It postdated most micrite and bladed calcite cement and predated radiaxial fibrous and equant calcite cements.

The texture and composition of crystalline internal sediment is distinct from that of the surrounding rock and other internal sediments (see below). Perched and inclined deposits suggest deposition from flowing water. It is temporally associated with formation of vugs and molds and with calcrete cementation. For these reasons the sediment is interpreted as vadose silt (Dunham 1969) formed through physical and chemical erosion within the pore network of a subaerially exposed limestone.

Internal clasts occur only in grain-supported limestones. Vugs containing internal clasts have irregular margins which grade into interparticle pores, whereas vugs without internal clasts generally are more smoothly bounded (compare Fig. 18.10 with Fig. 18.11). This suggests that internal clasts also were derived by physical and chemical erosion in lightly cemented host rocks (Dunham 1969, Minero 1983a).

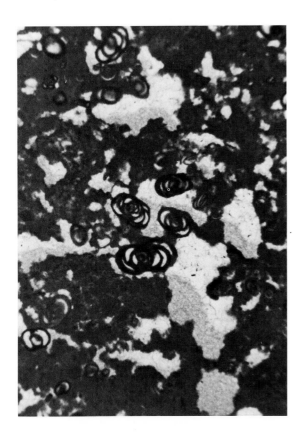

FIGURE 18.9. Photomicrograph, 8 mm across, of crystalline internal sediment reducing vuggy porosity in a peloid-miliolid wackestone. Note the varied degree of inclination of the sediment/equant cement interface. VIII 51.3 m.

Discussion of Syndepositional Meteoric Diagenesis

Tidal flats and islands permitted intermittent to nearly continuous introduction of meteoric water into the sediment. Early diagenesis in subaerially deposited rocks included episodes of dissolution, cementation, and internal sedimentation in the vadose and shallow phreatic zones. Discontinuity surfaces formed during prolonged periods of subaerial exposure. Their physical extent and repetition testify to the importance of penecontemporaneous meteoric diagenesis. Subaerial surfaces were covered by soil or regolith, as indicated by the presence of calcrete profiles and rounded microkarst morphologies, or dissolution locally acted on bare limestone, resulting in jagged microkarst morphologies. Tidal-flat sediments were locally dolomitized.

Islands were storm built and therefore only several meters above sealevel. As porosity and permeability were initially high in rudistid-skeletal limestone islands, the water table, where present, was near sealevel. Vadose diagenesis was widespread but meteoric phreatic diagenesis was largely restricted to small lenses within topographic highs such as beach ridges composed of rudistid-skeletal limestone. Some isopachous bladed cements formed under

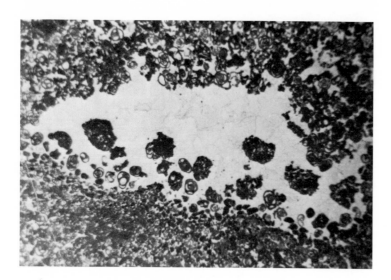

Figure 18.10. Photomicrograph, 14 mm across, illustrating a vug with a geopetal accumulation of internal clasts identical to the host peloid-miliolid grainstone. The vug is filled by equant calcite cement. Partially cross-polarized light. VIII 18 m.

temporary phreatic conditions, as they alternate with vadose micrite cements in the lower part of calcrete profiles. On the leeward tidal flats, the meteoric vadose zone may have passed directly into a saline phreatic zone.

The meteoric vadose zone was shallow. Microkarst channels and vugs, and vadose internal sediment and cements typically occur within 1 m below exposure surfaces. Horizontally elongate vugs and recessive weathering zones form horizons several decimeters below subaerial exposure surfaces (Fig. 18.6A). These may record dissolution along shallow paleowater tables (Land 1973). As the uppermost, unlithified part of weathering profiles was apparently eroded prior to renewed sedimentation, the original depth to the water table may have been somewhat greater than indicated on outcrop.

Secondary pores may be partially of vadose origin as they are reduced by vadose sediment and cement, but locally truncate these features. Secondary porosity could also have formed in the meteoric phreatic zone or in a mixing zone (Plummer 1975) at the margins of the meteoric realm. Episodic marine recharge of meteoric lenses by storm floods may have resulted in temporary mixing zones.

Early Marine Diagenesis

Fossiliferous Internal Sediment

Geopetal accumulations of fossil-bearing internal sediment (Fig. 18.11) reduce all pore types except fractures. Bioclastic lime mudstone is most common, but grain-supported textures also occur in larger vugs. Ostracodes, foraminifera, gastropods, bioclasts, and peloids are common. Internal clasts are rare. Flat or inclined textural lamination occurs in internal sediment of larger vugs. Bands of radiaxial fibrous mosaic alternating with fossiliferous internal sediment also impart a layered aspect.

Two generations of internal sediment are recognized, although their composition and texture are similar due to common provenance. The earlier generation is characterized by the presence of internal clasts and molds of bioclasts suggesting temporal association with meteoric diagenesis. Fossils, particularly foraminifera, indicate a marine source for internal sediments, although not necessarily deposition in a marine diagenetic environment. Apparently, storm-borne marine sediment occasionally entered the vuggy pore system developing in exposed limestones, diluting the background accumulation of internal clasts. Subsequent development of moldic pores in internal sediment is compatible with deposition in a subaerial setting coincident with vug formation.

A later, more abundant generation of internal sediment is interlayered with radiaxial fibrous calcite and postdates boring and encrustation of pore margins. It accumulated in the marine environment. Hydrodynamically layered coarse internal sediments suggest free exchange between seawater and the vuggy pore system.

Radiaxial Fibrous Calcite Mosaic

Isopachous crusts of radiaxial fibrous mosaic (Kendall 1985) are volumetrically important in occlusion of all pore types except fractures. It occurs only within approximately 2 km of the platform margin. Crusts are up to several millimeters thick and commonly overlie or are interlayered with fossiliferous, marine internal sediment (Fig. 18.11). Rarely, crystals are encrusted or bored and subsequently infilled by internal sediment. Radiaxial fibrous mosaic postdated early meteoric diagenesis and was succeeded gradationally by clear equant calcite cement (Fig. 18.6C).

Radiaxial fibrous mosaic contains numerous inclusions finer than 10 μm, some of which were identified as dolomite by etching and staining (Lohmann and Meyers 1977) and X-ray diffraction. Electron microprobe analysis indicates an average $MgCO_3$ content of 1.5 mole %. Stable isotope compositions (Fig. 18.12) average $\delta^{13}C$ (PDB) $= +2.2$‰ and $\delta^{18}O$ (PDB) $= -2.7$‰. Isotopic composition is similar to that of calcite bioclasts.

Kendall (1985) reported radiaxial fibrous mosaic to be a primary cement fabric. El Abra mosaics are interpreted as early marine cements precipitated from pore fluids in open exchange with seawater as it is interlayered with marine internal sediment, micritized, bored, and encrusted. Optical characteristics of radiaxial fibrous calcite (Kendall 1985) apparently developed prior to precipitation of epitaxial equant calcite cement. The isotopic composition is compatible with this interpretation (Woo et al. 1985). Lithoclasts cemented by radiaxial fibrous mosaic are present in rudistid-skeletal storm deposits and locally abundant in adjacent rudist

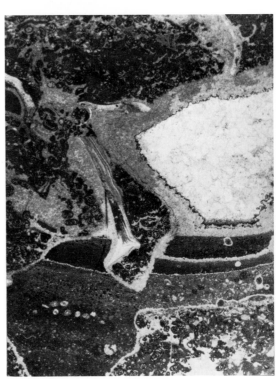

FIGURE 18.11. *A, B A.* A slab, 10 cm across, illustrating a vug reduced by laminated fossiliferous internal sediment. Grey banded cement is radiaxial fibrous calcite. Note that it is interlayered with internal sediment and does not line the lower part of the vug. Remnant porosity was occluded by equant calcite cement. III 41.6m. *B.* Photomicrograph, 2.5 cm. across, of a large vug reduced by fossiliferous internal sediment and radiaxial fibrous calcite and filled by equant cement. Partially cross-polarized light. IV 11 m.

reefs (Enos 1986). This indicates a synsedimentary origin of this mosaic. The presence of dolomite inclusions exclusively in the mosaic, Mg content, and the low concentrations of Sr by microprobe analysis imply it was originally Mg-calcite.

Discussion of Early Marine Diagenesis

The distribution of radiaxial fibrous cement was controlled by the effectiveness of mechanical pumping of seawater through the limestone (Matthews 1974). Whereas coarse pore networks were present throughout the SEA study area, marine cement was restricted to within 2 km of the agitated platform margin. Radiaxial fibrous mosaic commonly lines larger pores but decreases in abundance with distance from these permeability channels. It may be absent in interparticle pores within several centimeters of coeval larger pores.

Episodically, internal sediment entered the porous limestone while cementation proceeded. Interlayered transported sediment and cement and marine-cemented lithoclasts indicate that postdepositional marine phreatic diagenesis began near the sediment–water interface. Whereas early marine diagenesis postdated syndepositional meteoric diagenesis in a given rock, it was concurrent with meteoric diagenesis in younger depositional cycles. Diagenetic sequences in successive depositional packages were parallel rather than synchronous (cf. Enos and Freeman 1978).

Meteoric Phreatic Diagenesis (Stage 2 Karst)

Equant Calcite Cement

Equant cement (Figs. 18.10, 18.11) is present in all pore types and lithofacies. It is clear, low-Mg, nonferroan calcite forming inward-coarsening mosaics. Equant cementation was the final and locally the only major pore-occluding event. Commonly, early equant cement is epitaxial to radiaxial fibrous mosaic. Two families of equant calcite cement are discernible from stable isotope compositions (Fig. 18.12). Early equant cement has an average composition (PDB) of $\delta^{13}C = +1.7‰$ and $\delta^{18}O = -4.2‰$.

Late equant cement (sampled only from reef facies) has an average composition of $\delta^{13}C = +0.3‰$ and $\delta^{18}O = -9.9‰$.)

Four cathodoluminescent zones are present and correlative throughout the study area. Zone 1 luminesces dull orange and contains up to six bright internal bands, zone 2 is nonluminescent, zone 3 is dull orange with scarce bright bands, and zone 4 is nonluminescent. Zone 4 cement is crosscut by fractures and truncated by stylolites. A few fractures contain equant cement of zones 2, 3, and 4.

Equant cement postdated early meteoric and marine diagenesis but largely predated the major episode of fracturing attributed to Laramide deformation (see Mesogenesis, below). Isopachous outlines of luminescent zones indicate precipitation from fluid-filled pores.

Chemistry of the solution from which equant cement precipitated can be estimated from the minor element content of calcite by use of the homogeneous distribution coefficient, D (Pingitore 1978). Precipitation from an open (fluid-dominated) system near 25‰ C (shallow burial) is assumed. Sr and Mg were analyzed by electron microprobe; Mn, Fe, Zn, and Na concentrations are below the detection limit (<1000 ppm, as oxide) of the tool. Maximum cement values for Mn and Fe correspond to the minimum sensitivity of the cathodoluminoscope and combined stains, respectively. The calculated El Abra aquifer chemistry is very similar to that of recent limestone aquifers (Table 18.3).

Discussion of Meteoric Phreatic Diagenesis

Aguayo (1978) reported that the uppermost surface of the El Abra Formation in the SEA locally is a poorly exposed karst surface. Vugs and collapse breccias contain mid-Cretaceous limestone lithoclasts in a matrix of early Campanian pelagic limestone. As El Abra deposition may have continued into Late Turonian time, and as the platform margin was submerged by Campanian time, this karst surface largely formed during Coniacian and Santonian time (Aguayo 1978). Equant cement postdated syndepositional diagenesis but predated Laramide stylolites and fractures formed during Maastrichtian to Paleocene time. Temporal relations

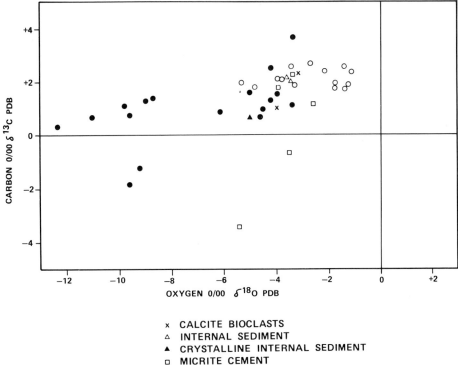

FIGURE 18.12. Cross-plot of δ ^{18}O vs. δ ^{13}C values for various components in the El Abra Platform. Stable isotope compositions in ‰ vs. PDB standard.

Legend:
- x CALCITE BIOCLASTS
- △ INTERNAL SEDIMENT
- ▲ CRYSTALLINE INTERNAL SEDIMENT
- □ MICRITE CEMENT
- ○ RADIAXIAL FIBROUS CALCITE
- ● EQUANT CALCITE CEMENT

TABLE 18.3. Estimated activity ratios for Late Cretaceous aquifer precipitating equant calcite cement.

Element	D	Method of detection	Analytic concentration	El Abra $aM_{aq}^{++}/aCa_{aq}^{++}$ (estimated)	$aM_{aq}^{++}/aCa_{aq}^{++}$ (limestone aquifers)[*,a,b]
Mn	5.4[c]	cathodo-luminoscope	20–40 ppm[d] (minimum)	6.7×10^{-6} 2.4×10^{-6} (minimum)	5.3×10^{-5}
	15[c]				
Fe	1[f]	combined stains	0.5 wt % FeO[g]	7×10^{-1} (maximum)	5.4×10^{-1}
Mg	0.033[h]	microprobe	1.3 mole % (maximum**) MgCO$_3$ (average)	0.40 (average)	0.45
Sr	0.055–0.27[i]	microprobe	1000 ppm (maximum**)	0.004–0.021 (maximum)	0.017

*Average of 10 analyses of shallow/intermediate meteoric aquifers.
**Maximum estimate corresponds to limit of detection of method applied if element is not detected.

$$D_{25°C} = \frac{aM_{calcite}^{++}/aCa_{calcite}^{++}}{aM_{aq}^{++}/aCa_{aq}^{++}}$$

Sources:
[a] Skougstad and Horr (1963)
[b] White et al. (1963)
[c] Michard (1968)
[d] Richter and Zinkernagel (1981)
[e] Pingitore (1978)
[f] Richter and Füchtbauer (1978)
[g] Lindholm and Finkelman (1972)
[h] Füchtbauer and Hardie (1976)
[i] Brand and Veizer (1980, after others)

and minor element chemistry suggest that equant calcite precipitated from a late eogenetic meteoric aquifer coincident with subaerial exposure and karst. Dissolution of limestone along the upper surface of the El Abra Formation was the probable source of calcite cement. Slight depletion of ^{18}O in the early equant cement relative to Cretaceous marine cement and sediment is compatible with precipitation from a shallow meteoric aquifer.

Mesogenesis

Dolomite and anhydrite precipitation and replacement, dedolomitization, fracturing, stylolitization, oil migration, and calcite cementation were minor late diagenetic features. Coarse saddle dolomite is a rare cement postdating most equant cement. Epitaxial calcite (dedolomite) has partially replaced saddle dolomite. Replacement anhydrite occurs in several samples. Stylolites and large fractures are mutually crosscutting. Bedding-plane stylolites have reduced section thickness by 1% or less. Microscopic extensional fractures extend from stylolite columns. Extension fractures are associated with stylolites formed during deformation (Nelson 1981). The main episode of stylolitization and fracturing thus coincided with Laramide deformation.

Oil-stained limestones and geopetal accumulations of tar are locally present. Impregnated beds typically are stylolitic. Locally, hydrocarbons migrated along stylolites as staining decreases away from stylolites over several millimeters.

Discussion of Burial Diagenesis

Burial diagenesis commenced in Campanian time with submergence of the platform margin. Mesogenesis culminated in the Laramide Orogeny during Maastrichtian–Paleocene time with gentle folding of the Sierra de El Abra. No major diagenetic products are attributed to this stage. Saddle dolomite and anhydrite probably were derived from pore fluids compacted from sulfate evaporites of the Lower Cretaceous Guaxcama Formation (Minero 1983a). Stylolites and associated extension fractures formed during Laramide deforma-

tion. Major fractures both postdate and predate stylolitization. Hydrocarbon migration, probably in post-Oligocene time (Enos 1983), postdates these features. The paucity of petrographically discernible compaction fabrics is attributed to extensive early cementation by calcrete, radiaxial fibrous mosaic, and equant cement prior to burial to 1 km or less.

Telogenesis (Stage 3 Karst)

The El Abra Formation is presently undergoing a third period of karst formation, which began with post-Laramide uplift and erosion of Upper Cretaceous cover. SEA topography is structurally controlled; typical subtropical karst forms are absent (Fish and Ford 1973). Drainage and dissolution occur in the subsurface because of well-developed cave and fracture systems. Sinkholes along Laramide joints and cavernous horizons are common in quarry outcrops (Fig. 18.13). Cave sediments include collapse breccia and colluvium, flowstone, bat guano, and organic detritus. Pleistocene rodent fossils have been recovered from reddish internal cave sediment in section VIII (Enos pers. comm.).

Phreatic caves occur at elevations up to 450 m, and some vertical segments are at least 300 m deep. Caves began developing when the coastal plain was 300 to 400 m above the present level (<100 m) (Fish and Ford 1973).

Small springs discharge along the eastern escarpment of the SEA. They are fed by local rainfall and stream capture (Fish and Ford 1973, Harmon 1981). Discharge and water chemistry vary seasonally, but small springs are undersaturated with respect to calcite at least during the rainy season (Harmon 1971). Large springs are typified by higher calcium, magnesium, and sulfate concentrations. At least part of their discharge is derived from deeply circulating groundwaters in Lower Cretaceous carbonates and evaporites in the Sierra Madre Oriental (Fish and Ford 1973).

Nonluminescent equant calcite cement similar to zone 4 cement locally postdated saddle dolomite and fills fractures which crosscut stylolites. This equant cement overlies geopetal tar accumulations in Taninul quarry (Enos 1983). The volume of this late-stage cement is unknown as only petrographic relations with rare

FIGURE 18.13. Stage 3 karst. A cave about 5 m high, with stalactite (right) and internal sediment at base, developed along a joint of Laramide age. VIII 10 m.

burial diagenetic features distinguish it from zone 4 calcite precipitated from the Late Cretaceous aquifer (i.e., that cement which predated Laramide fractures). It postdated hydrocarbon migration and degradation and therefore probably precipitated from oxidizing pore waters of the Late Tertiary to recent aquifer.

Summary and Conclusions

Rudist reefs developed along the windward margin of the Valles–San Luis Potosi platform in the Sierra de El Abra. Reefs passed rapidly bankward into islands with leeward tidal flats which prograded into the platform lagoon. Rudistid-skeletal limestone was deposited by storms on backreef islands. Lime mudstone, fenestral limestone, and cryptalgal laminites were deposited on tidal flats. Requienid and

peloid-miliolid limestones were deposited in the platform lagoon. Washover deposits which entered the platform lagoon were mixed with lagoonal sediment by bioturbation resulting in bioclastic limestones. Local beaches and channels are represented by laminated lime grainstones. Analogous modern carbonate sediments and environments occur in humid, tropical, shallow-water settings such as Andros Island, Bahamas (Hardie and Ginsburg 1977, Shinn et al. 1969) and Ambergris Cay, Belize (Ebanks 1975).

Subaerial discontinuity surfaces such as microkarst (Stage 1), calcrete profiles, and penecontemporaneously dolomitized horizons developed on backreef islands and intermittently on tidal flats. Discontinuities are correlative between measured sections, providing reliable horizons with which to identify laterally equivalent environments and to construct a depositional model (Figs. 18.2, 18.3, 18.4).

Calcrete-lithified islands of rudistid-skeletal limestone exerted the dominant control on paleogeography and the pattern of sedimentation. Coarse, subaerially deposited limestones near the platform margin grade rapidly bankward into cyclic marine to supratidal finer-grained limestones. Muddy sediments accumulated near the platform margin in the "energy-shadow" leeward of islands. Coalescing tidal flats filled the local platform lagoon during relatively stable sealevel stands, resulting in regressive and progradational shoaling-upward packages 3 m to 10 m thick. Cyclic depositional packages typically terminate with a subaerial discontinuity surface. The sedimentary record indicates rapid, episodic sealevel rises of up to 10 m net magnitude. Whereas island accretion rates equaled or exceeded that of sealevel rise, the tidal flats were repeatedly inundated. Depositional packages may have accumulated in as little as several thousand years. Discontinuity surfaces may record considerably longer periods of exposure and nondeposition or net erosion.

Relations between early diagenesis and the depositional environment are well illustrated by El Abra limestones. Diagenetic environments, although ephemeral, were recreated cyclically by depositional processes. Penecontemporaneous meteoric diagenesis on the perireef islands and leeward tidal flats was nearly continuous. It included mineralogic stabilization, dissolution, and cementation that transformed the sediment into a limestone, probably similar to many Quaternary limestones, prior to burial. Vadose cementation by calcrete formed a rigid framework at many horizons. Upon submergence, marine eogenesis was extensive as a consequence of the platform-margin setting and the coarse, solution-enhanced pore network. Mechanical pumping of seawater nourished Mg-calcite cementation and introduced internal sediment. Eogenetic lithification precluded significant compaction.

A meteoric aquifer developed during pre-Campanian time as a result of local subaerial exposure. The El Abra Formation acted as both a donor and a recipient limestone. Calcium carbonate derived from the karstic (Stage 2) upper surface was precipitated as late eogenetic equant cement occluding most remaining porosity.

Diagenesis was essentially completed at shallow depths prior to arrival of sulfate-rich compaction fluids which introduced minor saddle dolomite and anhydrite. Laramide deformation resulted in gentle folding of the Sierra de El Abra, fracturing, and ultimately, uplift and karst development (Stage 3) during the Cenozoic. Subsurface drainage is extensively developed in the SEA. Minor equant calcite cement precipitated from the regional Tertiary aquifer.

Acknowledgments. This paper is extracted from part of my dissertation research at the State University of New York at Binghamton. The work was supported by National Science Foundation grant EAR 7903960. Paul Enos supervised the original research, critically reviewed the manuscript, and provided isotopic analyses. Pecten International Company facilitated preparation of the manuscript through the typing services of Mary Dee Blaylock and Linda Davis.

References

Aguayo, J.E., 1978, Sedimentary environments and diagenesis of a Cretaceous reef complex, eastern Mexico: An. Centro Cienc. del Mar y Limnol., Univ. Nal. Auton. Mexico, v. 5, p. 83–140.

Brand, U., and Veizer, J., 1980, Chemical diagenesis of a multicomponent carbonate system—1: trace elements: Jour. Sed. Petrology, v. 50, p. 1219–1236.

Dodd, J.R., and Siemers, C.T., 1971, Effect of Late Pleistocene karst topography on Holocene sedimentation and biota, lower Florida Keys: Geol. Soc. America Bull., v. 82, p. 211–218.

Dunham, R.J., 1969, Early vadose silt in Townsend Mound (Reef), New Mexico: Soc. Econ. Paleontologists and Mineralogists Spec. Pub. 14, p. 139–181.

Dunham, R.J., 1971, Meniscus cement, *in* Bricker, O.P., ed., Carbonate cements: Baltimore, Johns Hopkins, p. 297–300.

Ebanks, W.J., 1975, Holocene carbonate sedimentation and diagenesis, Ambergris Cay, Belize: Am. Assoc. Petroleum Geologists Studies in Geology 2., p. 234–296.

Enos, Paul, 1974, Reefs, platforms, and basins of middle Cretaceous of northeast Mexico: Am. Assoc. Petroleum Geologists Bull., v. 58, p. 800–809.

Enos, Paul, 1983, Introduction, *in* Minero, C.J., Enos, P., and Aguayo, J.E., Sedimentation and diagenesis of mid-Cretaceous platform margin, east-central Mexico, Dallas Geological Society, p. 1–19.

Enos, Paul, 1986, Diagenesis of mid-Cretaceous rudist reefs, Valles Platform, Mexico, *in* Purser, B.H., and Schroeder, J.H., eds., Reef diagenesis: Berlin, Springer-Verlag, p. 160–185.

Enos, Paul, and Freeman, T., 1978, Shallow-water limestones from the Blake Nose, sites 390 and 392: Initial Reports of D.S.D.P., v. 44, p. 413–461.

Enos, Paul, and Perkins, R.D., 1979, Evolution of Florida Bay from island stratigraphy: Geol. Soc. America Bull., v. 90, p. 59–83.

Esteban, M., and Klappa, C.F., 1983, Subaerial exposure environment, *in* Carbonate depositional environments: Am. Assoc. Petroleum Geologists Memoir 33, p. 1–54.

Fish, J.E., and Ford, D.C., 1973, Karst geomorphology and hydrology of the Sierra de El Abra, S.L.P. and Tamps., Mexico: Proc. 6th Intl. Congress of Speleology (Prague), v. 2, p. 151–156.

Füchtbauer, H., and Hardie, L.A., 1976, Experimentally determined homogeneous distribution coefficients for precipitated magnesian calcites: applications to marine carbonate cements: Geol. Soc. Amer. Abs. Prog., v. 8, p. 877.

Graf, D.L., and Goldsmith, J.R., 1956, Some hydrothermal syntheses of dolomite and protodolomite: Jour. Geology, v. 64, p. 173–186.

Hardie, L.A., and Ginsburg, R.N., 1977, Layering—the origin and environmental significance of lamination and thin bedding, *in* Hardie, L.A., ed., Sedimentation on the modern carbonate tidal flats of northwest Andros Island, Bahamas: Baltimore, Johns Hopkins, p. 50–123.

Harmon, R.S., 1971, Preliminary results on the groundwater geochemistry of the Sierra de El Abra region, north-central Mexico, Nat. Speleological Soc. Bull., v. 33, p. 73–85.

Harrison, R.S., 1977, Caliche profiles: indicators of near-surface subaerial diagenesis, Barbados, West Indies: Bull. Can. Petroleum Geology, v. 25, p. 123–173.

Jennings, J.N., 1971, Karst: Cambridge, MA, MIT Press, 252 p.

Kendall, A.C., 1985, Radiaxial fibrous calcite: a reappraisal: carbonate cements: Soc. Econ. Paleontologists and Mineralogists Spec. Pub. 36, p. 59–77.

Kennedy, W.J., 1975, Trace fossils in carbonate rocks, *in* Frey, R.W., ed., The study of trace fossils: New York, Springer-Verlag, p. 377–398.

Land, L.S., 1973, Holocene meteoric dolomitization of Pleistocene limestones, North Jamaica: Sedimentology, v. 20, p. 411–424.

Lindholm, R.C., and Finkelmann, R.B., 1972, Calcite staining: semiquantitative determination of ferrous iron: Jour. Sed. Petrology, v. 42, p. 239–242.

Lohmann, K.C., and Meyers, W.J., 1977, Microdolomite inclusions in cloudy prismatic calcites: a proposed criterion for former high-magnesium calcites: Jour. Sed. Petrology, v. 47, p. 1078–1088.

Matthews, R.K., 1974, A process approach to diagenesis of reefs and reef associated limestones: Soc. Econ. Paleontologists and Mineralogists Spec. Pub. 18, p. 234–256.

McKee, E.D., 1959, Storm sedimentation on a Pacific atoll: Jour. Sed. Petrology, v. 29, p. 354–364.

Michard, C.R., 1968, Coprecipitation de l'ion manganeux avec le carbonate de calcium: Comptes Rendus Acad. Sci. Paris, Ser. D., v. 269, p. 1685–1688.

Minero, C.J., 1983a, Sedimentary environments and diagenesis of the El Abra Formation (Cretaceous), Mexico: Unpubl. Ph.D. thesis, SUNY Binghamton, 367 p.

Minero, C.J., 1983b, Back-reef facies, El Abra Formation, *in* Minero, C.J., Enos, P., and Aguayo, J.E., Sedimentation and diagenesis of mid-Cretaceous platform margin, east-central Mexico: Dallas Geological Society, p. 32–52.

Müller, G., 1971, "Gravitational" cement: an indicator for the vadose zone of the subaerial diagenetic environment, *in* Bricker, O.P., ed., Carbonate cements: Baltimore, Johns Hopkins, p. 301–302.

Multer, H.G., and Hoffmeister, J.E., 1968, Subaerial laminated crust of the Florida Keys: Geol. Soc. America Bull., v. 79, p. 183–192.

Nelson, R.A., 1981, Significance of fracture sets associated with stylolite zones: Am. Assoc. Petroleum Geologists Bull., v. 65, p. 2417–2425.

Perkins, R.D., 1977, Depositional framework of Pleistocene rocks in south Florida: Geol. Soc. America Memoir 147, p. 131–198.

Pingitore, N.E., 1978, The behavior of Zn^{2+} and Mn^{2+} during carbonate diagenesis: theory and applications: Jour. Sed. Petrology, v. 48, p. 799–814.

Plummer, L.N., 1975, Mixing of sea water with calcium carbonate ground water: Geol. Soc. America Memoir 142, p. 219–238.

Read, J.F., 1974, Calcrete deposits and Quaternary sediments, Edel province, Shark Bay, western Australia: Am. Assoc. Petroleum Geologists Memoir 22, p. 250–282.

Richter, D.K., and Füchtbauer, H., 1978, Ferroan calcite replacement indicates former magnesian calcite skeletons: Sedimentology, v. 25, p. 843–860.

Richter, D.K., and Zinkernagel, U., 1981, Zur anwendung der katodolumiszenz in der karbonatpetrographie: Geol. Rundschau, Bd. 70, p. 1276–1302.

Shinn, E.A., 1968, Selective dolomitization of recent sedimentary structures: Jour. Sed. Petrology, v. 38, p. 612–615.

Shinn, E.A., Ginsburg, R.N., and Lloyd, R.M., 1965, Recent supratidal dolomite from Andros Island, Bahamas: Soc. Econ. Paleontologists and Mineralogists, Spec. Pub. 13, p. 112–123.

Shinn, E.A., Lloyd, R.M., and Ginsburg, R.N., 1969, Anatomy of a modern carbonate tidal flat, Andros Island, Bahamas: Jour. Sed. Petrology, v. 39, p. 1202–1228.

Skougstad, M.W., and Horr, C.A., 1963, Occurrence and distribution of strontium in natural water: U.S. Geol. Survey Water Supply Paper 1496-D, p. 55–97.

Stoddart, D.R., and Steers, J.A., 1977, The nature and origin of coral reef islands, *in* Jones, O.A., and Endean, R., eds., Biology and geology of coral reefs: New York, Academic Press, v. 4, p. 59–105.

Sweeting, M.M., 1973, Karst landforms: New York, Columbia Univ. Press, 362 p.

Walkden, G.M., 1974, Paleokarstic surfaces in Upper Visean (Carboniferous) limestones of the Derbyshire Block, England: Jour. Sed. Petrology, v. 44, p. 1232–1247.

White, D.E., Hem, J.D., and Waring, G.A., 1963, Chemical composition of subsurface waters: U.S. Geol. Survey Prof. Paper 440-F, 67 p.

Woo, K.S., Sandberg, P.A., and Anderson, T.F., 1985, Radiaxial fibrous calcite in mid-Cretaceous rudist limestones, Geol. Soc. America Abstracts with Programs, v. 17, p. 754.

Wright, V.P., 1982, The recognition and interpretation of paleokarst: two examples from the Lower Carboniferous of South Wales: Jour. Sed. Petrology, v. 52, p. 83–94.

Index

The *italic* page numbers refer to figures.